Magnetic Techniques for the Treatment of Materials

Magnetic Techniques for the Treatment of Materials

by

Jan Svoboda

Formerly Research Programme manager and Principal Technology leader, De Beers Diamond Research Laboratory, De Beers Consolidated Mines (Pty.) Ltd., Johannesburg, South Africa

KLUWER ACADEMIC PUBLISHERS

DORDRECHT / BOSTON / LONDON

A C.I.P. Catalogue record for this book is available from the Library of Congress.

ISBN 1-4020-2038-4 (HB)
ISBN 1-4020-2107-0 (e-book)

Published by Kluwer Academic Publishers,
P.O. Box 17, 3300 AA Dordrecht, The Netherlands.

Sold and distributed in North, Central and South America
by Kluwer Academic Publishers,
101 Philip Drive, Norwell, MA 02061, U.S.A.

In all other countries, sold and distributed
by Kluwer Academic Publishers,
P.O. Box 322, 3300 AH Dordrecht, The Netherlands.

Printed on acid-free paper

Printed in the Netherlands.

Contents

Preface

In the years following the publication of *Magnetic Methods for the Treatment of Minerals* by Elsevier in 1987, many changes have taken place in magnetic technology. While fundamental and thorough, the above treatise reflected preferences and philosophy of research and the development and application of magnetic methods as they were practiced in the second half of the last century. Although demand for metals and minerals has not diminished, the dominant drivers of the early 21st century differ significantly from those of the late 20th. The priorities of nuclear power, defence, and energy-demanding and waste-generating beneficiation of mineral resources are being replaced by those associated with technology sustainability, environmental and knowledge management, recycling, and health care. Research priorities and product development of the last century cannot, therefore, satisfactorily meet criteria of the 21st century.

Considerable technological progress has been achieved in areas such as automation, computerization, sustainable material science, laboratory and plant practices and separation equipment. New permanent magnetic materials, advances in practical applications of superconductivity and availability of sophisticated modelling tools have changed the technological landscape. As a result, innovation and technology transfer in magnetic technology have been remarkably successful during the last two decades.

The title of this monograph reflects the fact that the book covers not only the application of magnetic techniques in the minerals industry, but also in recycling, environmental engineering and biomedical sciences. The goal of the book is to reflect the current technological trends and to re-position the research, development and practice of magnetic methods of material manipulation, according to current needs and strategic thinking. The book is intended for researchers, students, development workers and consulting metallurgists, as well as plant practitioners and equipment manufacturers who wish to acquire comprehensive knowledge of magnetic separation technology.

The book is based on work of numerous scientists and engineers. I am grateful to many of my colleagues and friends at universities, research organizations, and mining and manufacturing companies around the world, for their contribution in advancing magnetic separation technology. Interaction and collaboration with them over the years has been enriching and rewarding and they will be identified in this monograph by citation of their published work. I am also indebted to many individuals and companies who have given me invaluable help by providing data and photographs for reproduction. Their important contributions have been acknowledged directly in the text.

I am grateful to Mr. E. H. Roux and Drs. D. Slatter, G. Hill and N. Allen for their kindness in reading portions of the manuscript and suggesting improvements. I also wish to express my appreciation to my colleagues Drs. V. Murariu and L. Krüger and Mr. A. Sarelis for assistance with various aspects of this book.

Jan Svoboda
Johannesburg, South Africa
February 2004

Chapter 1

Principles of Material Treatment by Magnetic Means

1.1 Magnetism and the innovation milestones in magnetic separation

Magnetic phenomena have been known and exploited for many centuries. The earliest experiences with magnetism involved magnetite, the only material that occurs naturally in a magnetic state. Thales of Miletus stated that the magnetic interaction between lodestone, or magnetite, and iron was known for at least as long ago as 600 B.C. That magnetite can induce iron to acquire attractive powers, or to become magnetic, was mentioned by Socrates. Permanent and induced magnetism, therefore, represents one of man's earliest scientific discoveries.

Practical significance of magnetic attraction as a precursory form of magnetic separation was recognized in 1792, when W. Fullarton obtained an English patent for separating iron ore by magnetic attraction [D1]. Since that time the science and engineering of magnetism and of magnetic separation have advanced rapidly and a large number of patents have been issued. While separation of inherently magnetic constituents was a natural early application of magnetism, Wetherill's separator, devised in 1895, was an innovation of significant proportions. It demonstrated that it was possible to separate two components, both of which were commonly considered to be non-magnetic. In the ensuing time various types of disk, drum and roll dry magnetic separators were developed although the spectrum of minerals treatable by these machines was limited to rather coarse and moderately strongly magnetic materials. Since the end of the nineteenth century there has been a steady expansion of both the equipment available and the range of ores to which magnetic separation is applicable.

The development of permanent magnetic materials and improvement in their

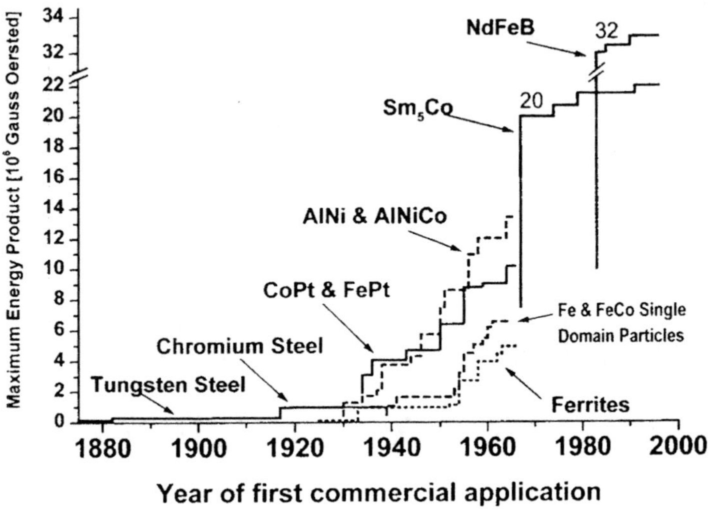

Figure 1.1: Development of permanent magnet materials.

magnetic properties have been main drivers of innovation in magnetic separation. Figure 1.1 illustrates the history of improvement of the maximum energy product $(BH)_{\max}$ of permanent magnets. Three innovation milestones can be identified in the graph. At the end of the nineteenth century very feeble steel-based magnets were employed, while in the 1940s permanent magnets that were able to compete with electromagnets, were developed. Probably the most important innovation step was made in the late seventies of the last century when rare-earth magnets became available. These magnets allowed new solutions for challenges that were not possible or feasible with electromagnets.

Another significant driver of innovation in magnetic separation was the introduction of ferromagnetic bodies (such as balls, grooved plates or mesh) into the magnetic field of a separator. In 1937 Frantz developed a magnetic separator consisting of an iron-bound solenoid packed with ferromagnetic steel ribbons [F1] and this proved to be an important milestone in the development of the present high-intensity and high-gradient magnetic separators. This innovation extended the range of applicability of magnetic separation to many weakly magnetic and even to diamagnetic minerals of the micrometer size.

Although the significance of the discovery of superconductivity has been equated with the invention of the wheel [K1], its importance for magnetic separation does not seem to represent a major breakthrough. The need for magnetic induction greater than 2 T has never been convincingly demonstrated in matrix separators [S1, S2] and the main advantage of superconducting magnets is, therefore, the reduced energy consumption and the possibility of generating a high magnetic force in large volumes, even without using matrices.

The concentration of various ferrous and non-ferrous minerals has been an

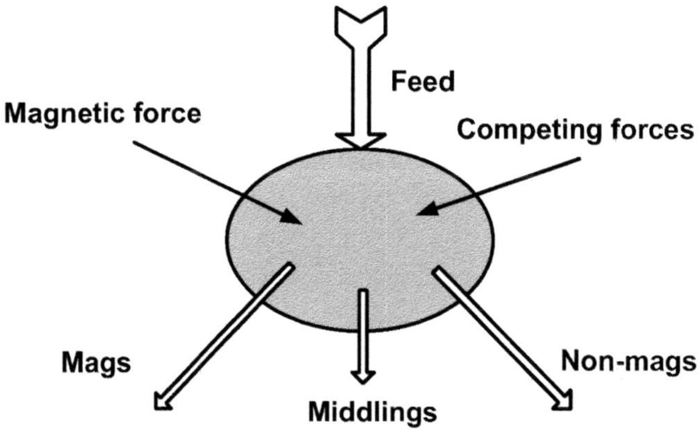

Figure 1.2: Schematic representation of the process of magnetic separation.

important application of magnetic separation, as has the removal of low concentrations of magnetizable impurities from industrial minerals. In recent years, as a result of numerous economic, environmental and social challenges, the recycling of metals from industrial wastes and the concentration or removal of biological objects in medicine and biosciences have become important areas of application of magnetic technology. Recent research and development of eddy-current separators, magnetic fluids and magnetic carriers illustrate the enormous effort that has been expended over the last twenty years in order to convert a wealth of novel ideas into workable techniques and introduce them into material manipulation operations.

1.2 Principles of magnetic separation

Magnetic separation is used for the concentration of magnetic materials and for the removal of magnetizable particles from fluid streams. The separation is achieved by passing the suspensions or the mixtures of particles through a non-homogeneous magnetic field. This process leads to preferential retention or deflection of the magnetizable particles. The same objective is often achieved in a very different fashion, the common features being a competition between a wide spectrum of forces of various magnitudes and ranges.

In magnetic separation the separating (differentiating) external force is the magnetic force. The separation of one material from another or the removal of magnetizable particles from streams depend upon their motion in response to the magnetic force and to other competing external forces, namely gravitational, inertial, hydrodynamic and centrifugal forces. Interparticle forces of electromagnetic and electrostatic origin contribute to the overall scenario, which is depicted in Fig. 1.2.

It is thus clear that a necessary (but not sufficient) condition for a successful separation of more strongly magnetic from less strongly magnetic particles in a magnetic field is that the magnetic force F^m_{mag} acting on more magnetic particles must be greater than the sum of all the competing forces F^{im}_{comp}. Simultaneously, a magnetic force acting on less strongly magnetic particles, F^n_{mag} must be smaller than the sum of the corresponding competing forces. Therefore, the following conditions must be met in a magnetic separator:

$$F^m_{mag} \geq \sum_i F^{im}_{comp} \quad \text{and} \quad F^n_{mag} \leq \sum_i F^{in}_{comp} \tag{1.1}$$

For different objectives eqs. (1.1) will have specific forms. For instance, in order to achieve high recovery of magnetic particles, the magnetic separating force must be greater than the sum of the competing forces as shown in eqs. (1.1). If, however, the magnetic force is much greater than the competing forces, i.e.

$$F_{mag} >> F_{comp} \tag{1.2}$$

selectivity of separation will be poor, as no distinction will be made between magnetizable species of different values of magnetic susceptibilities. High selectivity of the separation process can thus be obtained when the magnitudes of the magnetic and competing forces are of comparable magnitudes, compatible with conditions given by eqs. (1.1). The selectivity of the process will be, therefore, critically determined by the relative values of the magnetic and competing forces. And these are affected by the correct choice of a separator and its operating conditions. For instance, selective separation of a magnetizable material (1) from magnetizable material (2) will be achieved when the following relationship is met:

$$F^{(1)}_{mag} > F_{comp} > F^{(2)}_{mag} \tag{1.3}$$

If the non-magnetic fraction is the valuable product, the "brute force" approach to separation, expressed by relationship (1.2), will result in poor yield of the non-magnetic product, as a result of mechanical and magnetic entrainment of the non-magnetic material in the magnetic tailings.

In general, a mixture of particles introduced into the magnetic separator will be split into two or more components. However, in any real separation, both magnetic and non-magnetic particles can be found in the magnetic fraction, non-magnetic fraction and the middling fractions. The efficiency of separation is usually expressed by the recovery of the magnetic component, and by the grade of the magnetic product. However, these criteria are not unique and they must be selected on the basis whether the useful product is the magnetic or non-magnetic fractions.

Since all materials are magnetic to some extent, methods of separation that use magnetic force offer a unique approach to material manipulation in a wide array of industries. One of the main advantages of material treatment in a magnetic field is that the magnetic force can be applied in a controlled manner, in a wide range of values. Moreover, this force can be superimposed on other physical forces and several physical properties of materials can thus be exploited simultaneously. In addition, magnetic separation can treat a wide range of types of materials, ranging from colloidal to large sizes and from "non-magnetic" to strongly magnetic. Magnetic separation is an environmentally friendly technique and it can operate in wet and dry modes, making it a technique of choice for arid or arctic regions.

1.2.1 Magnetic force on a magnetizable particle

The density of the magnetic energy w_m in a linear isotropic medium is given by:

$$w_m = \frac{1}{2}\vec{H} \cdot \vec{B} \tag{1.4}$$

where H and B are, respectively, the magnitudes of the magnetic field and the magnetic flux density (magnetic induction).

It transpires from eq. (1.4) that the magnetic energy U_{mp} of a magnetizable particle of volume V_p placed in the magnetic field is:

$$U_{mp} = \frac{1}{2}\mu_p V_p H^2 \tag{1.5}$$

while the magnetic energy of a fluid of the same volume is given by

$$U_{mf} = \frac{1}{2}\mu_f V_p H^2 \tag{1.6}$$

In eqs. (1.5) and (1.6), μ_p and μ_f are magnetic permeabilities of the particle and the fluid, respectively. The energy increment U of the system (particle + fluid) is given, to first order, as the difference between the energies given by eqs. (1.5) and (1.6). For weakly magnetic particles this is a good approximation [G1]. Thus:

$$U = \frac{1}{2}(\mu_f - \mu_p)V_p H^2 \tag{1.7}$$

In general, a force can be expressed as $\vec{F} = -\nabla U$, where ∇ is the operator of the gradient. Taking into account that $\mu_j = \mu_0(1+\kappa_j)$, where κ_j is the volume magnetic susceptibility of material j and μ_0 is the magnetic permeability of vacuum, the magnetic force can be written (in SI units):

$$\vec{F}_m = \frac{1}{2}\mu_0(\kappa_p - \kappa_f)V_p \nabla H^2 \tag{1.8}$$

In practical situations the magnetic flux density B is frequently used, rather than the magnetic field strength H, and eq. (1.8) can then be expressed as:

$$\vec{F}_m = \frac{1}{\mu_0}(\kappa_p - \kappa_f)V_p B \nabla B \tag{1.9}$$

For sufficiently strongly magnetic particles ($\kappa_p >> \kappa_f$) it is advantageous to re-write eq. (1.8) as:

$$\vec{F}_m = \mu_0 V_p M \nabla H \tag{1.10}$$

where M is the magnetization of the particle.

Researchers as well as practitioners of magnetic separation sometimes use old *cgs* system of units, in which eq. (1.9) can be written

$$\vec{F}_m = (\kappa_p - \kappa_f) V_p B \nabla B \tag{1.11}$$

In practice it is easier to measure specific (mass) magnetic susceptibility χ rather than volume susceptibility and eq. (1.9) can be re-written, assuming that $\chi_f << \chi_p$:

$$\vec{F}_m = \frac{1}{\mu_0}(\chi_p - \chi_f) m_p B \nabla B \tag{1.12}$$

where m_p is the mass of the particle.

It can be seen from eqs. (1.8) to (1.12) that the magnitude of the magnetic force is proportional to the product of the magnetic flux density and its gradient. This force is in the direction of the field gradient, not of the magnetic field. It is also clear that in a homogeneous magnetic field, i.e. when $\nabla B = 0$, the force on a particle is zero. It can also be seen from the above equations that

$$F_m \approx b^3 \tag{1.13}$$

where b the particle radius.

1.2.2 Competing forces in a magnetic separator

As has been mentioned above, magnetic force in a magnetic separator competes with various external forces such as forces of gravity and inertia, centrifugal force and hydrodynamic drag. The importance of each force is a function of the type of the separator and of the mode in which it operates.

For a spherical particle of density ρ_p the gravitational force is given by:

$$\vec{F}_g = (\rho_p - \rho_f) V_p \vec{g} \tag{1.14}$$

where ρ_f and g are the density of the fluid medium and the acceleration by gravity, respectively.

The centrifugal force can be expressed as:

$$\vec{F}_c = (\rho_p - \rho_f) \omega V_p \vec{r} \tag{1.15}$$

where r is the radial position of the particle and ω is the angular velocity.

The hydrodynamic drag force can be obtained from Stokes's equation:

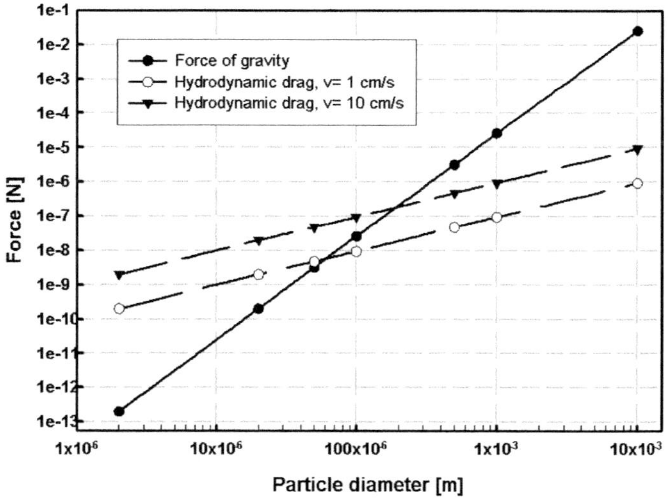

Figure 1.3: Hydrodynamic drag and force of gravity, as a function of particle diameter, for particles of density 5000 kg/m^3 (e.g. hematite), moving in water ($\eta = 10^{-3}$ Pa×s), with relative velocity v..

$$\vec{F}_d = 6\pi\eta b(\vec{v}_f - \vec{v}_p) \tag{1.16}$$

where η is the dynamic viscosity of the fluid and v_f and v_p are velocities of the fluid and the particle, respectively.

It can be seen from the last three equations that:

$$F_g \approx b^3, F_c \approx b^3 \text{ and } \quad F_d \approx b^1 \tag{1.17}$$

Because some of the forces have different dependence on particle radius, their relative importance will vary with particle size. The dependence of the force of gravity on the third power of the particle radius means that the force of gravity will be significant for large particles, On the other hand, the hydrodynamic drag, which in the Stokes regime depends on the first power of particle size, will be important for small particles. This situation in depicted in Fig. 1.3 and demonstrates that the force of gravity, for this specific case, is dominant for particle diameter greater than 500 μm, while for particle diameter smaller than 50 μm the hydrodynamic drag is the main competing force.

The magnetic force depends on the cube of particle radius and its relative importance in the process of magnetic separation is determined by the magnetic properties of the particles and by technical parameters of the magnetic separator (i.e. magnetic field strength and its gradient). Figure 1.4 illustrates the dependence of the magnetic force on a hematite particle of magnetic susceptibility $\kappa = 9.4\times10^{-3}$ (SI) ($\chi = 188\times10^{-8}$ m^3/kg), as a function of particle size,

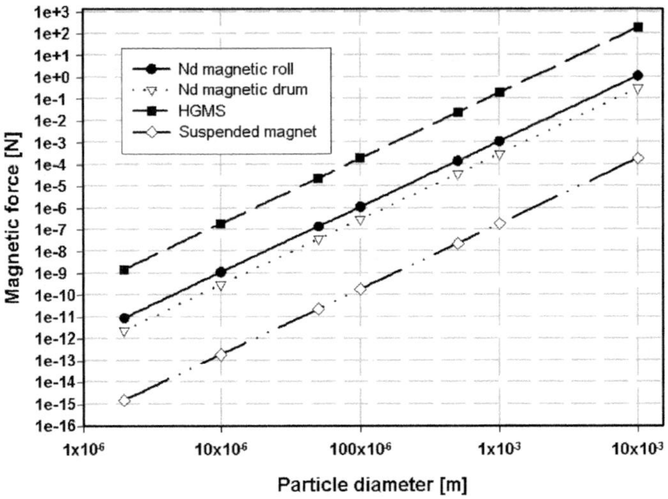

Figure 1.4: Magnetic force on a hematite particle in several types of magnetic separators, as a function of particle size.

Table 1.1: Typical values of B∇B for selected magnetic separators.

$B\nabla B[\mathrm{T}^2/\mathrm{m}]$			
Suspended magnet	Nd drum magnetic separator	Nd roll magnetic separator	HGMS with steel wool matrix
0.05	80	300	5×10^4

for various types of magnetic separators. The values of the product $B\nabla B$ used in construction of the graph are summarized in Table 1.1.

It is clear from eq. (1.9) and from Fig. 1.4 that the particle size is a more important discriminating factor in magnetic separation than the magnetic susceptibility. Since $F_m \approx \kappa b^3$, even a large difference in magnetic susceptibilities of components of a poorly classified mixture to be separated will not necessarily result in selective separation. As can be seen in Table 1.2, a particle characterized by magnetic susceptibility of 1 a.u. (arbitrary unit) and size 10 a.u. will be acted upon by the same magnetic force as a particle with susceptibility of 1000 a.u. and size 1 a.u. Selectivity of separation will be determined by the balance of the magnetic and competing forces acting on these two components.

It is evident from the above discussion that the subject of magnetic separation cannot be treated without first defining the physical quantities usually encountered in magnetism, and the units in which these quantities are measured. The next sections will be devoted to these issues.

Table 1.2: Magnetic force vs particle size.

Particle size [a.u.]	Magnetic susceptibility [a.u.]	Magnetic force [a.u.]
10	1	1000
1	1000	1000

1.3 Fundamental quantities of magnetism and their units

1.3.1 Magnetic field and magnetization

When we describe a magnetic field, we employ two different quantities, namely magnetic field strength H and magnetic flux density (or magnetic induction) B. H and B are both vector quantities having direction as well as magnitude. In vacuum B and H are not independent and are related by the equation

$$\vec{B} = \mu_0 \vec{H} \tag{1.18}$$

where the magnetic permeability of vacuum μ_0 is numerically equal to $4\pi \times 10^{-7}$ H/m. In a magnetic material of magnetization M, the total magnetic induction becomes, in Sommerfeld convention, adopted by IUPAP:

$$\vec{B} = \mu_0(\vec{H} + \vec{M}) \tag{1.19}$$

In the *cgs* system, since $\mu_0 = 1$, eq. (1.19) becomes

$$\vec{B} = \vec{H} + 4\pi\vec{M} \tag{1.20}$$

Magnetization is defined as the total magnetic moment μ_M of dipoles per unit volume V, i.e. $M = \mu_M/V$. The factor 4π arises from the fact that a unit pole produces a unit field everywhere on the surface of a sphere of unit radius (1 cm) enclosing the pole, and the surface of this sphere is 4π cm^2.

In the Kennelly system, which is traditionally favoured by electrical engineers, B is given in SI by:

$$\vec{B} = \mu_0\vec{H} + \vec{J}$$

where J is the magnetic polarization. J and M are related by equation:

$$\vec{J} = \mu_0\vec{M} \tag{1.21}$$

The magnetic induction thus includes contributions from the magnetization M, which is defined as the magnetic dipole moment of a body per unit volume, or polarization J defined by eq. (1.21).

1.3.2 Magnetic susceptibility and permeability

The magnetization of a material, in general, depends on the magnetic field acting on it. For many materials, M is proportional to H (at least when H is not too large) and we may write:

$$\vec{M} = \kappa \vec{H} \tag{1.22}$$

where κ, the volume magnetic susceptibility, is a property of the material. Since M and H have the same dimensions, κ is dimensionless. We can combine eq. (1.22) with eq. (1.19) to get

$$\vec{B} = \mu_0(1+\kappa)\vec{H} = \mu_0\mu_r\vec{H} = \mu\vec{H} \tag{1.23}$$

where

$$\mu_r = 1+\kappa \quad \text{and} \quad \mu = \mu_0(1+\kappa) \tag{1.24}$$

The quantity μ_r is called relative magnetic permeability and is dimensionless, while μ is called magnetic permeability and has a unit of H/m.

While eq. (1.19) is general, eqs. (1.23) and (1.24) are based on assumption that the material is both isotropic and linear. In other words, that M is proportional to H and in the same direction. This assumption is never completely true for ferromagnetic materials.

Either μ_r or κ may be used to characterize a material. Volume magnetic susceptibility κ ranges from values close to 0, both positive and negative, to positive values greater than 1, for different materials. For materials that have very small susceptibilities, it is much more convenient to use κ than μ_r . For instance for $\kappa = 10^{-6}$, $\mu_r = 1.000001$, and for $\kappa =$ - 10^{-6}, $\mu_r = 0.999999$.

Magnetic susceptibility can also be expressed with respect to the unit mass of material of density ρ, and then

$$\chi = \frac{\kappa}{\rho} \tag{1.25}$$

where χ is the mass, or specific, magnetic susceptibility. Since a sample mass m is usually better known than its volume, mass magnetic susceptibility is very commonly employed. For the same reason the mass magnetization σ rather than magnetization M is often used to characterize materials. The mass magnetization is defined as

$$\sigma = \frac{\mu_M}{m} \quad \text{thus} \quad M = \rho\sigma \tag{1.26}$$

1.3.3 Magnetic units

The units and symbols used in this book are those of the Système International d'Unités, designated SI. This system, which is based on the meter, kilogram, second, ampere, kelvin and candela, is essentially the same as the Giorgi, or MKSA, system in the field of electromagnetism.

Essentially all students today are taught in SI units and most scientific journals require their contributors to use SI units. Nevertheless, many papers on magnetism - and magnetic separation is no exception - are still published in *cgs* units. Occasionally, the *cgs* system is polluted with the *emu* "unit", which is not really a unit to which dimensional analysis can be applied [S3]. Therefore, even today the field of magnetism is not blessed with a lucid system of quantities and units and numerous panel discussions and articles on the subject illustrate the reigning chaos [C1, S4, G2]. Those authors who follow the current trend and convert *cgs* units to SI units find that the task of conversion is gigantic and that errors almost always creep in.

Table 1.3 lists the SI units in use for most important quantities, together with their conversion factors to electromagnetic *cgs* units.

Conversion of magnetic induction in the *cgs* unit of Gauss into the SI unit of Tesla should present no difficulty, with the values in Gauss divided by 10^4. When the magnetic field strength is used, Oersted multiplied by $10^3/4\pi$ (approximately 79.58) will give the value in the SI unit of A/m.

Magnetic susceptibility is a more complicated quantity. It is the volume magnetic susceptibility κ that enters most equations used in magnetic separation. Although dimensionless in both systems of units, volume magnetic susceptibility differs by a factor of 4π in the systems of units, as can be seen in Table 1.3. Experimentally, it is easier to determine the mass magnetic susceptibility and therefore most values of magnetic susceptibility that can be found in the literature are those of mass susceptibility, often in *cgs* units. Various names are used for this unit, e.g. *emu/cm, emu/g×Oe* and *cgs*, but all of them usually represent the same unit, namely cm^3/g. Thus in SI the unit is m^3/kg and conversion factor from *cgs* to SI is $4\pi\times10^{-3}$. The following equations summarize relationships between volume and mass magnetic susceptibilities, in both SI and cgs systems of units (density is expressed in g/cm^3).

$$\kappa(SI) = 4\pi\kappa(cgs) \tag{1.27}$$

$$\kappa(cgs) = \rho\chi(cgs) \quad \text{and} \quad \kappa(SI) = 10^3\rho\chi(SI) \tag{1.28}$$

$$\kappa(SI) = 4\pi\rho\chi(cgs) \tag{1.29}$$

$$\chi(SI) = \frac{4\pi}{10^3}\chi(cgs) \tag{1.30}$$

$$\chi(cgs) = \frac{10^3}{4\pi}\chi(SI) \tag{1.31}$$

1.4 Magnetic properties of materials

It has been stated at the outset that all materials display certain magnetic properties, regardless of their composition and state. According to their magnetic

Table 1.3: Units of important magnetic quantities. The quantity is cgs units is obtained by multiplying the quantity in SI by the conversion factor. For instance, 1 G = 10e-4 T.

Quantity	**Symbol**	**SI unit**	**cgs unit**	**Conversion factor**
Magnetic field strength	H	A/m	Oersted (Oe)	$10^3/4\pi$
Magnetic induction (flux density)	B	Tesla (T)	Gauss (G)	10^{-4}
Magnetization	M	A/m	Oe, erg/Gcm^3	10^3
Mass magnetization	σ	Am^2/kg	emu/g	1
Magnetic flux	Φ	Wb (Weber)	Mx (Maxwell)	10^{-8}
Polarization	J	Tesla (T)	G	10^{-4}
Magnetic moment	μ_M	Am^2	erg/G	10^{-3}
Volume magnetic susceptibility	κ	dimensionless	dimensionless, emu/cm^{3}Oe	4π
Mass (specific) magnetic susceptibility	χ	m^3/kg	cm^3/g	$4\pi \cdot 10^{-3}$
Demagnetization factor	N	dimensionless	dimensionless	$1/4\pi$
Energy product	BH	J/m^3	GOe	$1/40\pi$
Permeability of vacuum	μ_0	H/m	1 (dimensionless)	$4\pi \cdot 10^{-7}$
Bohr magneton	μ_B	9.274×10^{-24} Am^2		
Nuclear magneton	μ_N	5.051×10^{-27} Am^2		
Anisotropy constant	K	J/m^3	erg/cm^3	10^{-1}

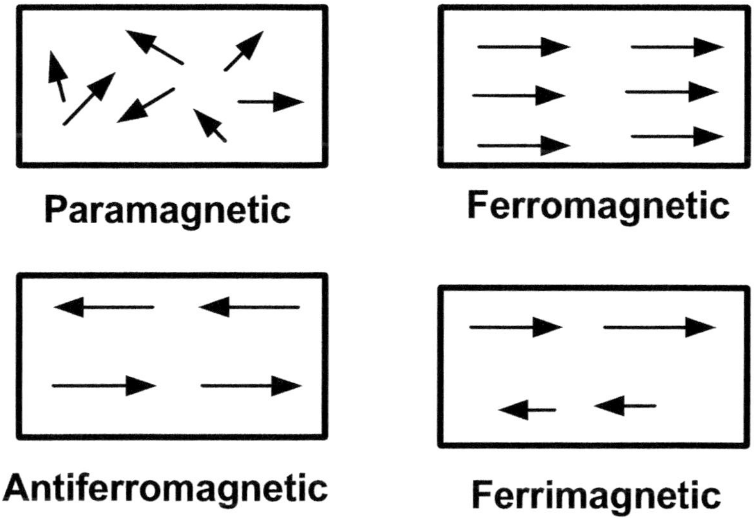

Figure 1.5: Schematic diagram of the alignment of magnetic moments.

properties, materials can be divided into five basic groups: diamagnetic, paramagnetic, ferromagnetic, antiferromagnetic and ferrimagnetic. The last three groups have generally higher magnetic susceptibilities than the other groups and are frequently termed "ferromagnetic" *sensu lato (s.l.)*. This broad definition must be distinguished from one of its subdivisions, ferromagnetic *sensu stricto (s.s.)* [T1] as discussed below. The alignment of magnetic moments in each type of material is shown in Fig. 1.5. The classification of magnetism in terms of magnetic permeability is given in Table 1.4.

1.4.1 Diamagnetism

Diamagnetism has its origin in the modification of the electron orbit magnetic moment by an external magnetic field. The currents thus induced give rise to an extra magnetic moment. However, according to Lenz's law, the resulting magnetic moment is in the opposite direction to the field that has induced the current. Therefore, approaching a source of the magnetic field, a diamagnetic material is repelled from it. This effect is present in all materials, independently of temperature. Diamagnets have a very small and negative volume magnetic susceptibility, of the order of -10^{-5} (SI).

While many inorganic and almost all organic molecules are diamagnetic, in many materials their diamagnetism is hidden by a much stronger paramagnetic/ferromagnetic *(s.l.)* effect. This is, for instance, the case of metals that are normally associated with small negative susceptibility, often overwhelmed by positive paramagnetic or ferromagnetic contribution. The values of selected diamagnetic metals and non-metallic compounds are given in Table 1.5. Table

Table 1.4: Magnetic susceptibility in various types of magnetism.

Parameter	**Diamagnetism**	**Paramagnetism Antiferromagnetism**	**Ferromagnetism Ferrimagnetism**
Susceptibility	< 0	≥ 0	$>> 0$
Relative permeability	< 1	≥ 1	$>> 1$

Table 1.5: Magnetic susceptibilities of diamagnetic materials.

Material	κ**[SI]** $\times$ **10^5**	χ**[m^3/kg]** $\times$ **10^9**
Copper	-1.0	-1.1
Zinc	-1.4	-1.96
Gold	-3.6	-1.86
Lead	-1.56	-1.38
Diamond	-2.0	-5.6
Water	-0.9	-9.0
Ethylalcohol	-0.73	-9.3

1.6 summarizes the magnetic susceptibilities of various diamagnetic minerals.

1.4.2 Paramagnetism

In paramagnetic materials the magnetic atoms or ions have permanent intrinsic magnetic moments and occur in low concentrations. Susceptibility arises from the competition between the aligning effect of the applied magnetic field and the randomizing effect of thermal vibrations. In zero applied magnetic field, the moments point in random directions. When a field is applied, a small magnetization develops, but the magnetic susceptibility is very small and is inversely proportional to the absolute temperature.

In the presence of the external magnetic field, and under standard practical conditions (i.e. when $\mu_M H/kT << 1$, where μ_M is the magnetic moment, k is the Boltzmann constant and T is the absolute temperature) the magnetization of a paramagnetic material is

$$M \cong \frac{c\mu_0 n\mu_M^2 H}{kT} \tag{1.32}$$

where c is a constant and n is the number of magnetic dipoles. The susceptibility of paramagnetic materials is then given by

$$\kappa = \frac{c\mu_0 n\mu_M^2}{kT} = \frac{C}{T} \tag{1.33}$$

Table 1.6: Mass magnetic susceptibilities of assorted diamagnetic minerals.

Mineral	**Chemical formula**	**Mass magnetic susceptibility $[m^3/kg]\times 10^9$**
Quartz	SiO_2	- 6.0
Calcite	$CaCO_3$	- 4.8
Sphalerite	ZnS	- 3.2
Galena	PbS	- 4.4
Baddeleyite	ZrO_2	- 1.4
Barite	$BaSO_4$	- 3.8
Magnesite	$MgCO_3$	- 6.4
Apatite	$Ca_5[PO_4]_3(F,Cl)$	- 3.3
Fluorite	CaF_2	- 7.9
Halite	NaCl	- 9.0
Kaolinite	$Al_4Si_4O_{10}(OH)_8$	- 20.0

where C is the Curie constant.

Equation (1.33) is called the Curie law, which states that magnetic susceptibility of the paramagnetic substance is inversely proportional to the temperature. The paramagnetic susceptibility is always small ($\kappa \approx 10^{-5}$ to 10^{-3} (SI)), although paramagnetism can be several orders of magnitude larger than diamagnetism.

As follows from eq. (1.32), the dependence of the magnetization of a paramagnetic material on the field strength is expressed by a straight line. In most cases, in standard conditions such as $T = 300$ K and $B = 1$ T, the above formulae apply. In extreme conditions, such as at cryogenic temperatures and very high magnetic field, a saturation effect appears. Taking into account eq. (1.23), it transpires from eq. (1.32) that for paramagnetic materials the magnetic susceptibility is independent of the magnetic field. Table 1.7 summarizes the values of magnetic susceptibility for selected paramagnetic metals and compounds, while Table 1.8 lists susceptibilities of assorted paramagnetic minerals.

Compilations of susceptibility of minerals are numerous [A1, S1, K2, D2] and large variations can be observed. Conflicting data can be explained by the inherent variability in mineral composition and the presence of inclusions and contaminants. Different measuring techniques tend to give differing values of susceptibility [M1] and the usage of different systems of units also contributes to frequent inconsistency of the values of magnetic susceptibility found in the literature.

1.4.3 Ferromagnetism

In ferromagnetic materials *(s.s.)* interaction between neighbouring atoms is strong so that the magnetic moments of all atoms are aligned parallel to each other against the randomizing force of thermal motion. The strong internal

Table 1.7: Magnetic susceptiblities of paramagnetic materials.

Material	**Mass magnetic susceptibility $[m^3/kg]\times10^9$**	**Volume magnetic susceptibility $[SI]\times10^5$**
Aluminium	7.67	2.1
Chromium	42.0	30.2
Titanium	42.2	19.0
Vanadium	73.0	44.5
Platinum	12.3	26.4
Zirconium	16.7	10.8
$CuSO_4{\cdot}5H_2O$	76.7	17.5
$NiSO_4{\cdot}7H_2O$	201	39.8
MnO	860	445.5

Table 1.8: Magnetic susceptibilities of paramagnetic minerals.

Mineral	**Chemical composition**	**Mass magnetic susceptibility $[m^3/kg]\times10^9$**
Hematite	$\alpha{-}Fe_2O_3$	250 - 3800
Siderite	$FeCO_3$	350 - 1500
Goethite	FeOOH	250 - 400
Ilmenite	$FeTiO_3$	200 - 1500
Rutile	TiO_2	10 - 50
Wolframite	$(Mn,Fe)WO_4$	350 - 1200
Pyrite	FeS_2	3 - 200
Limonite	$Fe_2O_3{\cdot}nH_2O$	100 - 400
Manganite	$MnO_2{\cdot}Mn(OH)_2$	100 - 500
Pyroxene	$(Mg,Fe)_2Si_2O_6$	40 - 1000
Dolomite	$CaMg(CO_3)_2$	5 - 20

fields, which align the spins, are called molecular or Weiss fields. These fields are the result of a quantum mechanical process called exchange interaction. Weiss assumed that the molecular field is very large, its magnitude independent of external magnetic field, and its direction not fixed, but always parallel to the magnetization. The Weiss hypothesis predicts that a material can be spontaneously magnetized even in the absence of an applied magnetic field at temperatures below the Curie temperature, at which the thermal agitation overcomes the molecular field. This tendency aids the external field in producing saturation, i.e. complete alignment.

There is an apparent contradiction between theory, which explains that magnetic materials are fully magnetized even in the absence of an external field, and practice, which generally shows that such materials exhibit no magnetization or magnetization much smaller than saturation. In fact, theory is correct at a mi-

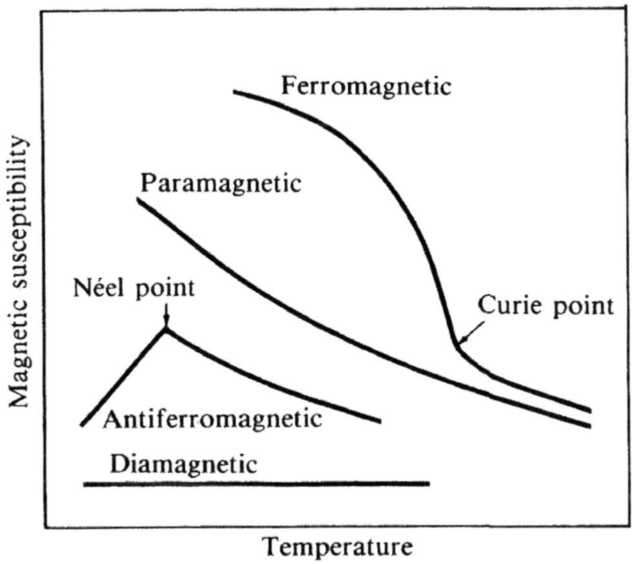

Figure 1.6: Temperature behaviour of various classes of magnetic materials.

croscopic level, but at a macroscopic scale the ferromagnetic body is subdivided into Weiss domains, each of which is spontaneously magnetized to saturation, but not in the same direction, by virtue of a strong exchange magnetic field. Inside a domain the magnetization is at a maximum (saturation), but as all domains compensate each other, on macroscale the mean magnetization is nil.

The value of the saturation magnetization varies with temperature, decreasing from a maximum value at $T = 0$ K, becoming zero at the Curie temperature T_c. Above T_c, the behaviour is similar to that of a paramagnetic material, with the magnetization being proportional to the field and the susceptibility decreasing with increasing temperature. However, the susceptibility does not follow the Curie law, eq. (1.33), but varies as

$$\kappa = \frac{C}{T - T_c} \tag{1.34}$$

Equation (1.34) is called the Curie-Weiss law. The temperature behaviour of various types of magnetism is shown in Fig. 1.6.

In contrast to diamagnetic and paramagnetic materials, magnetization of ferromagnetic materials (*s.l.*) depends not only on the field strength, but also on the shape and the magnetic history of the sample. For instance, a ferromagnetic material can remain magnetized after removal of the external field. The ferromagnetics *(s.l.)* are thus characterized by hysteresis as is illustrated in Fig. 1.7.

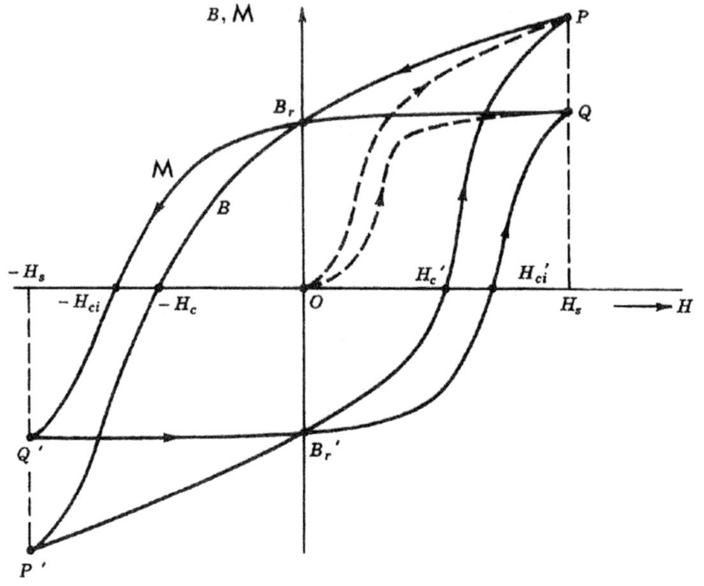

Figure 1.7: Hysteresis loop of a ferromagnetic *(s.l.)* material.

It can be seen in Fig. 1.7 that two kinds of the magnetic curve (or hysteresis loop) can be determined. The induction curve is $B = B(H)$, while the magnetization (or intrinsic) curve illustrates the $M = M(H)$ dependence. Since $B = \mu_0(M + H)$, and since M is much greater than H in most parts of the initial magnetization curve (curves OP and OQ, respectively, in Fig. 1.7), both curves are comparable. The difference occurs mainly in the saturation region where $M(H)$ is asymptotic to M_s, while $B(H)$ approaches $\mu_0(M_s + H)$.

It can be seen in Fig. 1.7 that after the initial (or virgin) magnetization reaches its saturation value B_s or M_s, (points P and Q, respectively), irreversibility, resulting from imperfections in the material, prevents the magnetization (or the induction) from returning to zero when the external field is removed. Magnetization and induction at $H = 0$ are called remanent magnetization, or remanent induction, respectively.

By reversing the direction of the external field H, the residual induction and magnetization decrease to zero for particular fields H_c and H_{ci} called coercive force and intrinsic coercive force, respectively. As can be seen in Fig. 1.7, H_c is not the same for $M(H)$ and $B(H)$ curves. For low-H_c materials (soft magnetic materials) the difference is generally negligible, while for hard material or permanent magnets (high H_c) the difference becomes significant.

Magnetic permeability, defined as $\mu = B/H$, can vary widely at various points on the magnetizing and demagnetizing portions of the hysteresis loop. The initial permeability μ_i and the maximum permeability μ_{max} are important

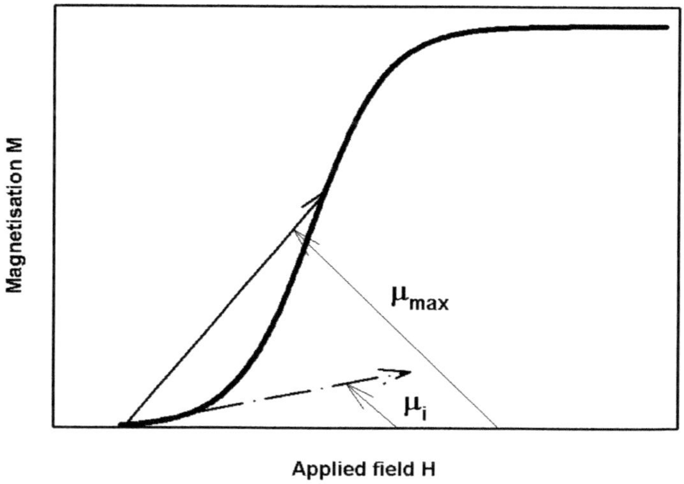

Figure 1.8: The initial magnetization curve and permeabilities of a ferromagnetic material.

parameters that can be determined as the slopes of the magnetization curve as shown in Fig. 1.8. The initial permeability illustrates the behaviour of the permeability when the external magnetic field approaches zero. With increasing magnetic field the permeability of the ferromagnetics *(s.l.)* increases and reaches its maximum. With further increase of the magnetic field permeability decreases and the relative permeability asymptotically approaches unity. This trend is shown in Fig. 1.9 which illustrates the dependence of the relative permeability on the external magnetic field for decarburized steel used for magnetic yokes [B1]. In a similar way we can define magnetic susceptibility of a ferromagnetic material, taking account relationship (1.24). Magnetic susceptibility has a similar trend as permeability, and at a high magnetic field it asymptotically approaches zero.

Ferromagnetism occurs in nine elements: three transition metals, iron, cobalt and nickel, and six rare earth metals, gadolinium, terbium, dysprosium, holmium, erbium and thulium. Most alloys consisting of the three transition metals are ferromagnetic, and so are many of their alloys with non-magnetic elements. Many alloys of manganese with non-magnetic elements are also ferromagnetic, although manganese is not ferromagnetic in the pure state. On the other hand, there is no mineral that would possess ferromagnetic properties. The values of magnetization and of the Curie temperature for three ferromagnetic elements (*s.s*) and other ferromagnetic materials are given in Table 1.9.

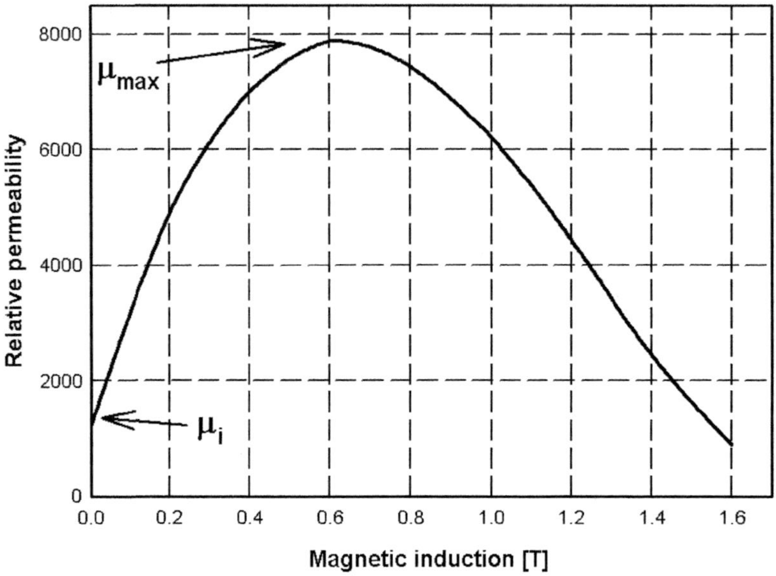

Figure 1.9: Permeability of decarburized steel used for manufacture of magnet yokes [B1].

Table 1.9: Magnetization and Curie temperatures of ferromagnetic materials.

Material	**Magnetization** $\mu_0 M_s$ **[T] at 0 K**	**Curie temperature** $\mathbf{T}_c$ **[^{0}C]**
Iron	2.18	770
Nickel	0.64	358
Cobalt	1.81	1120
Armco	2.11	
Si-Fe	1.97	
Permalloy 45	1.6	
Permendur	2.45	
Mumetal	0.8	

1.4.4 Antiferromagnetism

Antiferromagnetic materials were originally thought of as a class of anomalous paramagnets, since they have small positive susceptibilities of similar magnitude to many materials of the latter class. However, their magnetic susceptibility does not increase steadily as the temperature decreases all the way to absolute zero. At high temperatures they follow the relationship

$$\kappa = \frac{C}{T + T_N} \tag{1.35}$$

where T_N is the so-called Néel temperature. At this critical temperature the susceptibility ceases to obey eq. (1.35) and below T_N the susceptibility generally decreases with decreasing temperature. The situation is depicted in Fig. 1.6.

In the original Néel's theory, an antiferromagnetic substance was envisaged as composed of two sublattices, one of whose spins tend to be antiparallel to those of the other. Néel assumed the magnetic moments of the two sublattices to be equal, so that the net moment of the material was zero, as is schematically shown in Fig. 1.5. Since Néel's original hypothesis the term antiferromagnetism has been extended to include materials with more than two sublattices and those with triangular, spiral and canted spin arrangements.

Antiferromagnetism occurs mostly in transition metal oxides, such as MnO, CoO and NiO, as well as in other similar compounds such as sulphides and selenides. Antiferromagnets have small (10^{-10} to 10^{-8} m^3/kg) positive susceptibilities and hematite is one of the minerals that possess antiferromagnetic properties.

1.4.5 Ferrimagnetism

In a ferrimagnetic material magnetic moments are ordered regularly in an antiparallel sense, but the sum of the moments pointing in one direction exceeds those pointing in the opposite direction. This situation is illustrated in Fig. 1.5. The magnetic properties of ferromagnets and ferrimagnets are generally similar: both exhibit saturation and their magnetization is much greater than that of other magnetic classes. Ferrimagnetism occurs mainly in ferrites, mixed oxides of iron and of other elements, and also in two of the oxides of iron, namely magnetite Fe_3O_4 and maghemite $\gamma{-}Fe_2O_3$.

The shapes of magnetization curves of various classes of materials are shown in Fig. 1.10 and a summary of magnetic properties of materials is given in Table. 1.10.

1.4.6 Demagnetization

It was previously stated that the magnetic properties of ferromagnetic (*s.l.*) particles also depend on the shape of the body. The magnetic field H_i inside a magnetized body is not equal to the field applied to it (H_0). The magnetization of a ferromagnetic body of a finite size will produce a magnetic field inside the

Table 1.10: Classification of materials according to their magnetic properties.

Class	Critical temperature	Magnitude of κ [SI]	Temperature variation of κ
Diamagnetic	None	Approximately - 10^{-6} to - 10^{-5}	Constant
Paramagnetic	None	Approximately + 10^{-5} to + 10^{-3}	$\kappa = C/T$
Ferromagnetic	Curie temperature T_c	Large, e.g. > 1 (below T_c)	Above T_c $\kappa = \frac{C}{T-T_c}$
Antiferromagnetic	Néel temperature T_N	As paramagnetic	Above T_N $\kappa = \frac{C}{(T \pm T_i)}$ with $T_i \neq T_N$ below T_N decreases, anisotropic
Ferrimagnetic	Curie temperature T_c	As ferromagnetic	Above T_c $\kappa \approx \frac{C}{(T \pm T_i)}$ with $T \neq T_i$

body in a direction opposite to that of the magnetization. The induced field is called the demagnetizing field H_d because it always tends to oppose the applied field H_0. The situation is depicted in Fig. 1.11. The magnitude of H_d is proportional to that of magnetization M:

$$\vec{H}_d = -N\vec{M} \tag{1.36}$$

where N is the dimensionless constant of proportionality called the demagnetization factor. The value of N varies from zero to unity, depending upon the shape of the material and the direction of magnetization. The resulting magnetic field inside the body is given by

$$\vec{H}_i = \vec{H}_0 + \vec{H}_d = \vec{H}_0 - N\vec{M} \tag{1.37}$$

In general, the calculation of N is quite complex and numerical values for various shapes can be found e.g. in [B2, B3, D3]. With increasing aspect ratio, with the major axis parallel to the field, the demagnetization factor decreases. For a sphere $N = 1/3$, while for the infinitely long cylinder whose axis coincides with the direction of the field, $N = 0$. For a thin disc, perpendicular to the field, the demagnetization factor achieves its maximum value, $N = 1$.

For a linear soft magnetic material, for which $M = \kappa H$, eq. (1.37) can be re-written

$$\vec{H}_i = \frac{1}{1 + N\kappa}\vec{H}_0 \quad \text{and} \quad \vec{B}_i = \mu_0 \frac{1}{1 + N\kappa}\vec{H}_0 \tag{1.38}$$

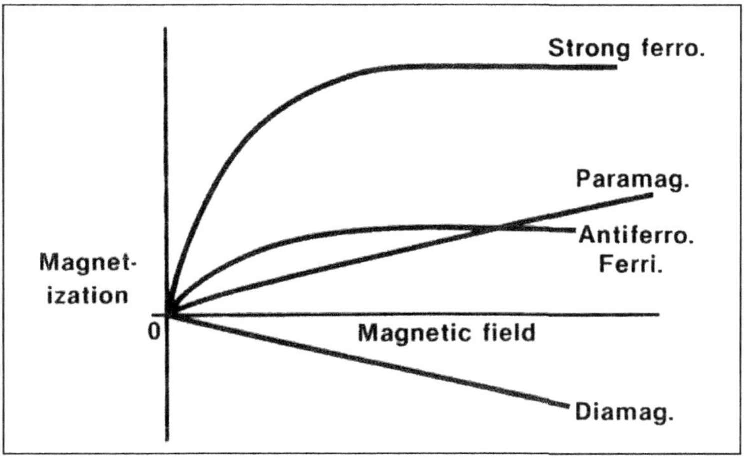

Figure 1.10: Shapes of the magnetization curves of various classes of materials (not in scale).

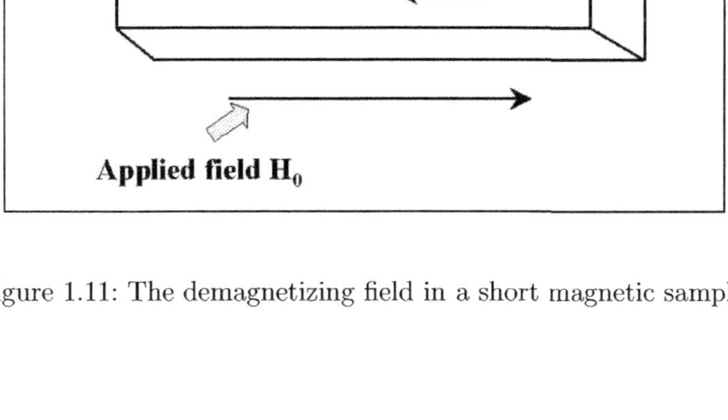

Figure 1.11: The demagnetizing field in a short magnetic sample.

In the studies of mineral magnetism it is important to distinguish extrinsic susceptibility κ_e, which is the susceptibility that is conventionally measured by most instruments, from intrinsic susceptibility κ_{in}, which is the true susceptibility after removal of the effects of internal demagnetizing fields [T1]. These susceptibilities are related by the following equation:

$$\kappa_e = \frac{\kappa_{in}}{1 + N\kappa_{in}} \tag{1.39}$$

1.4.7 Mixtures containing ferromagnetic *(s.l.)* materials

In magnetic separation, many weakly magnetic materials contain ferromagnetic (*s.l.*) impurities. As a result of a high magnetic permeability of most ferromagnetic (*s.l.*) materials, the bulk magnetic susceptibility of a weakly magnetic mineral will be determined by the concentration of ferromagnetic impurities. For such a mixture the observed magnetic susceptibility is given by

$$\kappa_e = \frac{c\kappa_{in}}{(1 + N\kappa_{in})} \tag{1.40}$$

where c is the volume concentration of the ferromagnetic (*s.l.*) material. Contributions of various ferromagnetic (*s.l.*) and paramagnetic minerals to the total susceptibility of a mixture is shown in Fig. 1.12 [T1].

Since the magnetic permeability and susceptibility of a ferromagnetic (*s.l.*) material is a function of the magnetic field, the definition of magnetic susceptibility is determined by measuring techniques used to determine the susceptibility. The most frequently quoted value is the initial susceptibility (or permeability) κ_i as shown in Fig. 1.8. For strongly susceptible materials, such as magnetite, $\kappa_e = 1/N$, and the initial susceptibility can be written:

$$\kappa_i = \frac{c}{N} \tag{1.41}$$

For spherical particles $N = 1/3$ and thus

$$\kappa_i \approx 3c \tag{1.42}$$

while for ovoid particles, with the ratio of long and short axes of 10, the susceptibility is [L1, T1]:

$$\kappa_i \approx 18c \tag{1.43}$$

Although the magnetic susceptibility of paramagnetic material is independent of the magnetic field, minerals displaying various forms of magnetic order (i.e. ferromagnets (*s.l.*)), or paramagnetic minerals containing the admixtures of the magnetically ordered materials, exhibit a field-dependent magnetic susceptibility. This phenomenon can seriously alter the predictions of the results of magnetic separation based on the field-independent magnetic susceptibility.

The measured magnetic susceptibility, in a relatively low applied magnetic field ($B < 0.6$ T), can be represented by the Owen-Honda equation [O1, H1]:

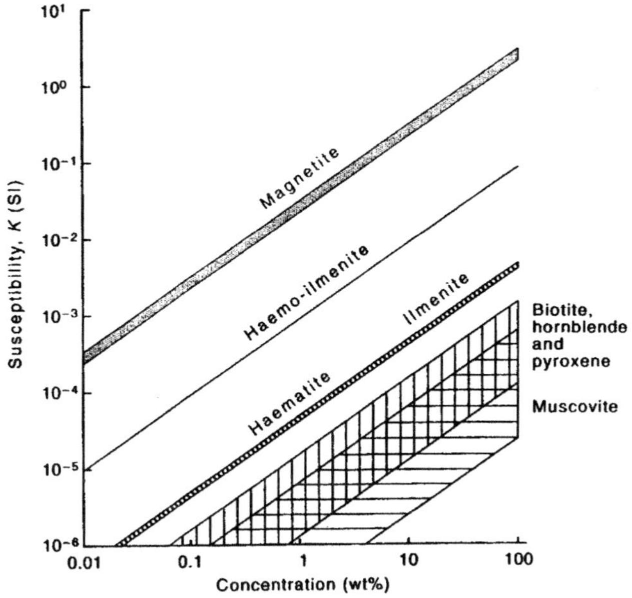

Figure 1.12: Contribution of various minerals to the total susceptibility of a mineral mixture [T1].

$$\kappa = \kappa_p + \kappa_d + \frac{M_s}{H} = \kappa_\infty + \frac{M_s}{H} \tag{1.44}$$

where κ_p and κ_d are paramagnetic and diamagnetic susceptibilities, respectively, κ_∞ is the magnetic susceptibility extrapolated to the infinite magnetic field, and M_s is the saturation magnetization of the ferromagnetic (*s.l.*) component of the mixture.

It has been experimentally determined ([N2, P2]) that κ_∞ is characteristic of a mineral, and is nearly constant for samples of the same mineral from different localities. On the other hand, saturation magnetization M_s, which indicates the degree of weak ferromagnetism, varies widely between samples from differing localities and between size fractions of the same sample.

1.5 Magnetic properties of minerals

1.5.1 The classification of minerals according to their magnetic properties

Although all matter responds to the presence of the external magnetic field, the degree of this response varies widely. For practical reasons the technical classification of materials, and particularly of minerals, differs from physical categorization discussed in previous sections. For the purposes of magnetic separation, materials can be divided into three basic groups:

Strongly magnetic materials can be recovered by a magnetic separator with the use of a relatively weak magnetic field, up to, for example, 0.15 T, with a modest field gradient of the order of 0.5 T/m. The magnetic susceptibility that responds to such a low magnetic field is generally greater than 10^{-4} m^3/kg and ferromagnetic and ferrimagnetic materials like mild steel, iron, magnetite, maghemite and pyrrhotite fall under this group.

Weakly magnetic materials can be recovered by a magnetic separator that generates magnetic induction up to 1.0 T, and the field gradient in the range from 50 to 500 T/m. This broad group of materials includes antiferromagnetic, paramagnetic and some ferrimagnetic minerals. The group comprises iron and manganese oxides and carbonates, ilmenite, wolframite and other materials. Magnetic susceptibility of these materials ranges from 10^{-7} m^3/kg to 5×10^{-6} m^3/kg.

"Non-magnetic" materials are those materials that cannot be easily recovered by conventional magnetic separators. They include very weakly paramagnetic and antiferromagnetic materials with magnetic susceptibility smaller than, for example, $10^{-7}m^3/kg$. Austenitic stainless steel, aluminium, rutile, pyrite, garnet and red blood cells belong to this group. Diamagnetic materials, with their negative magnetic susceptibility are also included.

This classification is only conditional as it does not take into account such an important variable as particle size, which, as we have seen in Section 1.2.2, is as important as magnetic susceptibility in determining the efficiency of magnetic separation. Recent developments and innovation in magnetism extended the range of materials that can be treated by the magnetic means to such a degree that all classes of matter are included, down to the submicrometer or even nanosize range.

1.5.2 Magnetite

Magnetite $\alpha-Fe_3O_4$ is the most fundamental and important magnetic mineral. It has a cubic crystalline structure in which Fe^{3+} and Fe^{2+} cations form two distinct lattices. Each of the lattices is magnetized in opposite directions and magnetite behaves as a ferrimagnetic mineral .

The saturation polarization of magnetite at room temperature ranges from 0.55 T to 0.61 T and the Curie temperature is 575 ^{0}C. The density of pure Fe_3O_4 is 5200 kg/m^3, calculated from the cell dimensions [N1].

As has been mentioned earlier, magnetic properties of magnetically ordered mineral grains depend, among other things, on the grain size. The most important distinction is between very small particles, each of which is a single domain, and large grains, which are multidomain. For any material there is a maximum size for spherical simple domain, referred to as the critical size. For magnetite this critical size is about 0.05 μm [S5], while for hematite, for instance, this value is much higher, of the order of 200 to 300 μm [N1].

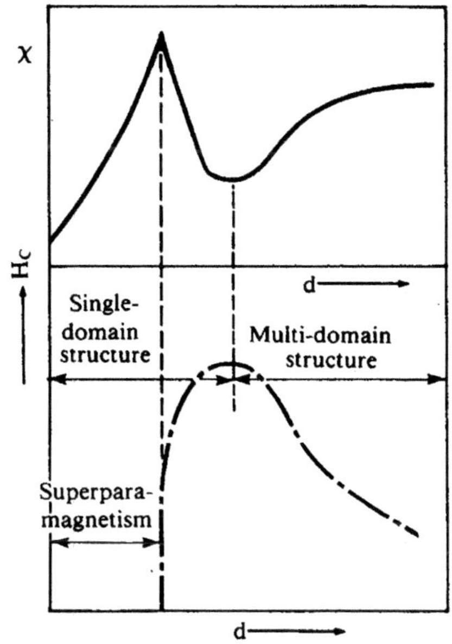

Figure 1.13: Magnetic susceptibility and coercive force of ferromagnetic particles, as a function of grain size.

The variation of the coercive force H_c with particle diameter d, in granular magnetic materials and over a wide range of grain sizes, is well represented by the relationship

$$H_c \approx d^{-n} \tag{1.45}$$

The value of n is usually in the range between 0.25 and 1. The effect of transition from single domain to multidomain structure, on magnetic properties of a ferromagnetic grain, is illustrated in Fig. 1.13. This Figure shows that, at the transition between multidomain and single domain arrangement, magnetic susceptibility achieves its minimum, while the coercive force is maximum. A simplified dependence of the coercive force and mass magnetic susceptibility on particle diameter is shown in Fig. 1.14.

Fig. 1.15 illustrates the dependence of the mass magnetic susceptibility of magnetite on the external magnetic field, ranging from 0 to 240 kA/m (0 to 3 kOe). The data reported in [P1] were obtained with size fraction - 74 + 53 μm.

1.5.3 Titanomagnetites

Titanomagnetites $Fe_{3-x}Ti_xO_4$ are one of the most common magnetic minerals. Magnetic susceptibility, saturation magnetization and Curie temperature vary in a regular manner with the composition parameter x from $x = 0$ (magnetite)

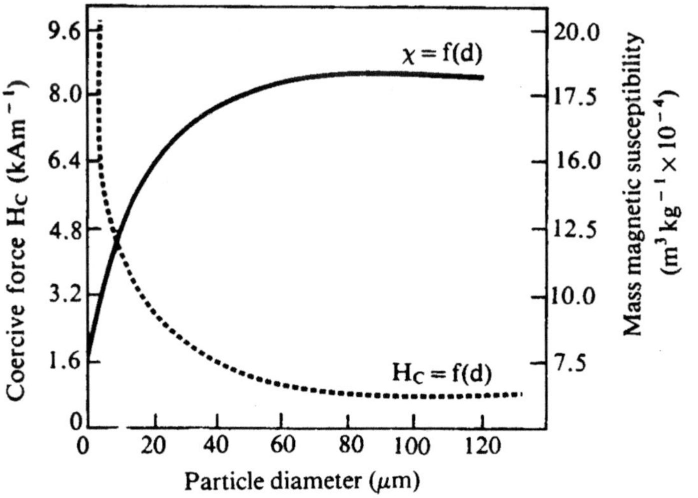

Figure 1.14: Dependence of the coercive force and magnetic susceptibility of magnetite on the particle size (after Derkach [D4]).

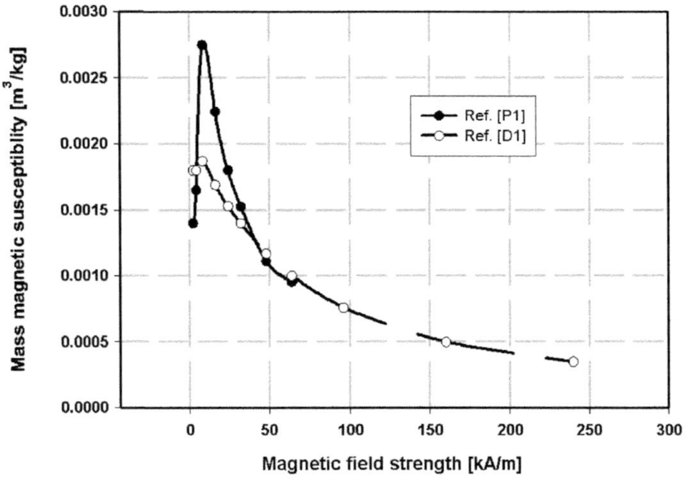

Figure 1.15: The field dependence of the magnetic susceptibility of magnetite (adapted from Derkach [D4]).

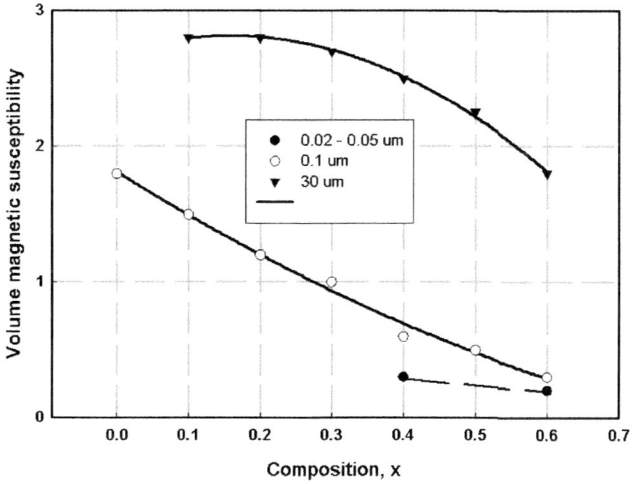

Figure 1.16: Variation of the room temperature magnetic susceptibility with composition, for titanomagnetites (after [O1]).

to $x = 1$ (ulvöspinel). The variation of the initial magnetic susceptibility with composition, for three size fractions, is shown in Fig. 1.16 [O1]. Fig. 1.17 illustrates the dependence of the initial magnetic susceptibility on particle size, for assorted compositions of titanomagnetites.

1.5.4 Ilmenohematites

Titanium-rich ilmenohematites $Fe_{2-y}Ti_yO_3$ are significant magnetic minerals, as their susceptibilities are more than two to three orders of magnitude greater than those of paramagnetic minerals. Ilmenite $FeTiO_3$ is antiferromagnetic below the Néel temperature of 55 K. Above the Néel temperature ilmenite is paramagnetic [N1] and its magnetic susceptibility (approximately 5×10^{-3} (SI) or 10×10^{-7} m^3/kg) is somewhat lower than for other members of the series [T1, D2], and was found to be essentially independent of the magnetic field strength in the range from 0.1 T to 9 T [D2].

Magnetic susceptibility measured below the Néel temperature (at 4.3 K) was found to be equal to 150×10^{-7} m^3/kg at magnetic fields smaller than 0.6 T. At higher magnetic fields the value of the susceptibility dropped to about 100×10^{-7} m^3/kg [D2]. Figure 1.18 illustrates the variation of the saturation magnetization of ilmenohematite samples of particle size of 9.8 μm, at room temperature, with composition [O2]. It can be seen that ilmenite ($y = 0$) is one of the magnetically weakest members with saturation polarization of approximately 0.012 T.

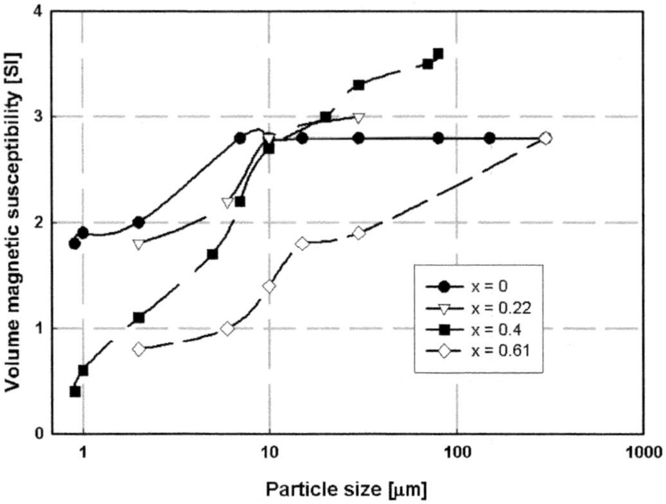

Figure 1.17: Variation of the initial magnetic susceptibility, at room temperature, with grain size, for titanomagnetites (after [O1]).

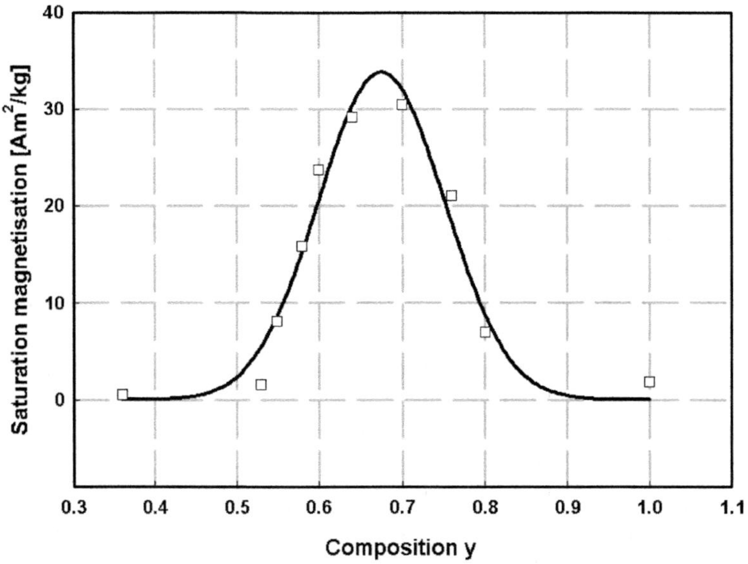

Figure 1.18: Saturation magnetization of ilmenohematites as a function of their composition (after [O2]).

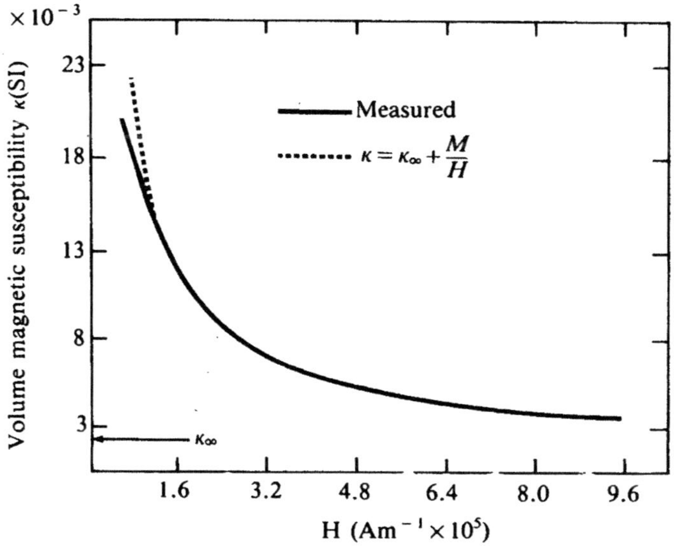

Figure 1.19: Magnetic susceptibility of hematite as a function of the applied magnetic field (after [N2]).

1.5.5 Maghemite

The magnetic properties of maghemite (γ-Fe_2O_3) are very similar to those of magnetite (albeit slightly less pronounced). Both minerals have fundamentally the same spinel structure, although the composition of maghemite is essentially that of hematite. Maghemite is metastable and converts to hematite at temperatures in the range from 300 ^{0}C to 350 ^{0}C. The saturation polarization at room temperature ranges from 0.48 T to 0.51 T.

1.5.6 Hematite

Pure hematite α-Fe_2O_3 has antiferromagnetic structure above the Néel temperature of approximately 680 ^{0}C. Between the Néel temperature and the Morin temperature T_M = - 100 ^{0}C hematite is a canted antiferromagnet ("weak ferromagnet"). At room temperature, the magnetic polarization is about 0.5 % that of magnetite, i.e. between 15×10^{-4} to 30×10^{-4} T. The weak ferromagnetism (*s.l.*) of hematite is not fully understood because of its complex and contradictory behaviour. The properties vary from sample to sample and depend on their origin. For instance, the ratio of the remanent polarization to saturation polarization J_R/J_S for fine hematite, ranges from close to zero to about 0.8 [A2]. The coercive force of fine hematite is very high, up to 2.5×10^5 A/m, in contrast to the low values of the order of 8×10^2 A/m [S5, D5], observed with crystals, which were a few millimeters in size.

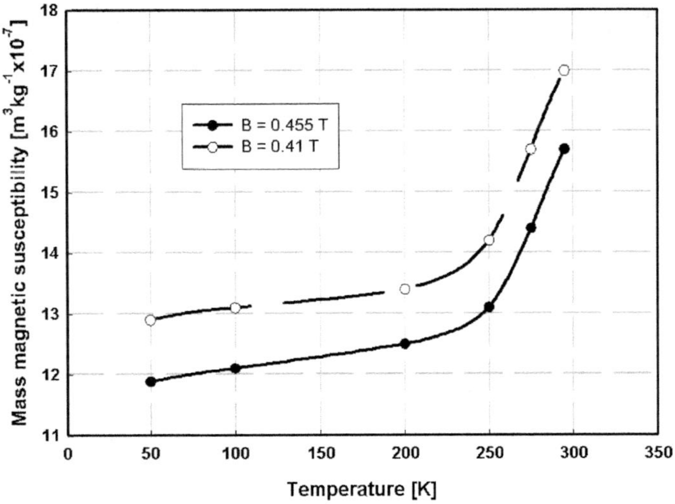

Figure 1.20: The effect of temperature and magnetic field on the magnetic susceptibility of specularite (after [P2]).

Hematite is also strongly anisotropic, i.e. crystals can be magnetized by magnetic fields applied along the basal plane, but are almost unaffected by fields perpendicular to this plane, unless extremely strong fields are employed [T1]. Hematite also tends to flake parallel to its basal plane and the flakes are often termed specularite.

The magnetic susceptibility of pure hematite is a complicated function of particle size, magnetic field, and temperature, determined by the interplay between contributions from weak ferromagnetism and antiferromagnetism [D5]. The effects of magnetic field strength and temperature on magnetic susceptibility are shown in Figs. 1.19 and 1.20, respectively.

A significant change in magnetic susceptibility for magnetic fields smaller than 8×10^5 A/m (10 kOe) can be seen in Fig. 1.19. The field-independent magnetic susceptibility κ_∞ was found to be equal to 1.43×10^{-3} (SI) [N2]. Pastrana and Hopstock [P2] investigated the effect of temperature and magnetic field on magnetic susceptibility in terms of eq. (1.44). They found that κ_∞ increased only slightly with temperature but the effect of temperature on κ was dramatic, as illustrated by Fig. 1.20.

1.5.7 Goethite

Goethite (α−FeOOH) is an antiferromagnetic mineral, which ideally should have perfectly compensated spins, and hence no net spontaneous magnetic moment. It appears, however, that one of the sublattices is usually canted; as a result the mineral behaves ferrimagnetically. It has a magnetization greater

Table 1.11: Some physical properties of ferromagnetic (s.l.) minerals.

Mineral	**Composition**	**Curie/Néel temperature** [^{o}C]	**Saturation polarization** [T]	**Density** [kg/m^3]
Magnetite	$\alpha-$ Fe_3O_4	575	0.59 - 0.61	5200
Maghemite	$\gamma-$ Fe_2O_3	350	0.48 - 0.51	4800
Hematite	$\alpha-$ Fe_2O_3	680	10×10^{-4} to 30×10^{-4}	5300
Goethite	$\alpha-$ FeOOH	120 - 130	50×10^{-7} to 50×10^{-4}	4300
Pyrrhotite	FeS	320	0.1	4600
Chromite	$FeCrO_4$	- 84	1.9×10^{-5}	5090

than that of hematite but weaker than that of magnetite. Goethite also has a high coercivity, comparable to that of hematite. The mass magnetic susceptibility ranges from 2.5×10^{-7} m^3/kg to 6×10^{-7} m^3/kg [A1, P2]. The temperature and the magnetic field dependence of the magnetic susceptibility of goethite, as reported by Pastrana and Hopstock [P2] is shown in Fig. 1.21. Physical properties of selected ferromagnetic ($s.l$) minerals, including goethite, are summarized in Table 1.11.

1.5.8 Siderite

Siderite $FeCO_3$ undergoes an antiferromagnetic transition at the Néel temperature of about 14 K. When heated under oxidizing conditions, it decomposes into minerals such as maghemite, magnetite and iron hydroxide. The final decomposition product is hematite [B4]. Results of extensive measurements of magnetic susceptibility of samples of siderite (- 150 + 105 μm) at room temperature [P3] show that the susceptibility does not vary with the applied magnetic field. The susceptibility ranges from 8×10^{-7} m^3/kg to 11×10^{-7} m^3/kg and similar values were reported in [T1, M2].

1.5.9 Iron sulphides

Iron sulphides exhibit magnetic properties over a narrow range of specific compositions, but these are strongly affected by impurities, grain size and similar factors. Pyrrhotite (FeS_{1+x} for $0.10 < x < 0.14$) is the most common ferromagnetic ($s.l.$) iron sulphide. Most of these minerals are chemically unstable on heating and the Curie temperature of pyrrhotite, 320 ^{0}C, is almost the same as the temperature at which is decomposes to magnetite [T1].

As in hematite, the spin structure consists of ferromagnetic order, the magnetic moments of adjacent planes being coupled parallel. Unlike hematite, the moments of the planes differ in magnitude and a strong ferrimagnetism results [O2]. Pyrrhotite is generally more strongly magnetic than hematite and the

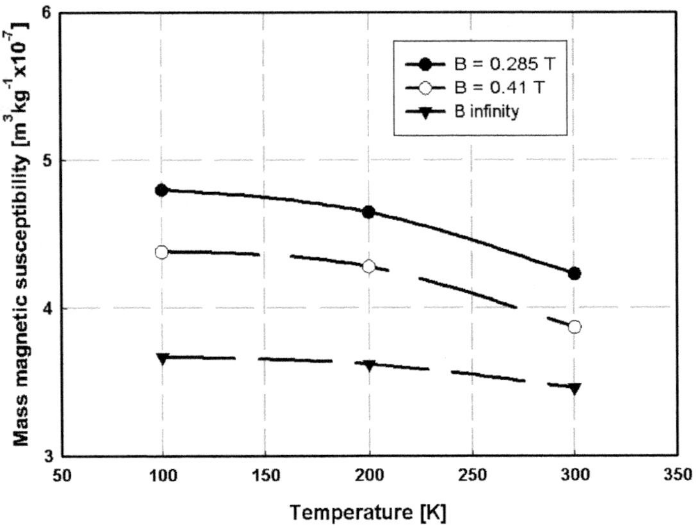

Figure 1.21: Magnetic susceptibility of goethite as a function of magnetic field strength and temperature (after [P2]).

saturation polarization, at room temperature, is about 0.12 T [O2], although smaller values are often found in literature (e.g. [T1]). The magnetic field dependence of the mass magnetic susceptibility is shown in Fig. 1.22 [D4].

1.5.10 Chromites

The magnetic properties of natural chromium-bearing spinels or chromites (Fe, Mg)(Cr,Al,Fe)$_2$O$_4$ are highly variable, the variation being closely related to the variation in their chemical composition. It is frequently assumed that the magnetic susceptibility and the hysteresis behaviour of chromites is controlled by the end-member magnetite [H2, S6]. The magnetic susceptibility is reported to range from 6×10^{-7}m^3/kg to 30×10^{-7} m^3/kg at 4.3 K [D2] and from 5×10^{-7} m^3/kg to 12×10^{-7} m^3/kg at elevated temperatures (78 and 136 K) [D2, S6].

1.6 Measurement of magnetic properties

This section briefly describes experimental techniques used for the measurement of magnetic quantities. There are hundreds of different techniques and their modifications, each having distinct merits as well as limitations. There is no single technique which is universally applicable to all types of measurements. There are several comprehensive sources of information [Z1, F2, B5, B6, C2] that provide a good starting point in the pertinent literature. It is interesting

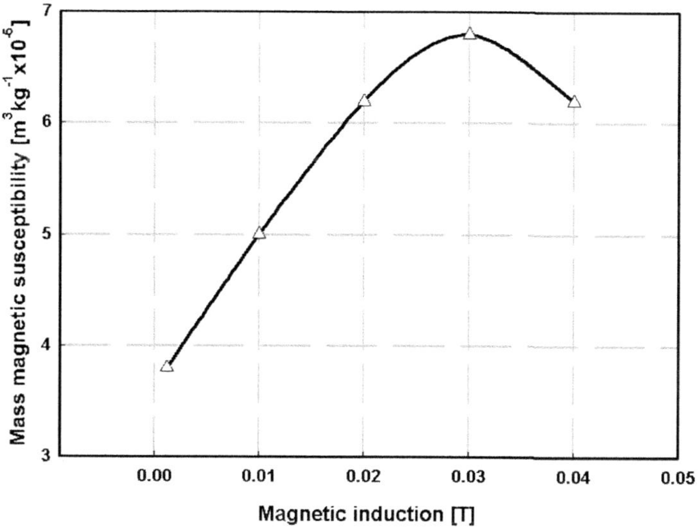

Figure 1.22: Mass magnetic susceptibility of pyrrhotite as a function of the magnetic induction (after [D4]).

to note that while the measurement methods have remained virtually unchanged for a long period, the equipment has been subjected to continuous development.

Practitioners as well as researchers who use magnetic techniques as a tool for mineral beneficiation, material concentration or removal will be interested primarily in the measurement of the magnetic field strength and its gradient and of magnetic susceptibility. There is a wide range of measuring instruments commercially available, some of them easily affordable and suitable for field applications.

1.6.1 Measurement of the magnetic field strength

There are several ways of measuring the magnetic field strength or magnetic induction. The choice of the measurement method depends on several factors. The field strength, its non-homogeneity, as well as the required accuracy all need to be considered. Figure 1.23 shows the accuracy of absolute measurements of magnetic field strength for a few commonly used methods.

The fluxmeter method

This method, which is based on Faraday's induction law, is one of the oldest, and still very accurate, techniques of the field measurement. It consists of placing a coil in the magnetic field so that part of that field is enclosed as a magnetic flux through the coil area. Switching off the field or removing the coil from the field gives rise to an induction voltage across the coil terminals. The flux is

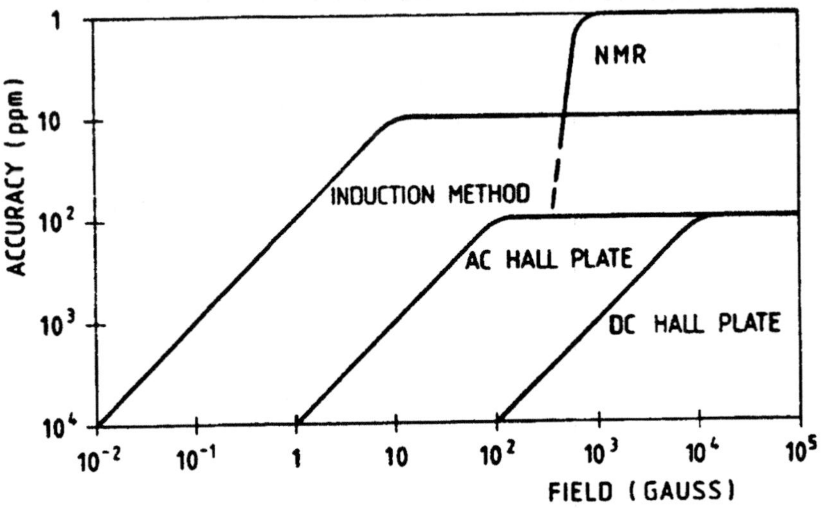

Figure 1.23: Measurement of the magnetic field strength: accuracies and ranges of various methods [H3].

then measured as the time integral of the voltage produced. Measurements are performed either by using fixed coils in a dynamic magnetic field, or by moving the coils in a static field.

The Hall Effect method

E.H. Hall discovered in 1879 that a metal strip immersed in a transverse magnetic field and carrying a current developed a voltage mutually at the right angle to the current and the voltage that opposed the Lorentz force on the electrons. The Hall-probe meters, known as gaussmeters, use very simple electronic instrumentation and provide an instant measurement. The small size of the probe sensing area (1 mm^2 and less), its static physical characteristics, as no motion of the probe is required, and there are no moving parts, make gaussmeters very useful instruments.

The basic premise of the Hall probe is that when a current I is flowing along one axis of a conductor or semiconductor, and when the magnetic field B is applied perpendicularly to the plane of the probe (or generator), then the Hall voltage U_H given by eq. (1.46) is generated perpendicular to both the current and the magnetic field. The situation is depicted in Fig. 1.24.

$$U_H = R_H \frac{I \cdot B}{d} \tag{1.46}$$

where d is the thickness of the Hall generator and R_H is the Hall constant.

The Hall voltage is thus, in the first approximation, proportional to the

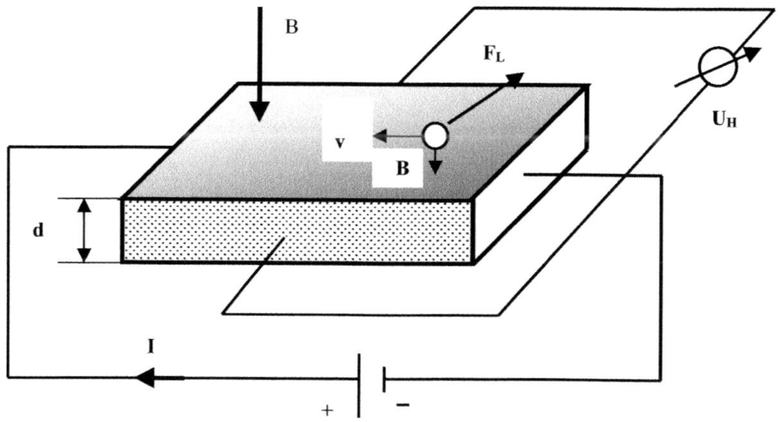

Figure 1.24: Schematic representation of the Hall-effect probe.

flux density. The Hall constant depends on various material parameters as well as temperature and, to a small extent, on the flux density B. The resulting linearity error is kept as small as possible by applying an appropriate ohmic resistor to the Hall voltage. This allows the construction of measuring devices that are linear to within ±1 % for a flux density of about 1 T. Equally important for accuracy is the effect of the temperature coefficient of the Hall voltage.

The Hall probes, usually made using indium arsenide, are very fragile. The Hall element is generally only a few hundredths of a millimeter thick and needs mechanical support. Aluminium is usually used as a substrate.

The Hall probes can be used to measure d.c. fields, a.c. fields and pulsed magnetic fields. The probes are available in various shapes and their use depends on the direction of the magnetic field to be measured. The probes are either transverse or axial. With the transverse probe, the flux density is measured perpendicular to the probe rod, while with the axial probe the measured magnetic field is in the direction of the probe axis. The overall views of hand-held and bench-scale gaussmeters are shown in Figs. 1.25 and 1.26.

In order to generate a high magnetic force on particles to be separated, the magnetic field in magnetic separators is generally very non-uniform. The value of the magnetic field thus changes considerably, often over a scale of several hundreds of micrometers, particularly in high-gradient type magnetic separators such as HGMS, or permanent magnet roll separators. In view of the finite size of the Hall sensing element, it is the mean magnetic field over the relatively large cross-sectional areas of the sensor, and not the maximum field, that is being measured by the Hall probe. With common commercially available Hall probes the measured magnetic field is usually much less than the maximum field. For instance, magnetic induction usually measured in an NdFeB permanent magnetic roll separator rarely exceeds 1 T, while a special Hall microprobe can

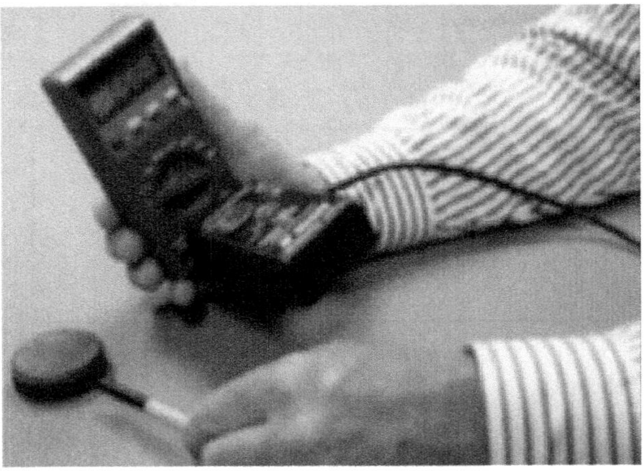

Figure 1.25: A portable hand-held gaussmeter (courtesy of Walker LDJ Scientific, Inc.).

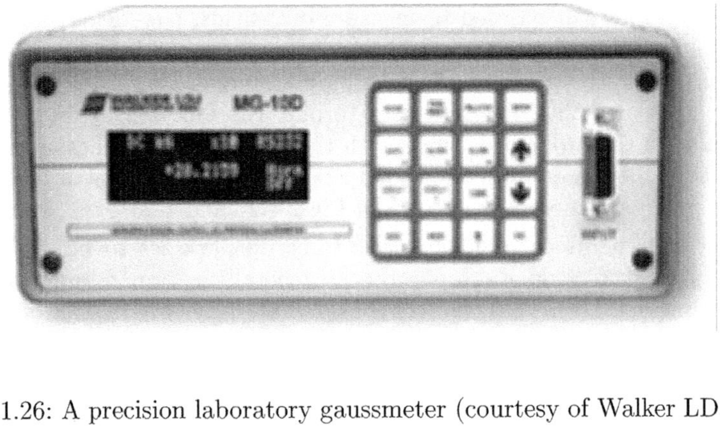

Figure 1.26: A precision laboratory gaussmeter (courtesy of Walker LDJ Scientific, Inc.).

measure the field as high as 1.9 T. Such values agree with the field strength obtained in computer modelling.

The above discussion indicates that it is rather difficult to measure accurately the magnetic field gradient in a magnetic separator. Although the knowledge of the spatial distribution of the magnetic field and its profile is important for serious understanding of the process of magnetic separation and of the design of magnetic separators, commercial Hall probes are usually unsuitable for this exercise. However, commercially available gaussmeters can be used for measurements of the magnetic field gradient in a low-gradient class of magnetic separators, such as magnetic pulleys and suspended magnetic separators. Accuracy of the measurement can be enhanced by devising a measuring rig [S7].

1.6.2 Measurement of magnetic susceptibility and magnetization

All techniques for the determination of magnetic susceptibility fall into two principal categories. The first is based on the measurement of a force that acts on a sample when it is placed in a non-uniform magnetic field. The other method makes use of the electromagnetic force induced in a coil when the magnetic field changes, or when there is a relative motion of the sample and the coil.

The force methods

The most familiar of the force techniques for measurement of magnetic susceptibility is the Gouy method. The sample under investigation is packed into a long cylindrical container, which is positioned so that one end is in a strong uniform magnetic field B_1 (near the centre of the magnet gap), while the other is in a much weaker field B_2, e.g. outside the gap. The magnetic force acting on the sample is equal to

$$F_m = \frac{\kappa_s - \kappa_m}{2\mu_0} A(B_1^2 - B_2^2) \tag{1.47}$$

where κ_s and κ_m are volume magnetic susceptibilities of the sample and of the surrounding medium, respectively, and A is the cross-sectional area of the sample. This force is detected as a weight differential between the field-on and field-off conditions. The principal merit of the Guoy method lies in its simplicity - one needs only a balance, a sample container and the magnet [D6].

The Faraday method rests in the measurement of the force exerted on a small sample by a non-homogeneous magnetic field. This can be achieved, for instance, by the use of special pole pieces. In this case the force on the sample can be written

$$F_m = \frac{\kappa_s - \kappa_m}{\mu_0} VB\frac{dB}{dz} \tag{1.48}$$

where z is the direction of the magnetic field gradient and V is the volume of the sample.

Figure 1.27: Magnetic susceptibility meter KLF-4 (courtesy of Agico Instruments).

Magnetic induction methods

In the simplest application of the induction principle, a sensing coil is wound around a sample. A second winding is used to generate variable magnetic field, which produces a magnetization. There are several instruments based on the induction principle available commercially. They are essentially induction bridges that measure magnetic susceptibility of a sample of any shape and form while the mass of the specimens can range from a few grams to several tens of grams. Figure 1.27 illustrates such an automatic computer-controlled inductivity bridge, which is suitable for rapid laboratory measurement of magnetic susceptibility of rocks, soils, minerals and other materials. Sensitivity of the instrument is 1×10^{-6} (SI) and the measurement is conducted at the magnetic field range from 10 A/m to 300 A/m (0.12 G to 3.8 G). A similar instrument with resolution of 2×10^{-6} (SI) is shown in Fig. 1.28.

An even more sensitive instrument is the KLY-3 Kappabridge shown in Fig. 1.29. It employs a fully automatic zeroing system with autoranging and automatic compensation of the thermal drift and its sensitivity is specified as 2×10^{-8} (SI).

While the above described laboratory susceptibility meters could be of considerable interest and use for practitioners and researchers in magnetic separation, they might not be easily affordable and sufficiently portable in plant operation. For field operation there are several types of portable susceptibility meters commercially available. Figure 1.30 shows such an easily affordable pocket-size susceptibility meter. The instrument weighs 0.18 kg and has the measurement sensitivity of 1×10^{-7} (SI). The field data can be stored in 250 memory registers and recalled later.

The advantages of the induction bridge susceptibility meters are their high accuracy, fast measuring rate and automated computer-controlled mode of operation. High sensitivity enables one to measure the magnetic susceptibility of

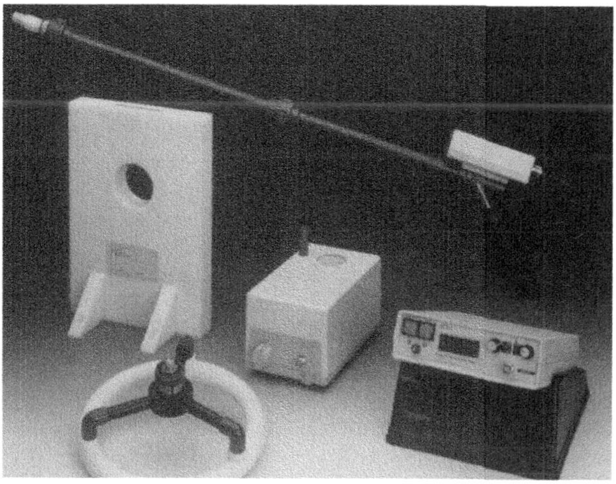

Figure 1.28: Magnetic susceptibility meter MS2 (courtesy of Bartington Instruments).

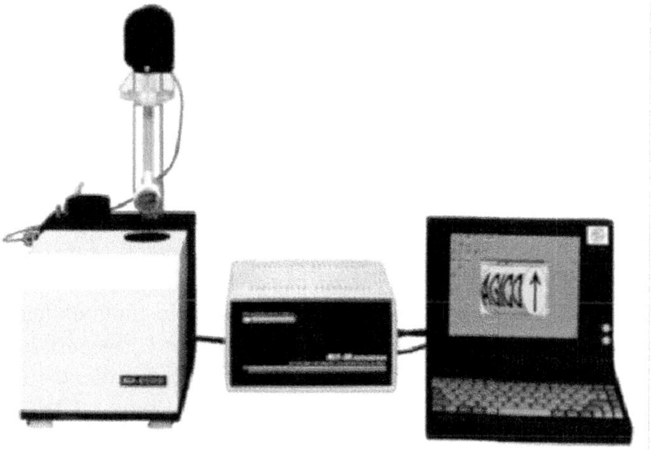

Figure 1.29: Magnetic susceptibility meter Kappabridge KLY-3 (courtesy of Agico Instruments).

Figure 1.30: Pocket-size magnetic susceptibility meter SM-30 (courtesy of ZH Instruments).

Table 1.12: Magnetic susceptibility of pyrrhotite measured by an induction bridge [S8].

Magnetic field [G]	**Volume magnetic susceptibility ([SI]x10^3)**
0.0256	186
0.9	233
5.36	357

materials ranging from diamagnetic to ferromagnetic.

The main drawback from the magnetic separation point of view is that these instruments operate at very low magnetic field and do not allow measurement of the susceptibility as a function of the magnetic field strength. The measurement of susceptibility at a low magnetic fields is only acceptable for paramagnetic and diamagnetic materials for which the susceptibility is field-independent. For materials with ordered structures, such as ferromagnetic (*s.l.*) materials, magnetic permeability and magnetic susceptibility are strongly dependent on the external magnetic field, as has been demonstrated in Figs. 1.9 and 1.15. Table 1.12 illustrates the dependence of the initial magnetic susceptibility on the magnetic field strength, as measured by an induction bridge.

It can be seen from Figs. 1.9 and 1.15 that the initial permeability or susceptibility is generally smaller than the effective permeability at the magnetic field at which magnetic separation takes place, i.e. $\mu_i < \mu_{eff}$. If the external magnetic field is sufficiently high then $\mu \rightarrow 1$ and $\kappa \rightarrow 0$, and the magnetic force on a particle will be determined by the saturation magnetization M_s as follows from eq. 1.44. The value of M_s can be high or low, depending on the

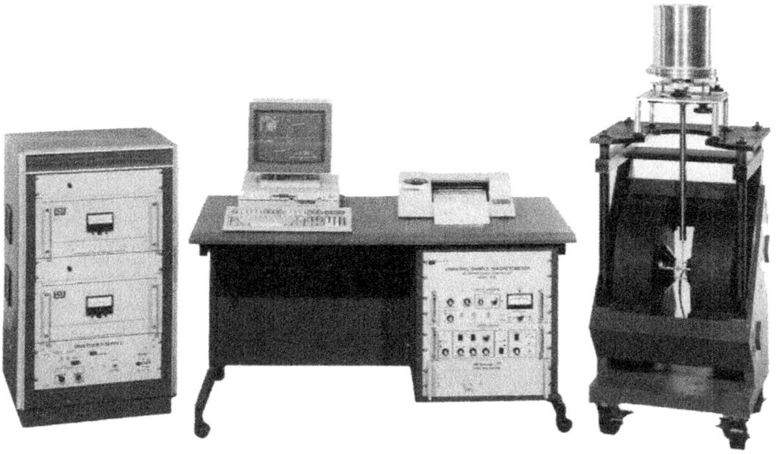

Figure 1.31: Vibrating sample magnetometer (courtesy of LDJ, Inc.).

type of material, irrespective of what the value of the initial permeability (or susceptibility) is. It is thus clear that evaluation of separability of ferromagnetic (*s.l.*) materials based on the measurement of the initial magnetic susceptibility can lead to erroneous conclusions.

Vibrating sample magnetometer

A vibrating sample magnetometer (VSM), shown in Fig. 1.31, is one of the most popular techniques in magnetometry [F3]. The sample is vibrated at small amplitude with a known frequency and phase in a uniform applied magnetic field. The field distortion produced by the sample varies in an a.c. manner and is detected by a pick-up coil. Using a lock-in amplifier to filter out signals at frequencies other than the sample vibration frequency, VSM is able to make highly sensitive measurements. The magnitude of the signal is dependent on the magnetic properties of the sample. The external magnetic field can be produced by superconducting magnets, various types of electromagnets or Helmholtz coils, providing excitation fields from a fraction of milliTesla to many Tesla.

For all magnetizable materials, the VSM test results are displayed as magnetization curves. For ferromagnetic (*s.l.*) materials the results are in the form of the simple hysteresis loop, with values provided for all of the critical parameters, such as coercive force, remanent and saturation magnetization, applied magnetic field and others. VSM can be used to test magnetic properties of metal alloys, minerals, permanent magnets, thin films, magnetic fluids etc. A typical hysteresis curve, as obtained by VSM, is shown in Fig. 1.32.

For purposes of evaluation of magnetic separability of materials, VSM is a very useful, albeit rather expensive instrument. Since susceptibilities of diamagnetic and paramagnetic materials do not depend on the strength of the

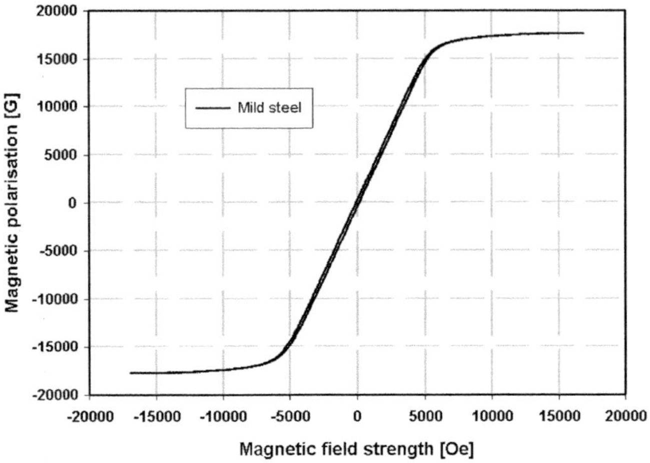

Figure 1.32: The hysteresis curve for mild steel, as obtained by VSM.

external magnetic field, information obtained using inductive bridges is sufficient to access separability of such materials. For ferromagnetic (*s.l.*) materials, however, the magnetic susceptibility is strongly dependent on the field strength and separability of such materials is determined by the magnetic properties at the operating magnetic field strength of a given magnetic separator.

Vibrating sample magnetometer allows the determination of the magnetic susceptibility at any point on the hysteresis curve (Fig.1.32), as the tangent to the curve. It is thus possible to determine the magnetic susceptibility at any value of the magnetic field. This procedure is useful only in the linear part of the magnetization curve. Once the material begins to saturate magnetically, the susceptibility approaches zero and the magnetic force on a particles is determined by the saturation magnetization. A typical field dependence of the magnetic susceptibility as determined from the magnetization curve obtained using VSM is shown in Fig. 1.33.

As has been stated earlier, in magnetic separation many paramagnetic and diamagnetic materials, and minerals in particular, contain ferromagnetic (*s.l.*) admixtures. These admixtures can significantly influence the values of magnetic susceptibility as a consequence of its dependence on magnetic field, temperature and particle size. A correct evaluation of magnetic susceptibility in terms of eq. (1.44), under given experimental or operational conditions, can only be effected with instruments such as VSM. However, VSM is only rarely accessible to a practitioner or even researcher in magnetic separation and detailed information about magnetic properties of minerals available in literature is limited. An exception is the work of Dahlin and Rule [D2] who investigated the magnetic susceptibility of numerous minerals as a function of the magnetic field strength.

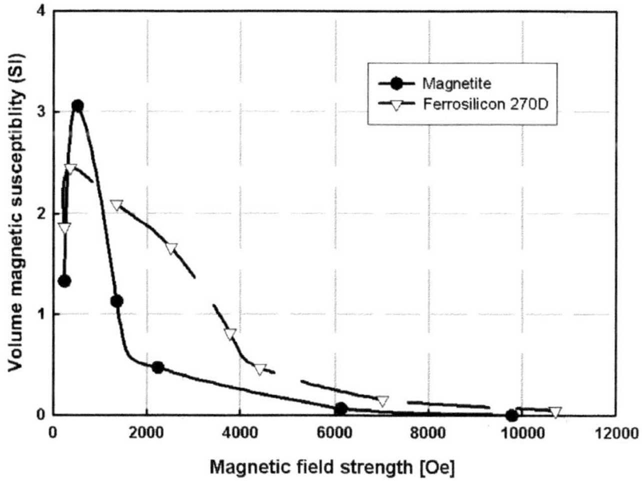

Figure 1.33: Volume magnetic susceptibility of magnetite and ferrosilicon 270D, as a function of magnetic field, as measured by VSM [S8].

A vibrating sample magnetometer with a superconducting magnet capable of producing a magnetic field of 7.16×10^6 A/m (90 kOe) was used. The results showed that magnetic susceptibility of most minerals that were investigated was essentially constant in magnetic fields above those needed to saturate the ferromagnetic constituents, in agreement with eq. (1.44). Figure 1.34 illustrates some of the results.

Hysteresisgraphs

Hysteresisgraphs are systems for testing and analyzing hysteresis properties (B-H loop) of hard and soft magnetic materials. The commercially available units are usually precision personal computer-controlled systems. Designers of permanent magnet-based magnetic separators can use such instruments to test, analyze and quality-control hard magnetic materials such as ferrites, Alnico, samarium-cobalt and neodymium-iron-boron alloys. Figure 1.35 illustrates such a hysteresisgraph.

1.7 Sources of magnetic field in magnetic separation equipment

The aim of magnetic separation is to recover or remove particles with sizes ranging from several tens of millimeters down to a fraction of micrometre, with a wide spectrum of magnetic susceptibilities, from ferromagnetic to weakly paramagnetic. Furthermore, it is frequently necessary to separate two or more ma-

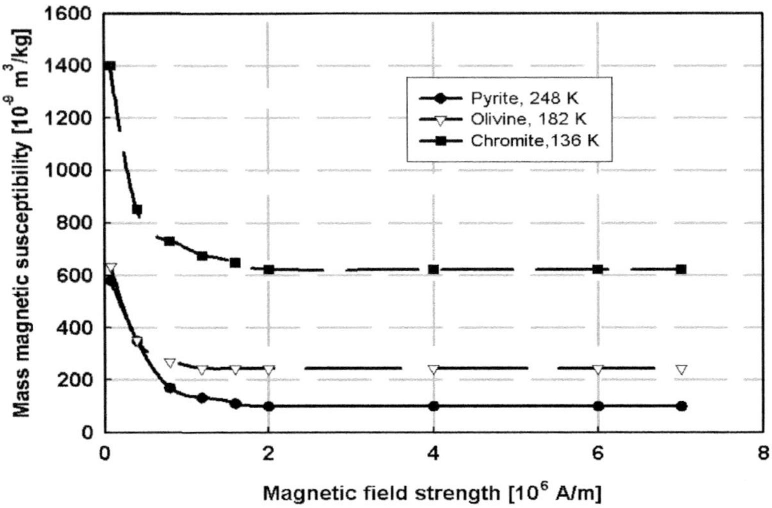

Figure 1.34: Magnetic susceptibilities of various minerals, measured by VSM, as a function of magnetic field strength (adapted from [D2]).

Figure 1.35: Hysteresisgraph Permagraph (courtesy of Magnet Physik GmbH) for quality control of permanent magnets.

terial species, the magnetic susceptibilities of which do not differ very much, of a wide range of particle sizes. The magnetic field strength needed to recover such particles, therefore, varies substantially from one application to another and different ways of generating the magnetic field are employed.

As can be seen from eq. (1.8), the magnetic force required to separate a magnetizable particle depends not only on the physical properties of a material, and on the magnetic field strength, but also on the value of the gradient of the magnetic field. Therefore, to achieve a selective separation of a valuable material from unwanted ones, a carefully selected combination of the magnetic field strength and its gradient must be used to warrant a high recovery and simultaneously a high grade of the concentrate. Magnetic separation can be considered successful only if it produces the right quality of the concentrate with the right recovery.

In some applications of magnetic separation, for instance in magnetic filtration, the "brute force" approach is often used to ensure a high degree of removal of magnetizable solids from suspension. Consequently an effort is usually made to maximize the external magnetic field or the field gradient, or both. This approach is also employed in some applications in mineral processing, for instance in removal of magnetic contaminants from industrial minerals. Such a combination of high magnetic field and field gradient often results in a high-quality non-magnetic product, at the expense of losses of the non-magnetic component into the magnetic fraction. Undue emphasis on the magnetic field strength and/or field gradient sometimes results in reduced recovery of the valuable components and reduced selectivity. Excessively high capital and operating costs are additional consequences of such an approach. In the majority of applications, only a moderate magnetic field, combined with an optimized field gradient, are required to achieve the optimum results.

The desired magnetic induction and the field gradient are thus a function of the type of application, required metallurgical performance and economic criteria. In the following section the basic methods of the generation of magnetic fields and their gradients, as used in magnetic separation, will be reviewed.

1.7.1 Permanent magnets

Recent years have seen a strongly increasing industrial demand for permanent magnet materials. Permanent magnets have become essential components in many electric, electronic and electromechanical devices, including magnetic separators. Although the earliest permanent magnet, lodestone, was known in ancient times, the greatest strides in magnet development have occurred only over the past hundred years. Each advance has been connected with the discovery of a new class of materials, characterized by ever more desirable properties. This evolution has been monitored generally by one figure of merit for a permanent magnet, the maximum energy product $(BH)_{\max}$, which provides a measure of the field that can be produced outside a unit volume of magnet material. The so-called theoretical maximum energy product, the largest value realizable in

Table 1.13: Magnetic properties of Alnico alloys.

Material	**B$_r$ [T]**	**H$_c$ [Oe]**	**H$_{ci}$ [Oe]**	**(BH)$_{max}$ [MGOe (kJ/m^3)]**
Alnico 5	1.27	640	645	5.5 (44.0)
Alnico 5-7	1.34	740	745	7.5 (60.0)
Alnico 5DG	1.33	685	690	6.5 (52.0)
Alnico 8B	0.9	1600	1640	6.75 (54.0)
Alnico 9	1.05	1500	1515	10.5 (84.0)

principle, is an intrinsic quantity defined by

$$(BH)_{\max} = \frac{1}{\mu_0}(\frac{J_s}{2})^2 \tag{1.49}$$

where J_s is the saturation polarization (in [T]).

This maximum energy product (in [J/m^3]) can be achieved only if the magnet retains J_s in a reverse field at least as large as $J_s/2$ [H4]. The maximum energy product is thus limited, to perhaps 1.2×10^6 kJ/m^3 (150×10^6 MGOe), by the maximum saturation polarization of elements and alloys of approximately 2.5 T. At the present time, one third of this maximum has been achieved with NdFeB material. The chronological trend of the maximum energy product achieved in various permanent magnet materials is shown in Fig. 1.1. It can be seen that there are four major families of permanent magnet materials: Alnico, ferrite, rare-earth-cobalt and rare-earth-iron.

Alnico

A major advance in permanent magnet technology began in 1932 with the discovery of excellent magnetic properties of an aluminium-nickel-iron alloy, Alnico. Numerous subsequent investigations resulted in many variations in composition and properties. Alnico magnets are expensive when compared to ferrite magnets since they contain 20% to 40 % of cobalt in addition to other components. Alnico alloys have desirably high remanent magnetization, good thermal stability, but comparatively low coercivities. Curie temperature is about 900 ^{0}C and the maximum recommended operating temperature is 500 ^{0}C. Alnico magnets are very resistant to corrosion, nearly as resistant as stainless steel. Only slight discolouration of the surface is encountered when a magnet is heated in air to 450 ^{0}C [P4]. The physical properties of Alnico alloys are rather poor. High coercivity is closely accompanied by extreme hardness and brittleness. Forming is by casting or sintering as close as possible to the desired size and shape. For close tolerance it is necessary to cut or wet-grind the magnet [P4]. Magnetic properties of several grades of Alnico are summarized in Table 1.13.

Table 1.14: Magnetic properties of the ferrite magnets.

Material	B_r [T]	H_c [Oe]	H_{ci} [Oe]	$(BH)_{max}$ [MGOe (kJ/m^3)]
Ferrite 5	0.395	2200	2230	3.6 (29.0)
Ferrite 7B	0.38	3250	3800	3.3 (26.0)
Ferrite 8A	0.39	2950	3000	3.5 (28.0)
Ferrite 8D	0.40	3100	3000	3.8 (30.0)
Ferrite 8C	0.43	4100	2200	4.3 (34.0)

Ferrite magnets

A new class of permanent magnetic materials was announced in 1952. The materials were based on the crystal anisotropy of barium oxide. This class of magnets is generally known as ferrite, but is sometimes referred to as oxide or ceramic magnets. Today, ferrite magnets are by far the most widely used magnets. The success of ferrites is due to several reasons. The raw materials are inexpensive and non-strategic. Their high coercive force combined with reasonably high magnetic induction has allowed permanent magnets to be incorporated into many types of devices. Ferrites are mechanically hard and brittle and their magnetic properties are characterized by relatively low remanent magnetization while the coercive force is satisfactory. The Curie temperature is 450 ^{0}C and the recommended maximum operating temperature is 250 ^{0}C. At the same time, surfaces of ferrite magnets are stable and are not subject to oxidation.

Ferrites are produced by powder metallurgy. Their chemical formula can be expressed as $MO{\cdot}6(Fe_2O_3)$, where M is Ba, Sr or Pb. Strontium ferrite has a higher coercive force than barium ferrite and is produced on a larger scale [P4]. Ferrite magnets are available in isotropic and anisotropic grades. Magnetic properties of various grades of ferrites are shown in Table 1.14.

Samarium-cobalt-based rare-earth magnets

The family of rare-earth permanent magnets (REPM) has evolved in the last 35 years. The materials quickly surpassed all earlier magnets, with currently best values of energy product and coercivity, both being 5 to 10 times those of Alnico and ferrites. While they are rather costly for universal use, the REPM, together with the hard magnetic ferrites, are rapidly broadening applications of permanent magnets in general [S9]. Two major subgroups are being distinguished: rare-earth Co-based magnets (particularly Sm-Co) and rare-earth Fe-based magnets represented by Nd-Fe-B. The latter third-generation REPM, in rapid development since 1983, use cheaper, more plentiful raw materials than Sm-Co and offer still higher energy density at room temperature.

Sm-Co-based rare earth magnets are of two types: the $SmCo_5$ group and Sm_2Co_{17} group. These two groups constitute the first and second generation rare-earth magnets, respectively [H5]. $SmCo_5$ alloys are the most basic materials that make up rare-earth magnets. More advanced Sm_2Co_{17} alloys were devel-

Table 1.15: Magnetic properties of Sm-Co magnets [W1].

Material	B_r [T]	H_c[kOe]	H_{ci} [kOe]	$(BH)_{max}$ [MGOe (kJ/m^3)]
$SmCo_5$	0.90	9.0	29.0	20.2 (162.0)
Sm_2O_{17}	1.10	10.0	33.0	37.5 (300.0)

oped following the success of $SmCo_5$, to obtain higher performance. In order to increase its rather low coercivity, the latter alloy is now based on Sm-Co-Cu composition. The general composition of this new class of Sm-Co magnets is $R(Co,M)_z$, where R stands for at least one rare-earth element, M for a combination of transition metals and/or copper and z is between 6 and 8.5. These multicomponent magnet alloys are generally termed the 2:17 systems. They have the advantage of higher magnetization and use less Sm and Co. As a results of their high Curie temperature of 700 ^{0}C to 800 ^{0}C, Sm-Co-based permanent magnets satisfy the needs for applications with operating temperatures up to about 300 ^{0}C. Further demand for operation at even higher temperatures has led to development of a new class of Sm-based alloys, which can operate at temperatures as high as 550 ^{0}C [W1]. Magnetic properties of Sm-Co magnets are listed in Table 1.15.

The samarium-cobalt materials are attractive replacements for electromagnets in many applications, but their use has been constrained by economic disadvantages. Samarium is the least abundant, and hence the most expensive, of the light rare earths, and this unfavourable circumstance is compounded by the fact that the price and availability of cobalt are subject to large, unpredictable excursions. The bulk of cobalt is mined in Congo and regional political instability in Africa resulted in diminished supplies and radically increased prices. This situation led to a continued effort to develop alternative high performance magnets not containing cobalt.

Neodymium-iron-boron magnets

Iron-based materials having the characteristics of Sm-Co had long been desired and in 1983 two different processing routes were announced. The standard powder metallurgy route (Sumitomo route) [S10] and the melt-spinning method (General Motors route) [C3] were proposed for the production of Nd-Fe-B permanent magnets. In the ensuing years production and application of this new type of magnet has been steadily increasing. It has been predicted that the NdFeB permanent magnet market will grow from the 2002 value of $2 billion to $4.8 billion in 2007 [B7].

Presently three common fabrication routes can be used to categorize NdFeB magnets: sintering, polymer bonding and hot deformation. The most successful procedure on the production scale is the powder metallurgical method. The reason is that sintered NdFeB magnets have outstanding magnetic properties, good productivity and superior energy efficiency and cost performance. On

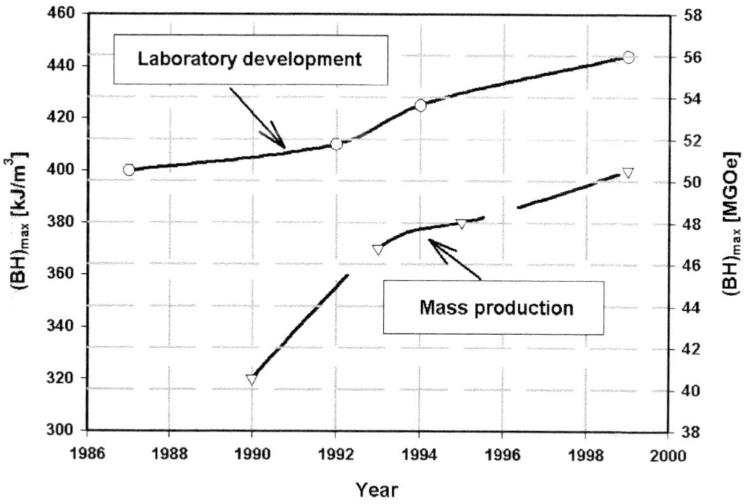

Figure 1.36: Progress in development of NdFeB sintered permanent magnets (after [K4]).

the other hand, sintered NdFeB magnets have certain disadvantages in terms of thermal and environmental stability. The theoretical value of the maximum energy product of the magnetic $Nd_2Fe_{14}B$ phase is 512 kJ/m^3 (64 MGOe) [S11]. So far, energy products exceeding 400 kJ/m^3 have been achieved, with the latest record of 444 kJ/m^3 (55.8 MGOe) [K4]. Progress in development of NdFeB sintered magnets is depicted in Fig. 1.36.

Bonded magnets have intermediate energy products of 80 to 145 kJ/m^3 (10 to 18 MGOe) and can be formed into intricate shapes. Hot deformed magnets possess intermediate to high energy product of 120 to 370 kJ/m^3 (15 to 46 MGOe), isotropic and anisotropic properties and the potential to be formed into net shapes. Comparison of magnetic properties of several types of NdFeB permanent magnets is shown in Table 1.16.

While NdFeB permanent magnets possess outstanding magnetic properties, they have insufficient level of thermal and environmental stability. The Curie temperature of the NdFeB alloys is about 310 ^{0}C. This proximity to the room temperature limits the practical range of applications of NdFeB magnets and causes appreciable changes in the flux density as the temperature varies. Operation of a permanent magnet at elevated temperatures can result in structural changes, which in turn lead to irreversible losses of magnetic properties. Table 1.17 compares the loss of remanent magnetization, at 20 ^{0}C, of various magnet materials, after exposure to the indicated temperature. Another study [F4] observed that after 30 minute exposure to the temperature of 220 ^{0}C NdFeB magnets lost up to 80% of their magnetization.

The maximum operating temperature of NdFeB magnets is about 150 ^{0}C.

Table 1.16: Magnetic properties of NdFeB permanent magnets [B7].

Magnet	$\mathbf{B}_r$ **[T]**	$\mathbf{H}_{ci}$ **[kOe]**	$(\mathbf{BH})_{\max}$ **[MGOe (kJ/m^3)]**
Sintered VCM	1.31	14	42 (336.0)
Bonded MQ1-B	0.69	9	10 (80.0)
Hot-pressed MQ2-E	0.825	17.5	15 (120.0)
Die-upset MQ3-F	1.31	16	42 (336.0)
Sintered (Sumitomo) [K4]	1.51	8.68	55.8 (445.0)

Table 1.17: Temperature-induced irreversible losses in permanent magnets [P4].

Magnet	**-60^0 C**	**+100^0 C**	**+200^0 C**	**+300^0 C**
Alnico	-1.45%	0.4%	0.8%	1.1%
Ferrite	-27%	0	0	0
$SmCo_5$	0	0.5%	1.7%	
NdFeB	0	6.5%	65.0%	

Comparison of the Curie and maximum operating temperatures for various permanent magnets is given in Table 1.18.

NdFeB permanent magnets are also sensitive to attack by both climatic and corrosive environments, resulting in a deterioration of the magnetic and physical properties. It appears that the susceptibility to corrosion is a problem that all NdFeB magnets experience, regardless of the fabrication process [H6]. It was found [C4] that the corrosion rate increases more with an increase in the humidity level than it does with an increase in the environmental temperature. It was also observed that nitrogen in the air retards the rate of corrosion. In practice, surface protection by epoxy resins or by Al coating, combined with chromating is applied. Typical coating thickness is from 10 μm to 24 μm for Al-chromated and about 25 μm for electrically deposited epoxy [H6]. The application of two separate coatings, shot-peened and chromated aluminium deposited by ion vapour

Table 1.18: Curie and maximum operating temperatures of various permanent magnets.

Magnet	**Curie temperature [^{0}C]**	**Maximum operating temperature [^{0}C]**
Ferrite	450	350
Alnico	900	500
$RECo_5$	700	250
NdFeB	310	150

deposition and electropaint deposited cathodically have been shown to resist a variety of climatic and corrosive atmospheres [M3].

Permanent magnets in magnetic separation

The early history of magnetic separation is closely associated with permanent magnets, which were used to remove strongly magnetic matter from wastes or to recover strongly magnetic (mostly iron) minerals from ores. With the development of powerful electromagnets the interest in permanent magnets as applied to magnetic separation temporarily diminished. In recent years, however, we have witnessed a remarkable renaissance of interest in permanent magnets, a phenomenon associated with the dramatic development of novel permanent magnets materials, based on rare earths.

Although the magnetic induction that can be generated by permanent magnets rarely exceeds 1 T, the value that can be easily produced by electromagnetic coils, in a much larger volume, permanent magnets offer several advantages over electromagnets. In small devices, permanent magnets have major advantages over electromagnetism. Even in large systems, however, permanent magnets are a good solution, as a result of their advantageous volume efficiency, low operating costs and simple support facilities. And very importantly, permanent magnets do not require a source of energy or a cooling fluid.On the other hand, permanent magnets possess a few disadvantages, among them the fact that the magnetic field is difficult to vary and that it cannot be switched off.

As a result of notable development of new permanent magnet materials during the last three decades, permanent magnets are currently used in a wide range devices for magnetic treatment of materials. In most applications, in which a relatively short reach of the magnetic force is required, permanent magnets replaced electromagnets. Permanent magnetic rolls, magnetic pulleys, suspended magnets, high-field permanent drum magnetic separators and eddy-current separators are typical examples. Even in those applications of magnetic separation where the magnetic force must be generated in a relatively large volume, permanent magnets are employed. Multiple-ring high-gradient magnetic separators are an example. In the application of magnetic techniques in biosciences permanent magnets are an essential part of biomagnetic separators and devices for drug targeting and hyperthermia.

1.7.2 Iron-core electromagnets

Of the several ways of generating a magnetic field in magnetic separation, the choice is dictated mainly by the required field magnitude, by the volume in which a given magnetic field is to be generated and by methods used to generate the field gradient. An effective separation of weakly magnetic particles of a small size often requires an increase of both the magnetic field and of its gradient beyond the values available with permanent magnets. Furthermore, as the magnetic field falls very rapidly from the surface of a permanent magnet, the effective volume in which a sufficiently high magnetic force can be created is limited.

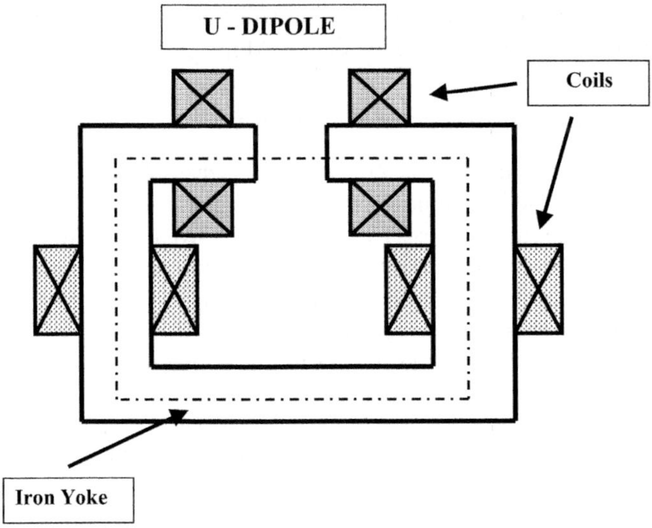

Figure 1.37: Schematic cross-section of the U-dipole iron-core electromagnet, with two possible positions of the magnetizing coils.

Therefore, in order to generate a sufficiently high magnetic field in a large volume that would guarantee a large throughput of a magnetic separator, an electromagnet must be used. Of the several ways of generating a magnetic field by an electromagnet, the choice is dictated by consideration of the required field magnitude, the volume in which the field is to be generated and by the duty cycle of the separation process. The relative importance of these considerations usually determines the method with little overlapping of choice. For example, if the separation process requires a magnetic field less than 2 T in a relatively small volume, an iron-core electromagnet is a good choice. For the field of 2 T or less, in a large volume, a water-cooled iron-clad solenoid should be used, although a superconducting magnet could also be a choice. Magnetic fields in excess of 2 T will undoubtedly dictate the use of superconducting solenoids.

In iron-core electromagnets, the electric current passing through the windings magnetizes a great mass of iron, which in turn produces a field in the working volume. Judicious location of the iron and of the windings may increase considerably the efficiency of the magnet. Although the field strength obtained in this way is limited by the saturation magnetization of iron, for most large-scale applications of magnetic separation for the recovery of weakly magnetic materials, the magnetic force generated by such iron-core electromagnets is usually sufficient.

Design of iron-core electromagnets is largely a matter of practical experience, and it is generally considered impossible to predict accurately the field which a particular magnet will produce [I1]. The situation has improved recently, when modern computer modelling software has been made available to magnet de-

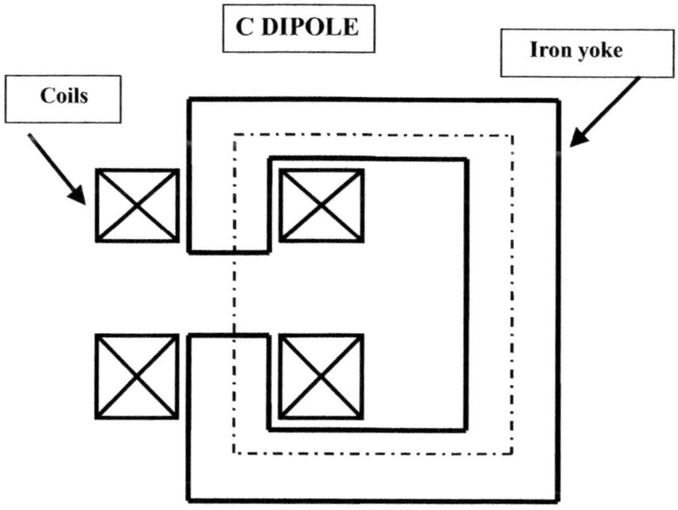

Figure 1.38: Schematic cross-section of the C-dipole iron-core electromagnet.

signers. The fact that the field magnitude and configuration is almost always determined by the iron pole pieces has both advantages and disadvantages. On the one hand, the presence of iron permits useful values of the magnetic induction (say, up to 2 T) to be reached with modest investments in magnetomotive force (ampere-turns). Iron also confines the stray magnetic flux to the immediate vicinity of the gap of the magnet. On the other hand, numerous drawbacks are due to the properties of iron such as nonlinearity, saturation, hysteresis and remanent field.

Figures 1.37 and 1.38 schematically illustrate cross-sections of the U-dipole and C-dipole electromagnets. These configurations are frequently used in the design of magnets for magnetic separators. The U-dipole design has been employed in conventional wet high-intensity magnetic separators (WHIMS) such as Jones (Humboldt Wedag), Eriez Magnetics or Boxmag-Rapid. On the other hand, the C-dipole configuration is used, for instance, in induced magnetic roll separators. A detailed description of such units will be given in Chapter 2, while the fundamentals of the design of iron-core magnets will be discussed in Chapter 4.

It is clear from Figs. 1.37 and 1.38 that as the working space of the magnet becomes larger, the iron core and the coils also become larger and therefore require increasing amount of power to generate the required magnetizing field. There is a definite crossover point at which a greater magnetic field can be generated for the same power in a solenoid of bore equal to the gap of the iron-core magnet [M4]. Iron can still be used in the form of a cladding surrounding the magnet.

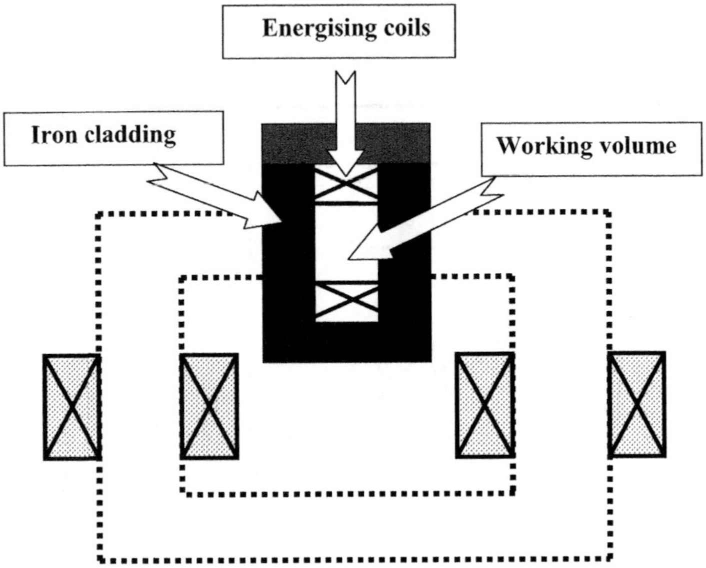

Figure 1.39: Comparison of a solenoid magnet (solid lines) and an iron-core electromagnet (dotted lines).

1.7.3 Solenoid electromagnets

As has been already mentioned, the magnetic induction obtained with iron-core electromagnets is limited by the saturation of the iron. To obtain still stronger fields, at even larger volume, one has to return to the original technique of the magnetic field generation, namely the construction of a solenoid. Instead of making use of the magnetization of iron, the field should be obtained directly from the current in the electrical conductor. Hence the magnet must be constructed in such a way that the windings are as close as possible to the field space, and the current through them must be as high as possible. Thus we come to a magnet consisting of a short coil, usually clad with a steel return frame which facilitates the passage of the magnetic flux. Comparison of iron-core electromagnet with an iron-clad solenoid both having the same width of the working space is shown in Fig. 1.39.

The total magnetic induction in the working volume of a magnet is equal to the sum of the contribution from the coil (B_c) and of the contribution from the magnetized iron (B_i)

$$B_{tot} = B_c + B_i \tag{1.50}$$

Approximately 75% to 90% of B_i comes from the iron in the proximity of the working volume. Therefore, approximately the same portion of iron of the yoke could have been eliminated without reducing B_i [M5]. Contribution from the coils to the field in the working gap in iron-core configurations (Figs. 1.37 and 1.38) is thus generally rather small. The coils are used mainly to magnetize

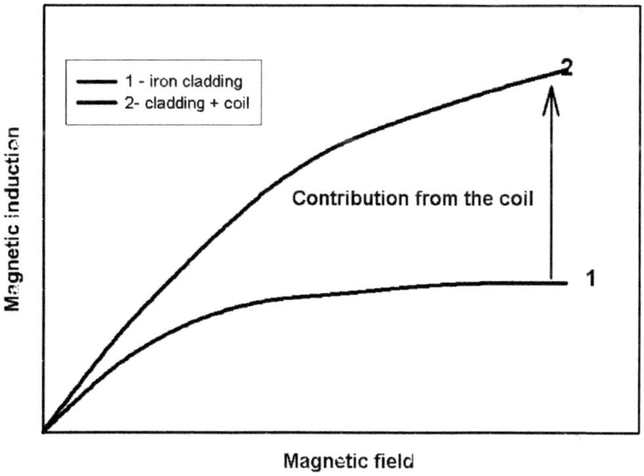

Figure 1.40: Contribution from iron cladding and from the coil to the total magnetic induction in the iron-clad solenoid.

iron and to generate high B_i. Nevertheless, if the iron-clad solenoid is used, the contribution from the coil to the magnetic induction B_c from the coil is much higher than in the iron-core magnet, while B_i remains the same. Comparison of the magnetization curves for these two magnetic circuits is shown in Fig. 1.40. It can be seen that B_i (iron-core magnet) $\approxeq$ B_i (iron-clad solenoid) and B_c (iron-core magnet) $<<$ B_c (iron-clad solenoid). It is thus clear that the iron-clad solenoid uses much less material and exhibits better parameters.

The concept of iron-clad solenoids has been used successfully in the design of magnetic separators. Design of conventional high-gradient magnetic separators that operate in a cyclic mode, such as those used to beneficiate kaolin, is based on an iron-clad solenoid. Several designs of continuous HGMS machines, such as Sala-HGMS carousel separator, VMS and SLON separators also use the iron-clad solenoid approach.

1.7.4 Superconducting magnets

Superconductivity is a remarkable phenomenon whereby certain metals and other compounds, when cooled to very low temperatures, become perfect conductors of electricity. Unlike the gradual changes in electrical resistance shown by all metals at more familiar temperatures, the superconducting state appears quite abruptly at a critical temperature T_c, which is characteristic of the metal in question. It is accepted that below this temperature the resistance of conventional class II superconductors suitable for magnet manufacture is indeed zero [H7]. Critical temperatures of these superconductors are very low, a few Kelvin (K), and the use of liquid helium as a refrigerant is essential. In recent

Table 1.19: Critical temperatures and magnetic induction of assorted superconductors.

Material	T_c [K]	B_c [T] at 0 K
Nb_3Sn	18.0	≈28
NbTi	10.6	≈15
V_3Ga	15.0	≈26
Nb_3Al	16 - 20	≈ 35
Y-Ba-Cu-O	92	140 (‖), 29 (⊥)
Hg-Ba-Ca-Cu-O	135 - 160	1150 (‖), 149 (⊥)

years, however, remarkable progress has been achieved in developing materials that become superconducting at temperatures as high as 160 K [M6, F5, C5].

After the discovery of superconductivity in 1911, it was recognized that large electromagnets, capable of generating high magnetic fields and consuming no power, could be built. However, these hopes were quickly dashed when it was discovered that superconductors, in addition to their critical temperature, have a critical field B_c, above which they revert from the superconducting state to a normal resistive state. Metals that were investigated at those times, i.e. mercury, lead and tin, the so–called class I superconductors, exhibited very low critical field, of the order of 0.05 T.

The era of superconducting magnets started only in 1960 after the discovery of the high-field properties of new class II superconductors, namely Nb_3Sn and NbTi. During the following years, the search for new materials of ever higher performance was actively carried out and a multitude of composites, such as V_3Ga and Nb_3Al, were investigated. Most of these materials proved to be difficult to prepare in the form of conductors and progressively disappeared [D7]. The only materials available commercially at present are the two very first Nb_3Sn and NbTi. The latter is by far the most widely used because of its excellent mechanical properties, while the former one, although difficult to work with, is the only choice for fields exceeding 10 T [D7]. Critical temperatures T_c and critical magnetic induction B_c of assorted superconducting materials are summarized in Table 1.19.

There are three obvious areas where superconducting magnets will have considerable impact. The first area covers all applications which will benefit from a significant reduction in the consumption of electrical energy. The second area includes those applications in which the reduction of the mass and size of the magnet is important. The third area is the market for very high field magnets where there is no viable alternative, or the magnets that can generate magnetic field in a large volume. An example of a superconducting solenoid is shown in Fig. 1.41.

As discussed above, conventional magnets make use of iron yoke or cladding to reduce magnetic reluctance, thereby minimizing the number of ampere-turns required. Superconducting magnets do not suffer from this constraint because additional ampere-turns are to be had relatively cheaply and do not significantly

Figure 1.41: A superconducting solenoid for high-energy experiments (courtesy of Ansaldo spa.).

affect the refrigeration power load. For this reason iron core or cladding are seldom used to augment the working field, but they are often used for shielding, to reduce stray magnetic fields [W2].

All these three areas play a role in the application of superconductivity to magnetic separation. With the use of a superconducting magnet in material separation it should be possible to generate very high magnetic fields at low input power, in large volumes. On the other hand, the need for these very high magnetic fields in material separation has not been sufficiently convincingly demonstrated. Although it was envisaged [K1] that magnetic separation will be the first industrial application of superconductivity, superconducting magnets have not yet made a significant impact in material processing.

Superconductivity has found application in cyclically operating solenoid-based high-gradient magnetic separators for beneficiation of kaolin. Further development resulted in industrial-scale operation of continuous reciprocating-canister HGMS. On the other hand, superconducting drum magnetic separators and open-gradient magnetic separators have not been, so far, successfully introduced into the material treatment industry.

1.7.5 Generation of the magnetic field gradient

As transpires from eq. (1.8), the magnetic force acting on a magnetizable particle in a magnetic separator is determined not only by the strength of the magnetic field but also by its gradient. A suitable combination of these parameters must be determined for each application. It is thus essential to provide

Figure 1.42: Pattern of the magnetic field around a suspended magnet.

not only a strong magnetic field but also a non-uniform field.

A large variety of designs have evolved over the years to render the magnetic field non-uniform and to produce suitable magnetic gradients. Separators based on these modes of the field gradient generation can be classified into three basic categories:

- Open-gradient separators
- Closed-gradient separators
- Matrix (polygradient) separators

In **open-gradient separators** the field gradient is generated in several ways:

a. As a result of the natural non-homogeneity of the field caused by the dependence of the field on the distance from its source. Typical examples are overband, suspended, plate and grate magnets. The pattern of the magnetic field around a suspended magnet as shown in Fig. 1.42 illustrates the formation of the field gradient in the separation zone below the magnet.

b. By special arrangements of permanent magnets or of the windings of electromagnets so that the patterns of alternate polarity are generated. Magnetic rolls, pulleys and drum separators use this type of the field gradient. Multipolar windings of conventional or superconducting coils with alternating polarity are employed in open-gradient magnetic separators (OGMS). Fig. 1.43 illustrates the open-gradient magnetic system of a drum magnetic separator while Fig. 1.44 shows the 3D pattern of the field around a permanent magnet roll. The magnitude of the gradient and the reach of the magnetic force are determined, among other things, by the dimensions of the permanent magnet elements or

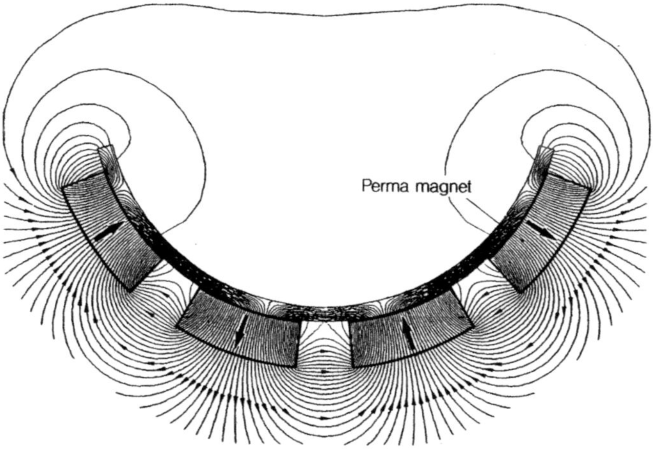

Figure 1.43: Pattern of the magnetic field around a magnetic drum [U1].

electromagnetic winding relative to the dimensions of the spacers (either air or steel) between these magnetic elements. A schematic view of a linear multipole OGMS system is shown in Fig. 1.45.

In **closed-gradient separators** a high field gradient is usually achieved by a suitable shaping of the pole pieces of the opposing magnetic poles. Magnetic field is thus focused on the working element of the separator and a high field gradient is created. Assorted geometrical arrangements of poles and counter-poles is shown in Fig. 1.46.

These different geometries, such as a hyperbolic pole facing a flat pole (case A), or a grooved pole (case B), triangular pole and multi-tooth arrangement generate various patterns of the magnetic field and its gradient [D4, S12]. The gradient of the field and the reach of the magnetic force are the function of the curvature of the poles and the counter-poles, by the pitch of the teeth and by the width of the separation gap. Induced magnetic roll and wet high-intensity magnetic roll separators employ this type of closed-gradient arrangement. Ferrohydrostatic separators with magnetic fluids also generate a closed gradient by a suitable shaping of the pole tips [R1].

Multipolar windings or multipolar arrangements of permanent magnets also generate closed gradient of the magnetic field. A quadrupole geometry was used to build a magnetic separator for biomedical applications [Z2, H8]. A sextupole arrangement is employed in Magstream separator with magnetic fluids [W9], while quadrupole magnets, either resistive [A3] or superconducting [C6] were also tested in magnetic separation. Figure 1.47 shows a schematic diagram of a pure quadrupole while the coil and pole arrangement in a conventional quadrupole magnet for physical experiments are illustrated in Fig. 1.48.

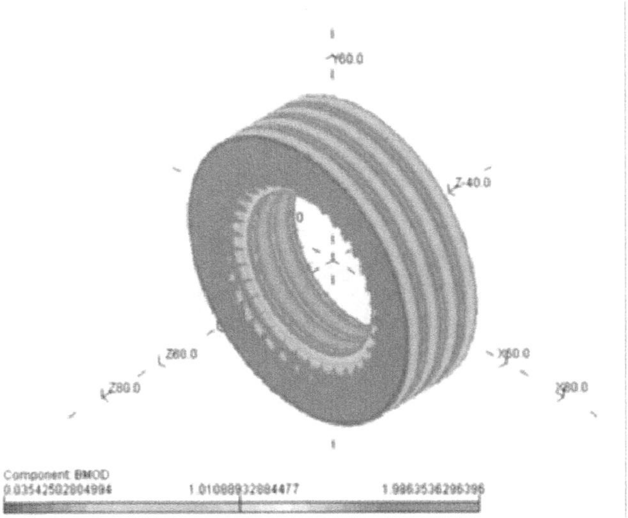

Figure 1.44: Pattern of the magnetic field on the surface of a permanent magnetic roll.

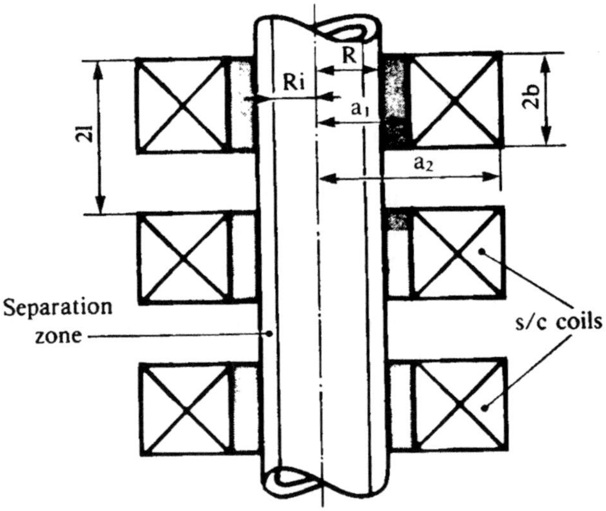

Figure 1.45: A multipole open-gradient magnetic system.

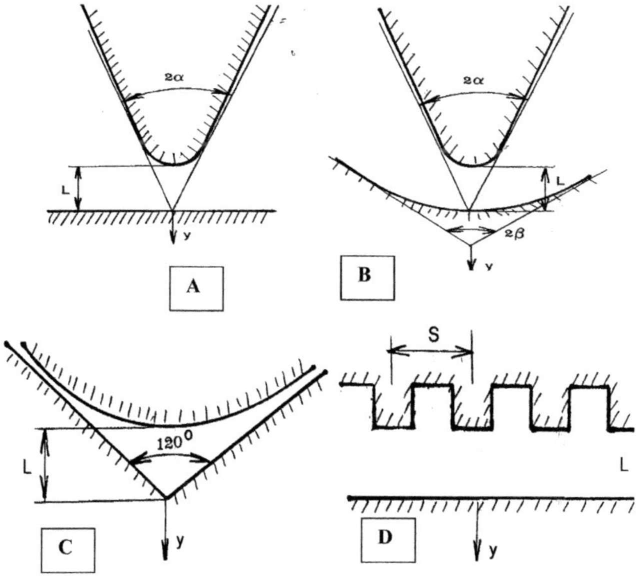

Figure 1.46: Shapes of poles and counter-poles in magnetic separators with closed gradient.

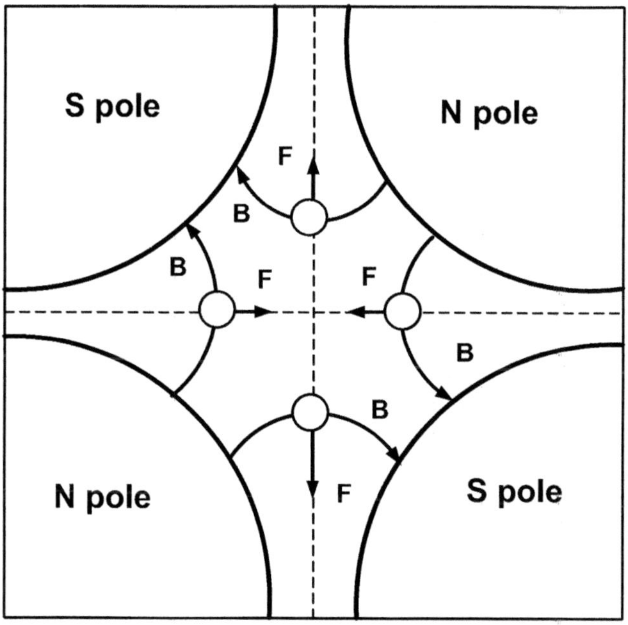

Figure 1.47: Magnetic field and forces inside a quadrupole magnet. Particles are entering the paper.

Figure 1.48: A quadrupole resistive magnet for high-energy physics experiments (courtesy of Danfysik A/S).

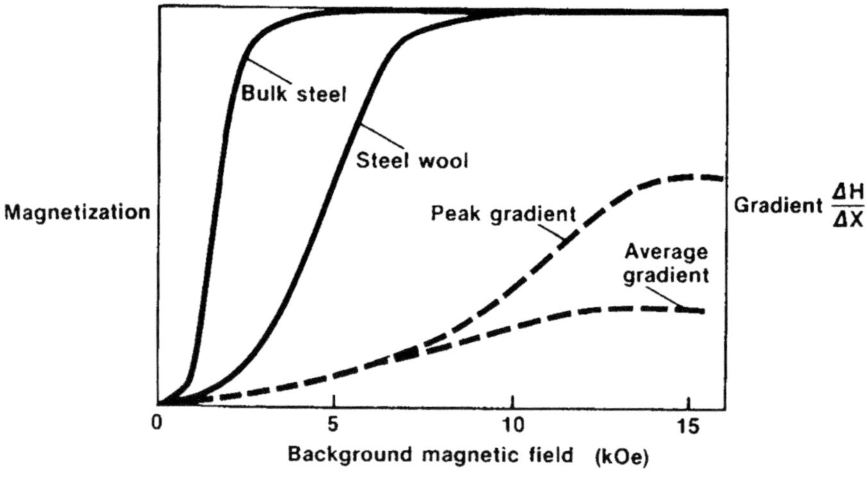

Figure 1.49: Magnetization curves for bulk steel and steel wool (after [K5]).

Table 1.20: Typical values of the gradient of the magnetic field in magnetic separators.

Separator	Gradient of the magnetic field	
	T/m	kG/cm
Suspended magnet	0.8	0.08
Nd magnetic roll	200	20
Nd drum separator	80	8
Davis tube	10	1
Frantz isodynamic	20	2
Belt separator	1 - 10	0.1 - 1
Steel wool (HGMS)	2.5×10^4	2.5×10^3
Steel balls (WHIMS)	1×10^3	100
Expanded metal	4×10^3	400
Grooved plates	2×10^3	200
Split pair OGMS	40 - 100	4 - 10
FHS	2	0.2
Magstream	30	3

In **matrix (polygradient) magnetic separators,** the gradient of the magnetic field is generated by means of induced magnetic moment in the ferromagnetic bodies (called the matrix) and placed in the external magnetic field. Strong magnetic field gradient is generated in the close vicinity of these bodies and its magnitude depends on the size and the shape of the matrix elements. Ferromagnetic bodies of various shapes have been used in high-intensity magnetic separators (such as balls, grooved plates, mesh, expanded metal, steel wool, rods etc.) and a proper choice of the matrix is one of the crucial prerequisites for successful separation.

Since the field gradient is inversely proportional to radius a of the matrix element, i.e. $\nabla B \approx J/a$, the magnetic force acting on the magnetizable particles is maximized by using as fine a matrix as possible. On the other hand, the dimensions of the matrix elements must be selected taking into account particle size of the material to avoid straining and to ensure the selectivity of separation. At the same time, the dimensions of the matrix bodies must be chosen in such a way that the magnetic force is sufficiently strong to recover the magnetic fraction. On the other hand, the magnetic force must not be too strong to cause the matrix blockage and retention of the "non-magnetic" component.

In order to magnetize a fine matrix, sufficiently strong external magnetic field is required. This is because the filling factor of the matrix is often low and the matrix does not conduct magnetic flux very well. Typically, the filling factor ranges from 0.04 (4%) for steel wool matrix to about 0.20 (20%) for expanded metal. The exception is the ball matrix where the filling factor can be as high 0.50, and grooved plates with the filling factor up to 0.80. Figure 1.49 illustrates that a background magnetic induction of at least 0.7 T is needed to saturate the steel wool matrix, compared to about 0.2 T that is necessary to

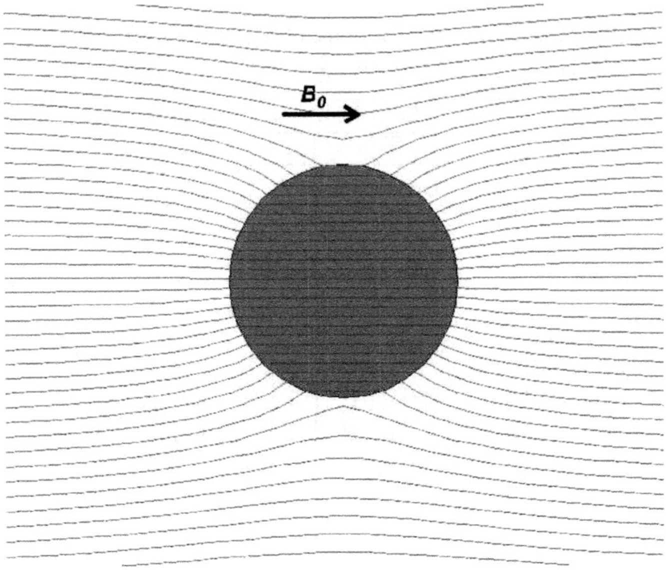

Figure 1.50: The magnetic field lines around a cross-section of a magnetized steel sphere or steel wire.

saturate the bulk material. The large background induction needed to saturate steel wool is a result of demagnetization effects. A large proportion of filaments in the randomly orientated steel wool is arranged along the difficult axis of magnetization and exhibits thus a large demagnetization factor. The pattern of the magnetic field around a magnetized steel wire or steel sphere is shown in Fig. 1.50. Typical values of the gradient of the magnetic field for various types of magnetic separators are given in Table 1.20.

Chapter 2

Review of Magnetic Separation Equipment and Techniques

In reviewing magnetic separators, it is necessary to examine hundreds of patents and papers and it is not possible to give adequate coverage to all of these. Moreover, there appear to be more models and modifications of magnetic separators than there are applications. Since each embodiment of a magnetic separation technique possesses distinct merits as well as limitations, no single technique is universally applicable to all types of magnetic separation.

Various classification schemes exist by which magnetic separator can be subdivided into categories. These schemes are based upon the dominant physical or technological feature. Magnetic separators can thus be grouped as follows:

a) Based upon the medium carrying the ore:

- dry
- wet

b) Based upon the requirement of the system:

- removal of iron and similar materials for the protection of machinery subjected to damage or wear
- extraction of valuable magnetic constituents
- removal of deleterious magnetic impurities
- separation of materials based on properties other than magnetic (e.g. density, conductivity)
- material handling

c) Based upon the way the magnetic field is generated:

- permanent magnets
- electromagnets with iron yoke
- resistive solenoids
- superconducting magnets

d) Based upon the magnitude of the magnetic field and its gradient:

- low-intensity magnetic separators
- high-intensity magnetic separators
- high-gradient magnetic separators.

Low-intensity separators are primarily used for the manipulation of ferromagnetic materials or paramagnetic materials of high magnetic susceptibility, and/or of sufficiently large particle size. These separators can operate either in dry or wet modes.

High-intensity separators are used for treatment of weakly magnetic materials, coarse or fine, in wet or dry modes.

Very fine, feebly magnetic materials can be treated in high-gradient magnetic separators.

From the practical point of view, classification according to the requirement of the system is the most illustrative. For example, the application of magnetic separation to iron removal is usually dictated by the material handling system involved. The most commonly used magnetic separators include:

- Magnetic pulleys, applied in belt conveyor systems as head pulleys
- Suspended magnets, applied over belt conveyor chutes or feeders
- Magnetic drums
- Plate magnets, applied in chutes
- Grate magnets, applied in product streams.

For concentration or removal of magnetizable particles, a wide spectrum of magnetic separators is used. These include:

- Magnetic pulleys
- Magnetic low-intensity or high-intensity drums
- Induced magnetic rolls
- Permanent magnet roll separators (IMR)

- Cross-belt and disk magnetic separators
- Magnetic filters
- Wet high-intensity magnetic separators (WHIMS)
- High-gradient magnetic separators (HGMS)

For magnetic material handling, the magnetic systems usually applied are:

- Magnetic conveyors
- Magnetic rolls
- Lifting magnets

Magnets are also used in numerous special applications:

- Separators with magnetic fluids, in which material is separated mainly on the basis of its density (FHS, MHS)
- Eddy-current separators for separation of non-ferrous metals (ECS)
- Dense-medium separators, in which a magnet is used to control magnetic heavy medium
- Demagnetizing coils
- Magnetic flocculators.

2.1 Dry low-intensity magnetic separators

Dry low-intensity magnetic separators are usually used to remove tramp iron, to concentrate coarse strongly magnetic iron ores, and to recover iron values from blast furnace and steel mill slag. Removal of strongly magnetic impurities from a variety of materials is also accomplished with dry low-intensity separators.

Tramp iron magnetic separators are used to protect handling and processing equipment such as crushers and pulverizers. These separators are usually used on dry material or on material which contains only surface moisture.

Tramp iron comes in many shapes and sizes. By conventional definition, tramp iron means that a particle has a significant size, which is sufficient to cause damage. Iron of abrasion, or particles smaller than 3 mm in diameter, usually do not cause equipment damage although they can discolour or contaminate the product [D7]. As far as the top size is concerned, large pieces as long as 2 m can be successfully removed. The size and shape of the tramp iron, together with the material handling system used must be considered in selecting the protective magnet.

Figure 2.1: Magnetic pulley (courtesy of Steinert GmbH).

2.1.1 Magnetic pulleys

An easy and simple way to remove tramp iron from material on a conveyor belt is by means of a magnetic head pulley. These are available both in permanent magnet and electromagnetic construction. They are relatively inexpensive and easy to install, and accomplish continuous tramp iron removal. The magnetic pulley, shown in Figure 2.1, usually replaces the head pulley or the tail pulley of a belt conveyor, converting the conveyor into a magnetic separator. Material carried on the conveyor passes over the magnetic pulley which holds the magnetic particles until they leave the region of the magnetic field while the non-magnetic material is discharged over the pulley. A typical installation is shown in Fig. 2.2, while Fig. 2.3 illustrates an actual process of separation by the pulley.

The permanent magnetic pulleys can be manufactured with axial poles in which the polarity alternates along the circumference, or radial poles, in which the polarity changes across the width the of the belt. For large particles and elongated shapes, the radial pole design is more suitable. For finely divided material, the uniformity of the field across the belt width makes the axial pole design more advantageous. Strontium ferrite magnets are usually used. Electromagnetic pulleys are used where control of the magnetic field is required.

The magnetic pulleys are manufactured in a range of diameters, from 500 mm to 1250 mm. The width of the magnetic pulley should be selected to match the belt width. The throughput of magnetic pulleys of different dimensions is shown in Figure 2.4. Standard belt speeds for which the throughputs were calculated are shown in Table 2.1. Inclined belts provide additional areas of contact with the magnetic field of the pulley and can tolerate higher capacities.

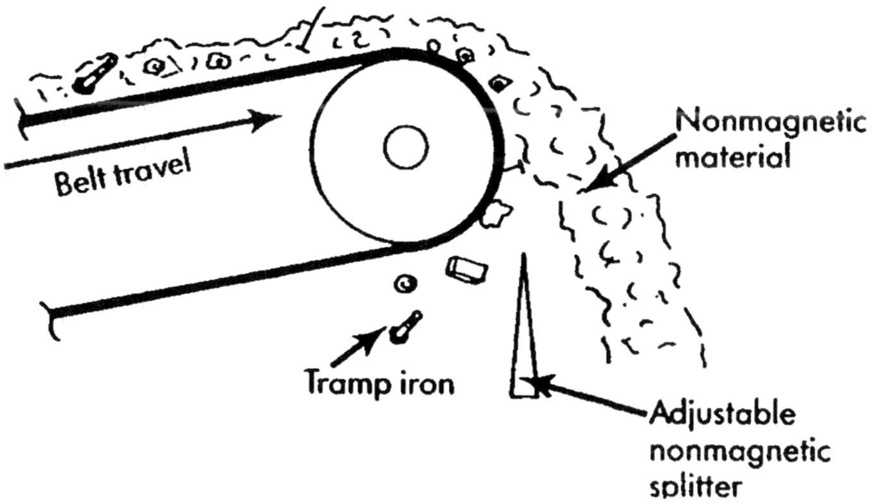

Figure 2.2: Principle of operation of a magnetic pulley.

Figure 2.3: Operation of a magnetic pulley (courtesy of Steinert GmbH).

Table 2.1: Standard belt speed for magnetic pulley operation.

Pulley diameter [mm]	Belt speed [m/s]
500	1.1
630	1.45
800	1.7
1000	2.0
1250	2.2

Table 2.2: Recommended diameters of magnetic pulleys.

Feed size [mm]	Pulley diameter [mm]	Throughput [t/h/m]
50 - 100	750 - 900	250 - 400
25 - 50	450 - 600	165 - 250
6 - 25	300 - 450	75 -135

Magnetic pulleys of special design are used in the concentration of magnetite and other ferromagnetic minerals. For the best results, the feed should be screened into sized fractions and each fraction treated on a separate pulley. Typical feed sizes would be - 100 mm + 50 mm, - 50 mm + 25 mm and - 25 mm + 6 mm. The magnetic pulley is not suited for the treatment of - 6 mm material. Pulley diameters recommended for various feed sizes are given in Table 2.2 [M7].

2.1.2 Plate magnets

Where the amount of tramp iron contained in a product is reasonably small and where automatic removal is not required, plate and grate magnets can be incorporated in chutes and ducts to remove both tramp iron and iron of abrasion. Such a plate magnet is shown in Fig. 2.5.

The plate consists of a series of alternating poles, uniform across the chute width and alternating along the chute length. As the material passes over the magnet face, tramp iron is trapped on the magnet face while the remaining non-magnetic material proceeds down the chute. The plate magnets must be cleaned and inspected daily since they are not effective if the accumulated tramp iron is allowed to remain on the face.

The magnetized plates are usually of permanent magnet type, either ferrite or rare earth. The effective depth of the magnetic field can be as high as 200 mm, depending on the type of the magnetic material and on the shape of particles to be removed. A chute angle not exceeding 45^0 is recommended [M7].

2.1.3 Grate magnets

The grate magnet shown in Figs. 2.6 and 2.7, consists of a series of magnetic tubes, which have poles alternating between one another and along their length.

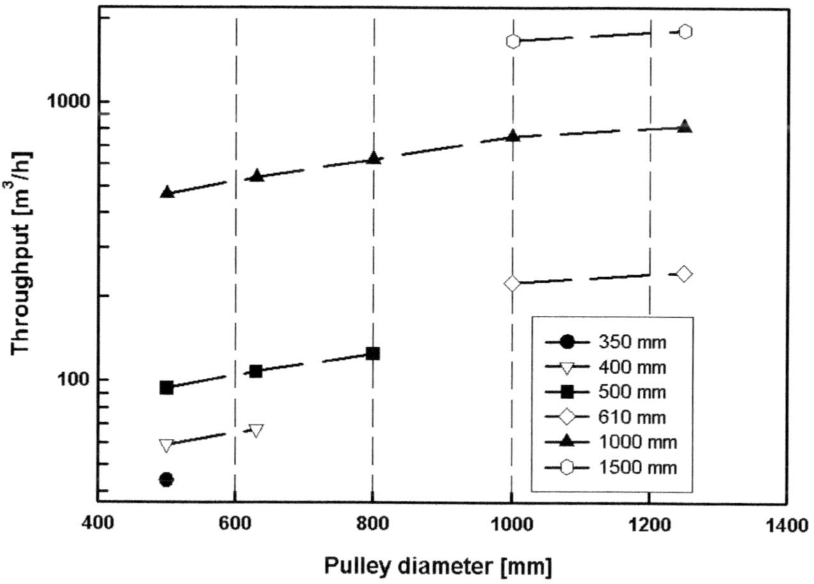

Figure 2.4: Throughput of magnetic pulleys of various diameters, for different widths of the conveyor belt. Standard belt speeds were assumed (after [M7]).

These units are used for granular free-flowing materials finer than 12 mm and are typically mounted at the discharge of a hopper. A number of rows of tubes provide several passes of finely divided material and can remove essentially all the magnetic particles. The permanent magnets used in the grate magnets are either ferrites, which are suitable for temperatures up to 150 ^{0}C, or Nd-FeB magnets, with operating temperature up to 90 ^{0}C. For higher temperature applications, up to 350 ^{0}C, the Alnico magnetic grates are also manufactured.

2.1.4 Suspended magnets

Suspended magnetic separators have been used over many decades to improve material purity and to protect processing machinery. Particularly when the belt speed is high and/or if large tramp iron objects are to be removed, suspended magnets are recommended. Suspended magnets are available in both circular and rectangular construction. The rectangular magnet is most commonly used and permits installation of self-cleaning construction. Units based on permanent magnets or electromagnets are manufactured, with the permanent magnet design being restricted to lighter burdens. Simple suspended magnets, designed to remove small amounts or occasional pieces of tramp iron, employ the manual cleaning mechanism. They must be periodically turned off in order to remove

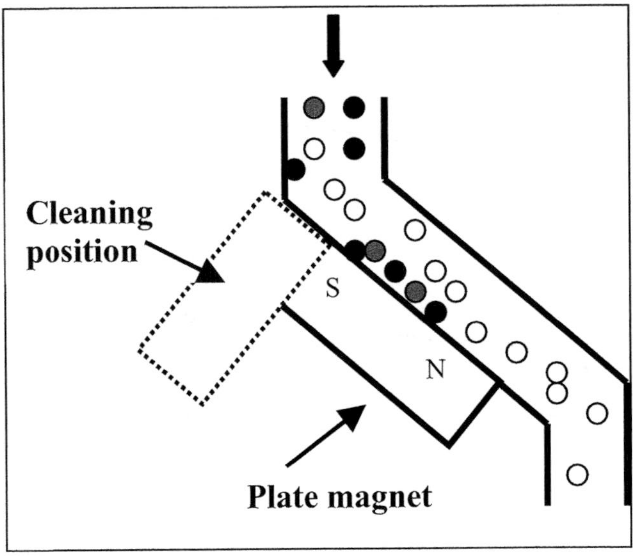

Figure 2.5: Plate magnet.

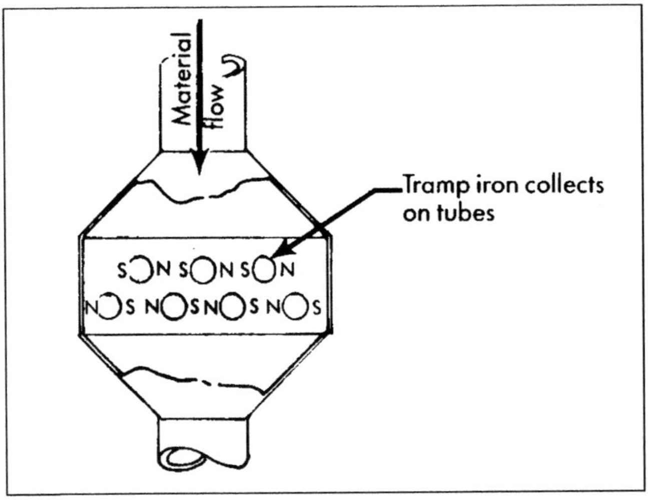

Figure 2.6: Grate magnet.

Figure 2.7: Grate magnet.

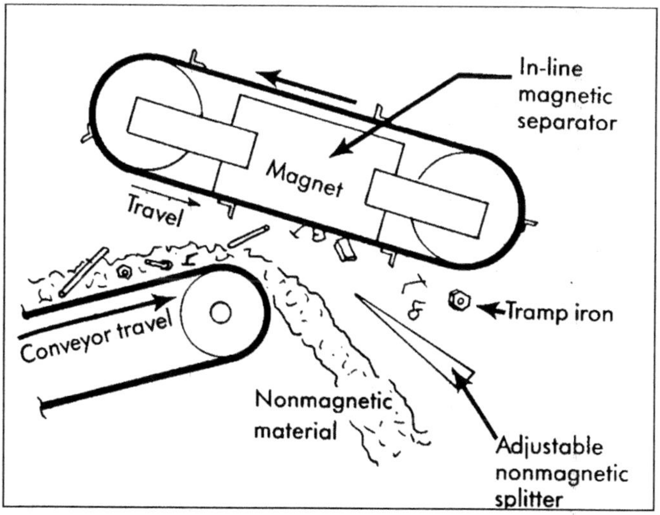

Figure 2.8: Principle of operation of an in-line suspended magnet.

Figure 2.9: Installation of a cross-belt suspended magnet (courtesy of Boxmag-Rapid Ltd.).

iron accumulated on the face of the magnet. Where a large amount of tramp iron is expected, or where access to the magnet for cleaning is difficult, self-cleaning magnets should be used. A belt driven across the magnet face allows continuous automatic removal of tramp iron.

Suspended magnets can be installed at many points in a material handling system: at any point along a conveyor belt, at the discharge end of feeders or screens and above chutes or launders. The preferred location for a suspended magnet is at an angle over the conveyor belt as shown in Fig. 2.8. In this orientation the tramp iron is moving into the face of the suspended magnet and the load is breaking open, allowing easier tramp iron removal. Self-cleaning magnets can be operated across the belt, as shown in Fig. 2.9, and as in-line units as illustrated in Fig. 2.10. In the cross-belt mode of operation, the magnet belt moves perpendicularly to the conveyor belt. Because the tramp iron must be attracted and turned by 90^0 from the movement of the conveyor belt, a cross-belt magnet is usually larger and stronger than an in-line magnet. Typical values of the magnetic field as a function of suspension distance for oil-cooled suspended electromagnets, for different lengths of the magnet, are shown in Fig. 2.11 [M7]. The selection process for suspended magnets will be discussed in Chapter 5.

Figure 2.10: Installation of an in-line suspended magnet (courtesy of Steinert GmbH).

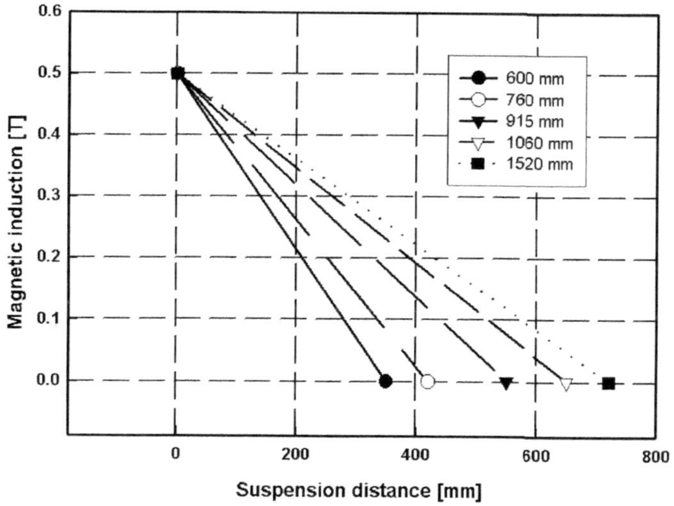

Figure 2.11: Magnetic field as a function of the suspension distance for oil-cooled suspended electromagnets of different lengths.

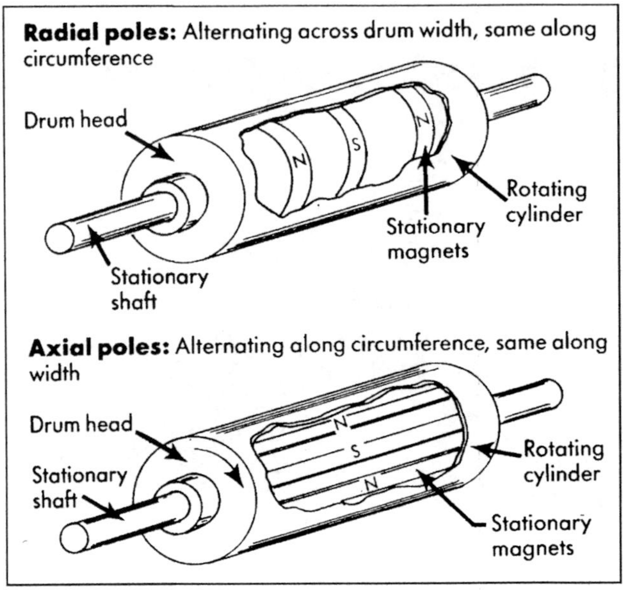

Figure 2.12: Pole configurations in drum magnetic separators [M8].

2.1.5 Drum magnetic separators

Drum magnetic separators are probably the most widely used type of magnetic separation equipment. The design of drum separators is predominantly based upon permanent magnet technology, although electromagnetic drums can also be occasionally found in industrial practice. Drum separators are applied to the treatment of particle sizes, which range from several centimeters down to several micrometers, and operate either in a dry or wet mode. While ferrite magnets are used in conventional low-intensity drum separators, recent progress in availability and affordability of rare-earth permanent magnets has resulted in the construction of powerful high-intensity drum separators.

The basic design of all drum magnetic separators is essentially the same. Blocks of permanent magnets are arranged in a stationary position inside a non-magnetic rotating drum, as shown in Fig. 2.12.

Drums are made in several configurations depending on the application. The magnet assembly, consisting of three to nine magnetic blocks of alternating polarity, covers an angle usually ranging from 90^0 to 120^0. As can be seen in Fig. 2.12, in the radial arrangement of the magnets, the polarity alternates across the width of the drum, and is the same along the drum circumference. Such an arrangement is suitable for the removal of elongated tramp iron or concentration of coarse particles when high recovery is required.

In the axial arrangement, the polarity alternates along the circumference of the drum and is uniform across the width of the drum. This arrangement

Figure 2.13: A pattern of the magnetic flux around a magnetic drum.

provides an agitating action enabling the release of entrained non-magnetic particles. The axial arrangement is used when a high quality magnetic concentrate is required, although sometimes at the cost of reduced recovery.

The development of improved permanent magnet materials resulted in increased magnetic induction available on the drum surface and at a suitable operating distance from the drum. Conventional ferrite-based drums generate up to about 0.22 T on the drum surface and 0.10 T at 50 mm. In NdFeB-based drums a magnetic induction as high as 1 T can be reached on the drum surface. A pattern of the magnetic flux around a magnetic drum is shown in Fig. 2.13.

Drums are manufactured with different diameters and widths. While diameters of ferrite drums range from 600 mm to 1500 mm, rare-earth drum are generally smaller, with diameters from 380 mm to 1000 mm. The most common width of the low-intensity drums is 1500 mm, although units 3000 mm wide are also manufactured.

Dry low-intensity drum magnetic separators are used mainly for tramp iron removal where magnetic pulleys and suspended magnets are not feasible. With rare-earth magnets, the drums are also able to separate stainless steel and iron-bearing minerals. These separators can operate with top feed and bottom feed, as shown in Figs. 2.14 and 2.15. The material to be separated is fed at the top of the rotating drum surface, as shown in Fig. 2.16; the non-magnetic fraction leaves the drum, while the magnetic fraction is retained by the magnetic force at the drum surface until carried outside the effective magnetic field. In bottom-feed drums, the magnetic component of the feed is picked up by the magnet while the non-magnetic fraction leaves the system unaffected. Drums with top feed produce the highest magnetic removal and are suitable for materials that contain relatively small amount of magnetic particles. However, the drum with

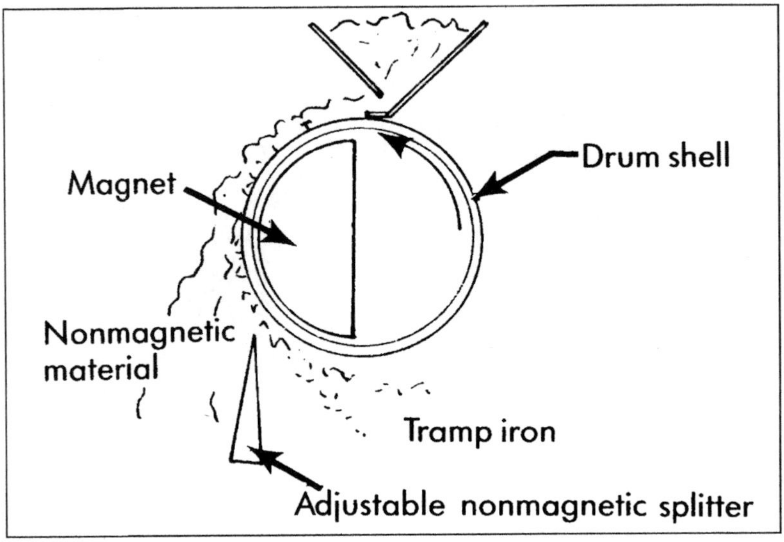

Figure 2.14: Dry drum magnetic separator with top feed [M8].

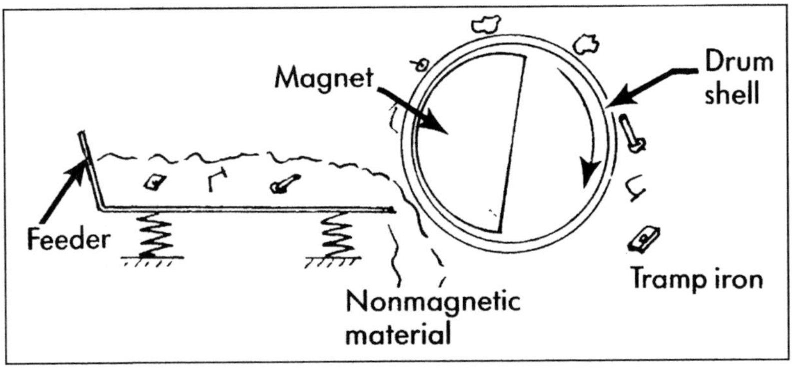

Figure 2.15: Dry drum magnetic separator with bottom feed [M8].

Figure 2.16: Operation of a dry drum magnetic separator (courtesy of Steinert GmbH).

bottom feed produces a clean magnetic concentrate with little non-magnetic carry-over and should be used to treat materials containing a high percentage of magnetic particles.

2.2 Wet low-intensity drum magnetic separators

Wet drum magnetic separators are used to recover ferromagnetic solids from a slurry feed. The major areas for their use are in media recovery in dense medium separation plants and in beneficiation of magnetite ore. Figure 2.17 shows the recovery of the magnetic particles by a wet drum magnetic separator. Physical construction of wet drum separators is slightly different for the two applications, with the ore concentrator being more rugged in construction and subject to more detailed specifications than the media recovery units.

The drum is partially submerged in a tank of water and carries the material to be treated across the face of the magnetic system and the magnetic concentrate out of the tank. The size of the feed treated in these separators rarely exceeds 6 mm while the bottom size of the feed can be as small as 20 μm.

The design of the tanks for the drum separator is very important and is determined by the objectives of the separation process. There are three basic tank designs, based on the slurry flow:

- concurrent
- counter-current

Figure 2.17: A wet drum magnetic separator.

- counter-rotation.

2.2.1 Concurrent tank

The concurrent tank design is shown schematically in Fig. 2.18. The feed is introduced at one end of the separator through the feed box and flows in the direction of drum rotation. Magnetic material is picked up by the magnet and the non-magnetic material is discharged at the bottom through the tailings discharge opening. Magnetic particles continue to adhere to the surface of the drum and pass through an opening between the drum shell and the tank wall. The magnetic fraction is then discharged down the discharge lip.

The concurrent design of the tank is used mainly for treating coarser material in the size range 5 mm and below. High throughput and high quality of the magnetic concentrate are characteristic features of this design.

2.2.2 Counter-rotation tank

In the counter-rotation design of the tank (Fig. 2.19), the feed is introduced through a special feed box to the drum, which rotates in the opposite direction to the slurry flow. Magnetic material is picked up by the drum and discharged almost immediately. Since the tailings flow along the entire magnetic arc of the drum before being discharged, losses are minimum. This tank design is used

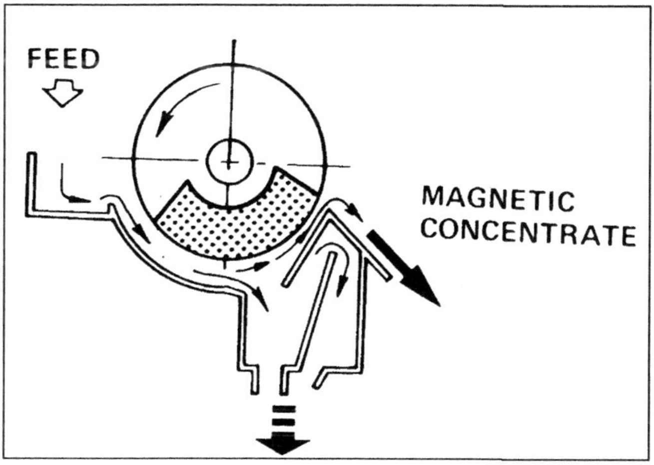

Figure 2.18: The concurrent tank.

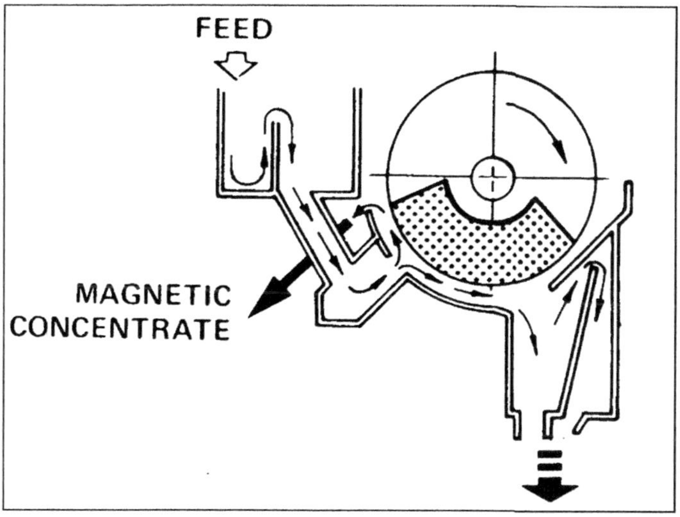

Figure 2.19: The counter-rotation tank.

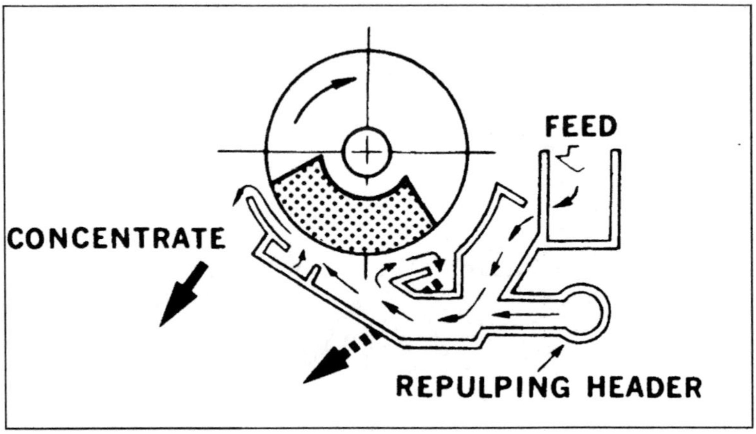

Figure 2.20: The counter-current (Steffenson) tank.

primarily where surges in throughput are experienced, possibly involving high content of the magnetic material. Very high recoveries can thus be achieved with this design, while the grade of the concentrate is of secondary importance only. The particle size can be up to 4 mm, but is preferably smaller than 0.5 mm.

2.2.3 Counter-current tank

The counter-current tank (Fig. 2.20) is used mostly for finishing where very thorough cleaning is required. The term "counter-current" refers to the fact that the tailings must flow counter to the rotation of the drum when leaving the separator. Feed is introduced near the bottom of the drum and agitated through wash-water jets. Tailings flow out from the opposite end of the tank through a tailings discharge overflow, which also acts as the slurry level control. This ensures a high recovery of magnetic material and a high quality of the concentrate. These tanks are most suited to finely ground material of 100 μm and less.

2.2.4 Design of the drum

The separating drum consists of two main parts: a stationary magnetic circuit and a shell, usually made of stainless steel and often protected by an anti-wear cover. The shell rotates freely around the magnets that cover an angle up to 150^0. The choice of the permanent magnet material and design of the magnetic circuit are the same as those for dry drum magnetic separators as discussed in Section 2.1.5. The drums are available in 760 mm (30") to 1200 mm (48") diameter; 900 mm (36") drum diameters being the most frequently used. The drum width ranges from 610 mm to 3050 mm, usually in increments of 300 mm.

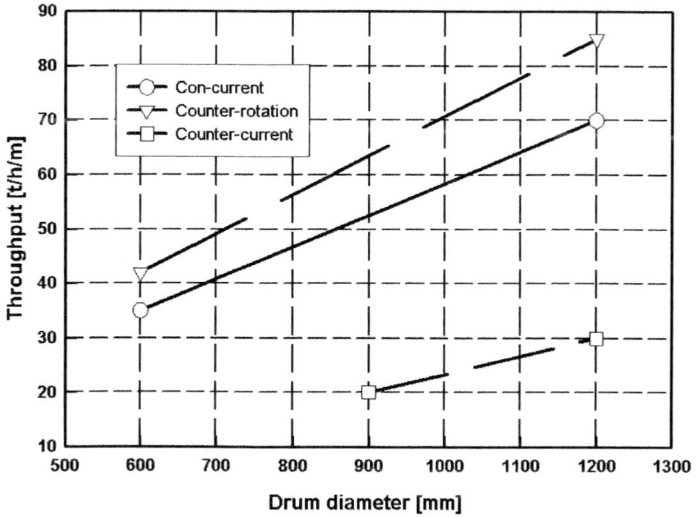

Figure 2.21: Throughput of a drum magnetic separator, as a function of the drum diameter (after [K6], [S13]).

With increasing diameter of the drum, the throughput also increases. This is illustrated, for different types of tank design, in Fig. 2.21. It appears, however, that the 1200 mm diameter is the optimum size, as a further increase in diameter has not apparently resulted in the corresponding increase in processing capacity [S13]. The increased throughput with increasing drum diameter can be expected as the residence time of particles in the magnetic field increases as the length of the magnetic arc increases.

This increase in residence time results either in improved recovery of the magnetic particles (possibly with a reduced grade of the concentrate due to increased entrainment of non-magnetic particles) or, at constant recovery, in increased throughput. Large diameter drums usually incorporate a larger pitch of the magnetic poles [S13]. Such a modification of the magnetic design then results in a deeper magnetic field around the drum and in the ability to treat higher volumes of material. However, a larger pitch of the poles reduces the field gradient and thus the magnetic force acting on the magnetizable particles. This could result in a reduced recovery.

Figure 2.22 illustrates the operation of wet-intensity drum separators in an iron-ore beneficiation plant. A more detailed discussion of practical aspects of operation and selection of wet drum magnetic separators is given in Chapter 5.

Figure 2.22: Wet drum magnetic separators in an iron-ore beneficiation plant.

2.3 Dry high-intensity magnetic separators

Weakly magnetic minerals can only be removed from a material stream in magnetic fields much greater than those available in low-intensity magnetic separators. Dry high-intensity magnetic separation has been used commercially since the beginning of the last century, while wet high-intensity high-gradient magnetic separators were developed only relatively recently.

Five types of dry high-intensity magnetic separators are in use:

- cross-belt separators
- induced magnetic roll separators
- permanent magnetic roll separators
- magnetic filters
- rare-earth drum magnetic separators.

2.3.1 Cross-belt magnetic separator

The cross-belt magnetic separator is one of the oldest types of separators used to concentrate moderately magnetic ores. The operating principle is shown in Fig. 2.23. Dry material is fed in a monolayer onto a conveyor belt and is carried between the poles of the magnetic system. The belt, with its load, passes

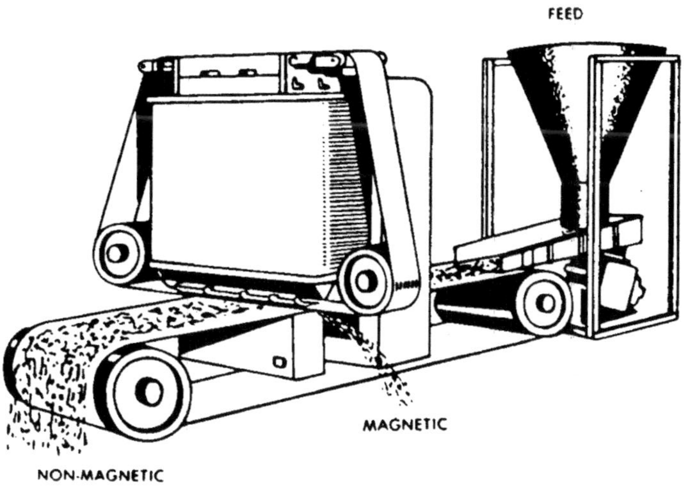

Figure 2.23: Cross-belt magnetic separator.

between the poles of two or more electromagnets. The lower pole-pieces are flat and immovable, while the corresponding upper pole-pieces are shaped and may be raised or lowered as required for the material undergoing treatment. An endless cross-belt runs around a series of pulleys and across each upper pole-face surface at right angles to the conveyor belt. As the material passes, the magnetic particles are attracted to the cross-belt, carried with it towards the discharge side and fall into a bin.

The material may be passed successively between several pairs of magnet poles when it is desired to make several products in one pass. Non-magnetic tailings discharge as the conveyor belt passes over its end pulley. Machines with two to eight pairs of poles are available and each pair can be adjusted independently by regulating the position of the upper poles and by altering the current through the coils. The height of the upper pole-pieces above the belt should be set at least 2.5 times the size of the largest particle in the feed.

The conveyor belt width ranges from 100 mm to 600 mm and particles ranging from 4 mm to 75 μm may be treated, preferably sized into several fractions. Depending on the gap between the poles and on the current setting, the operating magnetic induction ranges from 0.2 T to 2.0 T. The main advantage of the cross-belt separator is that several types of magnetic products can be recovered in one pass. However, low throughput is a major drawback. A five-pole cross-belt separator is shown in Fig. 2.24.

Figure 2.24: A cross-belt separator (courtesy of Roche Mining).

2.3.2 Induced magnetic roll separator

The induced magnetic roll (IMR) separator shown in Figs. 2.25 and 2.26 consists of a revolving laminated roll formed of alternate magnetizable and non-magnetic discs. The roll is placed between shaped pole-pieces of an electromagnet. The electromagnet induces a magnetic field in the magnetic lamination of the roll and generates regions of high magnetic field gradient.

Material to be treated is fed in a controlled thin stream by a vibratory feeder to the top of the roll. As the roll revolves, the material passes through a narrow gap between the pole of the magnet and the roll. Non-magnetic particles are discharged from the roll while the magnetic particles are attracted to the roll and report into a separate chute when they enter a region outside the sphere of influence of the magnet. Ferromagnetic material should be removed with a separate magnetic scalper before feeding to IMR, to prevent plugging the gap.

The gap between the feed pole and the roll is adjustable and also the setting of the splitter is of great importance. Depending on the width of the separation gap and the current in the magnetizing coils, the magnetic induction in the gap can approach 2 T. The material to be treated must be dry, free-flowing and for best results it should be in the size range - 2 mm + 100 μm. The gap should be set to at least 2.5 times the average grain size. The throughput can range from 1 to 3 t/h per meter length of the roll.

IMR separators have been extensively used for the beneficiation of beach sands and for the removal of weakly magnetic impurities from glass sand, andalusite, feldspar, wollastonite, magnezite and other industrial minerals. IMRs can also be used for concentration of materials such as chromite, monazite and wolframite.

IMRs possess several limitations, namely relatively low throughput, limited particle size range and large mass. In recent years, much simpler and easier to operate permanent magnetic roll separators have replaced IMRs in numerous

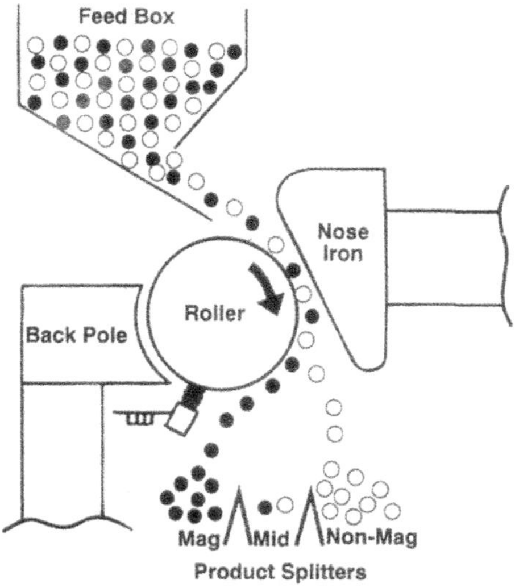

Figure 2.25: Schematic diagram of an induced magnetic roll separator (courtesy of Roche Mining).

applications.

An induced magnetic roll with bottom feed was developed by Carpco, Inc. It is offered as a laboratory or pilot-scale separator for treatment of particles up to 1 mm in size, in throughputs up to 50 kg/h. The separator, shown in Fig. 2.27, achieves highly selective separation by magnetically lifting weakly magnetic particles against gravity. As no entrapment of unwanted material takes place, the separator is suitable in those applications where the grade of the magnetic concentrate or the recovery of the non-magnetic concentrate are of prime importance.

2.3.3 Permanent magnetic roll separator

Although the idea of a permanent magnetic roll separator, consisting of permanent magnet disks or rings interleaved with mild steel disks, as shown in Fig. 2.28, dates back to 1968 [A4], it was only the advent of rare-earth permanent magnet materials that spurred the rapid development of these machines. A patent by E.L. Bateman Ltd. [Y1] became a starting point for a productive period of competitive development of a wide array of roll separators. Initially Sm-Co permanent magnets were used but they were quickly replaced by NdFeB magnets when they became commercially available.

The active part of this separator is a roll consisting of disks or rings of permanent magnets, sandwiched between mild steel rings. The adjacent per-

Figure 2.26: Reading induced magnetic roll separator (courtesy of Roche Mining).

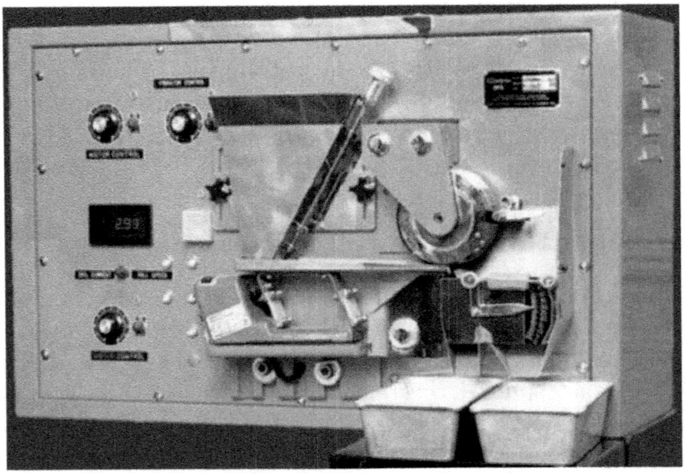

Figure 2.27: An induced magnetic roll separator with bottom feed (courtesy of Outokumpu Technology, Carpco Division).

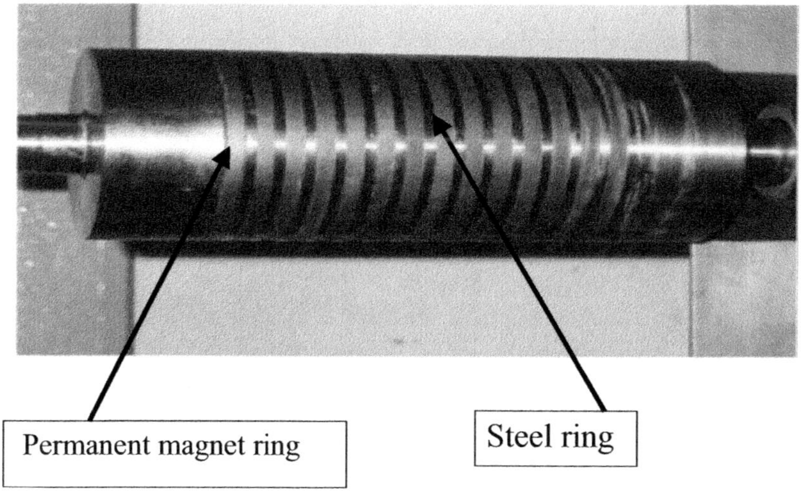

Figure 2.28: Details of a permanent magnet roll.

Table 2.3: Typical values of magnetic induction on the roll surface, as determined from modelling and by measuring using a conventional Hall-Effect probe.

Magnet	**B [T] - modelled**	**B [T] -measured**
Nd-Fe-B	1.9	1.0
Sm-Co	1.5	8.0
Ferrite	0.5	0.3

manent rings are arranged with the same polarity facing one another. Very high magnetic fields and field gradients are generated on the surface of the roll. As discussed in Section 1.6.1, it is difficult to measure the real magnitude of the field using conventional measuring equipment because the field varies over a scale of several hundreds of micrometers. Electromagnetic modelling software, however, allows determination the pattern of the field around a roll as shown in Fig. 2.29. Typical values of the magnetic field are shown in Table 2.3. The values of the magnetic field strength measured using a conventional Hall-Effect probe are also shown.

The values of the magnetic field strength and its gradient on the roll surface and around the roll are strongly dependent on the configuration of the magnet and steel rings. Thin magnet and steel rings generally produce a very high field, the field gradient and the magnetic force at the surface of the roll, while a long reach of the magnetic force can be achieved using thicker steel rings. The relative thicknesses of the magnet and steel rings generate a widely differing pattern of the magnetic fields around the roll. While the most common ratio of the widths of the magnet to the steel is 4:1, the correct choice of the roll

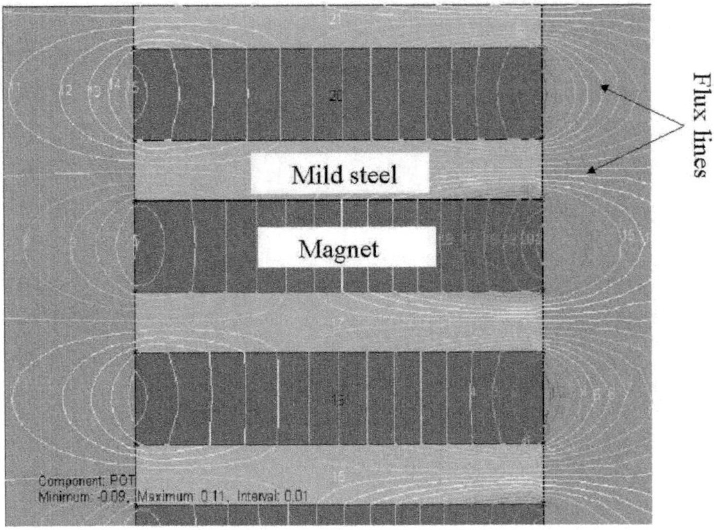

Figure 2.29: Pattern of the magnetic flux lines around a magnetic roll.

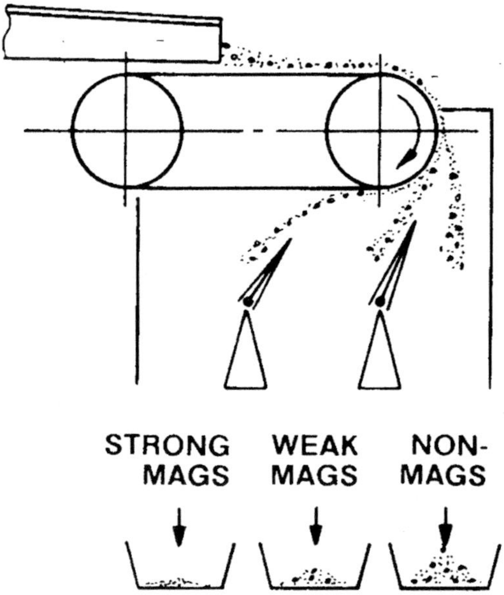

Figure 2.30: Operation of a permanent magnetic roll separator.

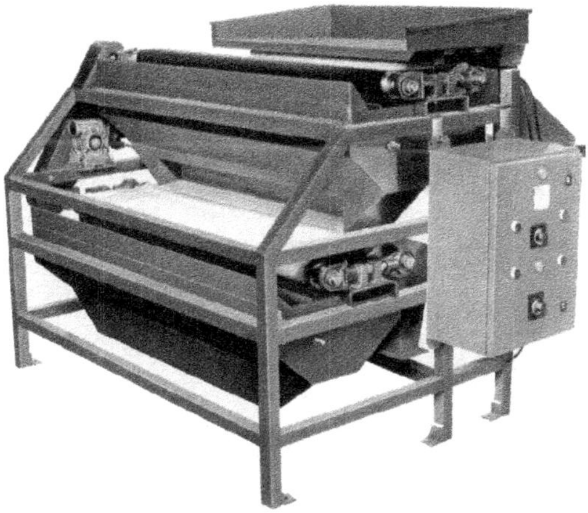

Figure 2.31: Permanent magnetic roll separator Permroll© (courtesy of E.L. Bateman Ltd.).

configuration is determined mainly by the magnetic properties and particle size distribution of the material to be treated.

Operation of permanent magnetic roll separators is schematically shown in Fig. 2.30. For easy removal of magnetic particles, the roll is covered by a belt, which can be as thin as 0.12 mm, to allow the feed material to be as close to the magnet as possible. The belt is supported by an idler roll. Below the conveyor is a hopper which collects the discharging material, while adjustable splitters divert the different fractions into collection bins placed beneath the hopper.

The most common diameter of the roll used to be 72 mm or 76 mm, while rolls 100 mm and 300 mm in diameter have now become commonly available from several manufacturers. The latter design is based on magnet segments rather than rings. With increasing roll diameter, the residence time of particles in the magnetic field, the field depth and the efficiency of separation increase. Alternatively, by increasing the speed of rotation, to maintain the same residence time as with a smaller roll, a higher throughput could be achieved with the roll of a larger diameter.

For production-scale applications, the width of the magnetic zone of the roll is either 1000 mm or 1500 mm, although for smaller scale applications 250 mm and 500 mm wide rolls are available. Throughput of roll separators is affected by many factors, which include particle size, particle density, proportion of the magnetic component in the feed and the desired quality of the final product. As a guideline, a throughput of 2 to 6 t/h/m can be achieved with material smaller than 2 mm, while larger material can be processed at a throughput up to 15 t/h/m. A modular design enables a number of passes to be arranged into a compact unit.

Figure 2.32: Installation of multistage Inprosys roll magnetic separators (courtesy of Outokumpu Technology Inc.).

A photograph of a two-stage production-size unit is shown in Fig. 2.31, while Fig. 2.32 illustrates the installation of multistage machines.

One of the most important issues that affects operational flexibility, ease of maintenance and cost efficiency is the belt tracking. Tracking, tensioning and changing the belts can be extremely time-consuming, in addition to a high cost of replacement of damaged belts. Several proprietary tracking and tensioning systems and roll support structure are available and it is claimed that it is possible to replace a belt in less 5 minutes [D8].

The magnetic force generated by the permanent magnetic roll is of the same order of magnitude as that in induced magnetic rolls. A spectrum of minerals that can be successfully treated by either magnetic separator is practically identical. Permanent roll separators, however, possess numerous advantages such as low energy consumption, and smaller mass and floor space. The absence of an air gap allows treatment of particles as large as 25 mm. A few drawbacks include belt wear and inability to control the magnetic field strength. While permanent roll separators are predominantly dry machines, wet units are being used, for instance, for the concentration of diamonds [R2].

2.3.4 High-intensity drum separators

Recent progress in the availability and affordability of rare-earth permanent magnets has resulted in design and construction of powerful high-intensity drum magnetic separators (REDS). Eriez Magnetics were probably the first to replace

Figure 2.33: Operation of a dry NdFeB drum magnetic separator (courtesy of Outokumpu Technology Inc.).

standard ferrite magnets with NdFeB permanent magnets and nearly all current suppliers of REDS adhere to this conventional design philosophy [A13]. Figure 2.33 illustrates the operation of a rare-earth dry drum magnetic separator.

A somewhat different approach [M13, W6, U1] to the design of the high-intensity drum separator was based on a plurality of small magnet blocks whose direction of magnetization changed in small increments, as can be seen in Fig. 2.34. This system which was claimed to require less magnetic material in comparison to a conventional system [W6] to achieve the magnetic induction of 0.7 T on the drum surface, was incorporated into the Permos separator of KHD Humboldt Wedag AG.

The density of the magnetic force (or the force index) $B\nabla B$ of a rare-earth drum magnetic separator is much smaller than the force index generated by most roll separators, usually not exceeding 20% of $B\nabla B$ for an NdFeB roll (see Table 1.1). The main reason is that the magnetic field gradient in REDS is much lower than that in a roll separator, as a result of the geometry of the magnetic system. On the other hand, the reach of the magnetic force can be extended to greater depths, allowing effective treatment of coarse particles. The current consensus is that roll magnetic separators are not only more powerful than drum magnetic separators, but they also seem to offer greater selectivity and flexibility [A13].

Generation of eddy currents in the rotating shell of the drum separator can be a serious problem because of limited temperature tolerance of NdFeB permanent magnets. The level of these currents depends on the magnetic induction, the number of magnetic poles, the speed of rotation, the shell thickness and the

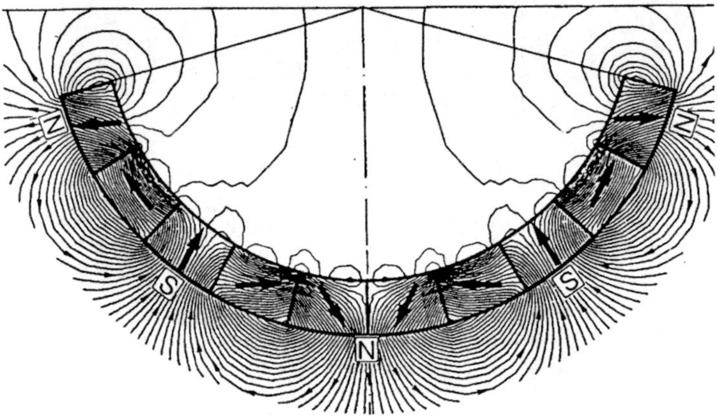

Figure 2.34: The magnet structure of the Permos separator [U1].

conductivity of the shell material. Heat will develop in proportion to the drive power that is required for the shell to overcome the braking effect of the eddy currents [A13]. The heat generation, sometimes aggravated by high temperature of the feed material represents, therefore, a problem for standard NdFeB materials, for which a maximum operating temperature of 80^0 is recommended. In such situations, more advanced magnet materials with a higher temperature tolerance must be used.

2.4 Wet high-intensity magnetic separators

2.4.1 Dry separation versus wet separation

Although dry high-intensity magnetic separators have been successfully applied to the beneficiation of a large spectrum of minerals, several basic disadvantages exist, which prohibit a wider use of a dry process. For dry magnetic separation to be successful, the ore must be completely dry and it almost invariably requires the ore to be sized into several fractions, each of which must be spread in a monolayer over the separators.

With dry methods, care must also be taken to ensure control of a possible dust hazard, an expensive precaution both in capital and operating costs. Furthermore, dry separators have considerably lower throughputs than wet machines. While dry separators frequently yield an excellent separation on material coarser than 75 μm, on unsized material containing a large portion of fines, the wet process is the only acceptable one. Considerable advantages must, therefore, be offered by dry separation to justify its use in preference to wet separation.

In order to extend the process of magnetic separation to fine weakly magnetic minerals, wet magnetic separators generating sufficiently high magnetic

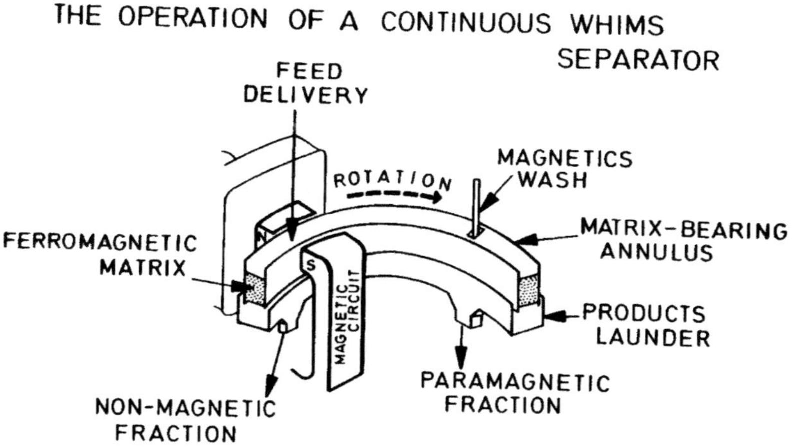

Figure 2.35: The principle of operation of a continuous wet high-intensity magnetic separator.

force in the working space must be used. A major development was achieved in this direction in 1955 by Jones [J1], who combined a high magnetic field strength with Frantz's idea of a magnetized matrix [F1]. By this approach, the magnetic force acting on magnetizable particles was increased by several orders of magnitude compared to dry magnetic separators. As shown in Table 1.1, a typical value of the product $B\nabla B$ for an Nd-Fe-B roll separator is 300 T^2/m, while with the magnetized steel wool matrix, this value is as high as 5×10^4.

Jones's wet high-intensity magnetic separator (WHIMS) became the starting point of a remarkable progress in this field and resulted in the development of a wide array of high-intensity high-gradient magnetic separators. This new type of separators extended the application of magnetic separation to classes of materials otherwise considered untreatable by magnetic means.

A schematic diagram of operation of a continuous wet high-intensity magnetic separator is shown in Fig. 2.35.

2.4.2 Iron core-based wet high-intensity separators

Jones high-intensity magnetic separator

The production-size Jones wet high-intensity separator is shown in Fig. 2.36 The magnetic field in the working gap is generated by a conventional iron-core electromagnet and the field gradient is enhanced by a matrix consisting of grooved plates of salient pole or of triangular groove plates, placed in the air gap. The plates, shown in Fig. 2.37, are vertically arranged in plate boxes that are placed around the periphery of the rotors.

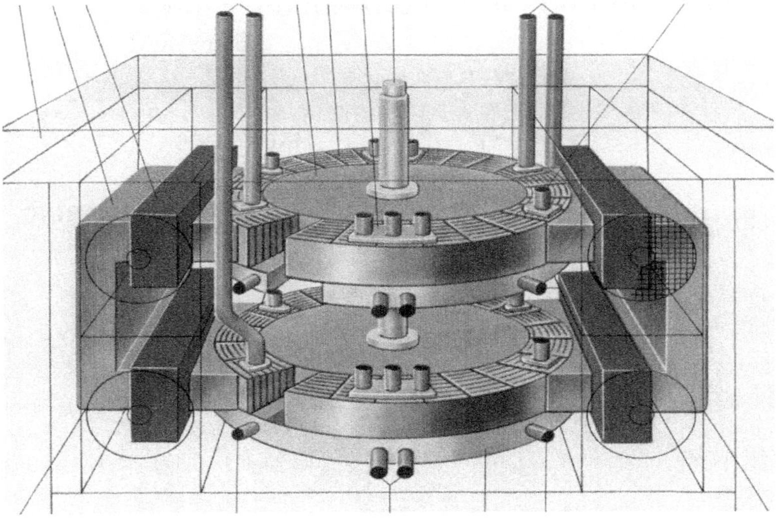

Figure 2.36: Perspective view of a Jones continuous wet high-intensity magnetic separator (courtesy of Humboldt Wedag AG).

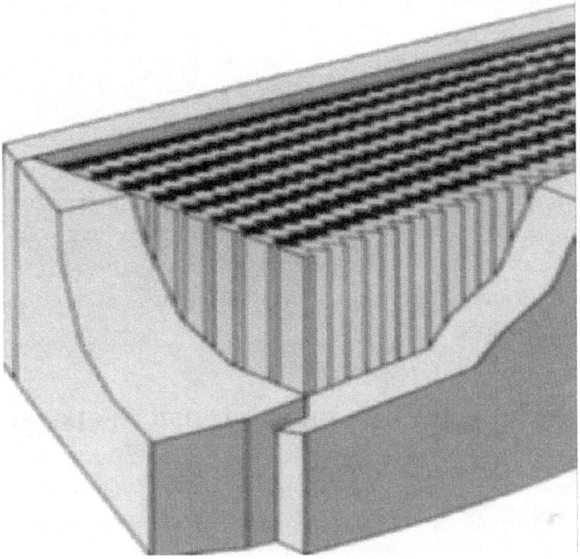

Figure 2.37: Grooved plates of Jones WHIMS.

Figure 2.38: Flush device for the magnetic fraction in the Jones WHIMS.

The slurry is gravity-fed onto the matrix where the magnetic particles are captured onto the plates, while the non-magnetic fraction passes through and is collected in a trough below the magnet. The feed points are at the leading edges of the magnetic field. Feeding is continuous due to the rotation of the plate boxes in the rotors. Each rotor has two symmetrically disposed feed points. Before a plate box with captured particles leaves the magnetic field, any entrained non-magnetic particles are washed out by low-pressure water and these middlings are collected in a separate launder under the rotor.

When the plate boxes reach the demagnetized zone half-way between the two magnetic poles, where the magnetic induction changes its polarity, and the magnetic field is essentially zero, the adhering magnetic particles are washed out with high-pressure water sprays. A flush device is shown in Fig. 2.38.

The particles in the slurry feed should be smaller than 1 mm and this determines the gap between the grooved plates. As a rule of thumb, the gap width, measured between two grooved plate tips, should be 1.5 to 2 times the maximum particle size [B8]. Particle size distribution also affects selection of the pitch of the grooves. Salient pole plates are used when a high grade of the magnetic concentrate is required, while triangular groove plates are used when high recovery into the magnetic fraction is important. The height of the plates is 190 mm and the filling factor of the plate box can range from 50% to 80 %, depending on the type of plates and the gap between them.

The maximum magnetic induction in the empty separation space does not exceed 1 T, although values up to 2 T are often stated. Such values can be

Table 2.4: Various models of Jones WHIMS.

Type	Capacity [t/h]	Mass [t]	Rotor diameter [m]
DP 335	180	114	3.35
DP 317	120	98	3.17
DP 250	75	70	2.50
DP 140	25	30	1.4
DP 90	10	16	0.9

Figure 2.39: Jones WHIMS DP 317 (courtesy of Humboldt Wedag AG).

achieved with the matrix installed in the plate box. The Jones separators are available in several sizes, some of them are summarized in Table 2.4. A photograph of a DP 317 separator is shown in Fig. 2.39.

The Jones magnetic separators have been widely used in the mining industry to beneficiate a wide spectrum of minerals. Although these separators have been used to upgrade industrial minerals such as glass sand, feldspar and bauxite, the main area of application has been the concentration of weakly magnetic iron ores. The success of Jones separators has been based on sound mechanical design and reasonably efficient performance. However, the grooved plates have proved to be an obstacle to long-term trouble free operation. As a result of the high filling factor of the matrix, several problems have been experienced. The frequently blocked matrix has to be removed from the plate boxes and the hard sediment must be laboriously removed. Also the effort to flush the matrix,

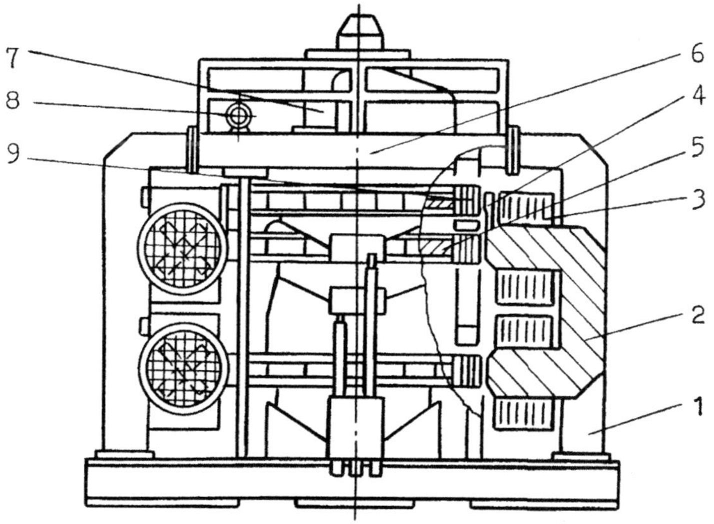

Figure 2.40: Schematic diagram of the three-stage 6-ERM-35/135 magnetic separator [U2].

which represents up to 80 % of the volume of the plate boxes, with high-pressure water, sometimes results in uncontrolled spillage.

6-ERM-35/135 high-intensity magnetic separator

The 6-ERM-35/135 WHIMS was developed in the former USSR in order to meet the increasingly demanding requirements for beneficiation of complex oxidized iron ores. It was observed [U2] that Jones separators could not ensure the efficient separation of ores with complicated mineral composition and fine dissemination, such as the deposits of Krivoy Rog and Kursk Magnetic Anomaly. While the design of the ERM separator is based on the Jones concepts, improved design of the matrix and of the feed mode, and incorporation of several stages of separation in one unit resulted in a greater resistance to matrix blockage, higher recovery and better selectivity of separation [U2, U3, T2].

A schematic diagram of the 6-ERM-35/135 separator is shown in Fig. 2.40. The three-rotor separator has two electromagnetic systems 1, each consisting of a magnetic circuit 2, two blocks of coils 3 and an additional upper pole 4. The rotor blocks 5, shown in Fig. 2.41, are positioned between the poles of the magnetic system. The upper rotor operates as a scalping stage to remove magnetite at a magnetic induction of 0.3 T. The lower rotors recover feebly magnetic particles at increasing magnetic induction, the lowest stage operating at a magnetic induction of 1.3 T. The throughput is specified as 100 t/h [U2]. The ERM separators are manufactured in the Ukraine by Magnis Co., Lugansk.

Figure 2.41: A rotor of the 6-ERM-35 magnetic separator.

Eriez high-intensity magnetic separator

Various designs of the magnetic circuits, differing from that of the Jones separators, were employed by other manufacturers of magnetic separators. For example, in the Eriez WHIMS, the carousel containing the matrix rotates between two poles of the electromagnet, as shown in Fig. 2.42. The background magnetic induction of up to 1.2 T in the air gap is generated by oil-cooled coils. Matrix types commonly used include expanded metal and steel balls. Commercial models range in capacity from 1 t/h to 120 t/h while the required input power ranges from 3 kW to 360 kW.

A similar design was used by Boxmag-Rapid in their HIW separators. The matrix in these separators was trapezoidal wire bars placed at an angle in the carousel. For treatment of particles smaller than 30 μm, the wedge wire was placed horizontally to increase the residence time. Figure 2.43 illustrates a single-head unit, while a photograph of two single rotor Boxmag-Rapid WHIMS operating in an ilmenite recovery plant is shown in Fig. 2.44..

Reading wet high-intensity separators

The Reading wet high-intensity magnetic separators are based on yet another concept of the magnetic circuit design. The magnetic poles of the electromagnetic system, shown in Fig. 2.45 are arranged along the circumference of the carousel with polarity illustrated in Fig. 2.46. The iron cores are magnetized by oil-cooled coils. Each two poles with opposite polarity form a unit in which a high magnetic field is generated and in which separation is taking place. The

Figure 2.42: Two-head Eriez CF-600 WHIMS, with capacity up to 60 t/h (courtesy of Eriez Magnetics, Inc.).

Figure 2.43: A single-head Boxmag-Rapid WHIMS (courtesy of Boxmag-Rapid, Ltd.).

Figure 2.44: Two Boxmag-Rapid WHIMS machines installed in an ilmenite recovery plant.

Figure 2.45: A detailed view of the magnet poles in the separation zones of the Reading WHIMS.

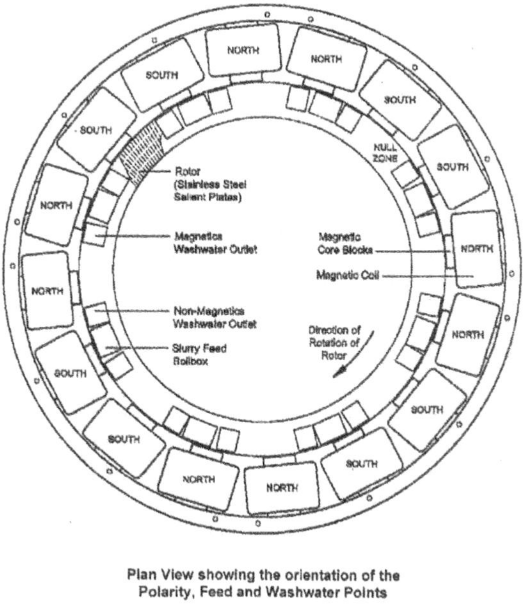

Figure 2.46: Schematic diagram of operation of the Reading wet high-intensity magnetic separator (courtesy of Roche Mining).

feed points are located close to the regions of the high magnetic field. While the non-magnetic fraction passes through the matrix unaffected, the magnetic fraction is flushed from the matrix at the zero field points that occur between the poles of like polarity.

Grooved plates of the salient type are used as the matrix, with an air gap of 2.5 mm. The radial length of the matrix is 68 mm although a wider version with 120 mm radial length is also available. The throughput of these two versions is 2 t/h to 3 t/h per feed point for the narrower version and 4 t/h to 5 t/h per feed point for the wider model. The Reading WHIMS, shown in Fig. 2.47, are manufactured in several models, namely 4 pole (2 feed points), 8 pole (4 feed points) and 16 pole (8 feed points).

The Reading wet high-intensity magnetic separators were developed by Reading of Lismore (Pty.) Ltd., Australia. They are well established in beach sand operations. Concentration of ilmenite, chromite and monazite are typical applications. The units are also applied to the beneficiation of fine weakly magnetic iron ores and to purification of various industrial minerals such as glass sand. Figure 2.48 illustrates the assembly of Reading WHIMS for application in the beach sand industry. The Reading product range was acquired by Roche Mining MT (formerly MD Mineral Technologies) in 2000.

Figure 2.47: The Reading 16-pole WHIMS (courtesy of Roche Mining).

Figure 2.48: Assembly of banks of Reading WHIMS.

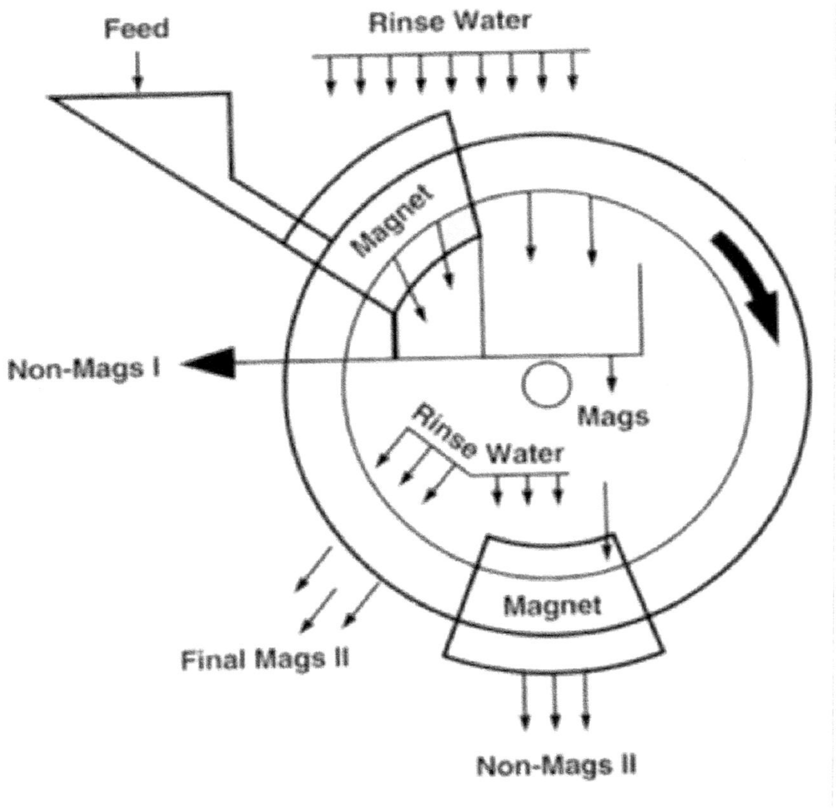

Figure 2.49: Schematic diagram of the Ferrous Wheel magnetic separator, operating in rougher/cleaner mode (courtesy of Eriez Magnetics, Inc.).

Ferrous Wheel magnetic separator

A Ferrous Wheel separator combines the basic concept of a high-gradient magnetic separator with the low-cost permanent magnet circuit. The principles of design and operation can be seen in Fig. 2.49. The vertical separating ring consists of two discs with the matrix situated in pockets between them, as shown in Fig. 2.50. The rings are assembled into a horizontal drum, the length of which is 5.5 m and outer diameter 2.4 m. Rotational speed of the drum is approximately 2 rpm.

The length of the matrix, fabricated from screen cloth, is 150 mm. The choice of the mesh size of the screen cloth is determined by the size distribution of the feed ore. The separating ring is open in the middle to allow the removal of the separated products. Barium ferrite permanent magnets are mounted on each side of the separating rings, as is shown in Fig. 2.51. Each of the two magnetic heads covers an arc of 60^0 and the magnetic induction generated in

Figure 2.50: The Ferrous Wheel separating ring with 125 mm wide matrix (courtesy of Eriez Magnetics, Inc.).

the gap is 0.1 T.

The Ferrous Wheel separator allows a two-stage separation. The separator can be configured as either rougher/scavenger or rougher/cleaner, the latter mode being illustrated in Fig. 2.49. The feed enters the matrix in the magnetic zone located at the top of the ring. The non-magnetic particles pass through the matrix and are channeled out of the separator. This fraction represents the rougher tailings. The magnetic particles, collected onto the matrix, are flushed from the matrix outside the magnetic field and then introduced, from inside the separating ring, into the second magnetic head, to remove the entrained gangue material. An assembly of Ferrous Wheel separators is shown in Fig. 2.52.

Roll-type wet high-intensity magnetic separator

A useful wet alternative to dry induced roll high-intensity magnetic separators was developed in the nineteen sixties in the former USSR [D4, K2]. These separators, suitable for treatment of weakly magnetic, relatively coarse materials (up to 10 mm), consist of profiled rolls that rotate against profiled stationary poles. The poles and the rolls are magnetized by electromagnetic coils wound around the iron yoke. Various profiles of the rolls and of the pole pieces, as shown in Fig. 1.46, have been employed, their selection depending upon the required magnetic force and the application. Triangular teeth facing grooved poles with cuts were found to be the most efficient profile [D4]. Magnetic induction as high as 1.6 T can be achieved in the working zone of the separator. A schematic

Figure 2.51: Magnetic system of the Ferrous Wheel magnetic separator. Permanent magnets are located on each side of the vertical separating ring (courtesy of Eriez Magnetics, Inc.).

Figure 2.52: Ferrous Wheel separators under construction at Eriez Magnetics (USA) (courtesy of Eriez Magnetics, Inc.).

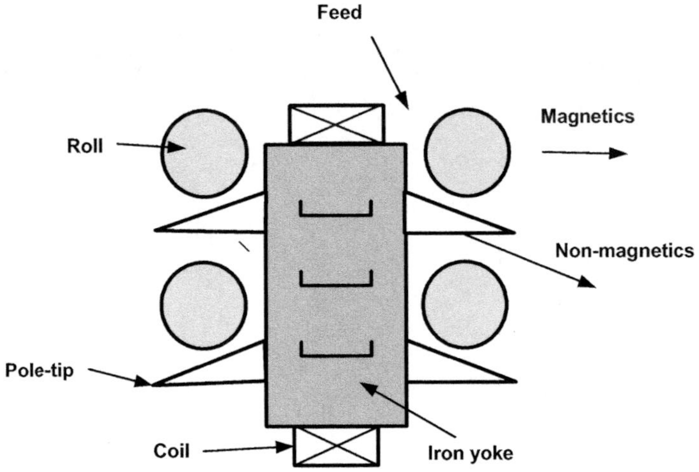

Figure 2.53: Schematic diagram of a four-roll wet high-intensity magnetic separator.

Table 2.5: Specification of wet high-intensity roll magnetic separators.

Parameter	5SVK	ERM3	2MSM5
Throughput [t/h]	2 to 6	5	3 to 6
Roll diameter [mm]	300	300	270
Roll length [mm]	2000	2000	2000
Number of rolls	2	4	2
Magnetic field [T]	1.2	1.6	1.2
Mass of separator [t]	4.7	10.9	4.9
Water consumption [m^3/h]	8 to 10	8 to 10	10
Input power [kW]	6		4

diagram of the four-roll separator is shown in Fig. 2.53 while an isometric view of the four-roll separator ERM-3 is shown in Fig. 2.54. A detailed view of the profiled roll can be seen in Fig. 2.55.

The lower pole tip contains slots through which the non-magnetic fraction leaves the separator. Magnetic particles from the feed are attracted to the teeth of the roll and are discharged when the roll leaves the sphere of influence of the magnetic field.

Typical specifications of various models of wet high-intensity roll magnetic separators are summarized in Table 2.5. A photograph of a 2-MSM-5 separator is shown in Fig. 2.56.

These separators were successfully applied to the beneficiation of coarse (- 5 mm) manganese ore, iron ores such as siderite and limonite, and to the removal of serpentine from magnezite. While these separators cannot treat efficiently

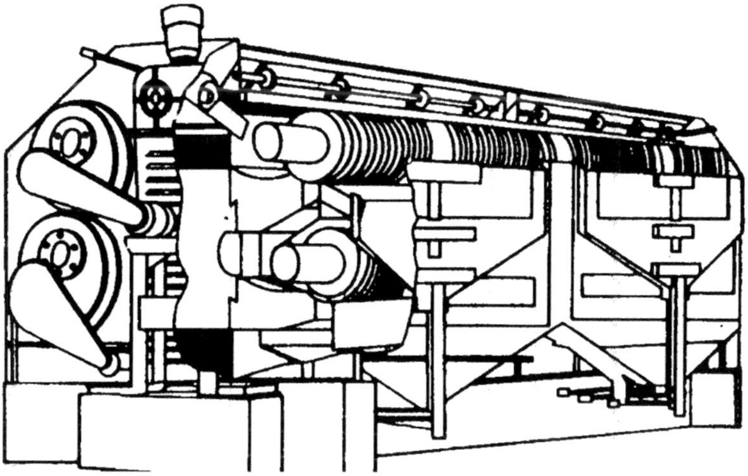

Figure 2.54: A four-roll wet high-intensity roll magnetic separator ERM-3.

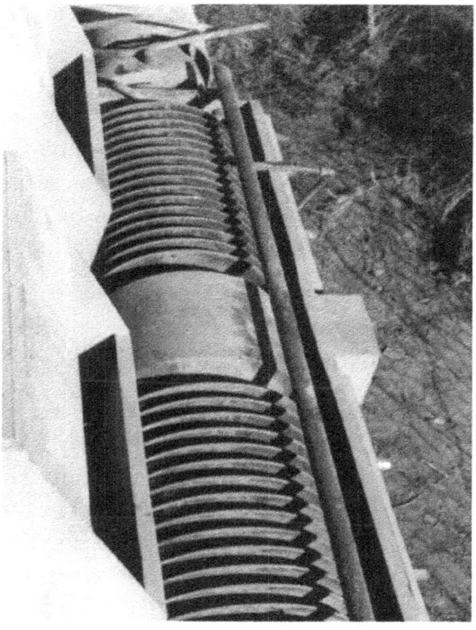

Figure 2.55: A detailed view of the profiled roll of a wet high-intensity roll magnetic separator.

Figure 2.56: 2-MSM-5 wet high-intensity roll magnetic separator (courtesy of ŽB, Czechoslovakia).

very feebly magnetic and very fine particles, their main advantage, compared to WHIMS, is the ability to treat coarse particles and the absence of the matrix blockage. The range of applicability is similar to that of the rare-earth drum magnetic separators, although the field generated by these separators is somewhat higher than that by the drum separators, with similar magnitude of the field gradient. Further advantage of wet induced roll separators is that it is possible to control the magnetic field strength. They can be efficiently applied to materials that fall between low-intensity magnetic separation and high-intensity high-gradient separation.

2.4.3 Solenoid-based magnetic separators

All wet high-intensity magnetic separators already discussed have produced the magnetic field by means of an electromagnet with iron yoke. For small working volumes and for moderately large magnetic induction, this constitutes an efficient and relatively cheap design, although the machines tend to be massive. The input power needed to saturate the iron poles is modest and reasonably high magnetic induction, of the order of 1 T, can be generated in the air gaps, which guarantees sufficiently high throughputs (tens of tonnes per hour) of fine weakly magnetic material. However, it follows from the discussion in Section 1.7.3 that this design suffers two serious drawbacks:

a. The maximum magnetic induction that is obtainable in the working volume is limited by the saturation polarization of iron (approximately 2 T). Since such an induction can be generated only in a narrow gap, much lower inductions

Figure 2.57: Frantz Ferrofilter magnetic separator.

are usually created in units designed for practical large-scale applications.

b. Scale-up is very difficult. In order to increase the throughput, the gap between the poles must be increased. This results either in a dramatic drop in the magnetic induction in the working volume, or in a substantial increase in the number of ampere-turns in order to maintain the same magnetic induction. The machine becomes extremely massive in relation to its capacity, magnetic induction becomes increasingly limited by the width of the gap, and the cooling of coils also becomes a problem.

To generate sufficiently large magnetic induction at large volumes, it is necessary to use an iron-clad solenoid, as has been analyzed in Section 1.7.3.

Frantz Ferrofilter

The first magnetic separator that used the iron-clad solenoid was the so-called Frantz Ferrofilter, introduced by S.G. Frantz in 1937 [F1]. This separator, shown in Fig. 2.57, became the conceptual basis of cyclic high-gradient magnetic separators. It consists of an iron-bound solenoid magnet generating a magnetic field of about 0.15 T, while the matrix is formed by screens fashioned from thin sharp ribbons of magnetic stainless steel, as shown in Fig. 2.58.

The device operates intermittently; the flow is interrupted, the magnetic field is reduced to zero and the magnetic particles are backwashed. Then the field is switched on again and the entire cycle is repeated.

Because of the low demagnetizing factor of the ribbon elements that are

Figure 2.58: A ribbon matrix used in a Frantz Ferrofilter.

oriented parallel with the direction of the magnetic field, the magnitude of the background magnetic induction was sufficient for the recovery of fine ferromagnetic particles. However, in order to magnetize the randomly oriented filamentary matrix with very high aspect ratio, a much higher magnetic field is needed. As a means of achieving a high degree of retention of quasicolloidal, feebly magnetic impurities from kaolin, Frantz's concept was extended so that more powerful iron-clad solenoids were used in conjunction with randomly oriented steel wool. The early developments of cyclic high-gradient magnetic separators are described in detail in references [O3, K5, K7, I2, I3].

Modern cyclic high-gradient magnetic separators

The first industrial high-gradient magnetic separator, which grew from Frantz's concept and from investigations into the removal of weakly magnetic discolouring agents from clays was built in 1969 [O4]. Figure 2.59 shows schematically the essential elements of operation of such a separator. The system comprises a canister filled with a matrix formed of compressed mats of magnetic stainless steel wool, although other types of matrices can be used. The canister is placed in an iron-clad solenoid which generates a magnetic induction up to 2 T. The matrix creates a high degree of non-homogeneity of the magnetic field and produces a large magnetic force acting on the magnetizable particles.

The slurry is fed to the canister either vertically downwards or upwards through the matrix with the magnetic field switched on. The magnetic particles from the slurry are trapped into the surface of the magnetized matrix, while the non-magnetic particles pass through the canister. When the matrix has been loaded with the magnetic particles, the flow is halted, the magnetic field

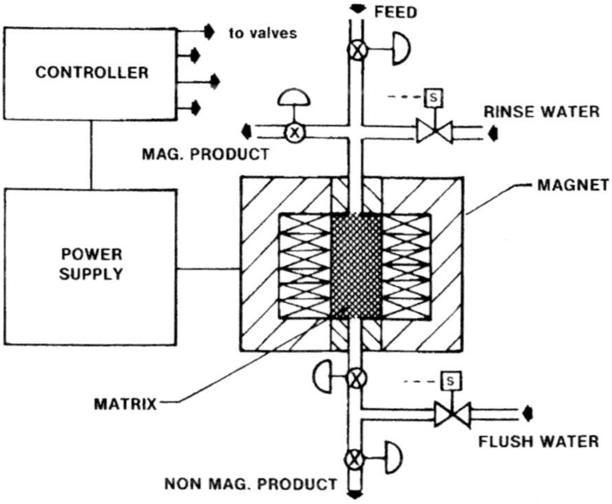

Figure 2.59: Schematic diagram of operation of a cyclic HGMS.

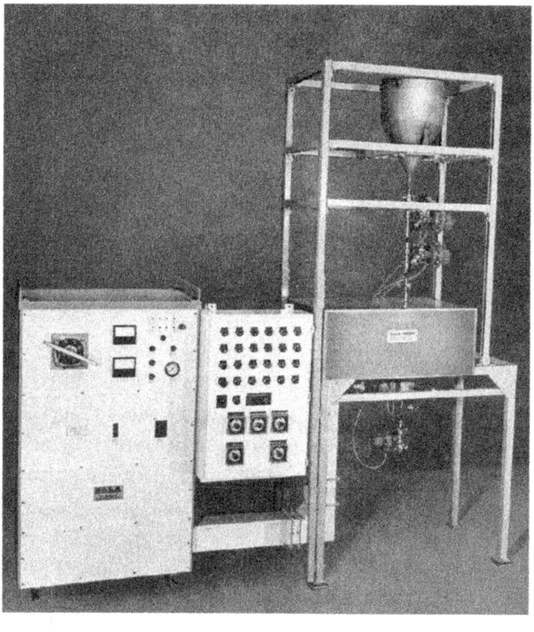

Figure 2.60: An automated small capacity cyclic Sala-HGMS^{TM}

Figure 2.61: High-gradient magnetic filter for effluent control in the steel industry (courtesy of Metso Minerals, Ltd.).

is reduced to zero and the magnetic fraction is flushed from the matrix.

Such cyclic devices are useful in applications where a slurry to be processed contains a small fraction of magnetic particles. These machines have found a widespread application in the kaolin industry and in waste water treatment.

A range of cyclic high-gradient magnetic separators is manufactured by several companies. The solenoid diameter ranges from 1 m to 3 m, input power from 100 to 500 kVA and throughput as high as 200 t/h is claimed. Figure 2.60 shows a small-scale cyclic HGMS while Fig. 2.61 presents a Metso Minerals cyclic HGMS for waste water filtration. A production-scale Eriez Magnetics HGMS for purification of kaolin is shown in Fig. 2.62.

Dry Allis-Sala HGMS Dry matrix high-gradient magnetic separation was developed by Allis Sala [J7] in order to extend applicability of dry separation to such a fine particle range that cannot be easily beneficiated by roll or drum dry separators. Schematic diagram of the separator is shown in Fig. 2.63. The magnetic field is generated by an iron-clad water-cooled solenoid and the matrix is placed in the bore of the coil. Particle transport through the matrix is accomplished by a compressed air-driven vacuum pump. The pump sucks the air through the heater into the separation circuit. The air inlet is positioned at the feed point, where the pre-heated air and particles are mixed before entering the matrix.

The economic expansion of high-gradient magnetic separation to numerous mineral processing applications requires the utilization of fully continuous mag-

Figure 2.62: Eriez Magnetics cyclic HGMS for kaolin beneficiation (courtesy of Eriez Magnetic, Inc.).

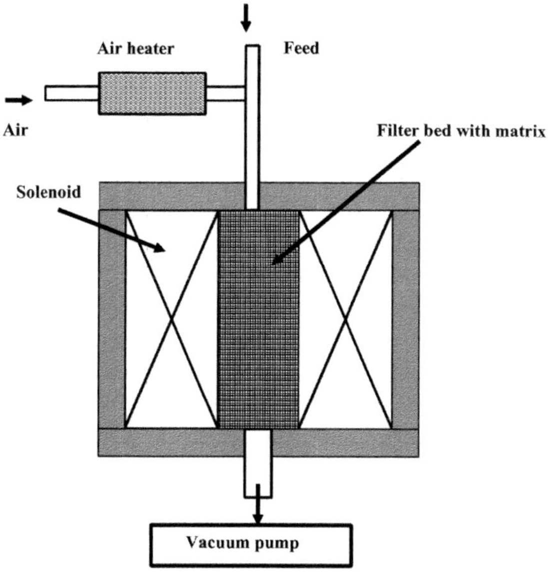

Figure 2.63: Schematic diagram of dry Sala HGMS.

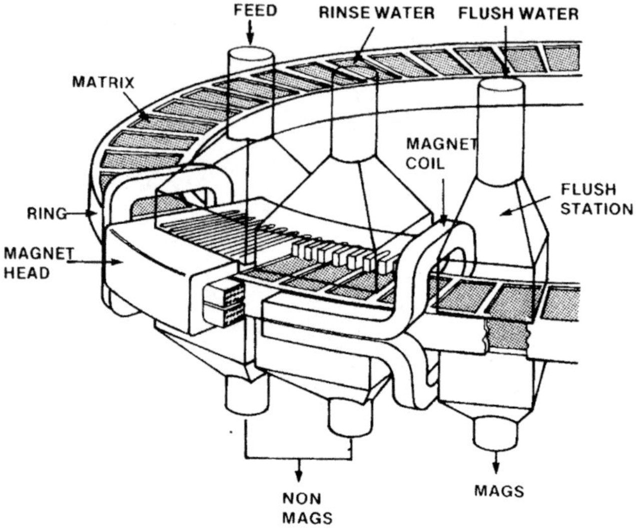

Figure 2.64: Continuous Sala high-gradient magnetic separator.

netic separation. Cyclic devices are limited in their use to separations in which the magnetic fraction is a small portion of the feed and to the feeds that must be passed through the device at controlled slow feedrates. Continuous solenoid high-gradient magnetic separators have eliminated the limitations of the cyclic machines and several advanced devices have been built.

Sala-HGMS carousel separator

The Sala-HGMS carousel separator was developed from the cyclic separator to meet the need for a continuous industrial machine capable of processing ores containing a high proportion of magnetics [O5, M9, A5]. As can be seen in Fig. 2.64, a double-saddle iron-clad coil surrounds the carousel ring. Seals mounted on the ring allow independent control of feed and flush velocity to enhance the separation and to increase the flushing efficiency of the magnetics. The operational sequence is similar to that used in conventional continuous iron-core-type high-intensity separators as is evident in Fig. 2.65.

Further advantage of such a magnetic separator is that it can be scaled up to treat high volumes of material. Magnetic field is parallel to the slurry flow. When increased throughput is required, the width of the magnet is increased, while its height remains constant. The surface area of the matrix can, therefore, be enlarged without increasing the width of the air gap and thus the number of ampere-turns to generate the required magnetic field.

The iron return frame reduces the magnetic flux leakage to places outside the magnetic head. In high-intensity magnetic separators with iron yokes, the stray

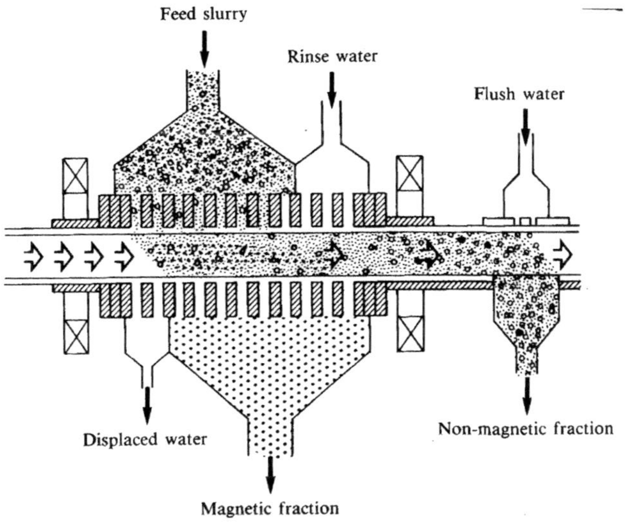

Figure 2.65: The operational sequence in a continuous Sala HGMS [A5].

magnetic field is so high despite a reversal of polarity of two adjacent magnetic heads that the magnetic induction in the flush station area usually amounts to about 20 mT. This is sufficiently high to hamper the efficient flushing of the matrix. With iron-clad solenoids, the magnetic induction outside the magnet is as low as 2 mT, depending on the matrix used.

Sala HGMS are manufactured under the Metso Minerals label in various models. The number of magnetic heads ranges from one to four and magnetic induction varies from 0.5 T to 2 T. Throughput of 200 t/h per magnet head is specified for the largest model 480 ABCD. The matrix is, typically, a compartmented stack of expanded metal screens. A photograph of a single-head Sala-HGMS unit, with nominal throughput of 15 t/h is shown in Fig. 2.66. The largest model that is presently being offered by Metso Minerals is the HGMS 350 shown in Fig. 2.67. This unit can be fitted with up to three magnetic heads and can reach a throughput up to 300 t/h.

VMS magnetic separator

A novel concept for replacement of a horizontal rotor or carousel by a vertical wheel was incorporated in the VMS magnetic separator. The VMS magnetic separator, the isometric view of which is shown in Fig. 2.68, was developed in the Ore Research Institute in Prague, in former Czechoslovakia [H9, H10, C7].

A magnetic field is generated by the horizontally oriented solenoid wound with a water-cooled hollow conductor. The coil is enclosed in steel cladding, the upper and lower pole-pieces of which are perforated to allow passage of the slurry. The slurry is fed into the separator via a feed box and it then enters

Figure 2.66: MK III Sala-HGMS carousel unit.

Figure 2.67: A three-head HGMS 350 carousel machine (courtesy of Metso Minerals, Ltd.).

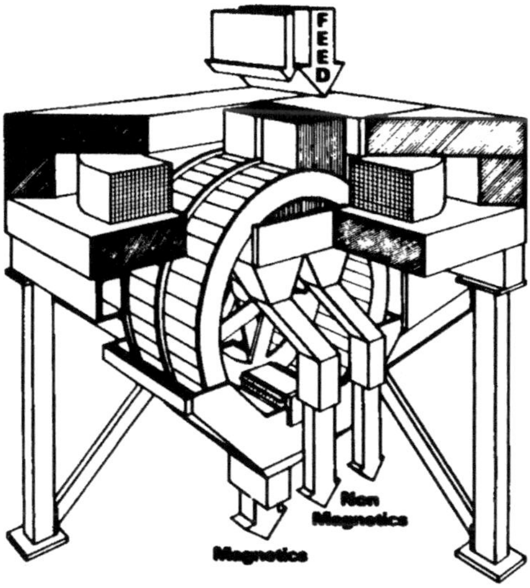

Figure 2.68: An isometric view of VMS.

a rotating drum, which carries separation chambers filled with matrix. The non-magnetic and the middling fractions are collected in launders underneath the magnet. Magnetizable particles are attracted to the matrix, placed in compartments along the periphery of the vertical wheel. When the matrix leaves the region of the magnetic field during the rotation of the wheel, the captured particles are flushed from the matrix. The magnetic product is thus washed off from the matrix at the flush station in a direction opposite to that of the feed and is collected in the trough at the bottom of the separator.

The maximum background magnetic induction generated by the magnet is 1.7 T, while stray magnetic induction outside the magnet is less than 5 mT. Mild steel rods of 3 mm and 5 mm in diameter, with spacing of between 1.5 mm and 2.5 mm, are used as a matrix. Woven wire mesh has also been used in some applications.

In order to increase the capacity of a magnetic separator, the width of the working space must be increased. The VMS concept allows this very simply by increasing the width of the wheel. Taking into account that the magnetic field in the VMS separator is oriented vertically, perpendicularly to the increase of the wheel width, the basic characteristics of the magnetic circuit remain unchanged, i.e. the vertical length of the air gap, which determines the magnetomotive force, or the number of ampere-turns needed to achieve the required magnetic induction, is the same for any width of the wheel. Because of the increase in the length of the conductor, energy consumption increases only linearly with the wheel width. A throughput of up to 150 t/h was specified for the largest separators VMS 100.

Figure 2.69: A vertical ring of the VMS HGMS, with matrix compartments (courtesy of ŽB Spišská Nová Ves, Slovakia).

Figure 2.70: VMS high-gradient magnetic separator (courtesy of Ore Research Institute, Czech Republic).

Figure 2.71: VMS-100 separators in a manganese ore beneficiation plant in the Ukraine (courtesy of Ore Research Institute, Czech Republic).

By replacing the conventional horizontal rotor by a vertically rotating wheel, Fig. 2.69, the VMS separator also addressed the problem of matrix clogging, an onerous problem for most carousel WHIMS or HGMS machines. The reverse flush and very low stray magnetic field in the flush section are additional features of this separator.

The VMS magnetic separator, shown in Fig. 2.70, was manufactured in several models and sizes by ŽB Co. in the former Czechoslovakia. These separators were earmarked for incorporation into the Krivoy Rog and Kursk Magnetic Anomaly iron ore beneficiation plants in the Ukraine. The proposed capacity of the plants was 30 Mt/a and it was envisaged that at least forty VMS separators, each with capacity of 100 t/h, would be required. At the same time, several units, Fig. 2.71, were installed in manganese ore beneficiation plants in the Ukraine.

In many demanding applications, particularly when feebly magnetic admixtures from industrial minerals are to be removed, it is necessary to control the residence time of the slurry in the separating zone and thus the flow velocity through the matrix. To cater for such applications, the VMS separator was modified in such a way that the magnetic system was situated below the rotation axis of the rotor. This VMKS separator, shown in Figure 2.72 and schematically in Figure 2.73, generates a magnetic induction of 2 T and has a rotor diameter of 3 m, width of 1.4 m and throughput of 10 t/h. It was successfully applied to purification of kaolin [Z3].

Figure 2.72: Continuous magnetic separator VMSK for purification of kaolin (courtesy of Ore Research Institute, Czech Republic).

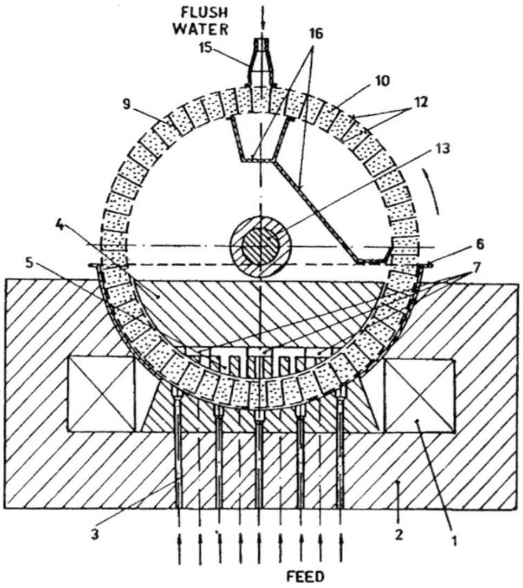

Figure 2.73: Sectional view of the VMKS separator for kaolin purification (Ore Research Institute, Czech Republic).

SLON magnetic separator

It became obvious in the late eighties, that the use of an iron-clad solenoid and application of the reverse flush of the matrix enhances the efficiency, scalability and availability of continuous high-gradient magnetic separators. The technical success of VMS separators prompted further innovative effort. The SLON HGMS, developed in 1988 by Ganzhou Non-ferrous Metallurgy Research Institute, Ganzhou, China [X1], and shown in Figure 2.74, is based on the VMKS concept. However, in order to improve the efficiency of separation even further, the pulsation of the slurry within the matrix was incorporated into the SLON separator.

The pulse generator induces vertical pulses in the slurry within the matrix. The particles are thus exposed to the pulsation force, which is much greater than the force of gravity and acts in both directions [Y2]. By the pulsation action, the mechanically entrained non-magnetic particles are removed from the matrix, which results in improvement of the quality of the magnetic concentrate. At the same time, the pulsation allows the magnetic particles in the slurry to be exposed to the entire depth of the matrix, which increases the recovery of the magnetic component.

The SLON separators are manufactured in several models, specifications of three of which are listed in Table 2.6.

The separator is used, on a large scale, to beneficiate feebly magnetic iron and ilmenite ores and appears to enjoy considerable commercial success in China.

Figure 2.74: SLON-2000 pulsating high-gradient magnetic separator (courtesy Ganzhou Non-Ferrous Metallurgy Research Institute, China).

Table 2.6: Parameters of SLON magnetic separators.

Parameter	**SLON-1000**	**SLON-1500**	**SLON-2000**
Ring diameter [mm]	1000	1500	2000
Maximum magnetic induction [T]	1	1	1
Input power [kW]	26	38	82
Pulsating stroke [mm]	0 to 30	0 to 30	0 to 30
Feed size [mm]	-1	-1	-2
Feed solid density (%)	10 to 45	10 to 45	10 to 45
Throughput [t/h]	4 to 7	20 to 35	50 to 80
Mass of the machine [t]	6	20	50

Figure 2.75: SLON-1500 magnetic separators operating in Pang Zhi Hua ilmenite processing plant in China (courtesy of Ganzhou Non-ferrous Metallurgy Research Institute, China).

It is set to make inroads into international markets. Figure 2.75 illustrates the operation of a bank of SLON-1500 separators in the Pang Zhi Hua ilmenite processing plant (China).

Other innovative designs of solenoid-based magnetic separators

Matrix blockage and insufficient throughput was the main limitation to recovery of uranium and gold from Witwatersrand ores and cyanidation residues by wet high-intensity magnetic separation [C8]. In order to satisfy all the requirements for a cost-effective and reliable machine that would be able to operate for a sufficiently long period in industrial conditions, the Council for Mineral Technology (Mintek), Randburg, South Africa, designed and constructed a prototype separator [S14] shown in Figure 2.76.

The design of the Mintek linear magnetic separator was based on a split-coil magnet that consisted of two parallel short race-track coils, internally water-cooled, separated by a gap through which the matrix could pass. This pair of coils was enclosed in a steel cladding. The matrix consisted of woven mesh screens held vertically on a flexible belt. The belt moved linearly in a horizontal direction and inverted back on itself by means of pulleys. As the belt moved round the pulleys, the matrix elements flared out, thus facilitating the removal of ferromagnetic material and wood chips. The residual mechanically entrained particles were then removed by a reverse flush on the lower side of the separator.

Figure 2.76: The Mintek linear high-gradient magnetic separator.

Scale-up of the machine could be accomplished by increasing the width of the belt and by the installation of two to four magnetic heads in a single machine along the belt. Although the separator performed well on pilot-plant scale, non-technical issues prevented the industrial application of the machine.

An innovative approach to the concept of the VMS separator was employed by Magnetic Technology Consultants (MTC) (South Africa) who designed and built a high-gradient separator (Figure 2.77) designed to treat weakly magnetic materials [A6]. Figure 2.78 illustrates the MTC separator operating in a glass sand plant. The feed material containing 0.08 % Fe_2O_3 was upgraded to 0.022% Fe_2O_3 in the first pass and to 0.011% Fe_2O_3 in the second pass.

Although the separator was designed primarily for wet applications, it was successfully tested in a dry mode. A vacuum system with vacuum seals was used to propel the pyroxenite ore through the matrix and high velocity of the ore and thus high throughput were achieved.

2.5 Superconducting magnetic separators

Although high-intensity and high-gradient magnetic separators that use resistive magnets meet, in most instances, the technological requirements of the mining and materials industries, their cost-effectiveness is impaired by high energy consumption, the need to cool the windings, limited working volume and considerable mass owing to the massive iron yokes or cladding. These disadvantages can, however, be overcome by using superconducting magnets, which

Figure 2.77: The MTC continuous HGMS separator (South Africa).

Figure 2.78: MTC HGMS separator at the glass sand plant. The feed into the separator is on the right-hand side while the non-magnetic product in the first pass is shown on the left.

offer a number of advantages:

a) Superconducting magnets can produce high magnetic induction in large volumes at low energy consumption. The power is required mainly for cooling rather than to compensate for ohmic losses in the coil of the magnet. Moreover, the cost of cooling is independent of the magnetic field generated, unlike resistive magnets where the cost of cooling is proportional to the square of the field intensity. The economic efficiency of the superconducting magnets thus increases with increasing field strength. In some applications the increased magnetic field allows the flow rate of the slurry to increase and thus the throughput of the separator increases.

b) Since iron cladding is generally not needed, the machine can be made physically smaller.

c) The superconducting coils can be arranged in such a way that a sufficiently large field gradient can be generated without using a matrix, while still maintaining high magnetic force.

Two types of magnetic separators that use superconducting coils have been introduced into industrial practice, namely the matrix separators and open-gradient magnetic separators (OGMS). The matrix separators fall into two distinct design categories, namely cyclic vertical flow units operating with liquid nitrogen and helium liquefier, and reciprocating HGMS.

As a simple extension of the matrix-based high-gradient magnetic separators, the conventional resistive solenoid magnet was replaced by a superconducting coil. These separators operating in the cyclic mode in the switch on/off regime are based on short, large diameter, coils, processing at slow slurry velocities with fairly infrequent flushing.

In order to minimize the energy consumption and the running costs of superconducting magnet systems, the magnet should be as passive as possible. The persistent mode option allows the magnet to remain energized with no additional power input. Elimination of the switch on/off regime would also result in a longer duty cycle and thus in higher throughput of the separator.

These issues of a passive magnet were addressed by the reciprocating concept [S15, W4, R3]. In this operation, one canister with the matrix processes the slurry while the second canister is cleaned in a region of zero magnetic field. This technology, first proposed by English China Clays, was not viable until the magnet technology caught up with the operational requirements in the early nineties [R4].

The second type of superconducting magnetic separators, namely OGMS, is based upon deflection of magnetizable particles in a non-homogeneous magnetic field, according to their magnetic susceptibility and size. Instead of employing a matrix to generate a magnetic field gradient, the geometry of the magnet winding is used to create the field gradients and magnetic force that deflects the magnetizable particles. These open-gradient magnetic separators include a linear multipole separator and a drum magnetic separator.

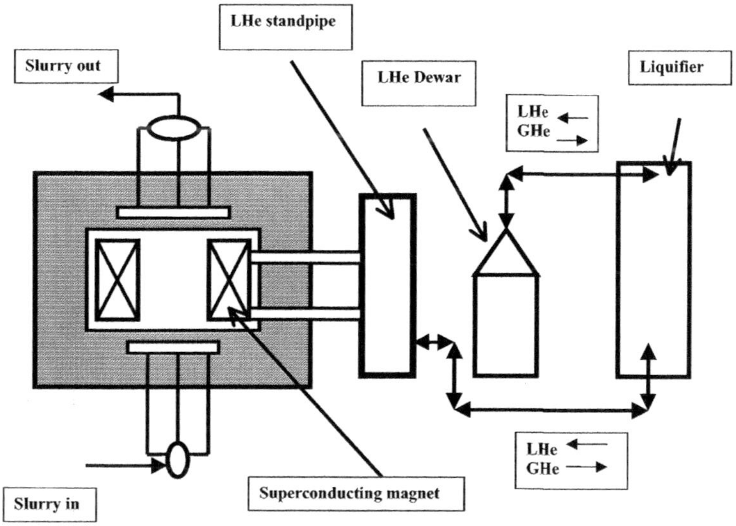

Figure 2.79: Schematic diagram of a cyclic superconducting HGMS.

2.5.1 Cyclic superconducting HGMS

The cyclic resistive high-gradient magnetic separators were exclusively designed for purification of kaolin. A resistive magnet of diameter of 2100 mm (84") generated the field of 1.8 T at a power input of up to 400 kW. Bigger units with a solenoid diameter of 3000 mm and weighing up to 625 tonnes consumed 600 kW. In order to reduce the energy consumption, the Huber Corporation (USA) decided to replace the resistive solenoid by the superconducting coil.

A schematic diagram of a cyclic superconducting HGMS is shown in Fig. 2.79. The first such unit, designed and built by Eriez Magnetics and shown in Fig. 2.80, was installed at Huber Corp. in May 1986 [S16]. The filling factor of the steel wool matrix is about 6% and the matrix is 500 mm deep. The clay slurry is pumped from holding tanks into the distributor header; it flows upward through the matrix to the return header. A typical flowrate is between 1 and 2 m^3/min, and typical retention time ranges from 50 to 90 seconds [W5].

After a period of time, the matrix accumulates enough magnetic material so that the efficiency of separation drops off. At this point the slurry feed is stopped, the magnet is de-energized, and the steel wool is washed in both directions with clean water. The magnet is then re-energized and the feed slurry is re-started. A typical duty cycle is 26 minutes separation, 1 minute of de-energizing, 2 minutes of flushing and rinsing, and 1 minute of re-energizing [W5].

This process means that two dead time periods occur during the ramp up and down of the magnetic field, along with a dead time while the flush of the magnetics is carried out. In order to reduce this dead time, fast ramp

Figure 2.80: A cyclic superconducting Eriez Magnetics HGMS for kaolin beneficiation (courtesy of Eriez Magnetics, Inc.).

superconducting magnets are used. The coil and power supply are subjected to rapid current cycles in order to apply and remove the magnetic field. This results in increased heating of the coil by eddy currents and larger and more expensive cooling power is required. A more robust power supply to drive the current quickly is also required. This results in significant cyclic stresses within the magnet itself [R4].

2.5.2 Reciprocating superconducting HGMS

Superconducting magnets can take advantage of a higher magnetic field compared to resistive magnets, and can increase the processing velocity. However, as the slurry velocity increases, the duty cycle decreases and the economics of separation is affected. Also, for the optimum operation of a superconducting magnet, taking into account coil stresses, field uniformity and power requirements, it is preferable to operate the superconducting HGMS in persistent mode. The issues of long duty cycles of separation and of optimum magnet operation are addressed by a reciprocating technology. In order to achieve economic operation, the captured particles must be flushed from the matrix, with the field on, and the capture zone must, therefore, be removed from the magnetic field.

The train of canisters, as shown in Figure 2.81, is moved by means of a ram that operates on a magnetically balanced canister, which houses a multi-section separation region. As one matrix canister is engaged in separation, the second is undergoing flushing and rinsing. Reciprocation can be accomplished in 10

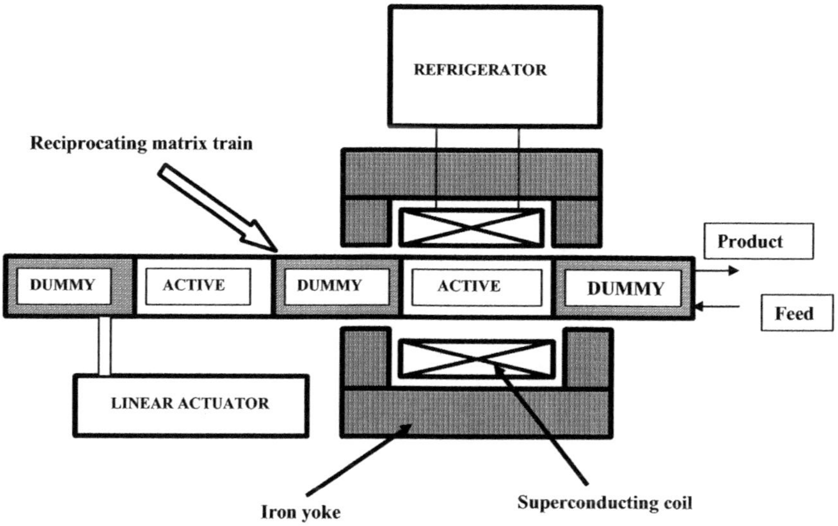

Figure 2.81: Schematic diagram of a reciprocating canister HGMS.

seconds [S17].

While economic design of resistive solenoids is based on short, large diameter coils, for superconducting coils it is far simpler to provide a coil whose length is greater than the diameter. Such a geometrical arrangement, used in Outokumpu/Carpco separators, allows the use of a radial slurry flow rather than an axial flow. As can be seen in Figure 2.82, the radial flow allows the feed area, and thus the throughput, to be increased.

The first industrial superconducting magnetic separator was designed by the Institute for Refrigeration Engineering in Prague, Czech Republic [F6, K8]. This 5 T unit, shown in Figure 2.83 was a reciprocating canister machine for the beneficiation of kaolin. Although, initially, the performance of the separator was encouraging, subsequent deterioration in efficiency and drop in throughput as a result of matrix blockage were given as the main reasons for closure of the separator [A7]. It was suspected that insufficient velocity of flush water and a strong stray magnetic field in the flush station were the main sources of the problems.

In order for a superconducting magnet to operate, it has to be cooled below its critical temperature. The most common way of cooling such a magnet is to immerse it in a bath of liquid helium at 4.2 K within a vacuum-insulated cryostat. The period between liquid helium additions can be further extended by using thermal insulation and liquid nitrogen. Mechanical coolers, based on the Gifford-McMahon refrigeration cycle, can extend the period between helium refills ever further. In this technique, a small volume of high-purity helium gas recirculates through a closed loop, generating a cooling effect at the point of

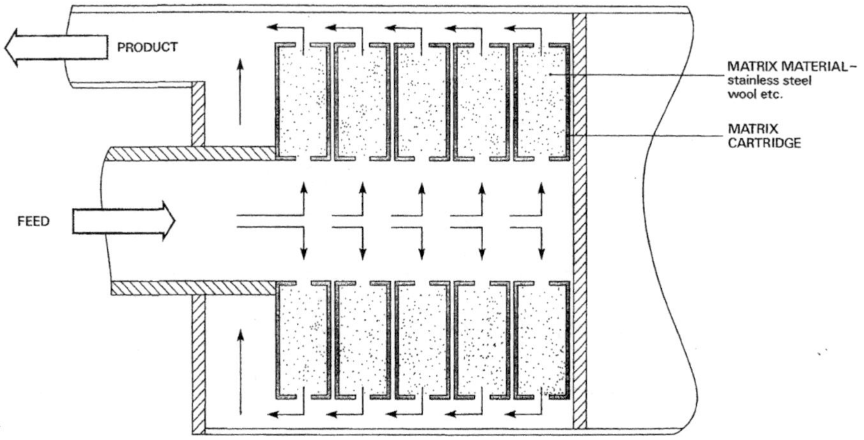

Figure 2.82: Diagram of an active matrix canister in reciprocating superconducting HGMS (Outokumpu Technology, Carpco Division).

Figure 2.83: The Czechoslovak reciprocating-canister superconducting magnetic separation for beneficiation of kaolin.

Table 2.7: Cryofilter HGMS models [S18].

Cryofilter Model	Application	Throughput
5T/200	Pilot	400 kg/h
5T/360	Industrial	up to 15 t/h
5T/460	Industrial	up to 50 t/h
5T/1000	Industrial	up to 150 t/h

Figure 2.84: Superconducting Cryofilter HGMS (courtesy of Outokumpu Technology, Carpco Division).

forced gas expansion. Carpco Cryofilter systems are using this concept of a combination of a liquid helium bath and mechanical cooling, which ensures a refill period of 1 year in normal processing operations [K9].

Further developments in the Oxford Instruments Cryofree® technology now allow superconducting magnet systems to be totally cooled by mechanical means, removing the requirement for helium completely [K9, A8].

Outokumpu Technology, Carpco Division, produce a range of the Cryofilter HGMS employed mainly in kaolin purification. Figure 2.84 illustrates the Outokumpu/Carpco Cryofilter HGMS, while Table 2.7 lists Cryofilter models.

While the Cryofilter supply package depends on the application, the primary items and options include [S18]:

- Basic Cryofilter with integral iron sheet, reciprocating canister and standard control package.

Table 2.8: Comparison of resistive and superconducting cyclic magnetic systems for HGMS, with reciprocating superconducting separators [K9].

Magnet type	Magnetic field [T]	Mass [t]	Power [kW]	Capacity
2100 mm resistive electromagnet	1.8	270	270 to 400	C
3000 mm resistive electromagnet	1.8	625	300 to 400	2C
2100 mm superconducting magnet	2.5	270	25	1.25C
3000 mm superconducting magnet	2.5	625	25	2.5C
5T360 Cryofilter	5	12	< 10	0.4C
5T/460 Cryofilter	5	24	< 10	1.3C
5T/1000 Cryofilter	5	120	<12	4C

- Slurry handling system.
- Advanced process control panel and SCADA monitoring system.
- Detergent matrix wash system.

Table 2.8 compares various types of matrix-based superconducting magnetic separators used for kaolin beneficiation [K9].

2.5.3 Open-gradient superconducting magnetic separators

The presence of a ferromagnetic matrix in high-gradient magnetic separators brings about a problem associated with keeping the matrix clean and passable under industrial conditions. One possible approach to eliminate the problems inherent in matrix machines is to employ a system in which magnetic particles are continuously deflected in a non-uniform magnetic field, according to their magnetic properties and size. These open-gradient magnetic separators, designed for the treatment of fine feebly magnetic materials, must be equipped with superconducting coils in order to provide sufficiently strong magnetic force. The field gradients generated by a suitable combination of the conductor filaments or by coils of varying polarity are much weaker compared to the gradients created by a matrix. Magnetic induction exceeding 2 T is thus required to achieve adequate magnetic force. Figure 2.85 illustrates two different modes of operation of an open-gradient magnetic separator.

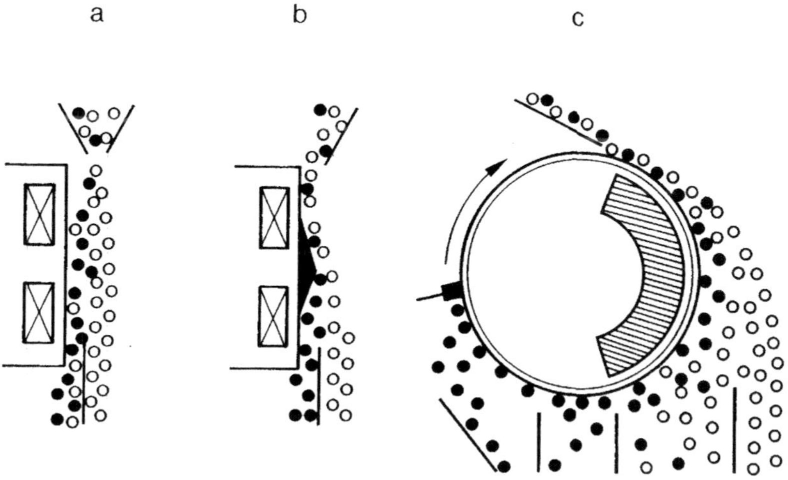

Figure 2.85: Different systems of open-gradient magnetic separators: linear multipole system [(a) and (b)] and drum separator (c).

Linear multipole OGMS

The design of the linear multipole system, as shown in Figure 2.85 (a) and (b), is based on the concept of the ancient Edison's deflection separator [T3]. It consists of an array of vertically positioned coaxial superconducting coils that are energized in opposite directions [S19]. In its simplest form, Figure 2.85(a), the dry ore is permitted to fall vertically as a curtain around the circumference of the magnet. The magnetic force directed radially inwards diverts the magnetic particles inwards while the non-magnetic particles fall vertically, unaffected by the magnetic force. The method failed to meet expectations mainly because the separation was taking place away from the region of the highest magnetic force. As a result of low depth of the magnetic field, its gradient and interparticle collisions, the metallurgical results were unsatisfactory even at low feedrates.

An improvement was achieved by bringing the feed point close to the magnet and introducing a ramp, as is shown in Figure 2.85(b). The performance of the system was sufficiently encouraging for Cryogenic Consultants Ltd. (CCL) to design and build a production-scale Cryofos OGMS unit for beneficiation of the pyroxenite ore in South Africa [G3]. In order to circumvent scale-up problems of magnets of cylindrical geometry used up to that time for OGMS, Good [G3] abandoned the cylindrical geometry in favour of a linear geometry. The split pair was replaced by a linear dipole of rectangular cross section, each side having a separation length of 3 m. The separator could be utilized along almost its entire length on both sides of its axis. Unfortunately, the magnet and the cryogenic system proved to be difficult to manufacture and the separator never reached operational status [S1].

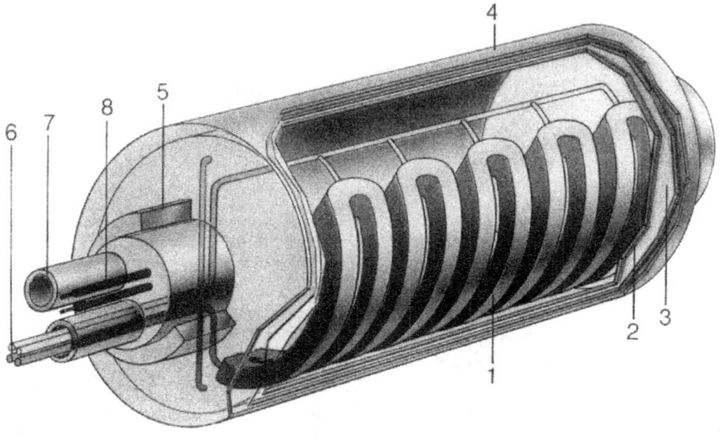

Figure 2.86: Superconducting drum magnetic separator Descos (Humboldt Wedag AG).

Further development at CCL and at Carpco Inc. addressed the issues of the optimized feeding close to the region of the strongest magnetic force and of the removal of the magnetic particles from the walls of the separator. The concept of a flat-faced magnet inclined at an angle with respect to the feeder evolved into the Cryostream® series of OGMS units marketed by Outokumpu Technology, Carpco Division [K9, S17]. In spite of considerable design progress, the concept of Cryostream has not yet enjoyed commercial success.

Descos OGMS

The drum magnetic separator developed by KHD Humboldt Wedag AG employed a multipole magnet and combined the advantages of a superconducting magnet with those of a conventional drum separator. As can be seen in Fig. 2.86, eight poles of the magnet system were arranged along one third of the circumference of the drum. Four race-track coils were placed in a liquid helium tank and generated 3.2 T on the surface. The drum of the separator was made of carbon-fibre reinforced plastic. The diameter of the drum was 1200 mm and its length 1500 mm. Commercialization of the product proved to be difficult as a result of its high cost, in spite of its reportedly satisfactory metallurgical performance.

2.6 Laboratory magnetic separators

A rule that applies to magnetic separation as much as to most physical separation techniques, is that the quickest and surest way of finding out how a material will behave in a separator is to run it through the machine. Theoretical models of the process, which would help to predict the efficiency of low-intensity magnetic separation are scarce and mostly too simplified to be of practical value. A wealth of theoretical descriptions of high-intensity and high-gradient magnetic separation applies mainly to idealized situations, with limited significance for practical problems. The information obtained from these models can sometimes even be misleading and this, combined with the fact that magnetic and some physical and chemical properties of the ores and other materials are difficult to ascertain, indicates that laboratory tests should be carried out before conclusions can be drawn about the efficiency of material treatment by magnetic means.

A wide spectrum of laboratory devices and bench-scale magnetic separators have been built to simulate a particular process in a particular production-scale machine. Generally, the usefulness of these small-scale machines is limited. For instance, in magnetic separators based on permanent magnets, such as magnetic pulleys, overband magnets and drum magnetic separators, the magnetic field generated is a function of the volume of the permanent magnets used in the manufacture of the separator. A small-scale separator will thus generate a lower magnetic field, often of different pattern, compared to a production-scale machine.

Moreover, physical dimensions of the separators, such as the diameter of a pulley or a drum, determine the residence time of material in the magnetic field and thus the efficiency of separation. Also, a small width of a magnetic roll or a drum introduces significant end effects into the results. In laboratory-scale tests of high-gradient magnetic separation, it is essential to keep the matrix depth and flow velocity of the slurry through the matrix the same as in the industrial conditions. Quite often it is not possible. Overall, tests on laboratory-scale units can thus be misleading and practical experience in evaluating the results of small-scale tests, on small, not necessarily representative samples, is essential.

Commercially available laboratory-scale magnetic separators, such as Davis Tube and Frantz isodynamic separator can, however, be used for a reliable analysis of the separability of materials.

2.6.1 Davis Tube

A Davis Tube (DT), shown in Figure 2.87, is a laboratory instrument designed to separate small samples of strongly magnetic ores into strongly magnetic and weakly magnetic fractions. It has become a standard laboratory equipment used for the assessment of the separability of magnetic ores by low-intensity magnetic separators [S20, S1]. It was developed in 1921 and no significant changes have been made in its design since then. The separator consists of an inclined tube 25 mm in diameter, placed between pole-tips of an electromagnet. Material to be separated is passed through the tube; the strongly magnetic fraction is held by the magnetic field while the weakly magnetic material is washed down the tube.

Figure 2.87: Davis Tube magnetic separator.

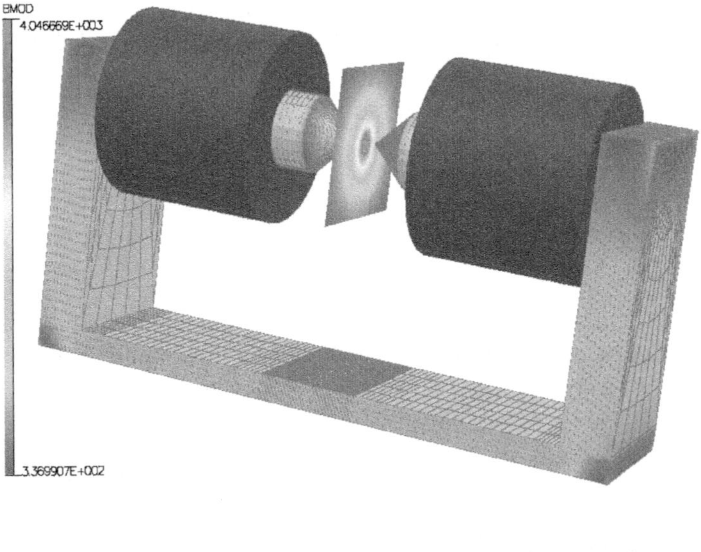

Figure 2.88: The 3D model of the Davis Tube electromagnet [M24, M10].

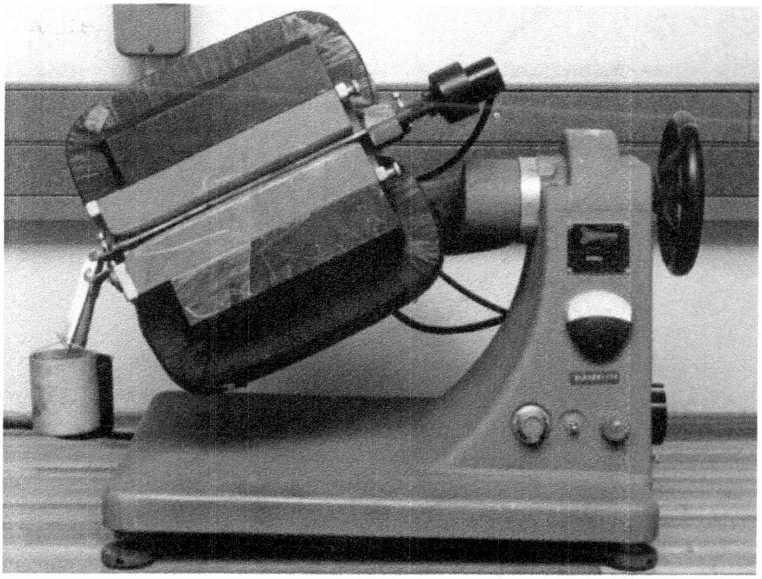

Figure 2.89: Frantz isodynamic separator.

Schultz [S20] suggested that a magnetic induction of 0.4 T or greater between the magnet poles should be used. However, Steinert and Boehm [S21] claimed that the current practice was to conduct the Davis Tube tests at a magnetic induction equal to that on the surface of the drum of the magnetic separator. Since it is the magnetic force or the force index, and not the magnetic field strength, that are decisive in the operation, it is unlikely that this practice would yield correct information.

Computer modelling of the pattern of the magnetic force generated by drum magnetic separators and a Davis Tube (shown in Fig. 2.88) was carried out by Murariu and Svoboda [M10]. It was found that in the case of conventional ferrite drum separators, with a standard gap of 25 mm between the drum and the tank, Davis Tube tests should be conducted at the magnetic induction of 0.1 T. On the other hand, correct assessment of performance of a rare-earth drum can be obtained by operating a Davis Tube at about 0.3 T or higher, depending on the design of the magnetic systems of the drum.

2.6.2 Frantz isodynamic separator

The Frantz isodynamic separator is a magnetic device that allows fractionation of materials based on their magnetic properties. This separator, shown in Fig. 2.89 consists of an electromagnet having two long pole pieces with a long narrow air gap between them. The pole design is such that the magnetic field profile is isodynamic, so that the magnetic force density across the width of the air gap

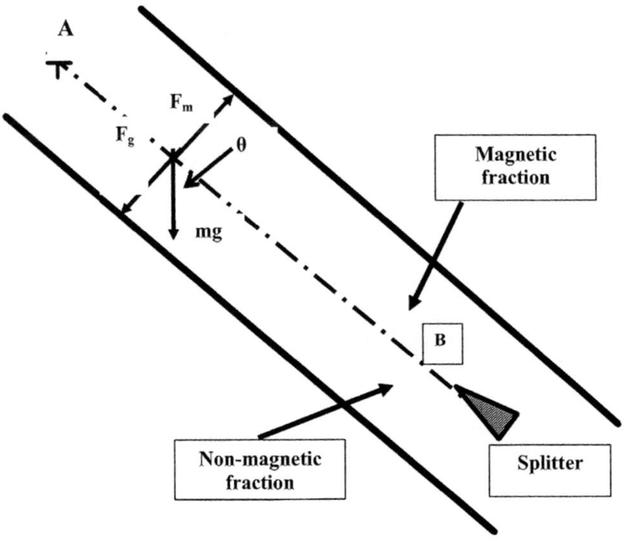

Figure 2.90: Forces acting on a particle in a Frantz isodynamic separator.

is constant, i.e.

$$B\nabla B = const. \tag{2.1}$$

Material to be separated enters the magnetic field at point A, as is illustrated in Fig. 2.90. If the force of gravity F_g and the magnetic force F_m on a particle are balanced in such a way that the particle moves along the central line of the air gap, this particle has a 50 % chance of reporting to a magnetic or non-magnetic fraction. The balance of forces can be expressed as:

$$m_p g \sin\theta = \frac{\kappa}{\mu_0} V_p B\nabla B \tag{2.2}$$

where m_p and V_p are the mass and the volume of the particle, respectively, g is the gravitational constant and θ is the transverse slope of the air gap. By re-arrangement we obtain:

$$\chi = \frac{\mu_0 g}{B\nabla B} \sin\theta \tag{2.3}$$

It can be seen that, because of the isodynamic pattern of the field, the action of the separator on a particle is determined solely by its mass magnetic susceptibility, and is independent of the size and the mass of the particle.

The magnitude of the term $B\nabla B$ is determined by the electric current I in the electromagnet so that

$$\chi = \frac{\sin\theta}{kI^n} \tag{2.4}$$

where n and k are constants [M11]. With an electric current of 1.8 A, the magnetic induction at the narrowest part of the gap is approximately 2 T. In order

Table 2.9: Chemical compounds used for calibration of magnetic susceptibility instruments.

Compound	Formula	χ [$10^7 \cdot m^3/kg$]
Mohr's salt	$Fe(NH_4)_2(SO_4)_2 \cdot 6H_2O$	4.1
Nickel sulphate	$NiSO_4 \cdot 7H_2O$	2.0
Copper sulphate	$CuSO_4 \cdot 5H_2O$	0.74
Cobalt sulphate	$CoSO_4 \cdot 7H_2O$	4.6
Ferrous sulphate	$FeSO_4 \cdot 7H_2O$	5.2
Manganese chloride	$MnCl_2$	15.4
Manganese pyrophosphate	$Mn_2P_2O_7$	12.9

to determine the exact values of these constants, each Frantz separator must be calibrated with compounds having well defined values of magnetic susceptibilities. Chemical compounds usually used for calibration are listed in Table 2.9, together with their mass magnetic susceptibilities.

Mass magnetic susceptibility χ can thus be obtained from eqs. (2.3) or (2.4) by plotting the cumulative per cent recovery of the material into the magnetic fraction, versus the current. Typical magnetic profiles, or partition curves, are shown in Fig. 2.91. The mean magnetic susceptibility of a sample can be defined as the value of χ for which 50 % of the sample reports into the magnetic fraction. While the sample represented by curve 1 in Fig. 2.91 has a narrow distribution of magnetic susceptibility, the sample described by curve 2 contains a wider spectrum of the values of magnetic susceptibility.

The Frantz magnetic barrier separator

While the Frantz isodynamic magnetic separator employs the isodynamic pattern of the magnetic force density, expressed by eq. (2.1), the operation of the barrier separator is based on the presence of the anadynamic and katadynamic force patterns [F7]. In the katadynamic region, the magnetic force increases with the increasing field strength, while in the anadynamic region the force decreases with the increasing magnetic field:

$$\frac{d}{dB} B\nabla B > 0 \text{ (katadynamic)}, \quad \frac{d}{dB} B\nabla B < 0 \text{ (anadynamic)} \qquad (2.5)$$

As can be seen in Fig. 2.92, the magnetic force density between the pole tips of the separator is bell-shaped and creates a barrier in the isodynamic plane, between the katadynamic and anadynamic regions [J2]. Depending upon whether paramagnetic or diamagnetic susceptibility is to be exploited, a mixture of particles is fed into the anadynamic or katadynamic region. As the mixture moves under gravity towards the isodynamic plane, the magnetic force slows the particles of the susceptibility to be exploited and accelerates the particles having susceptibility of the opposite sign. In the vicinity of the barrier, where

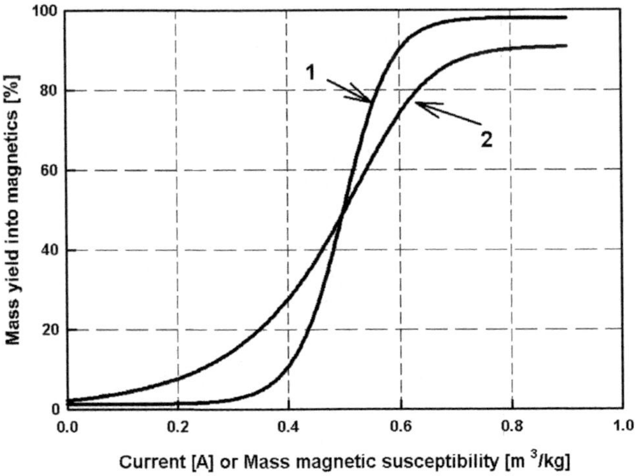

Figure 2.91: Magnetic profiles, or magnetic partition curves, of two types of materials, as obtained by Frantz isodynamic separator.

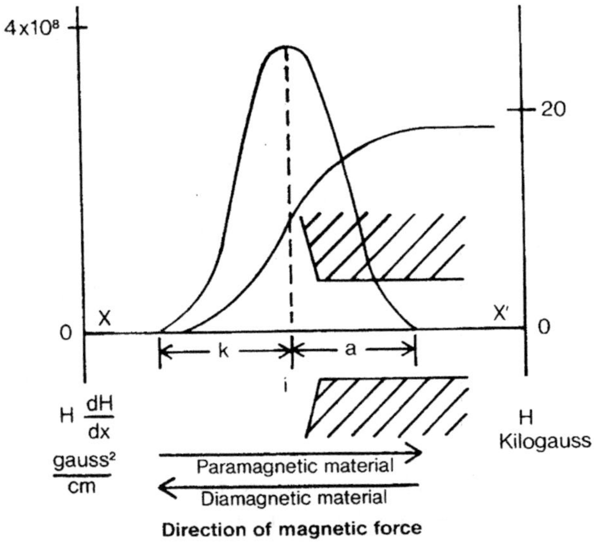

Figure 2.92: Superimposed graphs of the magnetic induction and magnetic force density in the barrier separator, showing katadynamic (k) and anadynamic (a) regions.

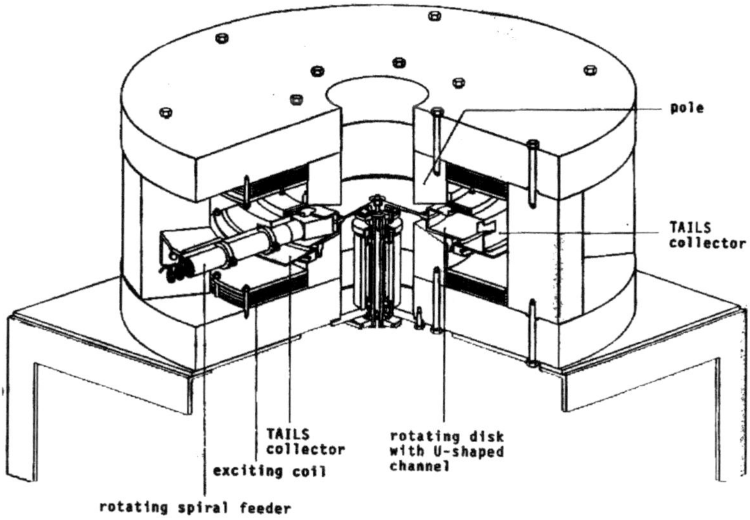

Figure 2.93: Rotating disk isodynamic magnetic separator.

the magnetic force has the maximum value, particles having susceptibility such that the magnetic force equals or exceeds the opposing force of gravity, are deflected in the direction parallel to the isodynamic plane. Particles having susceptibilities lower, or of the opposite sign, pass through the barrier. The barrier separator thus allows selective separation and fractionation of materials based on small differences in magnetic susceptibility [R5].

Disk isodynamic magnetic separator

Frantz isodynamic separator is a selective device that allows separation of mixtures based on differences in magnetic properties. This is in contrast to most other magnetic separators (such as roll, drum and matrix separators) which use the katadynamic field pattern and are thus non-selective from first principles. In these machines the separation is based on differences in magnetic properties as well as on particle size.

Frantz separator is a laboratory unit that does not lend itself to scale-up. In order to overcome limitations of the Frantz separator and, at the same time to retain its high selectivity, Gerhold et al. [G29, G30, G31] designed and built a rotating disk isodynamic magnetic separator shown in Fig. 2.93. The separator consists of a horizontally rotating disk and magnetic system with pole-pieces that generate an isodynamic magnetic field in the separation gap. The magnetic force created by the magnetic system acts radially inwards while the competing force is the centrifugal force. This force can be adjusted to the values ranging from five to ten times the force of gravity.

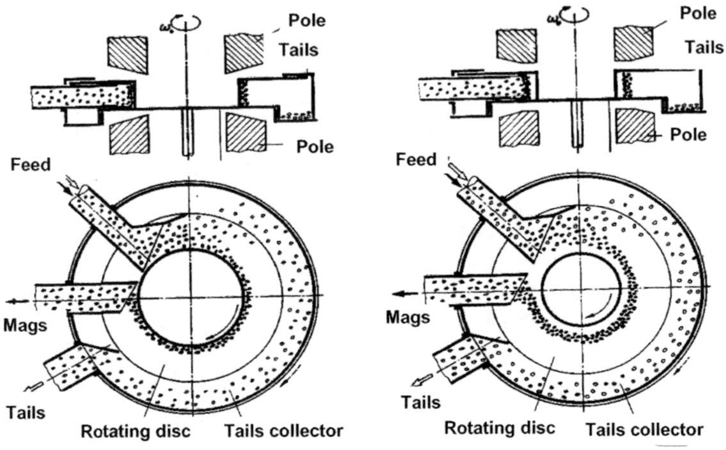

Figure 2.94: Schematic diagrams of operation of a dry rotating disk isodynamic separator, with two different profiles of the magnetic field (adapted from Gerhold [G30]).

Particles fed onto the disk assume the rotational velocity of the disk. The weakly magnetic particles are deflected outwards and are collected in a non-magnetic bin. More strongly magnetic particles are deflected inwards in a form of a rotating ring and are removed before they each the feed region. The operation of the separator, with two different profiles of the magnetic field, is shown schematically in Fig. 2.94.

A pilot-scale separator with a throughput of 2 t/h, maximum magnetic induction of 2 T and field gradient of 10 T/m, was built [G29] and selective separation of ferrimagnetic and strongly paramagnetic materials in the size ranging from 3 mm to 50 μm was obtained [G30].

2.6.3 Separators with a rotating magnetic field

Particle separation in rotating or alternating magnetic fields was proposed in the early years of development of magnetic separation technology [T3, D1]. It appears that this idea has not made its mark on the mineral processing industry. However, in many magnetic separators, the particle rotation is present as a side-effect rather then a deliberate attempt to rotate the particles. For instance, magnetic clusters in low-intensity magnetic separators are rotated slowly as they are carried past successive magnetic poles of opposing polarity. In doing so, the flocs release mechanically or magnetically entrained "non-magnetic" particles, improving thus the grade of the magnetic concentrate.

In some applications it is possible to use particle rotation alone to achieve separations not possible using conventional attraction-based magnetic separation, as has been shown by Allen [A9, A10, A11]. For instance, particles

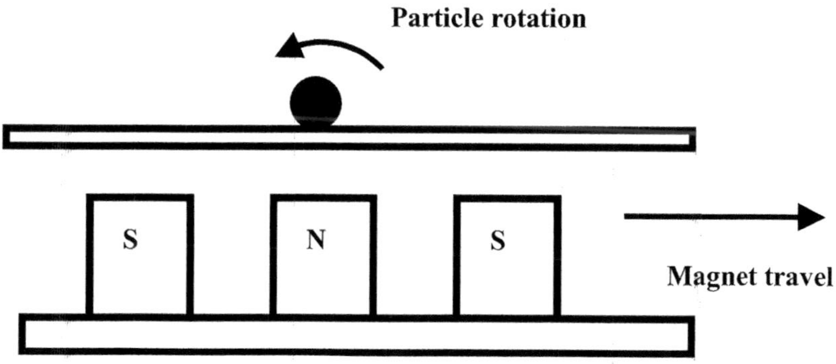

Figure 2.95: Generation of particle rotation.

having similar magnetic susceptibilities, but considerably different magnetic anisotropies, can be separated from each other by their selective rotation. As a result of magnetic anisotropy the external magnetic field exerts a torque on the particle in order to align its axis of easy magnetization with the direction of the external field. If the magnetic field generated, for example, by permanent magnets with alternating polarity, is moving, then the particles will rotate as is shown in Fig. 2.95.

The concept of a rotating magnetic field has been employed in rotating magnetic field (RMF) laboratory separators developed and manufactured by Ka (Pty.) Ltd., Australia. Figure 2.96 shows an RMF lift separator that can be employed for separation by rotating weakly ferromagnetic or even paramagnetic particles with magnetic anisotropy. Figure 2.97 illustrates an RMF separator for separation of paramagnetic minerals.

In these separators, the rotating magnetic field is generated by a rotating magnet rotor fitted with magnetic poles of alternating polarity around its circumference. The rotor in Fig. 2.96, rotating in a clockwise direction, produces a field that appears (to a particle on the outer drum surface) to be rotating in an anticlockwise direction. Therefore, a magnetically anisotropic particle dropped onto the drum will respond by rotating in an anticlockwise direction. Non-rotating particles will be carried by the drum and its belt in a clockwise direction.

RMF rotation separation is basically a separation of ferromagnetic *(s.l.)* particles from paramagnetic particles because of a large difference in magnetic anisotropies and thus rotabilities of these two classes of materials.

2.7 Eddy-current separators

Although the first patent for an eddy-current separator (ECS) was granted to T.A. Edison as early as 1889, a large-scale application of eddy-current separation started only recently, particularly in connection with the recycling of wastes

Figure 2.96: RMF lift separator (courtesy of Ka (Pty.), Ltd.).

Figure 2.97: RMF separator for separation of paramagnetic particles (courtesy of Ka (Pty.), Ltd.).

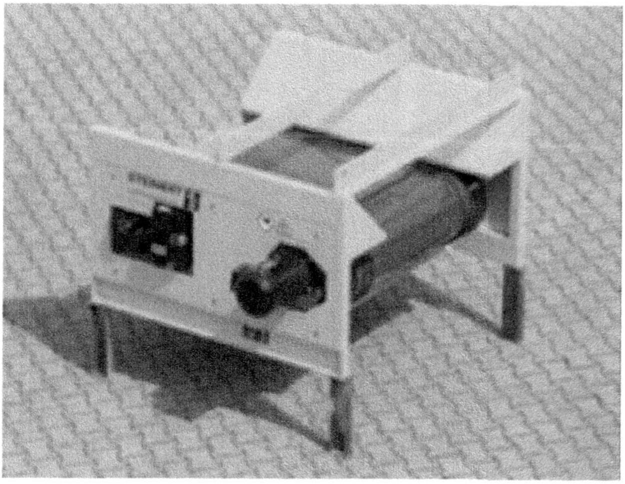

Figure 2.98: Eddy-current separator (courtesy of Steinert GmbH).

containing non-ferrous metals. The first industrial ECSs were based on an electromagnetic field induced either by single coils or by linear motor technique. As a result of the dramatic development of rare-earth magnets, the currently manufactured ECSs employ exclusively permanent magnets. Figure 2.98 illustrates such an eddy-current separator.

The eddy-current technique is based on the fact that if conductive particles are exposed to an alternating magnetic field, or move through a steady magnetic field, eddy currents are generated within the particles. These eddy currents are generated in such a way that the magnetic field they produce opposes the external magnetic field, according to Lenz's law. The opposite field then produces a repulsive force (Lorentz force) on the particle. Particles are thus deflected from the stream of material, as is shown in Fig. 2.99, depending on their conductivity, size, shape, and the field strength and the rate of change of the field.

The industrial eddy-current separators consist of two drums carrying a belt, the belt usually being driven by the rear steel drum. The front non-metallic drum contains a rapidly rotating magnetic system. The permanent magnets are arranged, with alternating polarity, along the circumference of the magnetic system. The separators usually have 6 to 18 poles which rotate at frequencies up to 3000 rpm. Depending on pole arrangement and the system speed, the frequencies of magnetic field change up to 1 kHz are achieved [N3, H11].

Two main designs of eddy-current separators, each with two different magnet configurations, are illustrated in Fig. 2.100, while Table 2.10 reviews the designs used by various manufacturers.

Type 2 design, which employs an eccentric magnetic system, is reported to have several advantages [N3, H11]. In concentric systems, strongly magnetic particles tend to accumulate between the belt and the magnetic drum and ulti-

Figure 2.99: Material separation in eddy-current separator (courtesy of Steinert GmbH).

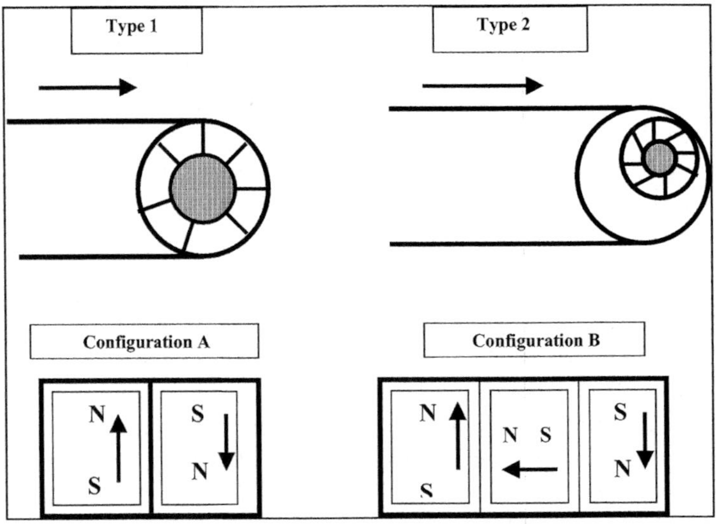

Figure 2.100: Two designs of drum eddy-current separators (after [N3]).

Table 2.10: Manufacturers of eddy-current separators [N3].

Manufacturer	Type	Configuration
Steinert (D)	2	A
Lindemann (D)	1+2	A
Eriez Magnetics (USA)	1	A
Huron Valley (USA)	1	A or B
Andrin (F)	1	A
Bakker Magnetics (NL)	1	A
Raoul Lenoir (F)	1	A
Newell Industries (USA)	1	A
Rutherford Light Metals	1	B

mately to damage the drum. In eccentric design, the ferromagnetic material is removed automatically from the region of the magnetic field generated by the rotating drum.

The repulsive force acting by eddy currents on a small particle [S22, Z4] is proportional to:

$$F_{rep} \approx m_p \frac{\sigma_p}{\rho_p} \frac{d}{f} B^2 \tag{2.6}$$

where m_p and d are the mass and diameter of the particle, respectively, B is the magnetic induction and f the frequency of its change, and σ_p and ρ_p are the electrical conductivity and density, respectively, of the particle. It can be seen that the efficiency of separation is determined, amongst other factors, by the ratio of material conductivity and density. The larger the difference in the values of this ratio for the materials to be separated, the more selective the separation will be. Table 2.11 lists this ratio for materials of interest in eddy-current separation. It is evident that aluminium and magnesium can be rather efficiently separated from semi- and non-conductors, while lead and stainless steel will not respond to eddy currents.

Selection of different models of eddy-current separators is usually dictated by particle size distribution of the feed material and the required throughput. The width of the belt ranges from 500 mm to 2000 mm and nominal throughput is of the order of 10 $m^3/h/m$. Magnetic induction on the surface of the shell ranges from 0.25 T to 0.38 T. Particles in the size range - 100 + 5 mm are usually treated in standard ECSs.

Separation of fine particles Although separation of particles smaller than 5 mm poses a problem [Z5, R6], several innovative designs have successfully reduced the minimum particle size to 0.1 mm. The Steinert model 6119 [H11] and Raoul Lenoir model Eddybox, are claimed to be able to separate efficiently particles as small as 1 mm. The Eddybox separator is based on free-fall of metallic particles through a magnetic structure, has no moving parts and is claimed to achieve a throughput as high as 1 t/h.

Table 2.11: The material parameter for eddy-current separation [N3, S22].

Material	σ/ρ $[\mathbf{10^3{\times}s{\times}kg^{-1}{\times}m^2}]$
Aluminium	13.0
Magnesium	12.2
Copper	6.3
Silver	6.0
Zinc and Zamac	2.4
Gold	2.1
Brass	1.3 to 1.8
Tin	1.2
Lead	0.45
Stainless steel	0.18
Glass	0
Plastics	0

Figure 2.101: Eddy-current separator Fourmi for separation of fine particles (Raoul Lenoir).

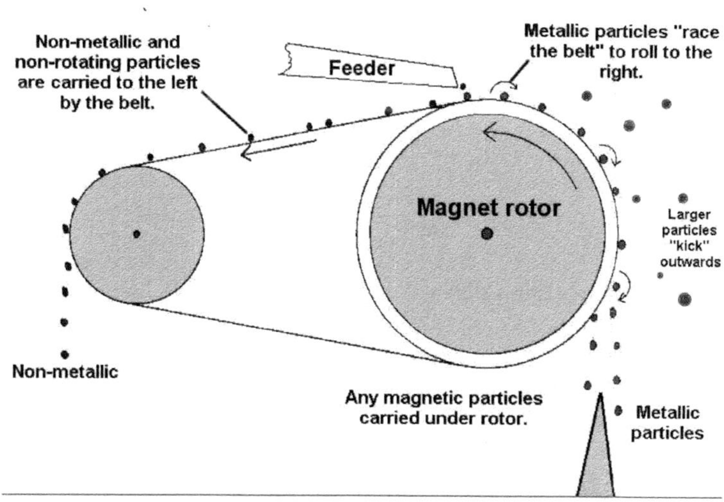

Figure 2.102: Schematic diagram of an eddy-current separator based on rotation of fine particles (courtesy of N. Allen).

The design of Fourmi® eddy-current separator (Raoul Lenoir), shown in Fig. 2.101, is based on a horizontally rotating polar disc with radially mounted NdFeB permanent magnets. An alternating magnetic field generates eddy currents in metallic particles that are fed on a stationary separation disc placed above the polar disk. The conductive particles, being acted upon by repulsive forces, become fluidized and are removed by air jets. Particles as small as 0.1 mm are claimed to be efficiently separated.

It will be shown in Section 3.9.3 that with decreasing particle size the torque on a particle becomes an important force. Particles in ECS then respond to the field changes by rotation rather than by repulsion. Allen [A21, A22] proposed the use of particle rotation as the separating mechanism of fine metal particles. A schematic diagram of such an ECS based on particle rotation is shown in Fig. 2.102.

2.8 Separators with magnetic fluids

2.8.1 Basic concept

Separation in magnetic fluids is a sink-and-float technique, which exploits differences in the densities of materials to be separated. In this technique, a magnetic fluid, placed in a non-homogeneous magnetic field, exhibits an apparent density different from its natural density [R7, R8]. This density can be controlled through a wide range of values, exceeding densities of all known elements and materials. While the scope of the conventional sink-and-float technique is restricted by a limited range of densities of the separation medium, the upper limit of ferrosilicon medium being approximately 3500 kg/m^3, and that of Clerici solution about 4200 kg/m^3 at room temperature, magnetic fluids allow densities as high as 20 000 kg/m^3 or even higher to be achieved.

The apparent density ρ_a of a magnetic fluid placed in a non-homogeneous magnetic field can be written (in SI units) as:

$$\rho_a = \rho_0 + \frac{\kappa_f}{\mu_0 g} B \nabla B \tag{2.7}$$

where ρ_0 is the physical (true) density and κ_f is the volume magnetic susceptibility of the fluid.

The apparent density of a magnetic fluid is thus dependent on the magnetic properties of the fluid, and the magnetic field strength and its gradient. A magnetic fluid, placed in a non-homogeneous magnetic field, is attracted to the regions of the highest magnetic field. This magnetic force produces an opposite force within the fluid that tends to expel an immersed non-magnetic body. This results in additional, magnetically generated, levitation, which causes particles to float.

In the early days, solutions of paramagnetic salts (such as $MnCl_2$) were used as a magnetic fluid [A12, A1]. As a result of their low magnetic susceptibility (of the order of 6×10^{-7} m^3/kg) very high magnetic fields and high gradient were required to achieve densities greater than those obtained using heavy liquids. For instance, in order to achieve an apparent density of the paramagnetic liquid equal to 3500 kg/m^3, the second term on the right-hand side of eq. (2.7) must be equal to 5×10^4 T^2/m. For a magnetic induction of 2 T, a gradient as high as 2.5×10^4 T/m is required. These are unrealistically high values, particularly for meaningful production-scale applications of this technique.

In the late sixties of the last century methods of preparation of stable suspensions of nano-sized magnetite particles in hydrocarbon carrier fluids were developed [P5, R10]. These ferromagnetic fluids (ferrofluids) are considerably more magnetic than the paramagnetic liquid and become magnetically saturated at moderate magnetic fields. Equation (2.7) can then be re-written as:

$$\rho_a = \rho_0 + \frac{M_f}{g} \nabla B \tag{2.8}$$

where M_f is the magnetization of the fluid (in [A/m]). If magnetic polarization J_f, (in [T]) is used, then eq. (2.8) is expressed in the form

$$\rho_a = \rho_0 + \frac{J_f}{\mu_0 g} \nabla B \tag{2.9}$$

In *cgs* units eq. (2.8) can be re-written as

$$\rho_a = \rho_0 + \frac{M_f}{4\pi g} \nabla B \tag{2.10}$$

where M_f and B are given in [G].

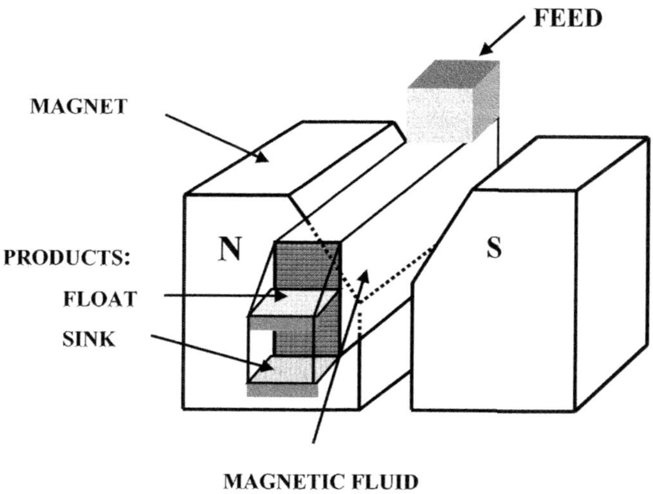

Figure 2.103: Schematic diagram of a separator with magnetic fluids.

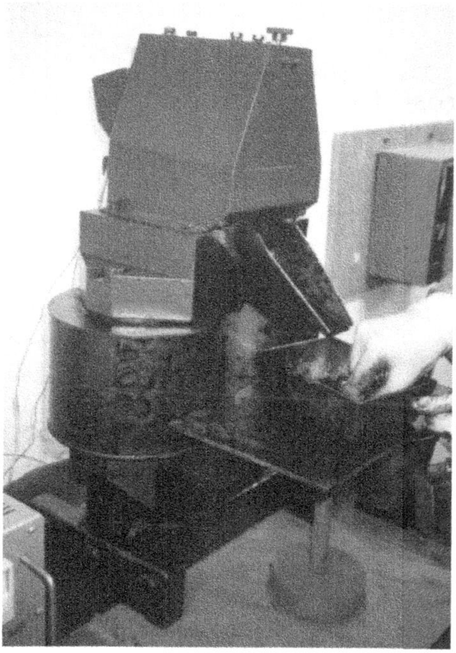

Figure 2.104: Laboratory ferrohydrostatic separator FGS-40 (courtesy of GMUO, the Ukraine).

Table 2.12: FHS separators manufactured by GMUO [G4].

Model	**Feed size [mm]**	**Throughput [kg/h]**	**Density range [kg/m^3]**
FGS-1A	- 50 + 0.2	60	1200 - 11 000
FGS-70	- 25 + 2.5	200	6500 - 10 000
FGS-40	- 10 + 0.2	5	4000 - 9000
FGS-40/3	- 5 + 0.2	2	2500 - 3500
FGS-200	- 100 + 10	4000	3000 - 6000

It can be seen from eqs. (2.8) to (2.10) that in order to achieve a constant apparent density of separation, in a magnetically saturated ferrofluid, the gradient of the magnetic field must be constant. This is achieved by suitably shaping the pole tips of the magnetic system. A ferrohydrostatic separator (FHS), which uses a ferrofluid as a working medium, thus produces two fractions, namely a float and a sinks products. Figure 2.103 illustrates the concept of a separator with magnetic fluids.

The efficiency of ferrohydrostatic separation decreases with decreasing size of particles. In order to overcome this inefficient separation of fine particles in gravity-based FHS, a rotation-based separation Magstream process was developed [W7].

2.8.2 Ferrohydrostatic separators

AVCO Corporation and NASA (USA) pioneered the FHS technique in 1973 by employing a kerosene-based ferrofluid to separate automobile scrap [M12]. Considerable effort to develop this technology was subsequently expended in Japan and the USA. Hitachi [N4], Nittetsu Mining Co. [S51], Tohoku University [S50] and US Bureau of Mines [R9] further confirmed the unique features of the FHS technology. Numerous pilot-plant-scale separators were designed and built, with throughputs up to 500 kg/h. Although extensive tests confirmed viability of this technique, particularly when applied to the recovery of non-ferrous metals from automobile scrap, production-scale application did not take place in the Western World at that time [F8]. Comprehensive review of the history of FHS in Japan and the USA was given by Fujita [F8].

Ferrohydrostatic separation was also researched and used, in approximately the same time period, on a production scale in the former Soviet Union. The Gipromashugleobogashcheniye Institute (GMUO) in Lugansk, the Ukraine, developed a comprehensive range of models of ferrohydrostatic separators for different applications and throughputs. These separators were used in such diverse applications as densimetric analysis of coal, recovery of gold from gravity concentrates, diamond concentration and separation of non-ferrous scrap. Table 2.12 summarizes various models of FHS units manufactured by GMUO [G4]. Figure 2.104 illustrates the versatile FGS-40 separator suitable for laboratory-scale separation of various minerals in the density range from 4000 kg/m^3 to

Figure 2.105: Production-scale ferrohydrostatic separator FGS-70 for separation of non-ferrous metals (courtesy of GMUO, the Ukraine).

Figure 2.106: Ferrohydrostatic separator Shlikh-1 with permanent magnets for recovery of alluvial gold (Too Geos Co.).

9000 kg/m^3. The FHS unit designed to separate non-ferrous metals in the density range between 6500 kg/m^3 and 10 000 kg/m^3, with particle size up to 25 mm, is shown in Figure 2.105.

As a result of geopolitical changes in the nineties of the last century, most markets for GMUO ferrohydrostatic separators in the former Soviet Union disappeared and the separators have not been innovated to meet the current standard of technology.

Ferrohydrostatic separators based on permanent magnets are produced by Too Geos Co. (Russia). These small-scale low-throughput machines have been designed for the recovery of alluvial gold from gravity concentrates. Throughput of the Shlikh-2 model is approximately 50 kg/h, while a laboratory model Shlikh-1, shown in Fig. 2.106 can treat about 10 kg/h.

All models of the Shlikh FHS separator are supplied together with a vibrating screen and disk and roll magnetic separators, as shown in Fig. 2.107. These units allow the classification of the material and the removal of strongly and medium magnetic materials from the feed into the FHS unit.

Since FHS is a unique technique for material manipulation in areas such as diamond-processing industry, De Beers Consolidated Mines (Pty.) Ltd. (South Africa) developed a range of selective FHS separators. These fully automated computer-controlled, hands-off units are capable of treating particles in the size

Figure 2.107: Shlikh-1 system comprising a vibrating screen, disk and roll magnetic separators and a ferrohydrostatic separator (Too Geos Co.).

range - 25 + 0.3 mm and can distinguish a density difference of at least 20 kg/m^3 (0.02 g/cm^3). Units with a throughput exceeding 1 t/h, depending on the feed density and size, have been built [S23]. Figure 2.108 shows a two-stage automated and secure ferrohydrostatic separator built by De Beers for the recovery of diamonds. The ferrohydrostatic separator Rhomag®, shown in Figure 2.109, is used for separating fine fractions of exploration samples.

2.8.3 Magstream separator

The main limitation of ferrohydrostatic separation, in which the important competing force is the force of gravity, is inefficient separation of fine particles. Efforts were made by Revnetsev [R11] and Andres et al. [A14] to replace the force of gravity by the centrifugal force. This proposal resulted in a rotation-based separation process Magstream [W7] developed by Intermagnetics General Corp.

The essential features of the Magstream separation process are shown in Fig. 2.110. Particles to be separated are combined with a ferrofluid and passed downward through a long, vertically oriented annular duct where they rotate with the flow guide and duct about the duct axis. The duct is surrounded by a multipole magnet that produces a non-homogeneous magnetic field. A radially inward displacement force, which results from the radially outward attraction of the magnetic fluid, is felt by all particles. This force is opposed (or aided) by a radially outward (or inward) magnetic force acting on paramagnetic (or diamagnetic) particles, and by the centrifugal force.

At a given speed of rotation, the inward displacement forces are greater on light particles, which then move radially inwards. Heavier particles experiencing

Figure 2.108: An automated hands-off two-stage ferrohydrostatic separator (courtesy of De Beers Consolidated Mines (Pty.) Ltd.).

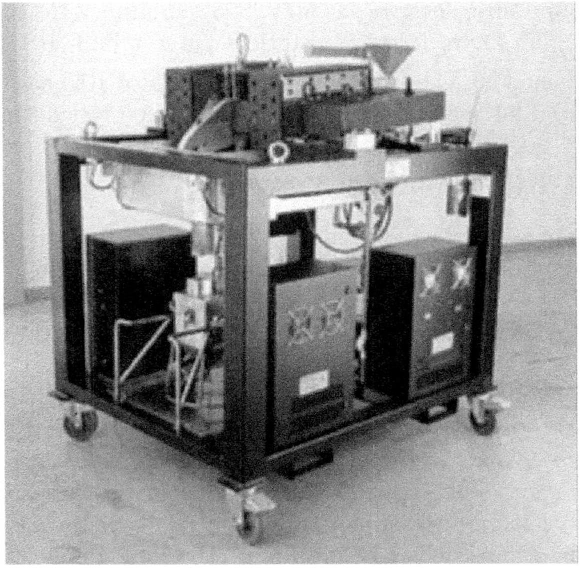

Figure 2.109: Ferrohydrostatic separator Rhomag® for separation of fine particles (courtesy of De Beers Consolidated Mines (Pty.) Ltd.).

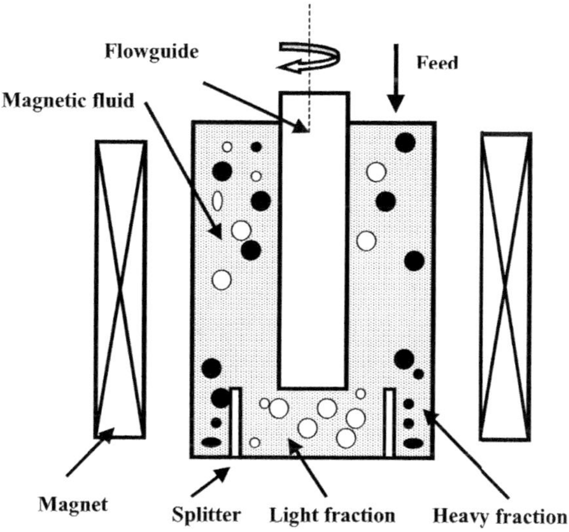

Figure 2.110: Schematic diagram of the Magstream separator.

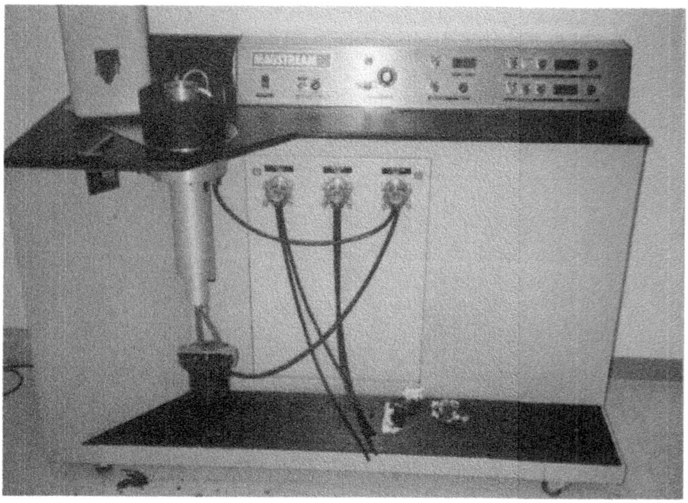

Figure 2.111: Magstream 200 separator.

higher centrifugal and magnetic forces, at the same speed of rotation, overcome displacement forces to move radially outwards. The separation density is then given by:

$$\rho_a = \rho_0 + const\frac{M_f B}{\omega^2} \tag{2.11}$$

where ω is the angular velocity. It can be seen that the additional degree of freedom, namely the speed of rotation, can be used to control the separating force acting on the particles.

The Magstream separators were available in several models, ranging from laboratory-scale portable units for batch separation of samples up to 300 g, to large continuous separators with a throughput up to 250 kg/h. Figure 2.111 shows the 200 Magstream model. Magstream separators can separate particles over a range of densities from 1300 kg/m^3 to beyond 21 000 kg/m^3. The accuracy of separation is claimed to be 100 kg/m^3 (0.1 g/cm^3). For the most efficient separation, the sample should be sized between 60 μm and 1 mm. Separation of samples up to 4 mm is possible with special accessories.

Chapter 3

Theory of Processes of Magnetic Separation of Materials

In view of the diversity of magnetic separation techniques the theoretical descriptions of the magnetic separation process are of limited general applicability. So many factors may affect the behaviour of a system of particles that many problems appear at first sight to be of bewildering complexity. In general, the particle motion in a non-homogeneous magnetic field is so involved as to prevent reasonable theoretical approach. On the other hand, however, a particular problem can often be simplified considerably and a realistic appraisal of the equations describing the motion of the particles is possible.

As has been mentioned in Chapter 1, a particle in a magnetic separator is acted upon by a wide array of forces of various magnitudes and ranges. The necessary (but not sufficient) condition for successful separation of more strongly magnetic particles from less strongly magnetic particles is expressed by eq. (1.1). In real situations, however, such a condition is difficult to evaluate. The general expression for the magnetic force (eq. (1.8)) acquires specific forms in different types of magnetic separators, depending on the pattern of the magnetic field and its gradient. Also, the importance of various competing forces acting on particles in a magnetic separator depends on the type of magnetic separator, on the properties of the particles, such as size, shape, magnetic properties, density, and on the type of fluid in which the particles to be separated are suspended.

Workable models of material separation must, therefore, be developed separately for individual types of magnetic separators. The situation is complicated by the fact that many branches of magnetic separation, such as separation by suspended magnets, magnetic pulleys or wet low-intensity drum magnetic separators still constitute highly empirical technology. Hesitant steps have been taken to develop theoretical models of dry separation in roll and drum magnetic separators. On the other hand, open-gradient magnetic separation, magnetic flocculation of weakly magnetic particles and wet high-gradient magnetic separation have received considerable theoretical attention. A notable number of papers dealing with the problem of particle capture in HGMS led to an understanding of the interaction between a particle and a matrix element. However, completely general treatment of the magnetostatic and hydrodynamic behaviour of an assembly of the ore particles in a system of matrix elements, in the presence of a strong magnetic field, is a theoretical problem of considerable complexity.

A brief survey of the current theoretical descriptions of separation of materials in various types of magnetic separation equipment will be given in the following sections.

3.1 The forces and the equations of particle motion

The equations of particle motion are relations between accelerations, velocities and coordinates of the particles. They are second-order differential equations for the function $r(t)$, where r is the radius vector and t is the time. Their integration makes possible, in principle, the determination of this function and so of the path of the particles.

The equation of particle motion in a magnetic separator can be written in a vector form as

$$m_p \vec{a}_p = \Sigma_i \vec{F}_i \tag{3.1}$$

where $m_p \vec{a}_p$ is the inertial force, m_p is the mass of the particle and a_p its acceleration. F_i are all the forces that may be present in a magnetic separator, such as the magnetic force, force of gravity, hydrodynamic drag, centrifugal force, the friction force, surface forces, magnetic dipolar forces and electrostatic forces among the particles and others. In order to be able to solve eq. (3.1), it is necessary to evaluate the components of all its terms in suitable coordinates.

To obtain the components of the inertial force in polar coordinates that are convenient for most types of magnetic separators, let us consider two mutually perpendicular unit vectors $\hat{r}$ and $\hat{\theta}$. We shall assume that these vectors have directions of increasing $\vec{r}$ and θ and their directions change with time. Since the derivative of a unit vector is perpendicular to the vector, the time rates of $\hat{r}$ and $\hat{\theta}$ can be expressed as [G1]

$$\frac{d\hat{r}}{dt} = \frac{d\theta}{dt}\hat{\theta} \tag{3.2}$$

$$\frac{d\hat{\theta}}{dt} = -\frac{d\theta}{dt}\hat{r} \tag{3.3}$$

Now, the position of the particle, in polar coordinates, is given by $\vec{r} = r\hat{r}$. Differentiating this equation and using eqs. (3.2) and (3.3), the components of the particle velocity $\vec{v}$ are found to be

$$v_r = \frac{dr}{dt} \tag{3.4}$$

$$v_\theta = r\frac{d\theta}{dt} \tag{3.5}$$

The subscripts r and θ indicate the radial and azimuthal components, respectively. By further differentiation, the components of the particle acceleration $\vec{a}_p$ are obtained and hence the components of the inertial force are [G1]

$$m_p a_{pr} = \rho_p V_p \left[\frac{d^2 r}{dt^2} - r\left(\frac{d\theta}{dt}\right)^2\right] \tag{3.6}$$

$$m_p a_{p\theta} = \rho_p V_p \left[r\frac{d^2\theta}{dt^2} + 2\frac{dr}{dt}\frac{d\theta}{dt}\right] \tag{3.7}$$

In Cartesian coordinates these equations can be written as

$$m_p a_{px} = \rho_p V_p \frac{d^2 x}{dt^2} \tag{3.8}$$

$$m_p a_{py} = \rho_p V_p \frac{d^2 y}{dt^2} \tag{3.9}$$

Complete equations of motion are obtained by combining eq. (3.1) with eqs. (3.6) and (3.7) or (3.8) and (3.9).

While it is fairly straightforward to express components of most of the competing forces, such as the force of gravity and the hydrodynamic drag, the magnetic force is more complicated. The general form of the components of the magnetic force, in polar coordinates are given by:

$$F_r = \frac{\kappa_p}{\mu_0} V_p (B_r \frac{\partial B_r}{\partial r} + B_\theta \frac{\partial B_r}{\partial r}) \tag{3.10}$$

$$F_\theta = \frac{\kappa_p}{\mu_0} V_p \frac{1}{r} (B_r \frac{\partial B_r}{\partial \theta} + B_\theta \frac{\partial B_\theta}{\partial \theta}) \tag{3.11}$$

In the Cartesian system the components of the magnetic force can be expressed as:

$$F_x = \frac{\kappa_p}{\mu_0} V_p (B_x \frac{\partial B_x}{\partial x} + B_y \frac{\partial B_y}{\partial x}) \tag{3.12}$$

$$F_y = \frac{\kappa_p}{\mu_0} V_p (B_x \frac{\partial B_x}{\partial y} + B_y \frac{\partial B_y}{\partial y}) \tag{3.13}$$

where κ_p is the volume magnetic susceptibility.

It can be seen that in order to use the above equations, we must be able to specify the components of the magnetic induction and its gradient. Each type of a magnetic separator generates its specific pattern of the magnetic field and the problem of particle motion has to be solved separately for each type of a magnetic separator.

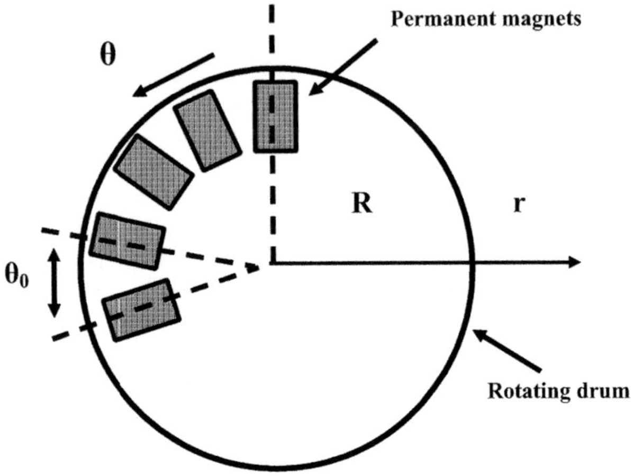

Figure 3.1: Arrangement of magnetic poles in a drum magnetic separator.

3.2 Particle motion in drum and roll separators

Drum and roll magnetic separators are characterized by periodic arrays of magnetic poles. Laurila [L2] made a fundamental investigation of the periodic magnetic field of the cylindrical array of magnets with radial and axial poles used in drum and roll separators. The general expressions for the components of the magnetic force derived in [L2] can then be simplified in specific geometrical arrangements of the magnetic poles.

Hopstock [H12] derived, based on mathematical and empirical analysis, approximate expressions for the radial and azimuthal components of the magnetic field around a magnetic drum with axial arrangement of the magnetic poles, as shown in Fig. 3.1.

Magnetic force in a drum separator

The approximate expressions for the radial and azimuthal components of the magnetic induction, derived by Hopstock [H12] are :

$$B_r = B_0 \cos(\pi \frac{\theta}{\theta_0}) \exp\left[\frac{-\pi(r-R)}{\theta_0 R}\right] \tag{3.14}$$

$$B_\theta = B_0 \sin(\pi \frac{\theta}{\theta_0}) \exp\left[\frac{-\pi(r-R)}{\theta_0 R}\right] \tag{3.15}$$

These equations were found to give a good fit to experimental data in a number of different magnet configurations [H12]. The magnetic induction around a drum, shown in Fig. 3.1, is then given by:

$$B = B_0 \exp\left[\frac{-\pi(r-R)}{\theta_0 R}\right] \tag{3.16}$$

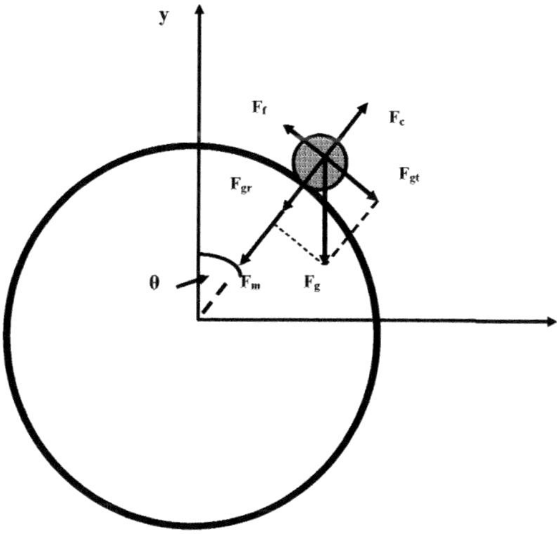

Figure 3.2: Forces acting on a particle placed on a magnetic drum or roll.

where B_0 is the magnetic induction on the surface of the drum. It can be seen that in this approximation B is independent of angle θ. The approximation is most accurate when the following condition is met

$$r - R \geqq \frac{\theta_0}{\pi} R \tag{3.17}$$

The term on the right-hand side of eq. (3.16) expresses the depth of the field, defined as a distance from the surface at which the field is reduced to $1/e$ of its surface value.

Equations (3.14) and (3.15) allow calculation of the magnetic field gradient and thus the magnitude of the magnetic force. As a first approximation, Hopstock found that the radial component of the magnetic force, acting on a magnetite particle in a drum magnetic separator is given by

$$F_{mr} = \frac{3\pi V_p B_0^2}{\mu_0 \theta_0 R} \tag{3.18}$$

Under Hopstock's approximations, the azimuthal component of the magnetic force is equal to zero, which is incorrect. In a more rigorous treatment by Birss et al. [B9], accurate expressions for the components of the magnetic field and force were derived in the form of a Fourier series.

Whether or not a particle is held on the drum and carried to the magnetic fraction depends on the relative magnitudes of the magnetic force and the gravitational and centrifugal forces. The forces acting on a particle situated on the surface of a drum or a roll are shown in Fig. 3.2. For a given angle of rotation

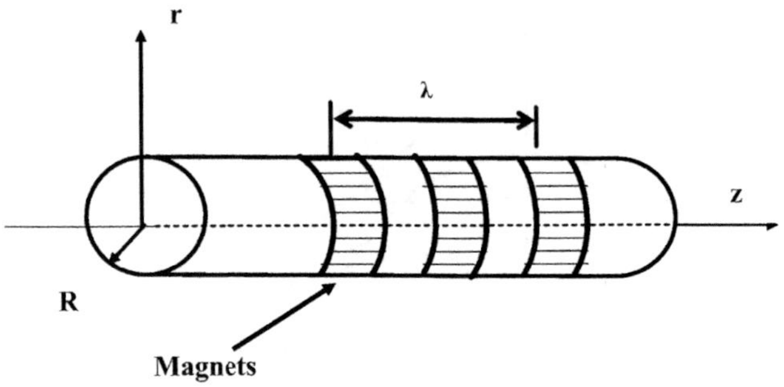

Figure 3.3: Arrangement of magnetic poles in a roll magnetic separator.

θ, the radial force acting of a particle can be expressed as:

$$F_r = \frac{\kappa_p}{\mu_0} V_p B_r \frac{dB_r}{dr} + V_p \rho_p g \cos\theta - V_p \rho_p \omega_p^2 R \tag{3.19}$$

where ω_p is the angular velocity of the particle on a roll or drum of radius R. For a specific case of a high-speed drum, when the force of gravity can be neglected, the centrifugal force equals the magnetic force acting on a magnetite particle when the drum is rotating at the critical angular velocity (in rad/s) given by [H12]:

$$\omega_c = \left[\frac{3\pi B_0^2}{\mu_0 \rho_p \theta_0 R^2} \right]^{1/2} \tag{3.20}$$

Expressions for the components of the magnetic induction for a roll magnetic separator, shown in Fig 3.3, can be written, in cylindrical coordinates, in a more rigorous form [M14]:

$$B_r = B_0 \sin(\frac{2\pi}{\lambda} z) \exp\left[-\frac{2\pi}{\lambda}(r - R) \right] \tag{3.21}$$

$$B_z = B_0 \cos(\frac{2\pi}{\lambda} z) \exp\left[-\frac{2\pi}{\lambda}(r - R) \right] \tag{3.22}$$

where B_0 is the magnetic induction on the surface of the roll, R is the radius of the roll and λ is the periodicity of the magnetic system. The above equations can be re-written in a more general form as [M14]:

$$B_r = B_0 \sin(k_1 z) \exp\left[-k_2(r - R) \right] \tag{3.23}$$

$$B_z = B_0 \cos(k_1 z) \exp\left[-k_2(r - R) \right] \tag{3.24}$$

where k_1 and k_2 are constants that describe the pattern of the magnetic field as a function of r and z. These constants can be determined, for instance, from the electromagnetic modelling of a specific roll separator.

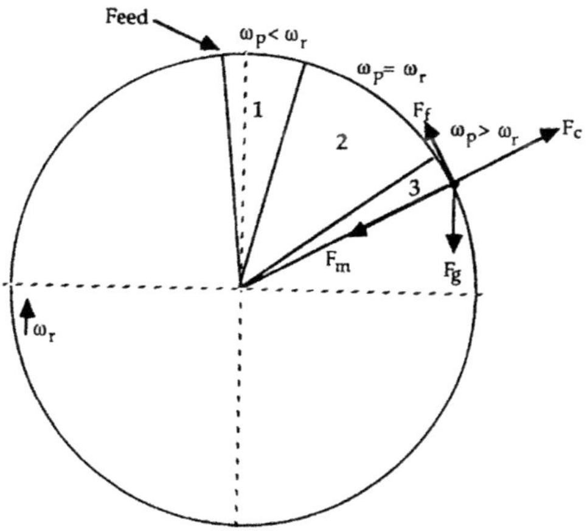

Figure 3.4: The forces acting on a particle in a drum magnetic separator (adapted from [C9] and [S24]).

Magnetic force acting on a spherical particle of radius b and volume magnetic susceptibility κ_p, on the magnetic roll or drum, can then be written as:

$$F_m = \mu_0 \kappa_p B_0^2 \frac{\pi b^2}{3} \left\{ \exp\left[-2k_2(r-b-R)\right] - \exp\left[-2k_2(r+b-R)\right] \right\} \qquad (3.25)$$

3.2.1 Motion of a particle on the drum or roll surface

The motion of a particle of radius b, situated on the surface of a drum separator of radius R, can be divided into three phases [C9], as shown in Fig. 3.4. In the first phase, the particle is accelerating under the influence of the force of friction F_f and the tangential component of the force of gravity F_{gt}, given by the following equation:

$$F_f = \mu_d (F_m + F_{gr} - F_c) \qquad (3.26)$$

$$F_{gt} = g V_p \rho_p \sin\theta \qquad (3.27)$$

where μ_d is the coefficient of dynamic friction and the angle θ is defined in Fig. 3.2. The end condition of phase 1 was defined by Cakir et al. [C9] as the point when the particle reaches rotor speed ω_r, i.e. $\omega_p = \omega_r$.

Once the particle acquires the speed of the drum, a static friction coefficient μ_s, replaces the dynamic friction coefficient in eq. (3.26). The particle then moves with the constant angular velocity equal to the speed of the drum. When the tangential component F_{gt} of the force of gravity (Fig. 3.2) becomes greater than the force of friction F_f , or when the particle leaves the drum, the motion of the particle enters the third phase. At the end of the second phase

$$F_f = \mu_s (F_m + F_{gr} - F_c) = F_{gt} \qquad (3.28)$$

In the third phase the particle begins to accelerate as it slips on the drum surface. The radial component of the force of gravity decreases in this region until the centrifugal force exceeds the radial force and the particle leaves the drum:

$$F_c = V_p \rho_p \omega_r^2 (R + b) = F_m + F_{gr} \tag{3.29}$$

Equations of motion of a particle on a magnetic drum in these three phases were developed by Murariu [M14]. The equation of particle motion in the first phase can be written

$$\frac{d^2\theta}{dt^2} = \frac{g}{R+b} \sin\theta + \mu_d \left[\frac{g}{R+b} \cos\theta - (\frac{d\theta}{dt})^2 + \frac{F_m}{\rho_p V_p (R+b)} \right] \tag{3.30}$$

In the second phase the equation of motion is:

$$\frac{d^2\theta}{dt^2} = 0 \text{ and thus } \frac{d\theta}{dt} = \omega_r \tag{3.31}$$

In phase three the equation of motion becomes:

$$\frac{d^2\theta}{dt^2} = \frac{g}{R+b} \sin\theta - \mu_s \left[\frac{g}{R+b} \cos\theta - (\frac{d\theta}{dt})^2 + \frac{F_m}{\rho_p V_p (R+b)} \right] \tag{3.32}$$

with initial conditions:

$$\theta = \theta_{01} \tag{3.33}$$

$$\frac{d\theta}{dt} = \omega_r \tag{3.34}$$

where θ_{01} is the angle of the slip point at the end of phase 2. At a certain value of the angle θ, denoted by θ_{02}, the particle leaves the drum. At that moment [M14]:

$$\rho_p V_p (R+b) (\frac{d\theta}{dt})^2 = \rho_p V_p g \cos\theta + F_m \tag{3.35}$$

It should be mentioned that in the case of the roll, the first phase is absent since a particle is deposited on the roll at the speed of the roll.

3.2.2 Motion of the particle after it leaves the roll

When the particle leaves the roll or the drum, it is being acted upon by the force of gravity and by the magnetic force. Schematic diagram of this situation is shown in Fig. 3.5. The equations of motion, in Cartesian coordinates, are [M14]:

$$\frac{d^2x}{d^2t} = -\frac{F_m}{\rho_p V_p} \frac{x}{(x^2 + y^2)^{1/2}} \tag{3.36}$$

$$\frac{d^2y}{dt^2} = -g - \frac{F_m}{\rho_p V_p} \frac{y}{(x^2 + y^2)^{1/2}} \tag{3.37}$$

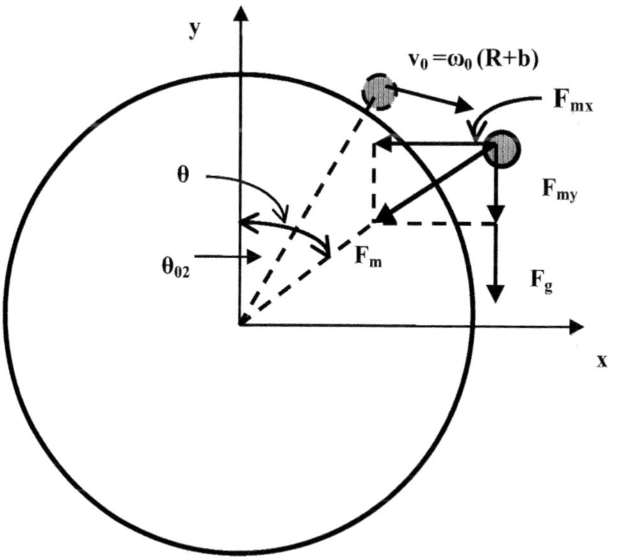

Figure 3.5: Forces acting on a particle when it has left the drum.

where the magnetic force is given by:

$$F_m = \mu_0 \kappa_p B_0^2 \frac{\pi b^2}{3} \left\{ \begin{array}{c} \exp\left[-2k_2((x^2+y^2)^{1/2}) - b - R\right] \\ -\exp\left[-2k_2((x^2+y^2)^{1/2}) + b - R\right] \end{array} \right\} \tag{3.38}$$

Differential equations (3.36) and (3.37) obey the following initial conditions

$$x(t=0) = (R+b)\sin\theta_{02} \tag{3.39}$$

$$y(t=0) = (R+b)\cos\theta_{02}$$

$$\frac{dx}{dt}(t=0) = \omega_0(R+b)\cos\theta_{02}$$

$$\frac{dy}{dt}(t=0) = \omega_0(R+b)\sin\theta_{02}$$

In equations (3.39) the term $\omega_0(R+b)$ represents the initial velocity of the particle at the moment when it leaves the drum, while ω_0 is the angular velocity of the particle at that moment. By solving the equations of motion (3.36) and (3.37), trajectories of particles can be obtained as a function of their magnetic susceptibility, size, density and of the coefficient of friction, magnetic field strength on the surface of the roll, the spatial distribution of the magnetic field and the angular velocity of the roll. An example of a family of particle trajectories is shown in Fig. 3.6.

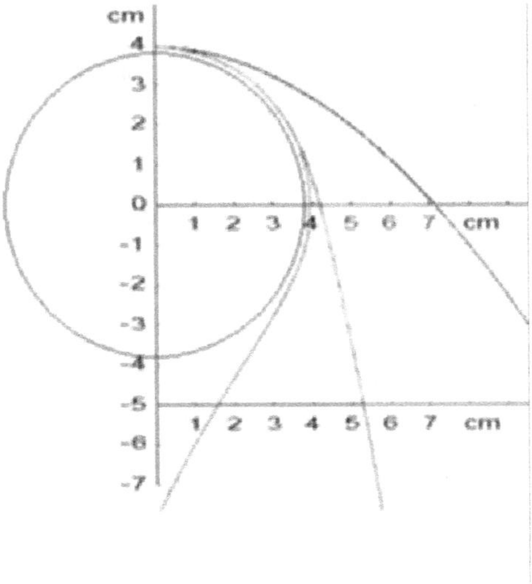

Figure 3.6: Particle trajectories in a roll magnetic separator. Roll diameter: 76 mm, speed of rotation: 200 rpm, particle size: 2 mm and particle mass magnetic susceptibilities (from right to left): 2.5×10^{-7}, 2.5×10^{-6} and 5×10^{-6} m^3/kg.

3.3 Separation of particles by a suspended magnet

Suspended (or overband) magnetic separators have been used over many decades to improve material purity and to protect processing machinery by removing unwanted iron and steel objects from conveyed bulk material. Although the design procedures of such magnetic systems, that use either electromagnets or permanent magnets are well established, the understanding of fundamental principles for extraction of metal is inadequate and standards for selecting suitable separators are inconsistent and often incorrect.

The common practice [A15, H13] is to base the evaluation of a magnetic separator on the so-called "force index", which is essentially the magnetic force density acting on a ferromagnetic body to be removed from a stream of material. Such an approach requires that, for the extraction to occur, the product of the magnetic induction and its gradient assumes a certain minimum value

Although such a concept is valid for certain special shapes of bodies, it does not, in general, describe correctly the behaviour of ferromagnetic particles in a magnetic field and can lead to erroneous design of a magnetic circuit.

3.3.1 Magnetic force in a suspended magnet

A starting point in the evaluation of the efficiency of a suspended magnet is the requirement that the magnetic force F_m acting on a particle must be greater than the sum of the competing forces, e.g. the force of gravity F_g, the drag of the environment F_d, the force of friction F_f and others. The magnetic force on a ferromagnetic particle can be conveniently written as

$$\vec{F}_m = (\vec{\mu}_M \cdot \nabla)\vec{B}_0 \tag{3.40}$$

where μ_M is the effective magnetic dipole moment of the particle and B_0 is the external magnetic induction.

When both the particle and the external medium are homogeneous and isotropic, and the body has an ellipsoidal shape, the effective dipole moment is [L3]

$$\vec{\mu}_M = V_p \vec{H}_0 \frac{\mu_p - \mu_m}{\mu_m + N(\mu_p - \mu_m)} \tag{3.41}$$

where H_0 is the external magnetic field strength, μ_p and μ_m are the relative magnetic permeabilities of the particle and the external medium, respectively, N is the demagnetization factor appropriate for the axis of the ellipsoid along which H_0 is directed.

Using eqs. (3.40) and (3.41) we get for the magnetic force [S25]:

$$\vec{F}_m = V_p \frac{\mu_p - \mu_m}{\mu_m + N(\mu_p - \mu_m)} (\vec{H}_0 \cdot \nabla)\vec{B}_0 \tag{3.42}$$

For a ferromagnetic body in a non-magnetic or weakly magnetic medium we have $\mu_p >> 1$ and $\mu_m \doteq 1$, so that eq. (3.42) becomes:

$$\vec{F}_m = V_p \frac{\mu_p - 1}{\mu_0 \left[1 + N(\mu_p - 1)\right]} (\vec{B}_0 \cdot \nabla)\vec{B}_0 \tag{3.43}$$

The relative magnetic permeability μ_p of the particle is the function of the magnetic field inside the particle, $\mu_p = \mu_p(H)$, where [L3]

$$\vec{H} = \frac{\mu_m \vec{H}_0}{\mu_m + N(\mu_p - \mu_m)} \cong \frac{\vec{B}_0}{\mu_0 \left[1 + N(\mu_p - 1)\right]} \tag{3.44}$$

for $\mu_m \cong 1$.

N is the demagnetization factor appropriate for the axis of the easiest magnetization (i.e. the axis of the smallest N), as the particle is assumed to be oriented so that its axis is parallel to the external field H_0.

The expression (3.43) for the magnetic force is a product of two factors: the first factor reflects the physical properties of particles to be removed, while the second factor, the magnetic force density

$$\vec{f}_m = \nabla(\vec{B}_0^2) \tag{3.45}$$

Table 3.1: Static force indices (FI) and their ratios for selected shapes [S25] (l - length, d-diameter).

Object	ℓ/d	**Force index** $[10^{-6}\ T^2m^{-1}]$	**Ratio** FI/FI_{rod}
Rod	1.0	29 889	7.4
	4.0	9 515	2.4
	10.0	4 037	1.0
Sphere		31 811	7.9
Plate			
13×50×50		20 300	5.0
13×200×200		11 130	2.8
25×200×200		15 300	3.8

describes the properties of the magnetic system. It is this magnetic force density that many designers call "force index" and use to evaluate magnetic separators.

Let us assume that a particle to be extracted by a suspended magnet rests on a stationary conveyor and that no burden of bulk material is present. In equilibrium $F_m = F_g$, where

$$\vec{F}_g = V_p(\rho_p - \rho_m)\vec{g} \tag{3.46}$$

and ρ_m is the density of the medium. Assuming that $\rho_p >> \rho_m$, and combining eqs. (3.43), (3.45) and (3.46) Svoboda obtained [S25]:

$$\vec{f}_m = \rho_p \frac{[1 + N(\mu_p - 1)]}{\mu_p - 1} \mu_0 \vec{g} \tag{3.47}$$

Equation (3.47) determines the minimum force density (or force index) required to lift a stationary ferromagnetic body of relative magnetic permeability μ_p, density ρ_p and demagnetization factor N, from its position on a belt, in the absence of a burden of bulk material.

It can be seen that this minimum magnetic force density is independent of particle size and depends solely on its shape (through the demagnetization factor), its magnetic properties (through the relative magnetic permeability) and its density. Equation (3.47) also shows that the ratio of force indices for various shapes is in the ratio of their demagnetization factors (provided that $N(\mu_p - 1) >> 1$, which is satisfied for all shapes but very long rods or thin plates, for which $N << 1$). Table 3.1 gives several examples of ratios of force index

The minimum static force index (or magnetic force density) as a function of the demagnetization factor, for a given material characterized by ρ_p and μ_p, in the absence of the burden has, therefore, general validity for any size of the object to be lifted. Figure 3.7 shows such a dependence for steel (ρ_p = 7800 kg/m^3, μ_p = 1000). It can be seen that, for instance, for a steel sphere (N = 0.33) FI = 0.03 T^2/m (80.8×10^3 G^2/inch).

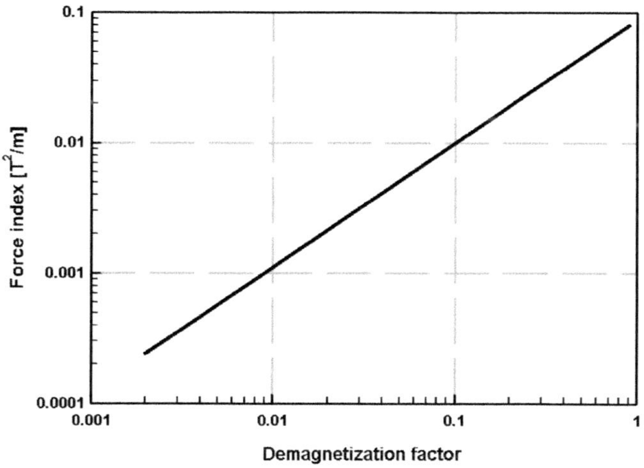

Figure 3.7: Magnetic force density (or force index), for steel, as a function of demagnetization factor (adapted from [S25]).

For majority of objects that are encountered in practice, the assumption that the external magnetic field is directed along the axis of easy magnetization of the body is invalid. Objects such as plates, rods or bolts usually rest on a conveyor belt with their axes of easy magnetization perpendicular to the direction of the external magnetic field. Demagnetization factors appropriate for such orientations are high and large force densities (or force indices) are required to lift the body towards the magnet.

3.3.2 Torque on a magnetizable body

Under the above mentioned circumstances, the role of the magnet should be first to re-orient the ferromagnetic body in such way that its long axis (the axis of easy magnetization) is parallel with the direction of the external magnetic field. This is accomplished by exerting a torque on the body by the externally applied magnetic field.

The magnetic torque is given [L3, L4] by:

$$\vec{T} = \vec{\mu}_M \times \vec{B}_0 \tag{3.48}$$

Since $\vec{\mu}_M = V_p \vec{M}_p$, where M_p is the magnetization of the particle, eq. (3.48) can be rewritten:

$$\vec{T} = V_p \vec{M}_p \times \vec{B}_0 \tag{3.49}$$

Assuming a body of spheroidal shape, the torque is perpendicular to the plane passing through the axis of symmetry of the body and the direction of

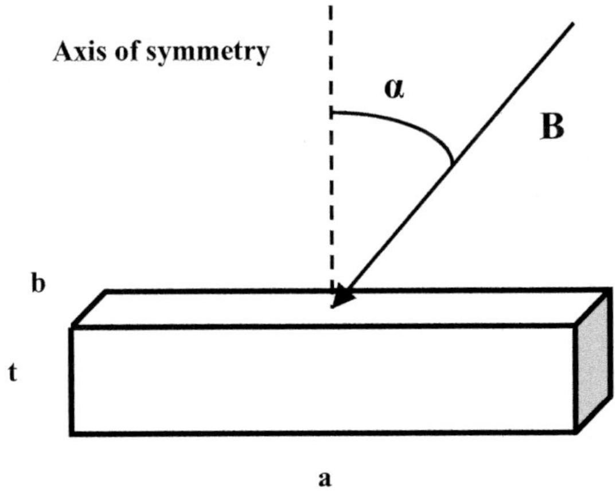

Figure 3.8: A ferromagnetic bar in a transverse external magnetic field.

B_0. The magnitude of the torque is given [L5] (by analogy with the torque on a dielectric spheroid in an electric field)[S25]:

$$T = \frac{(\mu_p - \mu_m)^2 \mid 1 - 3N \mid H_0 B_0 V_p \sin 2\alpha}{2\left[(\mu_m + N(\mu_p - \mu_m)\right]\left[\mu_p + \mu_m - N(\mu_p - \mu_m)\right]} \tag{3.50}$$

and for $\mu_m \cong 1$

$$T = \frac{(\mu_p - 1)^2 \mid 1 - 3N \mid B_0^2 V_p \sin 2\alpha}{2\mu_0\left[1 + N(\mu_p - 1)\right]\left[1 + \mu_p - N(\mu_p - 1)\right]} \tag{3.51}$$

Here α is the angle between the direction of B_0 and the axis of symmetry of the spheroid. It can be seen that the maximum torque occurs at $\alpha = 45^0$. The torque is directed in such a way that it tends to turn the axis of symmetry of a prolate (i.e. $N < 0.33$) or oblate ($N > 0.33$) spheroid parallel or perpendicular to B_0, respectively. Therefore, the torque tends to align the axis of easy magnetization of the body along B_0.

In equilibrium, the magnetic torque is balanced by the torque exerted by the weight of the particle G_p and by the weight of the burden G_b. We can thus write for a particle having a shape of a bar:

$$T = \frac{1}{2} ga(\rho_p + \rho_b \frac{h}{t})V_p \tag{3.52}$$

where a and t and dimensions of the bar, as shown in Fig. 3.8, h is the height of the burden of the material and ρ_b is its bulk density.

Combining eqs. (3.51) and (3.52) we obtain, in equilibrium, [S25]:

$$B_0^2 \sin 2\alpha = ag(\rho_p + \frac{h}{t}\rho_b)\mu_0 \frac{(\frac{1}{\mu_p - 1} + N)(\frac{\mu_p + 1}{\mu_p - 1} - N)}{\mid 1 - 3N \mid} \tag{3.53}$$

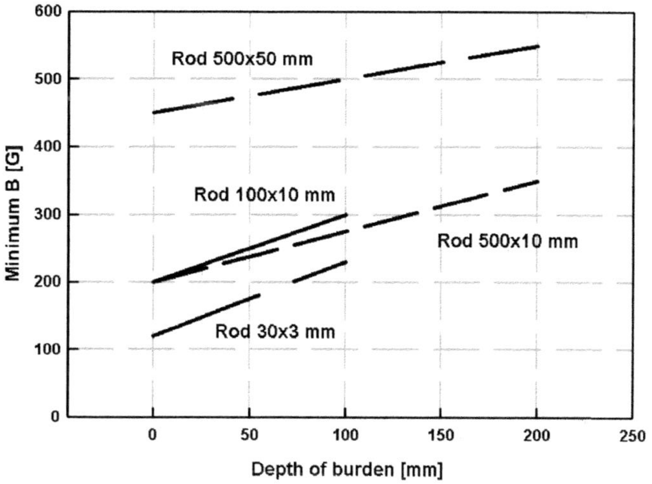

Figure 3.9: Minimum values of the magnetic induction, required to orient steel rods of selected sizes, as a function of the depth of the burden (adapted from Svoboda [S25]).

For $\mu_p >> 1$ we then have:

$$B_0^2 \sin 2\alpha \cong ag\mu_0(\rho_p + \frac{h}{t}\rho_b)\frac{N(1-N)}{\mid 1-3N \mid} \tag{3.54}$$

It can be seen that if the magnetic system generates the field that is perpendicular to the plane of the rod, the rod will not orient itself since $\alpha = 90^0$ and thus $T = 0$. It also transpires from eq. (3.48) that the gradient of the magnetic field does not contribute, in the first approximation, to the torque. Therefore, only the magnitude of the field determines whether the reorientation of the body takes place or not.

It is clear that in contrast to the conventional description of separation by a suspended magnet, which requires that the magnetic system is designed in such a way that the force index $B\nabla B$ achieves a certain minimum value, the physical picture based on the magnetic torque approach requires a minimum value of the magnetic field only. The physical content of such a description can have important implications for the design of the magnetic system.

It follows from the previous analysis that in a suspended magnet the important parameter is the magnetic field strength. If the field strength is high enough so that it can orientate the body along the axis of easy magnetization, the magnitude of the force index is required to be high enough to lift the particle in the direction of its axis of easy magnetization, only. Therefore, the force index is not the only parameter needed to evaluate a suspended magnet.

For instance, the force index $B\nabla B$ may be just large enough to lift an

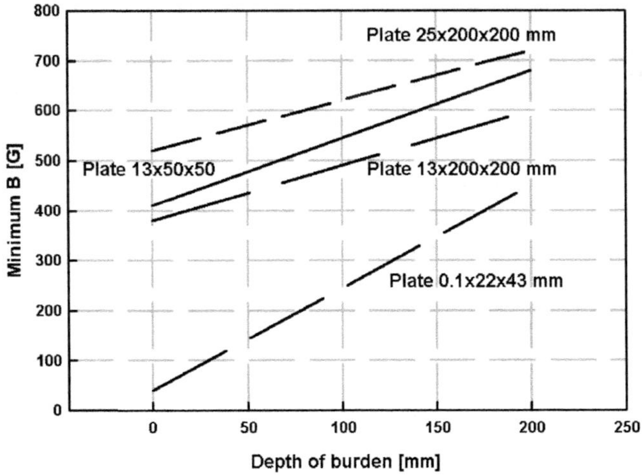

Figure 3.10: Minimum values of the magnetic induction needed to orient steel plates of selected sizes, as a function of the depth of the burden (adapted from Svoboda [S25]).

object in the direction of its axis of easy magnetization, but the magnitude of the magnetic field itself may still be too low for the condition of eq. (3.53) to be met. The object then will not re-orientate itself appropriately and will not be lifted.

When designing the magnetic system of a suspended magnet two conditions thus must be met: the magnitude and the direction of the magnetic field must satisfy eq. (3.54) and, at the same time, the force index must fulfill condition given by eq. (3.47).

Typical dependence of the minimum magnetic field on the depth of the burden, for selected sizes of steel rods and plates, is shown in Figs. 3.9 and 3.10. Similar diagrams can be constructed for other shapes and sizes using eq. (3.53), provided the demagnetization factors are known.

A comparison between theoretical values of the minimum magnetic induction $B_{\min}$ required to orient a body along its axis of easy magnetization, calculated using eq. (3.53), and experimental values is given in Table 3.2. It can be seen that the agreement is satisfactory.

3.3.3 Demagnetization factors

Calculation of the demagnetization factor for objects of shapes different from that of an ellipsoid is complex and the values found in literature, e.g. for ellipsoids and infinite cylinders, are often of limited practical use. Brown [B3] gives tables for circular cylinders of finite length, magnetized longitudinally, and for

Table 3.2: Comparison of theoretical and experimental values of the minimum magnetic induction required to orient steel objects, as a function of the depth of the burden [S25].

Object	$\mathbf{B_{min}}$ **[T]**		
	Burden depth [mm]	Exp.	Theor.
Rod 3×30 mm	0	0.0124	0.0113
	10	0.016	0.0132
Plate 0.1×22×43 mm	0	0.0033	0.0034
	10	0.0131	0.0114

Table 3.3: Demagnetization factors of ferromagnetic objects [S25]. Cylinders and plates are magnetized longitudinally(l - length, d - diameter).

Object	**Size** (ℓ/d)	**N**
Sphere		0.33
Finite cylinder	0	1.0
	0.4	0.528
	1.0	0.3116
	4.0	0.098
	10.0	0.041
	40.0	0.011
	100.0	0.004
	1000.0	0.0004
Plate	13×50×50 mm	0.19
	13×200×200	0.06
	25×200×200	0.11

rectangular rods magnetized transversely. These geometries, together with an ellipsoid, should be sufficient for analysis since other shapes usually encountered in the separation practice (plates, hexagons) can be, in the first approximation, converted into geometries for which the demagnetization factors are available. Values of demagnetization factors for selected shapes and sizes are given in Table 3.3.

3.3.4 The effect of the burden

It transpires from the above analysis that a ferromagnetic object to be lifted by the suspended magnet must be first re-orientated by the action of the magnetic field, in such a way that its long axis (i.e. the axis of easy magnetization) is parallel with the direction of the magnetic field. Once this is achieved, the magnetic force is needed to overcome the weight of the object. The object will then be lifted provided the force index acting on a particle is greater than the minimum force index given by eq. (3.47).

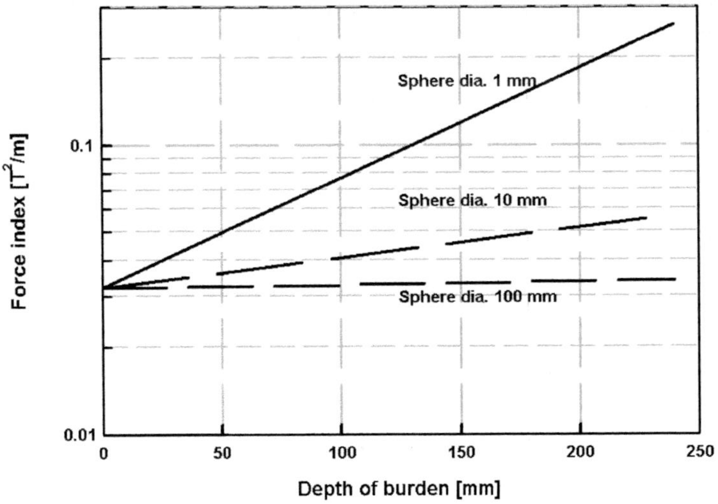

Figure 3.11: The static force index as a function of the depth of the burden, for various sizes of a ferromagnetic sphere ($\mu_p = 1000$, $\rho_p = 7800$ kg/m^3, $\rho_b = 800$ kg/m^3) (adapted from [S25]).

In practice, however, a ferromagnetic particle is usually buried in a layer of loosely packed material, e.g. coal. With the body at rest, the minimum magnetic force required to move the object is determined by its weight and by the force of pressure on the upper hemisphere of the body:

$$\vec{F}_p = \rho_b V_b \vec{g} \tag{3.55}$$

where V_b is the volume of the burden above the body. In equilibrium

$$\vec{F}_m = \vec{F}_g + \vec{F}_p \tag{3.56}$$

From eq. (3.56), taking into account that $\rho_p >> \rho_b$, a modified version of eq. (3.47) can be written [S25]:

$$\vec{f}_m = \mu_0 \frac{[1 + N(\mu_p - 1)]}{\mu_p - 1}(\rho_p + \rho_b \frac{V_b}{V_p})\vec{g} \tag{3.57}$$

It can be seen from eq. (3.57) that the static magnetic force density (or force index in the presence of the burden depends not only on the shape of the object, but also, in contrast to the situation when the burden is absent (eq. (3.47)), on the size of the object. A typical dependence of the static force index on the depth of the burden, for various sizes of a ferromagnetic sphere, is shown in Fig. 3.11.

Similar diagrams can be constructed, using eq. (3.57), for other shapes. It can be seen that the static force index increases steeply with increasing depth of the burden for small particles, while for larger objects the dependence is mild.

3.4 High-gradient magnetic separation

During the last three decades, high-gradient magnetic separation (HGMS) has attracted considerable attention, both experimentally and theoretically. It has established itself as a powerful technique for the manipulation of fine weakly magnetic particles. HGMS has also become a beloved subject of theoretical modelling and a notable number of publications dealing with the problem of particle capture led to an understanding of the interaction between a magnetizable particle and a matrix element.

The problem of particle capture in a matrix of a high-gradient magnetic separator can be looked upon from both the macroscopic and microscopic points of view. The macroscopic studies are aimed at the phenomenological description of the separation process and the prediction of its dynamic behaviour. The microscopic theories are intended to provide insight into the mechanisms of particle capture by the matrix. The complex problems facing the microscopic theories were considerably simplified by considering, in the first instance, the interaction between a single particle and a single matrix element. The theory of multi-particle and multi-collector magnetic separation was then developed as a direct extension of the single element model. Validity of such an approach has been questioned and theoretical analysis and experimental data confirm only limited applicability of such an extension to real-life multi-element scenario.

A detailed description of the theoretical models of high-gradient magnetic separation is given in a book by Gerber and Birss [G1] and other publications [B10, S1]. A brief survey of the theory of HGMS is given in the following section.

3.4.1 Theory of particle capture by a single-collector

The main emphasis in the early models of high-gradient magnetic separation was on the analysis of particle trajectories in the vicinity of a single magnetized matrix collector, usually a wire [W8, W9, L6, C10]. These studies were inspired by the early analysis by Zabel [Z6] who studied the mechanisms of dielectrophoretic capture on fiber matrix.

Watson [W8] considered a ferromagnetic wire of radius a placed along the z -axis in a diamagnetic fluid, as shown in Fig. 3.12. A uniform magnetic field H_0, strong enough to saturate the wire, is applied in the x direction. Spherical paramagnetic particles of radius b are carried by a fluid of viscosity η that flows with uniform velocity v_0 in the negative x-direction.

This orientation is one of the three possible arrangements between the directions of the magnetic field and particle flow velocity with respect to the ferromagnetic wire. It is always assumed that the wire is orthogonal to the external magnetic field; the fluid flow can be either parallel (longitudinal - L-

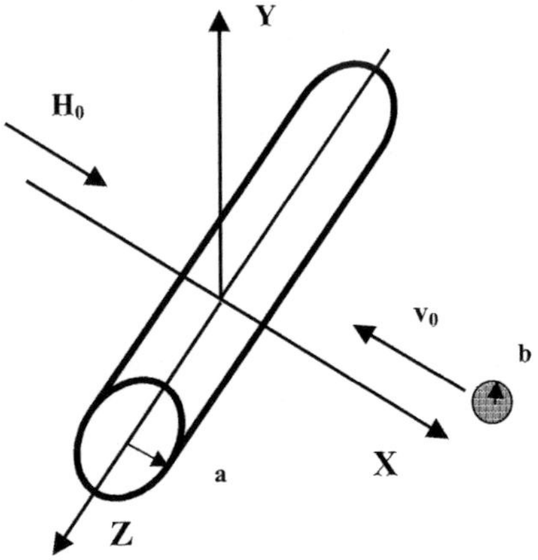

Figure 3.12: A ferromagnetic wire and a paramagnetic particle in a magnetic field.

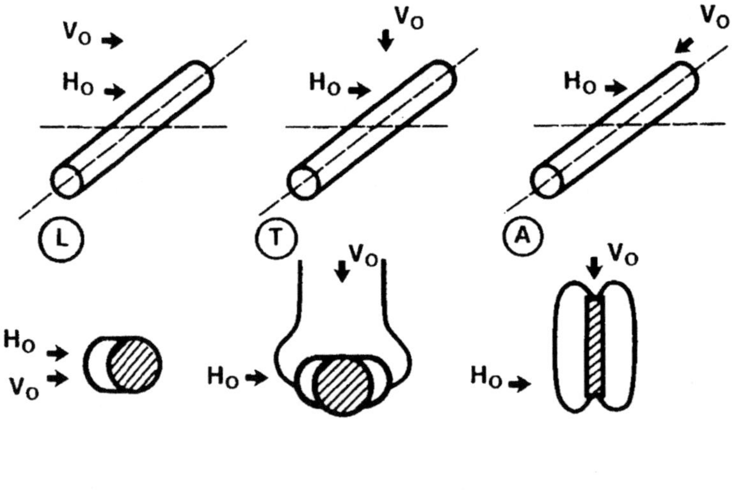

Figure 3.13: Graphic representation of three geometrical configurations in HGMS and of the profiles of the particle build-up on the ferromagnetic wire.

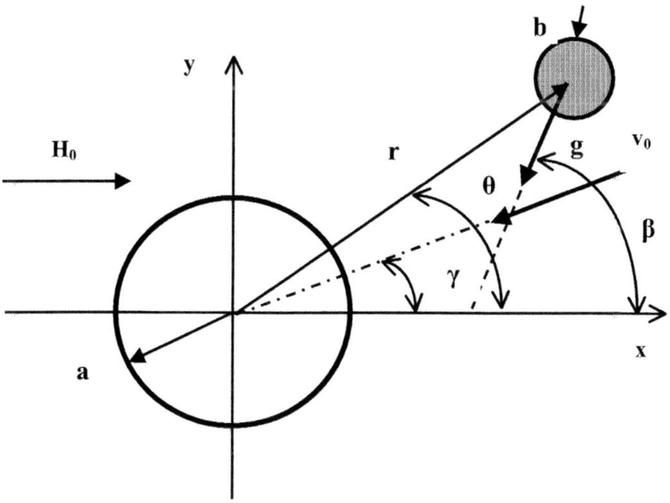

Figure 3.14: The configuration of particle capture in a single-wire approximation.

orientation) or perpendicular (transverse - T - orientation) to the direction of magnetic field. The third configuration, when the initial fluid flow is parallel to the wire, is known as the axial (A) configuration. These three geometrical configurations are illustrated in Fig. 3.13.

In order to describe the motion of a small paramagnetic particle travelling in a fluid stream in the presence of a magnetized wire, one assumes that the particle experiences magnetic, hydrodynamic and gravitational forces. The balance of forces which describes the particle motion is given by eq. (3.1), or in a more explicit form by eq. (3.58):

$$\vec{F}_i = \vec{F}_m + \vec{F}_d + \vec{F}_g \tag{3.58}$$

where F_i is the inertial force.

The general configuration of the particle motion problem in a single-wire approximation is shown in Fig. 3.14.

In order to obtain the magnetic force acting on a small paramagnetic particle in the vicinity of a magnetized wire, the magnetic field, external to and generated by the magnetized wire must be expressed as a function of the wire radius. The components of the magnetic field outside the wire, in polar coordinates, as defined in Figure 3.14, can be written [G1, S1] as:

$$H_r = H_0(1 + K\frac{a^2}{r^2})\cos\theta \tag{3.59}$$

$$H_\theta = -H_0(1 - K\frac{a^2}{r^2})\sin\theta \tag{3.60}$$

where

$$K = \frac{M}{2H_0} \tag{3.61}$$

For $H_0 < H_s$ $K = 1$ while for $H_0 > H_s$ $K = M_s/2H_0$. H_s and M_s are the saturation magnetic field strength and saturation magnetization of the wire material, respectively.

The magnetic force

By inserting eqs. (3.59) and (3.60) into eq. (1.8), components of the magnetic traction force on a magnetizable particle are obtained:

$$F_{mr} = -\frac{\mu_0 \kappa V_p M_s a^2}{r^3}\left(\frac{M_s a^2}{r^2} + H_0 \cos 2\theta\right) \tag{3.62}$$

$$F_{m\theta} = -\frac{\mu_0 \kappa a^2 V_p M_s H_0}{r^3} \sin 2\theta \tag{3.63}$$

$$F_{mz} = 0 \tag{3.64}$$

If the net volume magnetic susceptibility $\kappa = \kappa_p - \kappa_f$ is positive, then the radial component of the magnetic traction force between the particle and the wire is attractive provided that

$$\frac{M_s a^2}{r^2} + H_0 \cos 2\theta > 0 \tag{3.65}$$

The component F_{mr} becomes repulsive for $\kappa_p > \kappa_f$ in two symmetrical regions, the boundaries of which are determined by a critical angle θ_c [B10]

$$\theta_c = \arctan\left(\frac{1 + K\frac{a^2}{r^2}}{1 - K\frac{a^2}{r^2}}\right) \tag{3.66}$$

As shown in Fig. 3.15, the angle θ_c determines sections of the wire surface that are attractive or repulsive to paramagnetic or diamagnetic particles. It is apparent from eq. (3.65) that paramagnetic particles will be collected at angles smaller than θ_c and repelled at angles greater than θ_c. On the other hand, diamagnetic particles, with negative susceptibilities, will be attracted to the wire collector at angles $\theta > \theta_c$ and repelled in regions where $\theta < \theta_c$.

The situation is summarized in Fig. 3.16, which shows the contours of the magnetic force around a magnetized cylindrical wire. The hyperbolae indicate the lines of zero radial force. It can be seen from eq. (3.66) that the angle θ_c decreases with increasing magnetic field strength. Therefore, the area on the matrix collector, available for the capture of paramagnetic particles decreases with increasing field strength.

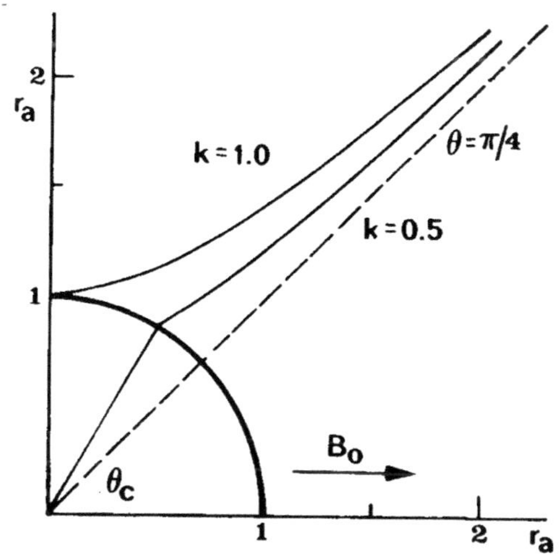

Figure 3.15: Attractive and repulsive regions around a wire collector ($r_a = r/a$).

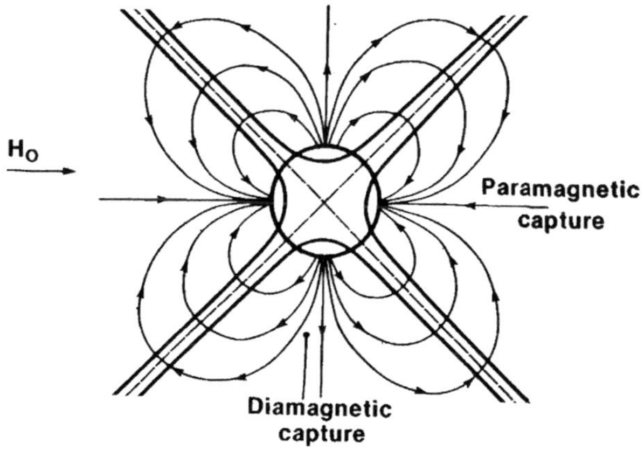

Figure 3.16: Pattern of the magnetic force around a magnetized cylindrical wire.

The fluid flow distribution and the viscous drag

The hydrodynamic drag acting on a particle travelling in a fluid stream is evaluated by calculating the flow field around a circular cylinder. The simplest approach to this difficult problem is to assume that the fluid is inviscid and incompressible. Then the fluid velocity can be written as the gradient of the velocity potential and by solving Laplace's equation it is possible to obtain the fluid velocity in polar coordinates [G1]:

$$v_r = v_0(1 - \frac{a^2}{r^2})\cos(\theta - \gamma) \tag{3.67}$$

$$v_\theta = -v_0(1 + \frac{a^2}{r^2})\sin(\theta - \gamma) \tag{3.68}$$

The potential flow gives good description of the flow conditions on the front side of wire for high Reynolds number, i.e. for Re > 1, where Re is given by

$$\mathrm{Re} = \frac{2v_0\rho_f a}{\eta} \tag{3.69}$$

For high Reynolds numbers the inertial forces exceed the viscous forces everywhere in the flow field except within a thin boundary layer at the wire surface and at the rear side of the wire.

For low Reynolds numbers, Re<1, the inertial forces are smaller than the viscous forces and the flow has a viscous (or creeping or laminar) character. The velocity distribution can then be obtained by Lamb's method [B11, G1]:

$$v_r = v_0 \frac{\ln(\frac{r}{a}) - 0.5\left[1 - (\frac{a}{r})^2\right]}{2.002 - \ln \mathrm{Re}} \cos(\theta - \gamma) \tag{3.70}$$

$$v_\theta = -v_0 \frac{\ln(\frac{r}{a}) + 0.5\left[1 - (\frac{a}{r})^2\right]}{2.002 - \ln \mathrm{Re}} \sin(\theta - \gamma) \tag{3.71}$$

As has been mentioned above, flow velocity at the wire surface becomes zero in the viscous flow model, while in the potential flow model it can reach values up to $2v_0$. These differences between the two models are particularly noticeable in the collection process. It was shown by Cummings et al. [C11] that particles in viscous (laminar) flow begin to move away from the wire much further upstream than those in potential flow. Because the viscous flow field "sees" the wire much further upstream, particle transport to the region of the highest magnetic force is greatly diminished.

Viscous drag force is obtained from Stokes's equation (eq. (1.16)) and the components of this drag force can be written:

$$F_{dr} = 6\pi\eta b(v_r - \frac{dr}{dt}) \tag{3.72}$$

$$F_{d\theta} = 6\pi\eta b(v_\theta - r\frac{d\theta}{dt}) \tag{3.73}$$

The components of the fluid velocity v_r and v_θ are given by eqs. (3.67) and (3.68) or eqs. (3.70) and (3.71), depending on the value of the Reynolds number.

The gravitational force

The gravitational force is given by eq. (1.14) and its components, in polar coordinates, can be written [G1, G5]:

$$F_{gr} = (\rho_p - \rho_f) g V_p \cos(\theta - \beta) \tag{3.74}$$

$$F_{g\theta} = -(\rho_p - \rho_f) g V_p \sin(\theta - \beta) \tag{3.75}$$

Equations of particle motion

The dynamic behaviour of the particle in a single-wire approximation of high-gradient magnetic separation is described by eq. (3.58). The complete equations of motion in polar coordinates were obtained by a number of authors, for instance [W8, G1, G5, C11, L6, L7]. If the analysis is limited to the motion of small particles (e.g. $2b < 75$ μm), for which the gravitational force is much smaller than the hydrodynamic force, and the condition

$$\frac{2\rho_p b^2}{9\eta} << 1 \tag{3.76}$$

is satisfied, the inertial and gravitational terms in eq. (3.58) can be neglected. The radial and azimuthal components of the equation of motion can be written [B10] as :

$$\frac{dr_a}{dt} = \frac{v_0}{a}(1 - \frac{1}{r_a^2}) \cos(\theta - \gamma) - \frac{v_m}{a}(\frac{K}{r_a^5} + \frac{\cos 2\theta}{r_a^3}) \tag{3.77}$$

$$r_a \frac{d\theta}{dt} = -\frac{v_a}{a}(1 + \frac{1}{r_a^2}) \sin(\theta - \gamma) - \frac{v_m \sin 2\theta}{a r_a^3} \tag{3.78}$$

where $r_a = r/a$ and

$$v_m = \frac{2\mu_0(\kappa_p - \kappa_f) M H_0 b^2}{9\eta a} \tag{3.79}$$

is the parameter introduced by Watson [W8]. It has the dimension of [m/s] and is termed magnetic velocity.

When $\gamma = 0$, eqs. (3.77) and (3.78) describe the motion of a particle in the longitudinal configuration, while when $\gamma = \pi/2$ they describe the motion in transverse configuration. By combining eqs. (3.77) and (3.78) the time-independent equation is obtained [W8]:

$$\frac{1}{r_a}\frac{dr_a}{d\theta} = -\frac{(1 - \frac{1}{r_a^2}) \cos(\theta - \gamma) - \frac{v_m}{v_0}(\frac{K}{r_a^5} + \frac{\cos 2\theta}{r_a^3})}{(1 + \frac{1}{r_a^2}) \sin(\theta - \gamma) + \frac{v_m}{v_0}\frac{\sin 2\theta}{r_a^3}} \tag{3.80}$$

Particle trajectories can thus be described by an equation of a general form

$$\frac{1}{r_a}\frac{dr_a}{d\theta} = f(\frac{v_m}{v_0}, \frac{M}{2H_0}, r_a, \theta) \tag{3.81}$$

The path taken by the particle depends, therefore, only on v_m/v_0, $M/2H_0$, and the initial position and the initial velocity of the particle.

The complete equations of motion (eqs. (3.77) and (3.78)) can be solved only numerically. The solutions (i.e. the particle trajectories) are of two types, depending on whether the particle is captured by the wire or not. The critical trajectory, for which the particle is just captured, is characterized by the capture radius R_c.

Approximate analytical solutions of the equations of motions in the case of the potential flow, under the condition that $\mid v_{ma}/v_{0a} \mid \geq \sqrt{2}$, yield the following expression for the capture radius in the longitudinal and transverse configurations [G1, G5]:

$$R_{ca} \simeq \frac{3\sqrt{3}}{4}(\mid \frac{v_{ma}}{v_{0a}} \mid)^{1/3} \left[1 - \frac{2}{3}(\frac{v_{ma}}{v_{0a}})^{-2/3}\right] \tag{3.82}$$

where the normalized quantities are defined as $R_{ca} = R_c/a$, $v_{ma} = v_m/a$ and $v_{0a} = v_0/a$. Thus, for a given $\mid v_{ma}/v_{0a} \mid$ ratio, both configurations have the same capture radius.

When $\mid v_{ma}/v_{0a} \mid < \sqrt{2}$, the critical capture radius in the longitudinal configuration ($\gamma = 0$) is given by

$$R_{ca} = -\frac{1}{2}\frac{v_{ma}}{v_{0a}} \left[(1 - K^2)^{1/2} + K(\pi - \arccos K)\right] \tag{3.83}$$

and in the transverse configuration ($\gamma = \pi/2$) [G1]:

$$R_{ca} = -\frac{1}{2}\frac{v_{ma}}{v_{0a}} \left[(1 - K^2)^{1/2} + K(2\pi - \arccos K\right] \tag{3.84}$$

It is apparent from eqs. (3.82), (3.83) and (3.84) that the dependence of R_{ca} upon the ratio v_m/v_0 can be divided into two categories: at low values of v_m/v_0, the capture radius increases linearly with this ratio

$$R_{ca} \sim \frac{v_m}{v_0} \tag{3.85}$$

while at large values it increases approximately as

$$R_{ca} \sim (\frac{v_m}{v_0})^{1/3} \tag{3.86}$$

Axial configuration

The analysis of a particle motion in the axial (A) configuration, where the wire axis and the fluid flow are mutually parallel, and the magnetic field is perpendicular to them (Figs. 3.13 and 3.17), does not require, in contrast to other cases, numerical computations. Since the results can be derived in an analytical form, this configuration became a favourite object of the modelling of particle capture. The dynamics of the particles and their build-up on the wire

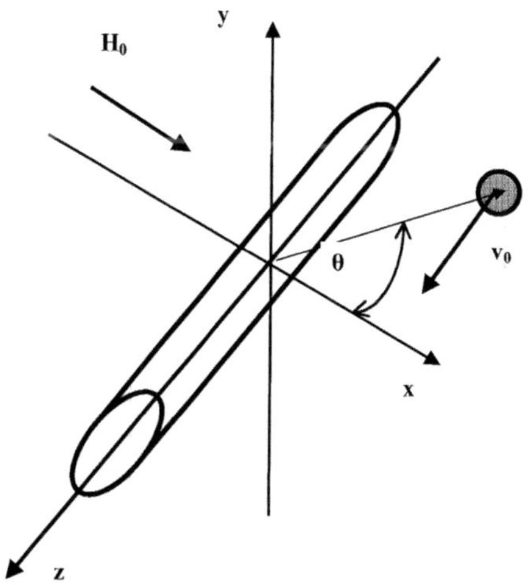

Figure 3.17: The axial configuration.

in A-configuration have been described in numerous papers, for example [G9, B12, U4, B13].

The equations of particle motion (eqs. (3.77) and (3.78)) are reduced in this configuration to the simplified forms

$$\frac{dr_a}{dt} = -\frac{v_m}{a}\left(\frac{K}{r_a^5} + \frac{\cos 2\theta}{r_a^3}\right) \tag{3.87}$$

$$\frac{d\theta}{dt} = -\frac{v_m}{a}\frac{\sin 2\theta}{r_a^4} \tag{3.88}$$

$$\frac{dz_a}{dt} = \frac{v_0}{a} \tag{3.89}$$

The elimination of time from eqs. (3.87), (3.88) and (3.89) produces a set of differential equations

$$\frac{dr_a}{d\theta} = \frac{K}{r_a \sin 2\theta} + r_a \cot 2\theta \tag{3.90}$$

$$\frac{dz_a}{d\theta} = -\frac{v_0}{v_m}\frac{r_a^4}{\sin 2\theta} \tag{3.91}$$

From these equations the particle trajectories can be determined analytically [G1,G9, B12].

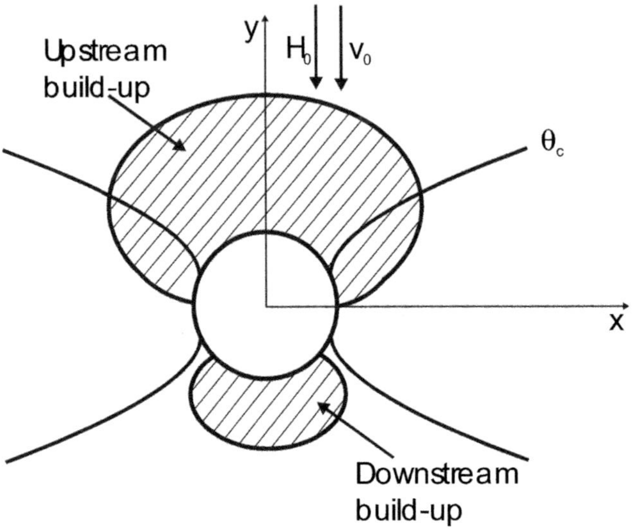

Figure 3.18: Build-up of paramagnetic particles on a magnetized wire in the longitudinal configuration.

Particle build-up

Previous analyses of particle trajectories in the vicinity of a magnetized wire considered the capture process in the initial stage, i.e. with no loading of the wire with the captured particles. The determination of the exact process of the particle build-up is a complicated problem. As the suspension flows through the matrix, some of the particles, under the influence of several forces, are deposited onto the matrix elements or onto already deposited particles. Consequently the geometry, the structure and the surface characteristics of the matrix may be significantly modified. These changes, in turn, affect the flow of the suspension through the matrix and the subsequent capture. Further complications may also arise as a result of dislodgment of the captured particles and their possible re-capture. As a result, the dynamic behaviour of the separation process varies with the time as well as with other variables.

The force that is principally responsible for the particle capture is the magnetic force. Particles are attracted towards the wire if the radial component of the magnetic force $F_{mr} < 0$. Thus the condition $F_{mr} = 0$ defines the critical angle θ_c given by eq. (3.66). The stability of a particle on the wire surface is determined by the balance of the tangential components mainly of the magnetic and hydrodynamic forces. Gerber found that the angle θ_p, for which a paramagnetic particle will be retained on the wire, is given as [G1]:

$$|\theta_p| \leqslant \arcsin\left[1 - (\frac{v_0}{v_m})^2\right]^{1/2} \qquad (3.92)$$

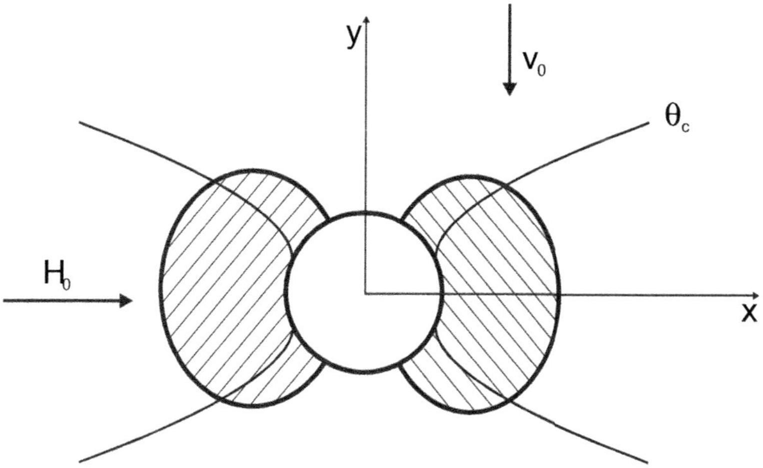

Figure 3.19: Build-up of paramagnetic particles on a magnetized wire in the transverse configuration.

It transpires from eq. (3.92) that retention will occur only if $v_m/v_0 > 1$ [W8, W9]. However, the experimental results show that a much less stringent condition can still lead to a reasonably high degree of capture.

Examples of the build-up of paramagnetic particles on a magnetized wire in three principal configurations are shown in Figs. 3.18, 3.19 and 3.20. Diamagnetic particles are retained in the complementary regions as follows from Fig. 3.16. The experimentally determined build-up of paramagnetic kimberlite particles in the transverse configuration is shown in Fig. 3.21.

Essentially all theoretical descriptions of particle capture by a magnetized wire assumed the idealized situation in which the collection wire axis was oriented orthogonally to the magnetic field direction. This assumption resulted from the belief that only these configurations would create sizable deposits. Birss et al. [B14, S26] considered the collection process on a cylinder with its axis oriented at an angle with the magnetic field direction. They found that the effect of the non-orthogonal wire orientation can be taken into account by replacing the coefficients v_m/v_0 and K by more general terms. These terms are the functions of the offset angle α, defined in Fig. 3.22.

Experimental observation by Svoboda et al. [S27], of particle collection at various offset angles α, demonstrated that even at the offset angle as large as 80^0 there is considerable collection of particles, as shown in Fig. 3.23. The length of the section of the wire where the diamagnetic capture takes place is a function of the offset angle. For the offset angle equal to $\pi/2$ (the magnetic field parallel with the wire axis) the paramagnetic capture takes place only at the end of the wire while the entire length of the wire is repulsive for paramagnetic particles, as shown in Fig. 3.24.

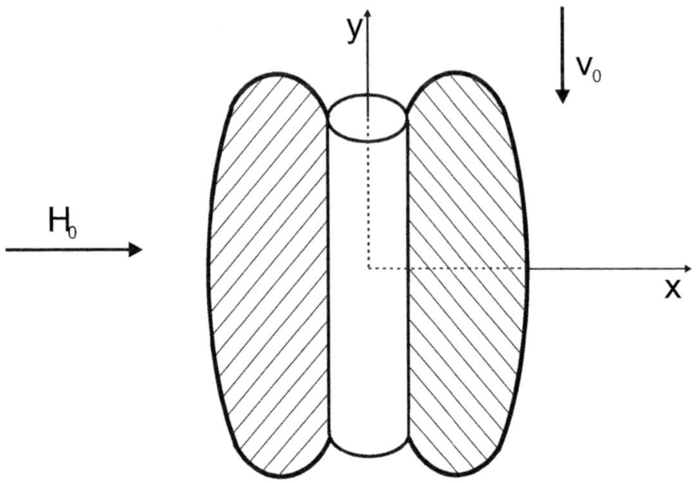

Figure 3.20: Build-up of paramagnetic particles on a magnetized wire in the axial configuration.

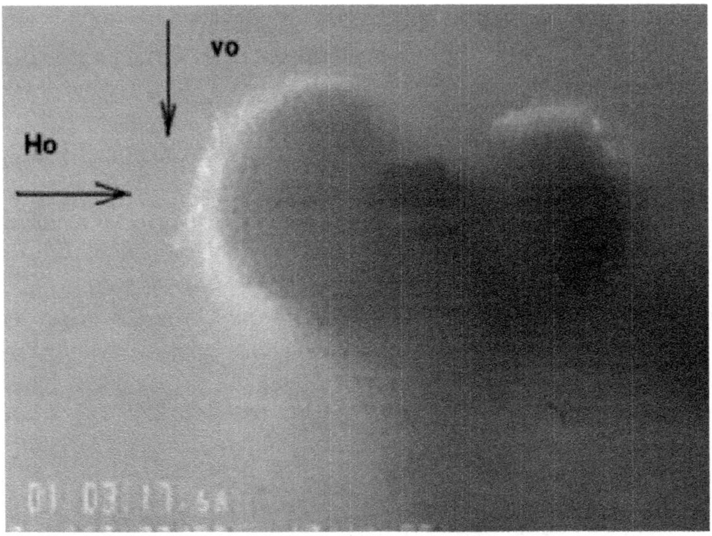

Figure 3.21: Build-up of paramagnetic particles of kimberlite (mean size 4.7 μm, magnetic susceptibility 1.7×10^{-6} m^3/kg) on a magnetized nickel wire in transverse configuration (B_0 = 1.7 T, v_0 = 0.25 cm/s) [S27].

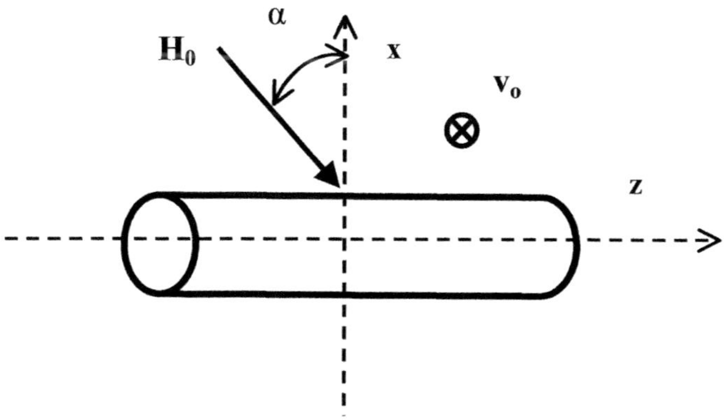

Figure 3.22: The plane view of the generalized transverse orientation with the magnetic field making an angle $(\pi/2 - \alpha)$ with the wire axis.

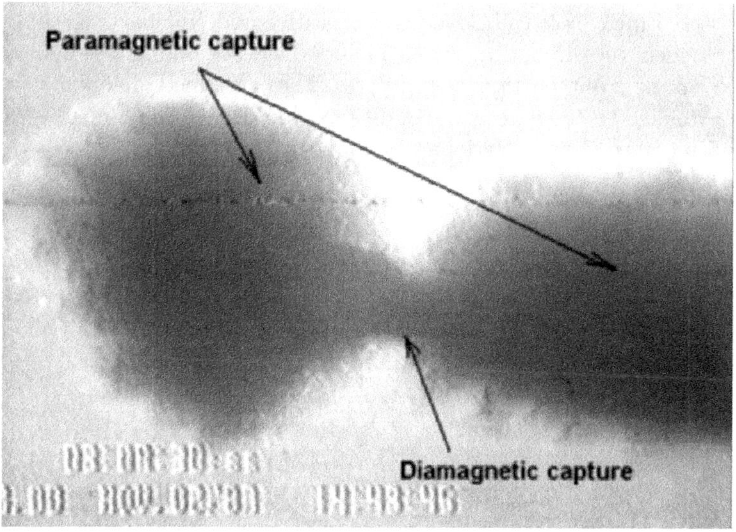

Figure 3.23: Collection of kimberlite particles on a nickel wire (diameter 125 μm) in the transverse configuration, with the magnetic field making an angle of 10^0 with the wire axis (offset angle $\alpha = 80^0$), ($B_0 = 1.7$ T, $v_0 = 0.25$ cm/s) [S27].

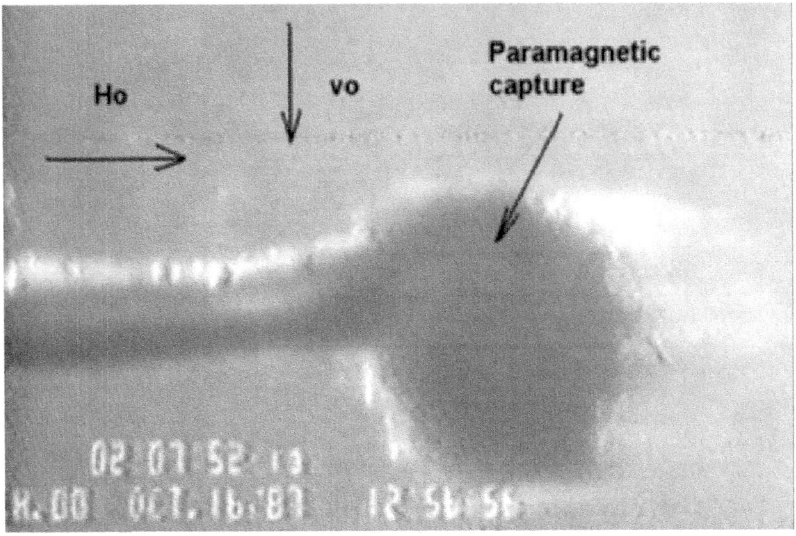

Figure 3.24: Collection of paramagnetic kimberlite particles on a nickel wire magnetized along its axis (offset angle $\alpha = 90^0$) [S27].

Nesset and Finch [N2, N5] developed a static model of particle capture in HGMS, in which the fluid drag force included the changing boundary layer thickness over the front portion of the wire. They assumed that for a particle to remain attached to the wire, two conditions would have to be met. First, the net radial force acting on a particle must be attractive, that is

$$F_{mr} + F_{gr} \leqslant 0 \tag{3.93}$$

where F_{mr} and F_{gr} are given by eq. (3.62) and (3.74), respectively.

The second condition is that the net tangential force must push particles towards the region of a greater attractive radial force. In the case of the longitudinal paramagnetic build-up, this condition can be expressed as:

$$F_{m\theta} + F_{g\theta} + F_{d\theta} \leqslant 0 \tag{3.94}$$

where $F_{d\theta}$ is the fluid drag force, which is proportional to the shear stress τ_0 in the boundary layer. Thus [N2]

$$F_{d\theta} = \frac{\pi^2}{2} b^2 \tau_0 \tag{3.95}$$

By solving equations (3.93) and (3.94), the outline of the build-up profile can be found computationally as intersecting loci of points $\{r, \theta\}$ satisfying these equations. The two retention criteria are satisfied only in the cross-hatched region shown in Fig. 3.25.

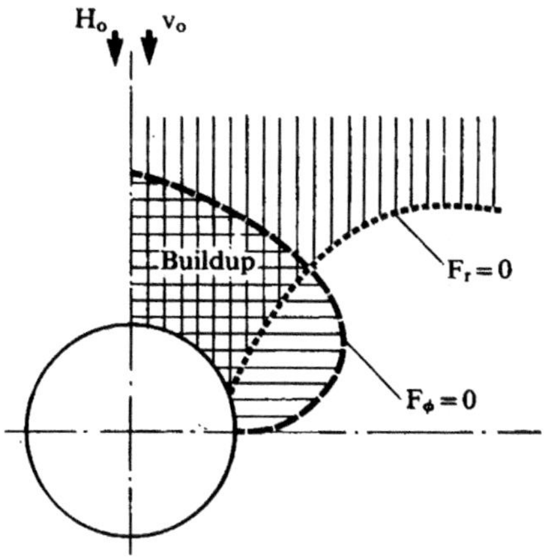

Figure 3.25: Loci of zero radial and azimuthal forces around the wire in the longitudinal orientation (adapted from [N5]).

Capture of diamagnetic particles

It transpires from eq. (1.8) that if the susceptibility difference $\kappa_p - \kappa_f$ is positive, then the magnetic force between the wire and the particle is attractive, provided that the radial component F_r is negative. This is the case of capture of paramagnetic particles, under the condition that eq. (3.65) is satisfied.

On the other hand, there are two symmetrical curved volumes, the boundaries of which are determined by the critical angle θ_c, given by eq. (3.66), in which the radial force acting on paramagnetic particles is repulsive, while the radial force acting on diamagnetic particles is attractive. These regions, in which collection of diamagnetic particles takes place, are complementary to the regions of collection of paramagnetic particles, as illustrated in Figs. 3.18, 3.19 and 3.20.

The value of θ_c depends on both M_s (or J_s) and H_0 (or B_0) as shown in Fig. 3.26. It can be seen that as the strength of the external magnetic field increases, the area of the wire collector available for the capture of diamagnetic particle also increases. The effective area for the capture of paramagnetic particles is, therefore, correspondingly reduced. This phenomenon indicates that an increase in the operating magnetic field can negatively affect the loading capacity of the matrix, the selectivity of separation and the overall performance of a magnetic separator. The first experimental confirmation of the diamagnetic capture was reported by Friedlaender et al. [F9]. They showed that by dissolving a paramagnetic salt in water, κ_f can be made positive and sufficiently large, so that $\kappa = | \kappa_p - \kappa_f |$ is large and the attractive magnetic force is sufficiently strong to recover particles of aluminium oxide. A theory of diamagnetic capture was developed and published by Birss and Parker [B15], simultaneously with

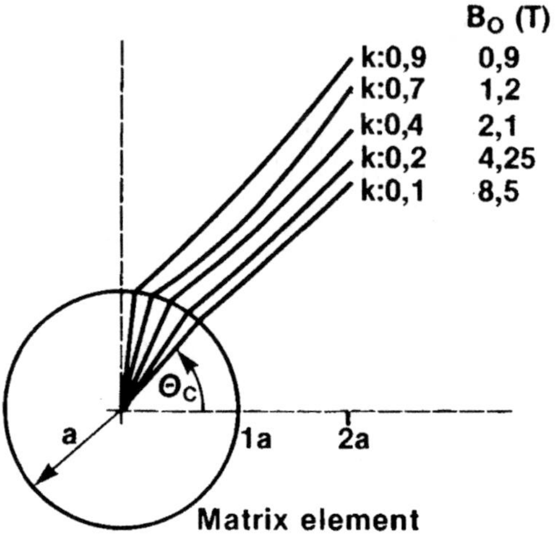

Figure 3.26: The critical angle θ_c as a function of the magnetic induction (for $J_s = 1.7$ T) [S1].

Friedlaender's experimental paper [F9].

Although the recovery of diamagnetic particles by magnetic separation is feasible in principle, simultaneous capture of paramagnetic particles prevents a magnetic separator from being a selective device for separation of diamagnetic materials. This limitation can be, however, overcome by magnetizing the wire collectors along their long axes (the offset angle $\alpha = \pi/2$). At this orientation, the paramagnetic particles are captured only at the end tip of the wire, as is shown in Fig. 3.24. The entire length of the wire is then attractive for diamagnetic particles. Collection of diamagnetic bismuth particles in this configuration was observed by Svoboda et al. [S27] and is shown in Fig. 3.27.

Magnetic separation at moderate Reynolds numbers

Experimental investigation by Friedlaender et al. into the particle capture on a single wire in the longitudinal configuration indicated [F10] that for moderate values of Reynolds number for the wire (Re 6 to 40), the material can be captured at the rear of the wire. Theoretical treatment by Watson [W10] showed that at higher velocities of the suspension a pair of stable standing vortices, which appear on the downstream side of a wire, can induce capture of particles on the trailing side of a matrix collector. This situation is illustrated in Fig. 3.28. The shaded areas on the upstream and downstream sides of the wire are attractive for paramagnetic particles. The unshaded portions around the y axis are repulsive, and particles arriving from the front part of the wire towards the wake must, therefore, pass through the repulsive regions.

It was shown in subsequent studies that the magnitude of the Reynolds num-

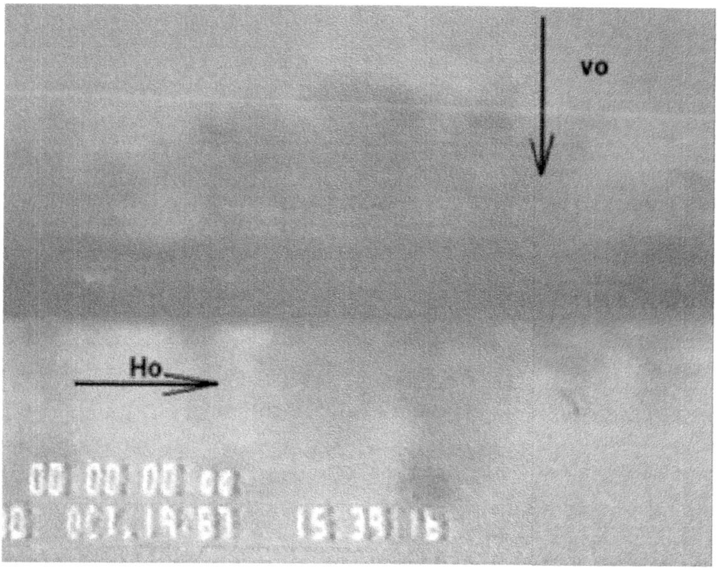

Figure 3.27: Collection of diamagnetic particles of bismuth (χ = - 16.3$\times 10^{-9}$ m^3/kg) on the upstream and downstream sides of the nickel wire magnetized parallel to its axis ($B_0 = 1.9$ T, $v_0 = 0.13$ cm/s) [S27].

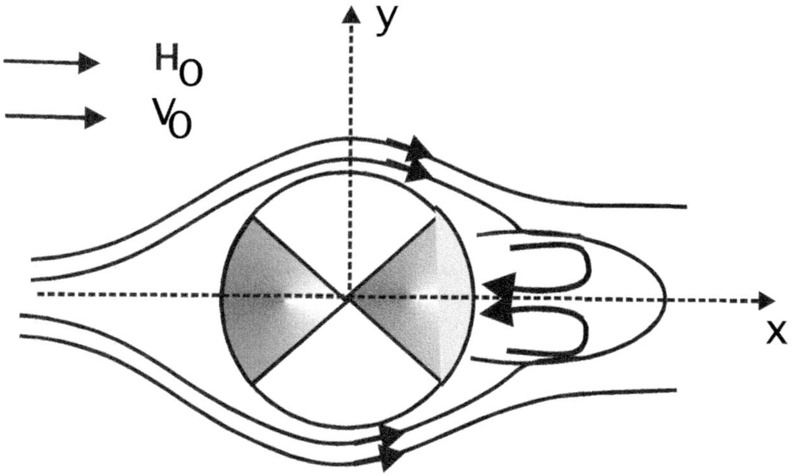

Figure 3.28: The downstream capture into a stationary wake formed by twin vortices.

ber is not the only condition for the vortex capture [W11]. The thickness of the boundary layer, the particle size and the field strength were found to be the main factors affecting the vortex capture. In particular it was determined that, for vortex capture to occur, it is necessary that $v_m/v_o < 1$. The separation process, taking place under such conditions, for which the downstream particle capture takes place, was termed vortex magnetic separation by Watson [W11]. The applicability of this phenomenon, investigated mainly in single-wire experiments, to real operational situations, has been a subject of controversy [S28, W12].

It was argued by Watson et al. [L8] that if the conditions for the downstream capture are met, the recovery, the grade of the magnetic concentrate and the throughput would be very high. However, it was shown by Svoboda [S28] that such conditions can be possibly met only when very fine (e.g. 125 μm) steel wool matrix is used and when the flow rate of the slurry through the matrix is close to the upper limit possible in such a matrix [S29]. Such conditions are usually used to remove fine weakly magnetic impurities from industrial minerals, such as kaolin. In those applications, however, the magnetic fraction is a waste and its grade is of no relevance to the efficiency of the process. On the other hand, in those applications where the magnetic fraction is the useful product, such as beneficiation of ferrous and non-ferrous ores, much coarser matrix (e.g. expanded metal or woven mesh) and much higher slurry velocities must be used in order to maintain meaningful economically viable throughput. Under these circumstances the conditions for the vortex formation are not met since the Reynolds number is at least an order of magnitude greater than the upper value required for downstream capture to take place.

3.4.2 Capture of particles in multi-collector matrices

The matrix of a real magnetic separator consists of a large number of individual collectors. There are essentially two possible ways of developing a theory of multi-collector magnetic filters: one way is to build the theory from first principles and the other is to incorporate the single-collector model into a multi-collector phenomenological description.

In the first approach, the particle trajectories and the capture efficiencies were investigated for the case of wire collectors arranged in regular arrays. Configurations of single and double layers of wires were analyzed and reported in numerous publications [H14, T4, E1, S30, G7, R12, R13]. As a result of considerable mathematical complexity of the problem the applicability of the theory to real matrix is limited.

Particle capture by ferromagnetic spheres. In spite of high efficiency of a ball matrix in some industrial applications of magnetic separation [S1], only a limited attention has been paid to the theoretical analysis of particle capture by spherical collectors. The character of the particle capture and of the build-up is inevitably different for spheres compared to the wires. The difference follows from the fact that the cross-sectional capture area of a sphere increases

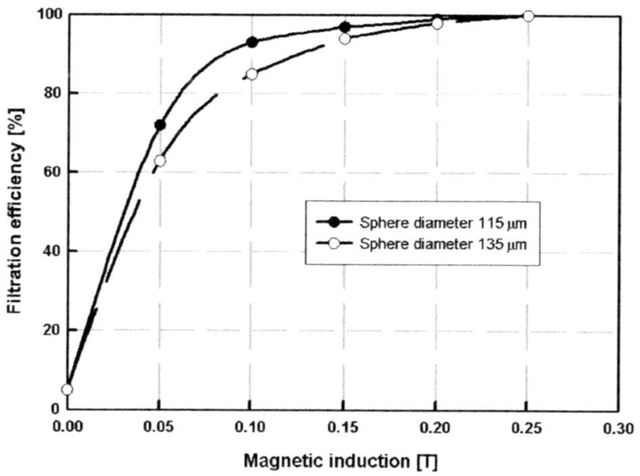

Figure 3.29: The efficiency of magnetic filtration of $\alpha-Fe_2O_3$ particles (diameter 0.08 μm) in a bed of steel spheres, as a function of the applied magnetic field (adapted from [A16]).

quadratically with the capture radius R_{ca}, whereas that of a wire increases only linearly. Friedlaender et al. [F11, F12] solved numerically the equation of motion (eq. 3.58) in transverse configuration and found that the capture of paramagnetic particles occurred in conically shaped surfaces defined by the critical angle

$$| \theta_c |= \arcsin \left\{ \frac{2\left[1+2K_s(\frac{a}{r})^3\right]}{3\left[1+K_s(\frac{a}{r})^3\right]} \right\} \tag{3.96}$$

which is equivalent to eq. (3.66) derived for a cylindrical wire. The parameter K_s given by

$$K_s = \frac{M}{3H_0} \tag{3.97}$$

is equivalent to eq. (3.61) for the wire. Capture of diamagnetic particles then takes place on conical surfaces complementary to those for paramagnetic capture.

It transpired from theoretical analysis of particle capture in a bed of spheres by Watson and Watson [W13] that the probability of particle capture increased with increasing background magnetic field until the field was high enough to saturate magnetically the ferromagnetic spheres.

However, Moyer et al. [M15, M16] argued that the observed saturation of the capture probability occurred at the magnetic field strengths far below those required to saturate magnetically the spheres comprising the matrix bed.

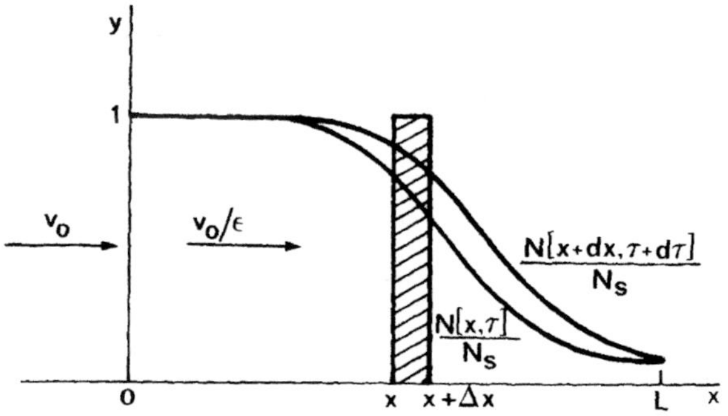

Figure 3.30: Schematic view of the normalized build-up N/N_s in a filter, as a function of the filter length.

This theoretical prediction was confirmed experimentally by Arajs et al. [A16]. As Fig. 3.29 illustrates, the filtration efficiency of submicrometer particles of $\alpha-Fe_2O_3$ in a bed of steel spheres levelled off at the magnetic induction of about 0.1 T, far below the saturation polarization of 1 T of the sphere material. Such a behaviour was not observed for an isolated collecting sphere [F12]. This is a very significant observation, since it indicates that the mechanisms of particle capture on a single collector and on an assembly of collectors in a packed bed are governed by different laws. It becomes clear again that it is not permissible to extend single-collector theoretical models (eqs. (3.83) and (3.84)) to multi-collector assemblies of matrix collectors.

Phenomenological model of HGMS

The alternative approach to multi-collector magnetic separation is a phenomenological model [W14, G1], which is less rigorous but at the same time less complicated and closer to practical needs of high-gradient magnetic filtration. It cannot, however, be easily applied to the process of separation of multicomponent mixtures of materials.

The fundamental equations describing the particle retention in a filter and the time-dependent performance of a filter are the macroscopic conservation equations and the rate equation. Let us consider particles travelling through an absorption filter of length L, as illustrated in Fig. 3.30. By equating the difference between the numbers of particles entering and leaving the volume Δx per unit time, to the time rate of increase of the number of particles, the conservation equation is obtained as

$$\frac{\partial N(x,\tau)}{\partial \tau} = v_0 \frac{\partial c(x,\tau)}{\partial x} \tag{3.98}$$

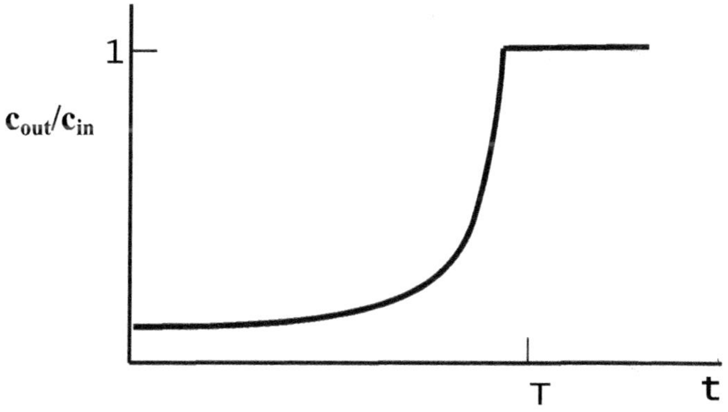

Figure 3.31: A breakthrough curve of a filter with a constant absorption coefficient λ. At time T the filter becomes fully saturated.

where $N(x,\tau)$ is the number of captured particles in the filter between x and $x+\Delta x$, at time τ. $c(x,\tau)$ is the number of suspended particles in a volume of the fluid $\epsilon\Delta x$, ϵ is the porosity of the filter matrix, v_0 the superficial velocity and τ the displacement time defined as $\tau = t - \epsilon x/v_0$.

The rate equation can be expressed in the form of the Iwasaki relation [R14]

$$-(\frac{\partial c}{\partial x})_t = \lambda c \tag{3.99}$$

where λ is the filter coefficient having the dimension of the reciprocal length. Equation 3.99 shows that the profile of the particulate concentrate in the liquid phase throughout a filter is logarithmic.

The solution of the rate equation can thus be written as

$$\frac{c(x)}{c_{in}} = \exp(-\lambda x) \tag{3.100}$$

$$\frac{c_{out}}{c_{in}} = \exp(-\lambda L) \tag{3.101}$$

where c_{in} and c_{out} are the influent and effluent concentrations.

Since nothing is known, in the phenomenological approach, about the physical processes taking place inside the filter, all information about the filter performance must be derived from its breakthrough curve, i.e. from the output concentration as a function of time, as shown in Fig. 3.31.

By incorporating the single-wire magnetic separation model into the above eqs. (3.100) and (3.101) Watson found [W8], the efficiency of the magnetic filter to be

$$\frac{c_{out}}{c_{in}} = \exp(\frac{-\alpha f L R_{ca}}{a\pi}) \tag{3.102}$$

where $\alpha = 2$ for ordered filters and $\alpha = 4/3$ for random filters. R_{ca} is the normalized capture radius given, for example, by eqs. (3.82) to (3.84), while f is the filling factor of the filter matrix ($f = 1 - \epsilon$).

Equation (3.102) clearly illustrates that it is not possible to extend the single-wire capture approach to multi-collector situation, as already demonstrated by Fig. 3.29. According to eq. (3.102), the efficiency of the filter depends on the capture radius calculated for the single-collector scenario. In the single-collector situation the particle capture is indeed determined by the capture radius and, therefore, by the magnetic velocity or by the strength of the applied magnetic field. In a multi-collector situation, there are several additional mechanisms that lead to particle collisions with the collectors and ultimately to capture. As will be shown in the next section, under suitable conditions, usually present in real operational situations, these mechanisms can be significantly more important than the magnetic traction mechanism, the only mechanism of particle capture considered in the single-collector approach. By inclusion of the single-collector capture radius R_{ca}, eq. (3.102), therefore, underestimates, by a wide margin, the efficiency of the filter and, at the same time, overestimates the role of the magnetic field in particle capture.

Particle capture in a real matrix

It transpires from the above discussion that it is not permissible to incorporate the single-collector model into a multiple-collector situation. In a single-collector situation the probability of a particle colliding with the matrix element is very small and the particle thus must be attracted to the collector by the magnetic traction force. On the other hand, in a real situation, where the matrix is essentially a bed of matrix elements, the efficiency of particle collision with the matrix elements is determined, to a great extent, by mechanical, i.e. "non-magnetic" mechanisms of collision, as has been argued by Svoboda [S31, S32]. It can, therefore, be assumed that HGMS is essentially a deep bed filter with an externally applied magnetic field.

High-gradient magnetic separation can thus be viewed as a stochastic process in which the chance of a particle reporting into the magnetic fraction is the product of two probability terms:

$$P_{recovery} = P_{collision} \times P_{retention} \tag{3.103}$$

where $P_{recovery}$ is the probability of recovery of a particle into the magnetic fraction, $P_{collision}$ is the probability of collision between a particle and the matrix and $P_{retention}$ is the probability of retention of a particle on the matrix.

While the first term on the right-hand side of eq. (3.103) is controlled by the hydrodynamic and geometrical conditions in the matrix, the second term depends on the magnetic and surface forces between the particle and the matrix element, and on the hydrodynamic regime.

Collision of particles with the matrix. Figure 3.32 illustrates typical collision mechanisms in a deep bed filter. These mechanisms also play a decisive role

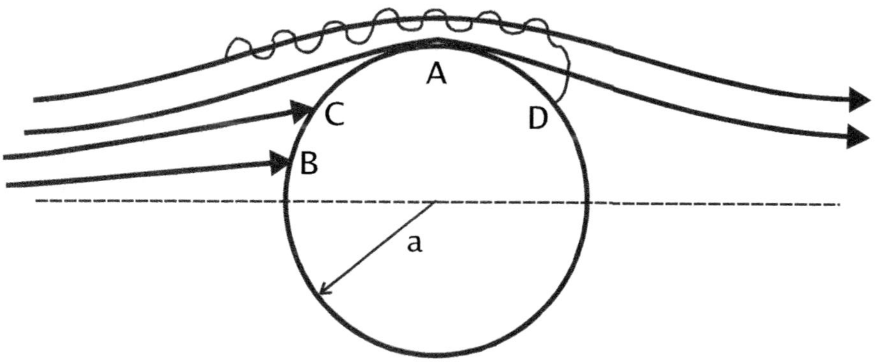

Figure 3.32: Collision mechanisms in deep-bed filtration.

in multi-collector HGMS. A particle following a streamline may come in contact with the matrix element by virtue of its finite size (interception or straining, case A), and as a result of inertia (case B), sedimentation (case C) or diffusion (case D).

The velocity of the liquid can be also greater on one side of the particle than on the other, causing the particle to rotate. As a result, a force that develops laterally to the direction of flow will drive particles across the flow field. This hydrodynamic effect also contributes to the collision of particles with collectors.

The collision probability can then be written, according to eq. (3.101):

$$\frac{c_{col}}{c_{in}} = 1 - \exp(-\lambda L) \tag{3.104}$$

where c_{col} is the concentration of particles which *collided* with the matrix collectors. The filter coefficient λ can be written in terms of the collision efficiency α of a single collector as [M17]

$$\lambda = -(\frac{3\pi}{4})^{1/3}\frac{p}{a}\ln(1-\alpha) \tag{3.105}$$

where $p = (1-\epsilon)^{1/3}$.

The total filter coefficient λ can be obtained from eq. (3.105) by calculating the collision efficiencies α_i of individual processes (e.g. [W15, I4, C12, H15]) that lead to collision. It was found [S33] that, for example, for steel wool and ball matrices, with element diameter of 200 μm and 6 mm, respectively, the dominant mechanism of collision is sedimentation. The hydrodynamic effect also contributes, particularly for small particles. On the other hand, wedging in the crevices of the matrix becomes significant for larger particles.

Equation (3.104) then can be used to calculate the ratio c_{col}/c_{in}, i.e. the ratio of the concentration of particles that collided with the matrix, to the concentration of particles that entered the matrix bed. Figure 3.33 documents that all particles as small as 4 μm, moving with superficial velocity of 0.07 m/s

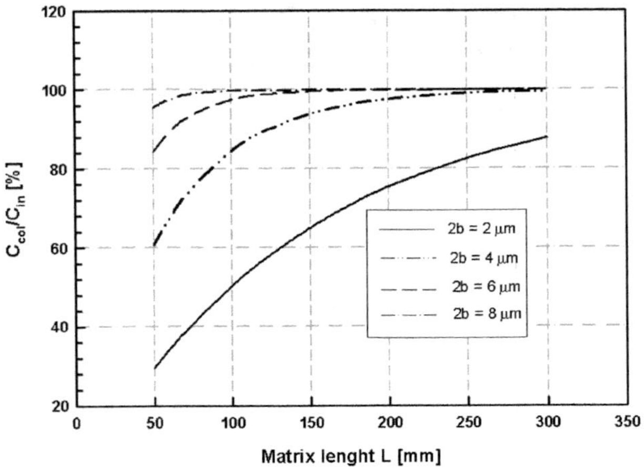

Figure 3.33: The proportion of particles colliding with steel wool matrix as a function of of the matrix length. $v = 0.07$ m/s, $2a = 400$ μm, $\epsilon = 0.95$ (adapted from Svoboda [S33]).

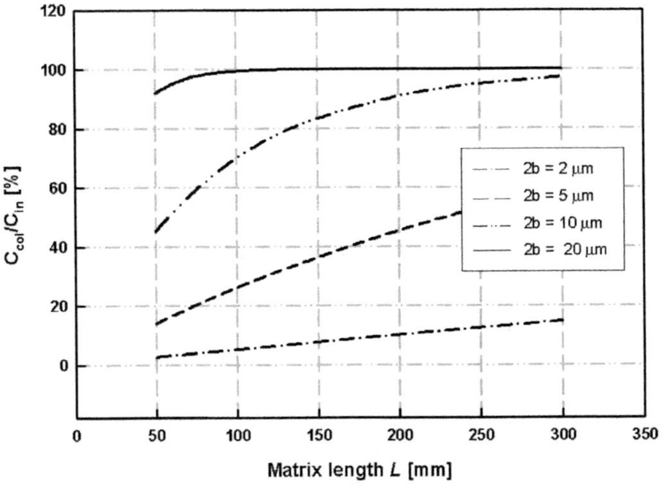

Figure 3.34: The proportion of particles colliding with a steel ball matrix, as a function of the matrix depth. $v = 0.07$ m/s, $2a = 6$ mm, $\epsilon = 0.4$ (adapted from Svoboda [S33]).

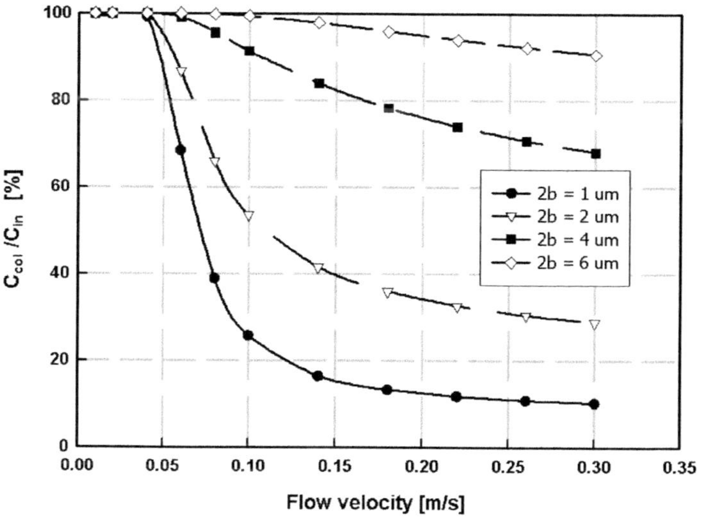

Figure 3.35: The proportion of particles colliding with steel wool matrix, as a function of superficial velocity. $L = 200$ mm, $2a = 400$ μm, $\epsilon = 0.95$ (adapted from [S33]).

or smaller, will have collided with the steel wool matrix, with the matrix bed depth of 200 mm. A similar trend can be observed for the steel ball matrix, as illustrated in Fig. 3.34. With decreasing velocity the probability of collision rapidly increases, as shown in Fig. 3.35.

The retention of particles For a particle to be retained on the matrix after a collision, the total interaction between the particles and the matrix and the fluid medium must be attractive. For sufficiently large particles the surface forces can be neglected and the magnetic dipolar and the hydrodynamic interactions are decisive. The mean free magnetic energy for the interaction of a spherical paramagnetic particle of susceptibility κ with a ferromagnetic sphere of saturation magnetization M_s, placed in an external magnetic field B can be written [S1] as

$$V_m = -\frac{4\pi\kappa B^2 b^2}{3\mu_0} - \frac{32\pi^2\kappa b^3 a^3 B M_s}{9r^3} \tag{3.106}$$

where r is the distance between the interacting bodies.

Since $a >> b$ and $r = a + b$, eq. (3.106) gives [S31]

$$V_m = -\frac{32}{9}\pi^2\kappa b^3 B M_s \tag{3.107}$$

The hydrodynamic potential energy of a particle captured on the surface of a collector is assumed to be Stokesian:

$$V_d = 6\pi\eta vab \tag{3.108}$$

A particle will remain attached to the matrix if $V_m + V_d < 0$. Therefore, for a particle to be retained by the matrix, the magnetic induction must be greater than the threshold magnetic induction that can be determined from the steady-state condition $V_m + V_d = 0$. Therefore, [S33]:

$$B > \frac{27\eta va}{16\pi\kappa b^2 M_s} \tag{3.109}$$

It is evident from eq. (3.109) that, for given conditions of separation, i.e. particle size and susceptibility, flow velocity and size of matrix elements, an increase in strength of the applied magnetic field beyond the threshold value will not increase the retention of particles on the matrix.

It might appear from eq. (3.109) that by increasing the magnetic field strength it is possible to increase the flow velocity and thus the throughput of a separator, and still maintain the required rate of particle retention. However, if the particle velocity is too high, the particles will never collide with the matrix bed and will thus not be available for retention. This scenario is illustrated in Fig. 3.35.

Furthermore, the selectivity of the separation process is be determined by the efficiency with which the "non-magnetic" particles are being swept away from the matrix by the erosive force of the fluid. Consequently, for particles that are supposed to report into the non-magnetic fraction, the inequality inverse to that given by eq. (3.109) should apply. This is in contrast to the trajectory models, in which the non-magnetic particles never reach the matrix surface.

3.5 Linear open-gradient magnetic separation

The concept of separation in a magnetic field with open gradient of the field is widely practised in drum and roll magnetic separators. Paradoxically, it is the linear multipole open-gradient magnetic separation (OGMS) that received considerable attention from the theoretical point of view, in spite of its failure to prove its advantages on the production scale. It has been explained in Section 2.5.3, that the linear multipole OGMS consists of an array of vertically positioned coaxial coils energized in opposite directions. The concept is schematically shown in Figs. 2.85(a) and (b) and in Fig. 3.36.

The magnetic force directed radially inwards diverts the free-falling magnetic particles inwards while the feebly magnetic particles fall vertically, unaffected by the magnetic force.

The main advantage of open-gradient separators, in comparison with HGMS, is the absence of matrix which allows the elimination of the blockage problem. Another advantage of OGMS is the use of superconducting magnets which lend themselves easily to scale-up. This is because in resistive magnetic systems the

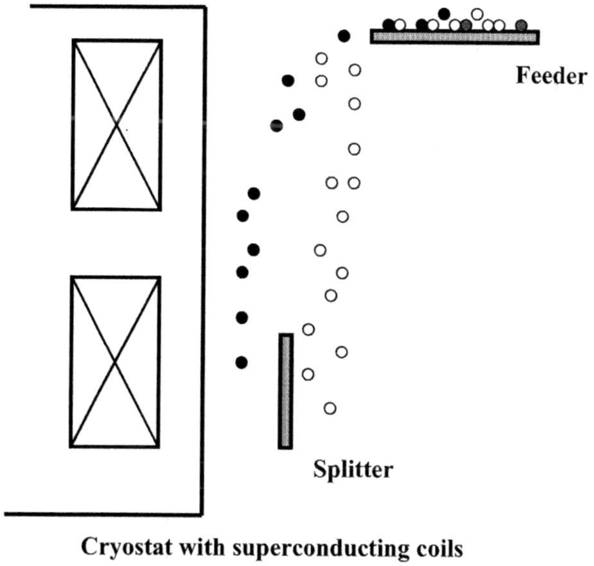

Figure 3.36: Schematic diagram of a linear multipole open-gradient magnetic separator.

magnetic field in the separation space is limited by the non-linear saturation effects in steel [G5]. On the other hand, systems consisting of superconducting magnets are linear as far as the magnetic field generation is concerned, and the electric power consumption, which is spent mostly for cryogenic purposes, increases only moderately with their size. It was also found by Gerber et al. [G8] that the metallurgical efficiency of the superconducting OGMS is superior to the conventional OGMS as a result of considerably larger particle deflections achievable in various designs of coil configurations.

A detailed analysis of particle trajectories in dry OGMS was made by Collan et al. [C13], while Kopp [K10, K11] developed a simple kinetic model of the separation process. These models were subsequently refined by numerous authors [B16 , G9, J3, B17, P6, G5]. Particle trajectories and particle deflections as a function of particle properties, particle collisions, electric current densities in the coils and the separator geometry were determined. Diamagnetic mode of OGMS was also described theoretically and confirmed experimentally by Krist et al. [K12]. Considerable effort was also devoted to the optimization of the design parameters of the superconducting coils for OGMS [G10, G11, K13, J4, P7].

Although theoretical analyses of particle motion in OGMS are of considerable mathematical elegance, their applicability to practical problems of material separation is limited. In view of a somewhat esoteric character of the subject and as a result of the failure of several efforts to industrialize the concept of

OGMS [G3, K9, A8, S17] and to achieve commercial success, detailed description is beyond the scope of this monograph. The interested reader is referred to the literature listed above.

3.6 Magnetic flocculation

The demand for mineral products is subject to two factors affecting the efficiency of beneficiation: a decrease in the quality of economically accessible ores and an increase in the degree of fineness of their valuable components. Since the recovery of fine particles is still very inefficient, vast quantities of valuable minerals are often lost in the form of slimes. Unless new technologies are developed, the increasing amount of finely-dispersed minerals that being mined and processed, will lead to increased volumes of waste. This will result in a substantial loss of valuable components and in environmental problems.

Methods for the treatment of fine particles are still in the early stages of development, and conventional methods are usually inefficient for particles smaller than 20 μm. Particles smaller than 5 μm can be manipulated only by HGMS, a rather specialized technique applicable to a limited range of materials. A more general solution to the problem may lie in the formation of aggregates comprising fine particles.

In the past, flocs were successfully formed by the addition of inorganic electrolytes or polymers, for instance for the removal of colloidal particles from effluents. Flocculation can also be induced by the application of external forces to a system of particles, e.g. by stirring, or by the application of magnetic or electrical fields.

Magnetic flocculation has been used for some time to create flocs of ferromagnetic particles. To induce this process, a magnetic induction of the order of 0.01 T is required. Experimental observations of magnetic flocculation of strongly magnetic minerals and its application to the mining industry have been described in numerous publications [B18, B19, B20, B21, L9, K14, K2]. The theoretical description of mutual coagulation of ferromagnetic particles is also well established [D9, J5, L10, L9].

Although investigations into the magnetic flocculation of paramagnetic and diamagnetic particles started much later [H16, W16, S34], systematic studies of the theoretical and experimental aspects [S35, P8, S36, P9, S37, S38, W17, W18, W19, P10, T5, S39, S40, S41] show that flocculation of fine, feebly magnetic, particles can be induced in magnetic systems that are presently available.

In the following sections the present status of the theory of magnetic flocculation of strongly and weakly magnetic materials will be briefly outlined.

3.6.1 Magnetic flocculation of strongly magnetic materials

The tendency of an assembly of ferromagnetic particles in the magnetic field of a magnetic separator to reduce its magnetostatic energy often results in the formation of aggregates of the particles. The effect of magnetic flocculation is the inherent part of any process of manipulation of finely ground strongly magnetic materials and directly affects the efficiency of material separation. In magnetic separation the process of magnetic flocculation can often play a positive role by increasing the recovery and/or throughput. In other processes, such as recovery of dense medium, magnetic flocculation is detrimental to the efficiency of dense medium separation.

Theory of magnetic flocculation

We shall consider a suspension of identical spherical ferromagnetic particles in a non-magnetic fluid. Under the influence of a sufficiently strong external magnetic field such particles tend to associate in long chains. Each particle of volume V_p and magnetization M_p carries a magnetic moment $\vec{\mu}_M = V_p\vec{M}_p$. The energy of magnetic interaction between particles (1) and (2) separated by a distance r is

$$V_M = \frac{1}{\mu_0}\{\frac{\vec{\mu}_M(1)\vec{\mu}_M(2)}{r^3} - \frac{3[\vec{\mu}_M(1)\vec{r}][\vec{\mu}_M(2)\vec{r}]}{r^5}\} \qquad (3.110)$$

If magnetization of the particles is sufficiently large, the electrical double layer interaction and the van der Waals interaction can be neglected. Total interaction energy, therefore, consists of the magnetic dipolar term only (eq. (3.110)). Effective magnetic interaction in the suspension can be obtained by averaging V_M over all orientations of the colloidal particles. Thus [C14, C15]

$$\bar{V}_M = -kT \ln < \exp(-\frac{V_M}{kT}) > \qquad (3.111)$$

where k is the Boltzmann constant and T is the absolute temperature and

$$< \exp(-\frac{V_M}{kT}) >= \frac{1}{64\pi^3}\int \exp(-\frac{V_M}{kT})d\Omega_1 d\Omega_2 d\Omega_3 \qquad (3.112)$$

where $d\Omega_i$ (i =1,2,3) represents an element of a solid angle describing the orientation of each of the dipoles (1, 2) and that of the external magnetic field (3). It was shown by Chan et al. [C14, C15] and by Scholten et al. [S42] that in the weak interaction limit, i.e. when $\mu_M^2/\mu_0 r^3 < kT$, the average interaction energy can be written as

$$\bar{V}_M = -\frac{\mu_M^4}{3\mu_0^2 r^6 kT} \qquad (3.113)$$

while in the strong interaction limit ($\mu_M^2/\mu_0 r^3 > kT$)

$$\bar{V}_M = -\frac{2\mu_M^2}{\mu_0 r^3} \qquad (3.114)$$

It can thus be seen that although the magnetic dipole-dipole interaction (eq. (3.110)) can be either attractive or repulsive, depending on the relative orientation of the dipoles, the orientational averaging process gives the energetically favourable (attractive orientation) heavy weighing.

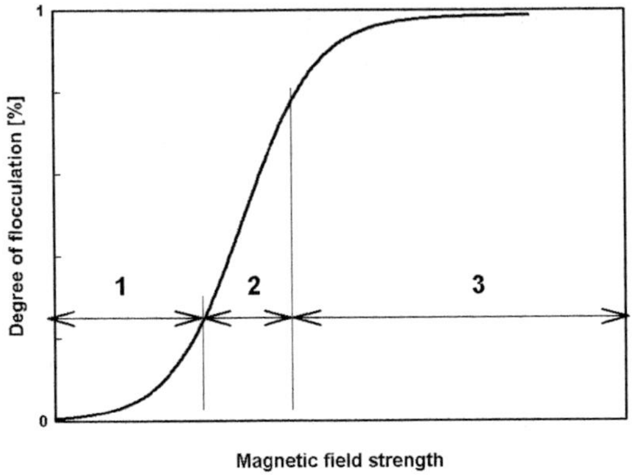

Figure 3.37: The effect of the magnetic field strength on the degree of flocculation of strongly magnetic particles (adapted from [K14, K2]).

A condition for magnetic flocculation to take place can be expressed as

$$F_m > F_d + F_e \tag{3.115}$$

where F_e is the electric force as a result of electrical double layer. Since $F_m >> F_e$ for ferromagnetic particles at reasonably strong magnetic field, the probability of flocculation is determined by the interplay between the magnetic and hydrodynamic forces. As a function of the magnetic field strength, the process of magnetic flocculation proceeds through several stages, as is shown in Fig. 3.37.

In stage 1, which corresponds to the steady-state phase of the process, in which $F_m \leqslant F_d$, the degree of flocculation ψ_f is proportional to F_m and, therefore, $\psi_f = kH^2$. At this phase the grade of the flocs is high and flocculation is selective.

With further increase in the magnetic field strength the magnetic interaction becomes dominant and the flocculation enters phase 2. The process proceeds non-selectively at a very high rate and the grade of the flocs is reduced. When the concentration of solids in a suspension is reduced below a certain threshold value, the mean distance between the particles and their characteristic flocculation time increase and the process enters an equilibrium phase 3.

3.6.2 Magnetic flocculation of weakly magnetic minerals

In contrast to strongly magnetic materials, the magnetic flocculation of paramagnetic and diamagnetic materials is a considerably more complex process. In spite of its significant potential, the process has not found any application on industrial scale as yet. Although several theoretical models that predict the behaviour of weakly magnetic particles of quasi-colloidal dimensions, in the presence of the external magnetic field, have been developed over the last twenty years, their experimental verification is still limited and is mainly of qualitative character.

Stability of colloidal suspensions

The criteria of the stability of colloidal suspensions are governed largely by the interplay of repulsive and attractive forces between the particles. A detailed description of this interplay is the basis of the theory of stability of hydrophobic colloids as proposed by Derjaguin and Landau [D14] and Verwey and Overbeek [V1] (DLVO theory). The DLVO theory views the stability of colloids as arising from electrical charges on identical particles and a consequent electrical repulsion between them. Opposing this repulsion are the attractive forces of the universal London-van der Waals type. If the repulsion force can be sufficiently reduced so that the attractive interaction becomes dominant then the particles colliding with one another will stick together to form quasi-stable aggregates.

The charge on a particle, together with the equal and opposite charge in a suspension, comprise the electrical double layer (EDL). Hogg et al. [H17] showed that the interaction of EDL between two dissimilar particles, under the condition of low surface potential and small thickness of EDL compared to the particle size, is given, under the assumption of a constant surface potential, by

$$V_R = \frac{\epsilon_r b_1 b_2 (\psi_1^2 + \psi_2^2)}{4(b_1 + b_2)} \left[\frac{2\psi_1\psi_2}{(\psi_1^2 + \psi_2^2)} \ln \frac{1 + \exp(-Xh)}{1 - \exp(-Xh)} + \ln(1 - e^{-2Xh}) \right] \quad (3.116)$$

where b_1, b_2 and ψ_1, ψ_2 are the radii and surface potentials of particles, respectively, ϵ_r the dielectric constant of the medium, X the Debye-Hückel reciprocal length parameter, and h their distance between the surfaces of the interacting particles. X is related to the ionic strength I_m of the medium by the expression

$$X = \frac{8\pi e^2 I_m^{1/2}}{\epsilon_r kT} \quad (3.117)$$

In the case of identical spherical particles eq. (3.116) reduces to

$$V_R = \frac{\epsilon_r b \psi^2}{2} \ln(1 + e^{-Xh}) \quad (3.118)$$

When the thickness of EDL is not small (i.e. when $bX < 1$), V_R can be expressed in a simple approximate form:

$$V_R = \pi \epsilon_r b^2 \psi^2 \frac{\exp(-Xh)}{(2b + h)} \quad (3.119)$$

Similarly, the London - van der Waals interaction for dissimilar spheres can be written as

$$V_A = -\frac{A}{6} \left[\begin{array}{c} \dfrac{2b_1 b_2}{h^2 + 2(b_1 + b_2)h} + \dfrac{2b_1 b_2}{h^2 + 2(b_1 + b_2)h + 4b_1 b_2} + \\ \ln \dfrac{h^2 + 2(b_1 + b_2)h}{h^2 + 2(b_1 + b_2)h + 4b_1 b_2} \end{array} \right] \quad (3.120)$$

where A is the Hamaker constant. For identical particles, eq. (3.120) simplifies to

$$V_A = -\frac{A}{6}\left[\frac{2b^2}{h^2+4bh} + \frac{2b^2}{h^2+4bh+4b^2} + \ln\frac{h^2+4bh}{h^2+4bh+4b^2}\right] \tag{3.121}$$

When the London forces operate over distances comparable to or larger than 0.1λ, where λ is the wavelength of the intrinsic electronic oscillations of the atoms, the interaction is reduced by the retardation effect [S43, R15].

Flocculation in a magnetic field

When a suspension of colloidal particles is placed in an external magnetic field there is, in addition to the interactions considered so far, a magnetic dipolar interaction, described by eq. (3.110).

In the strong interaction limit [C14, S34], the mean magnetic interaction energy can be written as

$$V_M = -\frac{8\pi\mu_0\kappa_1\kappa_2 b_1^3 b_2^3 H^2}{9(h+b_1+b_2)^3} \tag{3.122}$$

where κ_i are volume magnetic susceptibilities of particles and H is the magnetic field strength.

In agreement with the assumption of the DVLO theory that double layer and van der Waals forces are independent and additive, Svoboda [S34] supposed that the total energy of interaction of colloidal particles in the external magnetic field is, therefore, given by the sum of electrical double-layer energy, van der Waals interaction energy and magnetic dipolar interaction energy

$$V_T = V_R + V_A + V_M \tag{3.123}$$

Generally, the total interaction energy V_T can be either attractive or repulsive depending on the sign of the individual terms on the right-hand side of eq. (3.123). Electrical double layer energy is repulsive for interaction of particles of the same material, while it can be attractive or repulsive for interaction of unlike materials (e.g. between a valuable mineral and gangue).

The van de Waals interaction is usually attractive, but under certain conditions [V2], determined by relative values of the Hamaker constant of the interacting particle and the medium, the interaction becomes repulsive. Magnetic interaction can be either attractive or repulsive, depending on the signs of the magnetic susceptibilities of the interacting particles. It is clear that interaction between diamagnetic and paramagnetic particles is repulsive. On the other hand, magnetic interaction of identical materials, whether paramagnetic or diamagnetic, is always attractive. Magnetic flocculation of diamagnetic particles is, therefore, possible, in principle.

A curve of the total interaction energy V_T as a function of the interparticle distance is shown in Fig. 3.38. A potential barrier V_{max} separates two minima: a closer and deeper primary minimum and a more distant and shallow secondary

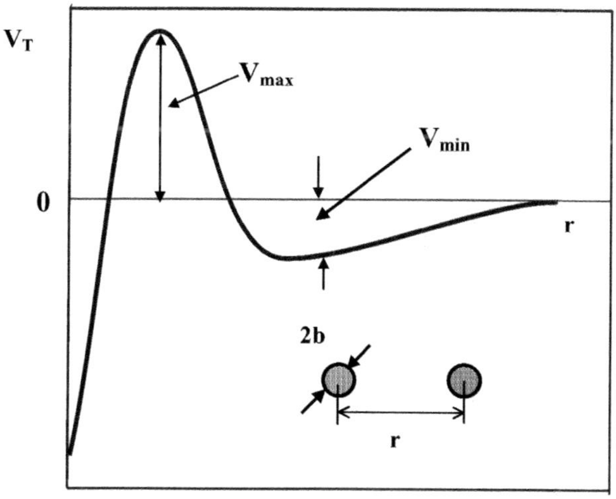

Figure 3.38: The general form of the total potential energy for a pair of interacting colloidal particles of radius b.

minimum of depth V_{min}. The individual potential energy contributions are affected by the properties of a suspension in different ways. The magnitude and range of the electric double layer interaction V_R are sensitive to ψ and X, which are controlled by the solution conditions (i.e. ionic strength and pH), whereas the van der Waals and magnetic contributions V_A and V_M are independent of these parameters.

The magnitudes of the individual potential energy components have a different dependence on the particle size: V_M is proportional to the cube of the radius while V_R and V_A are much less sensitive to particle size as they are proportional to the first power of particle size.

The height of the barrier determines the stability of a suspension. If the barrier is very low, there is nothing to prevent contact between the particles, and flocculation will take place at a rate governed by the frequency with which the particles collide. If there is a secondary minimum in the curve of a potential energy, particles can flocculate into this potential well. In this instance, the potential barrier need not be suppressed. Flocs thus formed will have a rather loose structure as a result of large interparticle distance at which the secondary minimum is created.

The existence and position of the primary and secondary minima can be affected by changes in the magnetic field, the magnetic susceptibility of the particles and the pH value of the suspension. The total interaction energy V_T is shown in Fig. 3.39 as a function of magnetic induction and particle size. In region I the potential barrier is present and the secondary minimum is sufficiently close, while in region II the secondary minimum is too distant. In regions III and IV, in which, owing to high magnetic induction, the potential

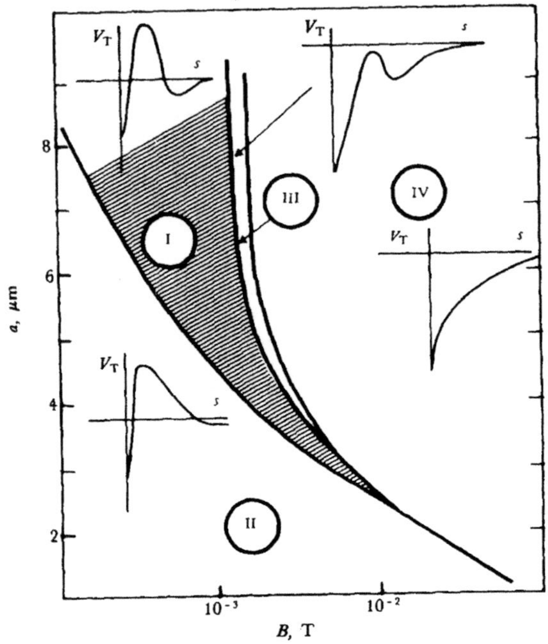

Figure 3.39: Regions in which primary and secondary minima are found in the (a,B) plane, for hematite ($\psi = 30$ mV, $X = 10^5$ m^{-1}), a - particle radius, B - magnetic induction (after [S36, S37]).

barrier is suppressed and the total interaction becomes attractive, and particles can flocculate into the primary minimum.

It follows from Fig. 3.39 that, for any particle size, a value of the magnetic induction can be found at which the potential barrier disappears. Wang et al. [W17] calculated the potential energy of interaction of particles of hematite and chromite in a magnetic field. Typical curves shown in Fig. 3.40 illustrate that at moderate magnetic field strength the potential barrier is eliminated and flocculation into the primary minimum could take place. A deep secondary minimum is created at even lower magnetic field.

A threshold magnetic field, at which the flocculation into the primary minimum starts, can be determined provided that the total potential energy of interaction V_T is negative [S34], or that the stability ratio W_p, which is given by eq. (3.124), is smaller than, or equal to, unity [S35]. The stability ratio is the ratio of the initial rate of flocculation to that which would occur if the process was only diffusion controlled (in the absence of an energy barrier) [P11].

$$W_p = 2 \int \exp\left(\frac{V_T}{kT}\right) \frac{ds}{s^2} \tag{3.124}$$

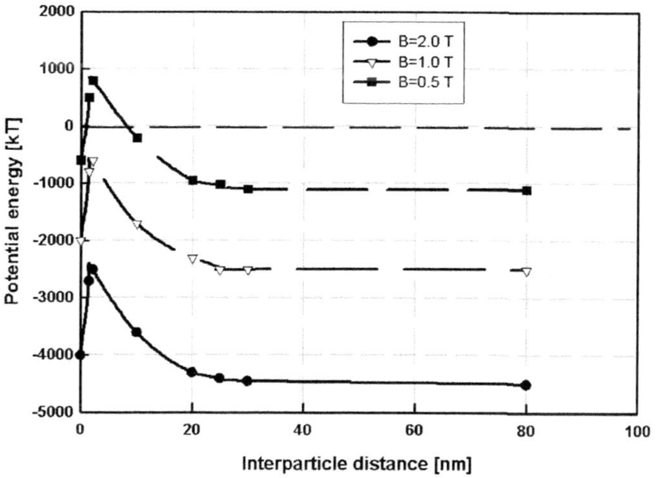

Figure 3.40: The potential energy as a function of interparticle distance for 4 μm hematite particles, $X = 10^8$ m^{-1} (adapted from Wang et al. [W17]).

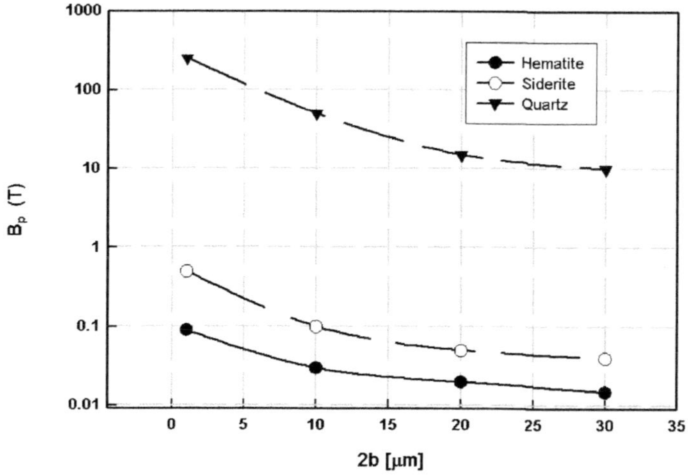

Figure 3.41: Threshold magnetic field, at which flocculation into a primary minimum begins, as a function of particle size (adapted from Svoboda [S35]).

The values of the threshold magnetic field B_p, at which magnetic flocculation into the primary minimum begins, are shown in Fig. 3.41, as a function of particle size, for various minerals. It can be seen that as the particle size and magnetic susceptibility increase, the threshold magnetic field decreases.

For fine particles with a volume magnetic susceptibility smaller than 10^{-3} (SI), the threshold magnetic field is too high for magnetic flocculation to be practicable. This is particularly true for the magnetic flocculation of diamagnetic materials, which require a magnetic induction of the order of tens of hundreds of Tesla.

Flocculation in the secondary minimum

As shown in Fig. 3.39, there is a secondary minimum (SM) in addition to the primary minimum in regions I, II and III. Although in regions I and II the potential barrier prevents particles from flocculating into the primary minimum, particles can flocculate into the secondary minimum. The secondary minimum in region II is too distant from the origin, and consequently, the aggregates will be packed too loosely. Furthermore, the secondary minimum is too shallow and the stability of the flocs will be low.

However, in region I, the secondary minimum can be deep enough and close enough to the origin to render it suitable for practical flocculation.

The mere existence of a secondary minimum does not imply an observable destabilization of a suspension. The other condition to be satisfied is sufficient depth of SM; the analysis indicates [S36] that, for $V_{\min} > 1$ kT (where kT is the unit of thermal energy), the value of the stability ratio W_s for flocculation in a secondary minimum, as given by eq. (3.125), will be close to unity, and the flocculation into SM will be rapid.

$$W_s = \frac{s_m \int_{s_m}^{\infty} \exp(\frac{V_T}{kT})\frac{ds}{s^2}}{1 - \exp(\frac{V_{\min}}{kT})} \tag{3.125}$$

where s_m is the position of the secondary minimum.

The threshold magnetic field B_s required to commence the flocculation into SM can be determined from the condition $W_s = 1$. The dependence on a particle radius of threshold magnetic field B_s in SM, which was determined from the condition of rapid flocculation ($W_s = 1$), is shown in Fig. 3.42.

A comparison of these values with those of the threshold magnetic field B_p for flocculation into the primary minimum (Fig. 3.41) indicates that the magnetic field needed to induce rapid flocculation in SM is substantially lower. This conclusion has been confirmed by detailed calculations by Wang et al. [W17], an example of which is presented in Fig. 3.40. Experimental confirmation of the magnetic field-induced flocculation is depicted in Fig. 3.43. Figure 3.44 illustrates the experimentally determined onset of rapid flocculation of hematite and chromite at threshold magnetic field of about 0.5 T [W17].

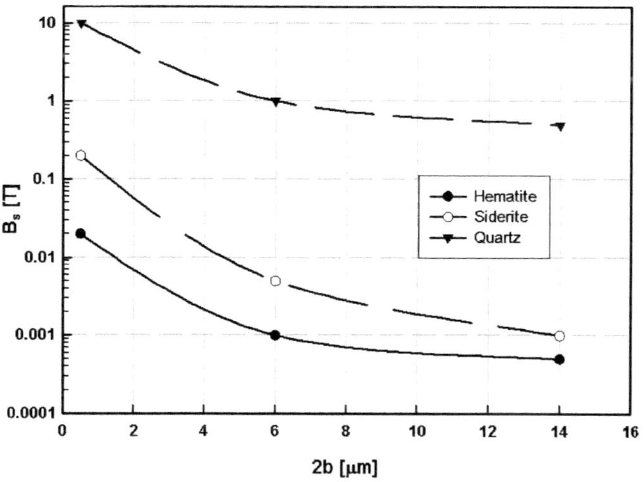

Figure 3.42: Threshold magnetic field for flocculation into a secondary minimum, as a function of particle diameter (adapted from Svoboda [S35]).

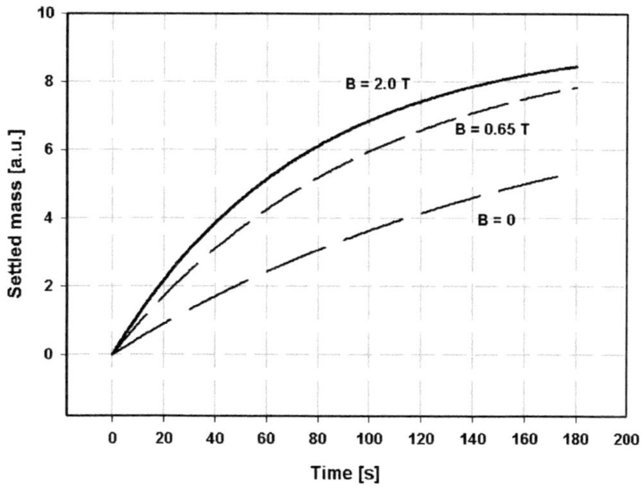

Figure 3.43: The magnetic field-induced flocculation of hematite particles at various values of magnetic induction. pH $= 7.8$, $\psi \approx 45$ mV, $d_{50} = 3.9\ \mu$m (adapted from Wang et al. [W17]).

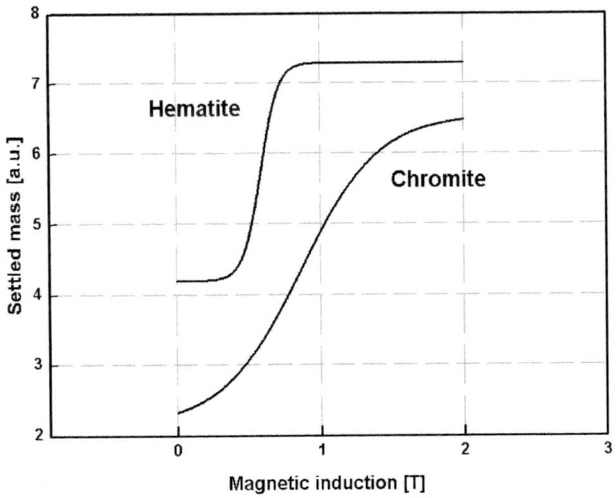

Figure 3.44: The onset of the magnetic field-induced flocculation of hematite (d_{50} = 3.9 μm) and chromite (d_{50} = 3.6 μm) particles (adapted from Wang et al. [W17]).

We have seen that under suitably chosen conditions, the interaction potential between a pair of colloidal particles can develop a significant ($\geqslant$ 1 kT) secondary minimum, while maintaining a large (>> 1 kT) primary maximum. Although the primary maximum will prevent irreversible coagulation into the primary minimum, the particles can undergo reversible flocculation into the secondary minimum.

The above treatment of the flocculation process is applicable only to colloidal particles. For particles greater than the colloidal range, the thermal motion and the surface forces play only a limited role, and the process of flocculation will be governed mainly by the gravitational and hydrodynamic forces.

The kinetics of flocculation

It has been shown that the stability of suspensions depends on the magnitude of the potential barrier and on the existence and depth of the secondary minimum. In addition, however, determination of how quickly flocculation will take place is necessary.

Fuchs [F13] showed that the time taken for the number of particles to be halved is given by the relation

$$t_{1/2} = \frac{3\eta W}{4kTN_0} \tag{3.126}$$

where η is the viscosity of the medium and N_0 is the initial concentration of the particles.

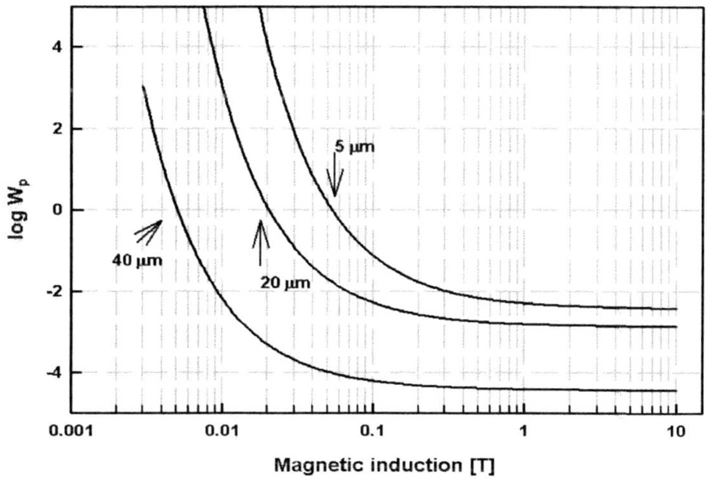

Figure 3.45: Dependence of the stability ratio W_p (as log W_p) on magnetic induction for assorted sizes of hematite particles (adapted from Svoboda [S35]).

It can be seen that the half-time of flocculation is directly proportional to the stability ratio and indirectly proportional to the initial concentration of the particles. To achieve a reasonably short flocculation time, low values of W and high values of N_0 must be used.

In chemical flocculation, the values of W smaller than 10^{-1} are reached with difficulty [V1]. This significantly restricts the possibility of accelerated flocculation. On the other hand, it was shown by Svoboda [S35], that in magnetic flocculation, the values of the threshold magnetic field determined from the condition $W = 1$ for rapid flocculation, allow reasonably short flocculation time to be achieved. As is shown in Fig. 3.45, the values of W_p, of the order of 10^{-3}, can be achieved at reasonably low magnetic fields so that the flocculation time into the primary minimum can be realistically short.

Magnetic-hydrophobic flocculation

In order to explain situations in which the original DLVO theory failed, numerous "non-DLVO" forces have been invented. A variety of forms of long-range forces such as hydrophobic/hydrophilic interaction have been introduced. For example, surface-active agents can adsorb at the mineral-water interface and render the solid hydrophobic. Particle-particle adhesion is favoured by the attraction between the hydrophobic coatings on different particles and the resultant reduction in the area of hydrophobic mineral-water surface.

A critical review of such extensions of DLVO theory to include "non-DLVO" interactions was published by Ninham [N6] who argued that confusion created

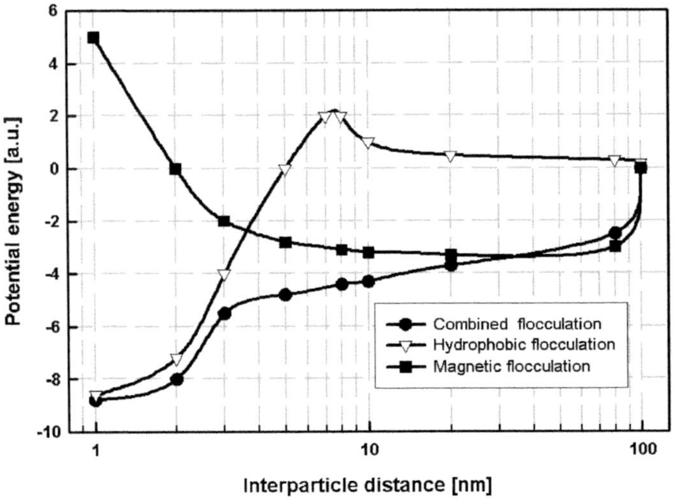

Figure 3.46: The potential energy curves for hydrophobic and magnetic flocculation of rhodochrosite ($2b_{50} = 2.25$ μm) hydrophobized by sodium oleate, at a magnetic induction of 0.9 T (adapted from [L11]).

by introduction of plethora of hydrophobic phenomena observed in the stability studies can, in principle, be explained by capillary condensation, i.e. thermodynamics.

Several authors have shown that the stability of a colloidal suspension can be controlled, in addition to the electrostatic surface potential and the magnetic interaction, by such a hydrophobic interaction. Lu et al. [L11] included the energy of hydrophobic interaction V_H as an additive term in the expression of the total interaction energy (eq. (3.123):

$$V_T = V_R + V_A + V_M + V_H \tag{3.127}$$

The hydrophobic interaction energy was expressed by two terms, each representing different mechanism of the hydrophobic interaction. The potential energy curves of three possible flocculation mechanisms for paramagnetic rhodochrosite were calculated and are shown in Fig. 3.46.

It can been seen that the potential energy of magnetic interaction exhibits a deep secondary minimum and a primary minimum barrier. In hydrophobic aggregation, a relatively low potential barrier partially prevents particles from rapid flocculation into the primary minimum. On the other hand, the potential energy curve for combined magnetic and hydrophobic flocculation shows deep primary and secondary minima, with no potential barrier.

The experimentally determined rates of flocculation in these three regimes are illustrated in Fig. 3.47. It is clear that the combined flocculation by a

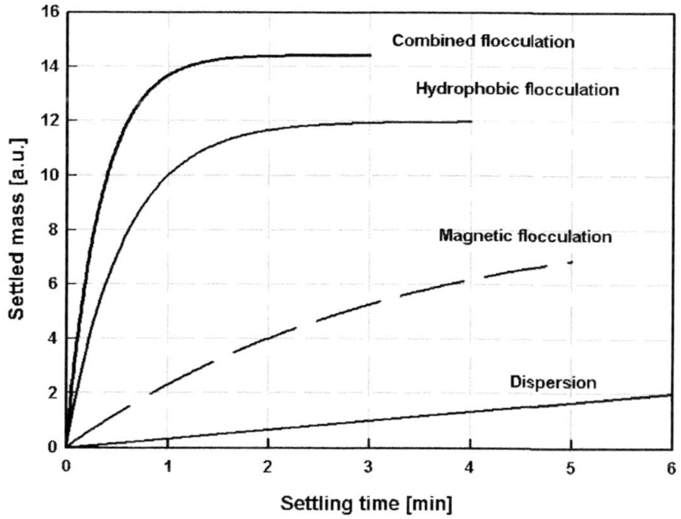

Figure 3.47: Flocculation of rhodochrosite by magnetic (B = 0.85 T), hydropohobic (sodium oleate + kerosene) and combined mechanisms (adapted from [L11]).

combination of magnetic and hydrophobic processes was more effective than aggregation induced by sodium oleate or magnetic field alone.

Škvarla and Zeleňák [S41] developed a simple model of magnetic-hydrophobic flocculation, as a variation of Lu's model [L11]. They assumed that the potential of the hydrophobic interaction was of the short range and it was approximated by a single exponential

$$V_H = -2\pi b \gamma_{SL} \lambda \exp(\frac{h_c - h}{\lambda}) \tag{3.128}$$

where γ_{SL} is the solid/water interfacial energy of the spherical particles, h_c is the contact separation distance of the spheres ($\cong 0.2$ nm), and λ is the empirical constant ($\cong 1$ nm). The model was verified by experimental observation of magnetic-hydrophobic flocculation of siderite. A satisfactory correlation between the measured and calculated potential energy maxima was observed for the hydrophobic process. The prediction of the model of magnetic flocculation alone was not satisfactory.

3.7 Magnetic separation by particle rotation

Motion of magnetizable particles in alternating or rotating magnetic fields is a well established research field. In 1925 Koizumi [K15] designed a magnetic separator with a rotating magnetic field, to eliminate entrapment of non-magnetic

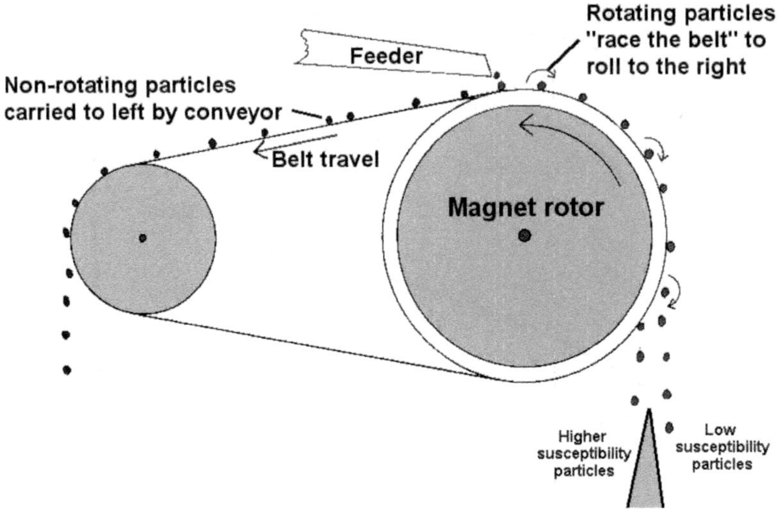

Figure 3.48: Schematic diagram of a separator with rotating magnetic field (courtesy of N. Allen [A17]).

particles in the magnetic concentrate. Similar approach was taken more recently by other researchers to improve the quality of magnetic concentration [T3, D1, F1, G12, G13, L9]. Although the results of these investigations had been positive, the technique did not appear to be accepted and developed further.

Recently, Allen [A9] employed the concept of the effect of a rotating magnetic field on magnetizable particles and developed rotating magnetic field (RMF) separators, described in Section 2.6.3. The rotation separation is based on magnetic anisotropy of particles, which is present in ferromagnetic as well as paramagnetic materials. However, in paramagnetic minerals the magnetic anisotropy is generally too weak to be of practical use for particle separation. In general, ferromagnetic particles will rotate if they are placed in a rotating external magnetic field, while paramagnetic particles will not.

An RMF is generated by a rotating magnetic drum fitted, around its circumference, with magnets, with alternate polarity. As is schematically illustrated in Fig. 2.95, the external magnetic field exerts a torque on a magnetically anisotropic particle and makes it rotate and move it in the direction opposite to that of the rotating field. A schematic diagram of such an RMF separator is shown in Fig. 3.48.

3.7.1 RMF rotation separation

In the RMF separator a particle is placed on a non-magnetic surface in a rotating magnetic field, as shown in Fig. 3.49. If the particle is capable of rotation in the

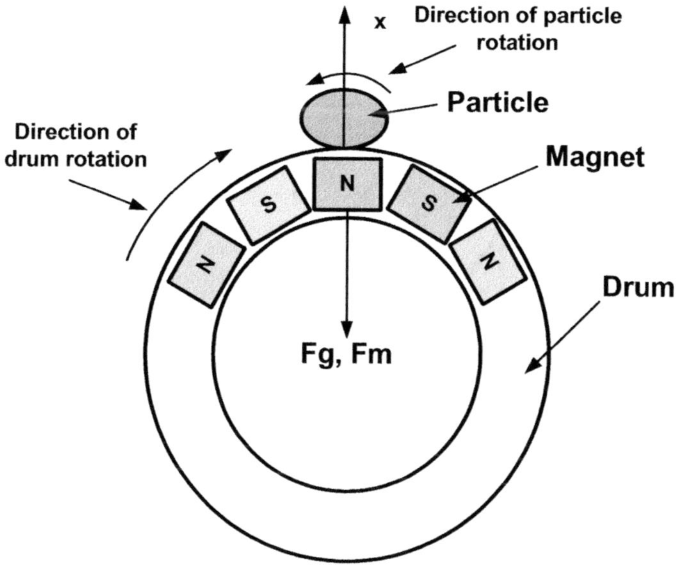

Figure 3.49: Schematic diagram of a particle placed on a rotating magnetic drum.

field, it will experience a torque in order to align its axis of easy magnetization with the direction of the external magnetic field. The torque, which thus arises when the direction of the magnetization of a particle and the external magnetic field do not coincide, will force the particle to roll. The rolling is opposed by the force of gravity and magnetic traction force acting on the particle.

The particles to be separated by rotation will, in general, contain ferromagnetic and paramagnetic regions. While the paramagnetic regions will make limited contribution to the particle rotation, they will contribute to the magnetic traction (holding) force. This traction force can, therefore, be written [A17] as:

$$F_m = \frac{1}{\mu_0}\kappa_p V_{pp} B\nabla B + M_{pf} V_{pf} \nabla B \tag{3.129}$$

where V_{pp} and V_{pf} are the volumes of the paramagnetic and ferromagnetic regions of the particle, respectively. M_{pf} is the magnetization of the ferromagnetic components while κ_p is the volume magnetic susceptibility of the paramagnetic contribution. The magnitude of the torque of the force F, required to rotate a particle, is the product of the magnitude of the force and the moment arm d; that is

$$T = Fd \tag{3.130}$$

The particle will not start rotating unless the initial magnetic rotational torque exceeds the initial gravitational and magnetic traction force torques (case (a) in Fig. 3.50). The cubic shape of the particle is usually used to demonstrate the

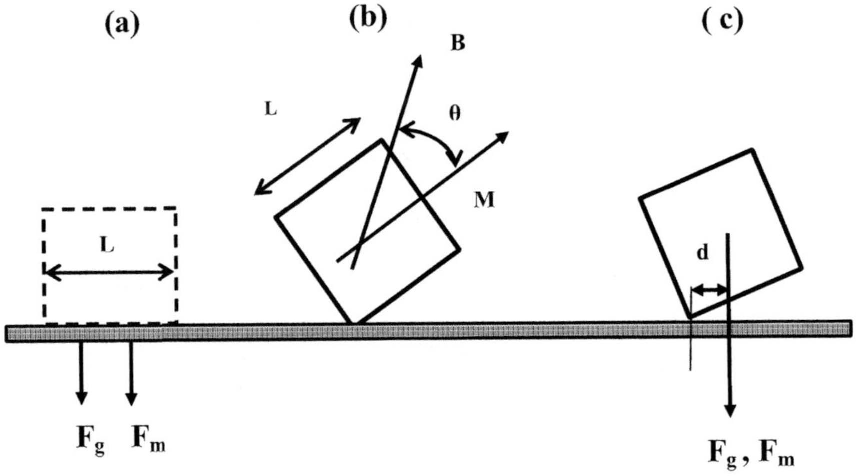

Figure 3.50: Forces acting on a cubic particle in traction and rotating magnetic fields.

case [A17, A18] and we have

$$T_{init} = \frac{L}{2}(F_g + F_m) \tag{3.131}$$

where L is the length of the cube edge and the force of gravity F_g is given by eq. (1.14).

For a cubic ferromagnetic particle, under the assumption that the first term on the right-hand side of eq. (3.129) can be neglected, the initial gravitational + magnetic holding torques (that is, when the particle is resting flat on the surface) can be written

$$T_{init} = L^3 \rho_p g \frac{L}{2} + M_p L^3 \nabla B \frac{L}{2} \tag{3.132}$$

3.7.2 Magnetic torque on a particle and condition for separation by rotation

The application of an external magnetic field to a ferromagnetic crystal, at a certain angle $\theta + \delta$ to a direction of easy magnetization, causes a deviation of the magnetization, as indicated in Fig. 3.51. The external magnetic induction B applies a torque to the particle magnetization M_p and causes it to deviate from the direction of easy magnetization x. As a result, the magnetocrystalline forces apply a torque on the magnetization to return it to the direction of easy

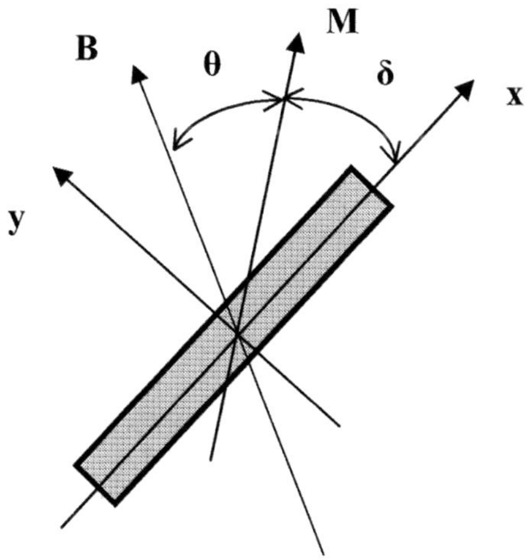

Figure 3.51: The torque experienced by an anisotropic magnetizable particle placed in an external magnetic field.

magnetization. This situation can be described by a relationship

$$M_p B \sin\theta = M_p A \sin\delta \tag{3.133}$$

where A represents the anisotropy field. If the external field is very low, then the angle δ is small. If the external field is very high, then θ is small [A17].

For a cubic particle, the magnetic torque given by eq. (3.48), can be rewritten as:

$$T_{mag} = V_p M_p B \sin\theta = L^3 M_p B \sin\theta \tag{3.134}$$

where θ is the angle between the directions of the magnetic field and particle magnetization, as can be seen in Figs. 3.50(b) and 3.51. Therefore, a particle will start rotating when

$$T_{mag} > T_{init} \tag{3.135}$$

or, explicitly,

$$B \sin\theta > \frac{L}{2}\left(\frac{\rho_p g}{M_p} + \nabla B\right) \tag{3.136}$$

Coercive forces are very low for many ferromagnetics *(s.l.)*, such as magnetite and steel, while anisotropy forces are high. If the external field is smaller than the coercive force ($B < \mu_0 H_c$), angular deviations from directions of easy magnetization are also small. Maximum torque is then applied when the external field is very close to 90^0 to the particle magnetization. [A17]. Although the angle between the magnetization and the external field has been maximized, the external field is very low for most materials and, therefore, the torque is small. However, for materials of very high coercive force, such as roasted ilmenite, this is an advantageous situation for magnetic separation by rotation.

In a magnetic field that exceeds the coercive force H_c of the material, the movement of domain walls within the solid substitutes for the movement of particles. The overall magnetization of a ferromagnetic particle is able to change direction in response to the direction of the external field. However, while the changes of the direction of the domain magnetization may occur freely between directions of easy magnetization within a crystal, they are restrained for intermediate angles by magnetocrystalline anisotropy. In this case the maximum torque angle is approximately one quarter of the angle between directions of the external field and the particle magnetization [A20, A17].

3.7.3 Rotation of particles in a liquid

It has been assumed, so far, that the particles to be rotated are suspended in air. If the particles are suspended in water, they are exposed, in addition to the force of gravity and the magnetic traction force, to hydrodynamic forces. These hydrodynamic forces are associated with a torque acting in the opposite direction [L9]. It was shown by Lantto [L9] that the hydrodynamic torque at a high Reynolds number is approximately given by

$$T_{hydro} = \frac{1}{2}\rho_f \omega^2 c^4 b^5 \tag{3.137}$$

where ω is the frequency of the rotating magnetic field, b the radius of the cylindrical particle, the length of which is cb (where $c >> 1$).

The magnetic torque is

$$T_{mag} = 2\pi c b^3 M_p B \sin\theta \tag{3.138}$$

Therefore, a cylindrical rod-like body will be brought into rotation when $T_{mag} > T_{hydro}$. This condition is met when

$$B > \frac{\rho_f \omega^2 c^3 b^2}{4\pi M_p \sin\theta} \tag{3.139}$$

It can be seen that rotation is more easily accomplished for small particles with large values of magnetization and large anisotropy angle θ. On the other hand, an increase in frequency of the rotating field hampers the ability of particles to rotate. It has to be stressed, however, that if the value of the magnetic induction given by eq. (3.139) exceeds that corresponding to the coercive force ($B > \mu_0 H_c$) for the particle material, the particles will either cease to rotate and may become demagnetized. If the particles are still rotating, the angle δ will be reduced.

In laminar flow, with small Reynolds numbers, the viscosity of the liquid will play an important role; otherwise the general character of the phenomenon is the same as that at high Reynolds numbers [L9].

3.8 Separation in magnetic fluids

The sink-and-float technique of gravity separation relies on selective levitation and sinking of materials based on their relative densities and that of the separating medium. The scope of the sink-and-float technique is restricted by a limited range of densities of the separating medium, the upper limit of ferrosilicon medium being approximately 3500 kg/m^3 and that of Clerici solution about 4200 kg/m^3 at room temperature.

In the mid sixties of the last century the application of magnetic fluids as a heavy medium, had been investigated, and it was established that by exposing a magnetic fluid to a non-homogeneous external magnetic field the fluid exhibited an apparent density exceeding densities obtainable with conventional heavy liquids. Two broad classes of magnetic fluids were investigated at that time, namely paramagnetic liquids [A1, A12, A19] and ferrofluids [R1, R7, R8, R16, G4, K16, F8].

Paramagnetic liquids are solutions of paramagnetic salts, such as $MnCl_2$ and $Mn(NO_3)_2$. These liquids are paramagnetic in their behaviour: their magnetization increases linearly with an increasing magnetic field and their susceptibility is low, of the order of 6×10^{-7} m^3/kg. On the other hand, a ferrofluid is a stable colloidal suspension of sub-domain ferromagnetic *(s.l.)* particles, for example magnetite, in a liquid carrier. The behaviour of ferrofluids is superparamagnetic, characterized by the absence of hysteresis and coercivity. Ferrofluids differ from paramagnets in having quite large magnetic susceptibilities and being able to become saturated in moderate magnetic fields. Saturation polarization of a ferrofluid can be as high as 0.1 T (1000 G)

The concept of separation of non-magnetic particles suspended in a magnetic fluid is based on the generalized Archimedes law whereby, in addition to the conventional force of gravity acting on the fluid, also a magnetically induced force acts on the fluid. This additional magnetic pull creates a magnetically induced buoyancy force on a particles immersed in the fluid. This buoyancy force can be controlled in a wide range of values and materials as dense as 20 000 kg/m^3 or higher can float in such a fluid.

A schematic diagram of the process of separation in magnetic fluids is shown in Fig. 3.52, while a general diagram of a magnetic circuit used in separation in magnetic fluids in shown in Fig. 3.53. Permanent magnets or electromagnets are used to generate a non-homogeneous magnetic field in the separation gap. The desired pattern of the magnetic field and its gradient is achieved by shaping the pole tips. A separation chamber placed between the pole-pieces of the magnetic circuit is filled with a magnetic fluid.

3.8.1 Apparent density of a magnetic fluid

There are two dominant forces acting on a volume of the magnetic fluid, placed in an external non-homogeneous magnetic field, namely the force of gravity and the magnetic traction force. The situation is schematically depicted in Fig. 3.52.

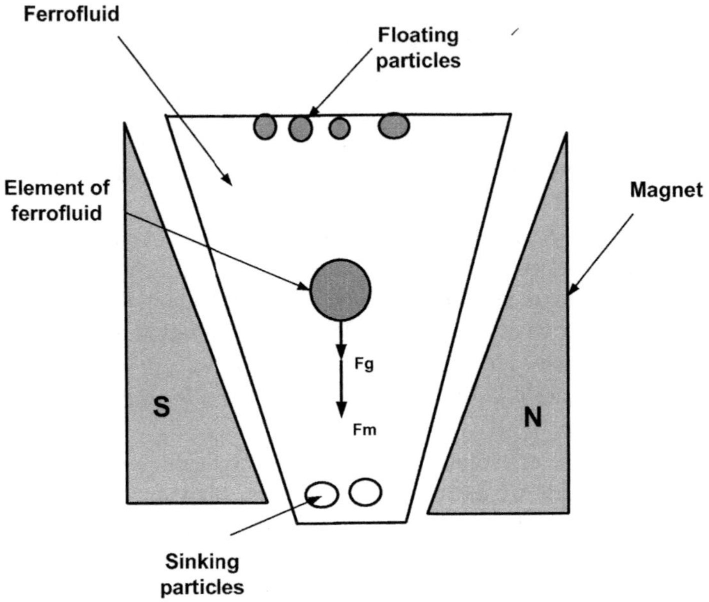

Figure 3.52: Schematic diagram of the process of separation in magnetic fluids.

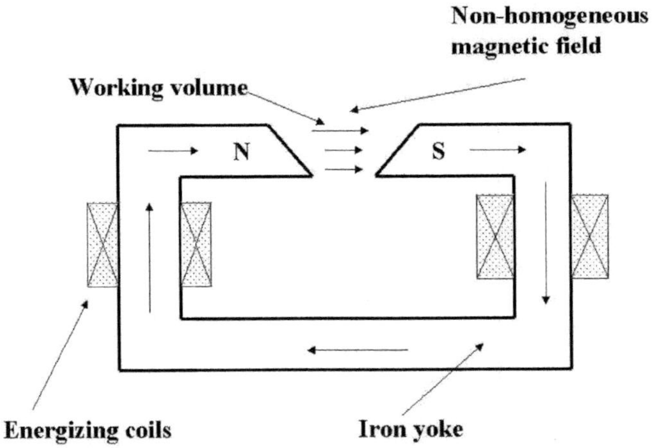

Figure 3.53: Schematic diagram of a magnetic circuit used in separators with magnetic fluids.

The total force on the magnetic fluid of volume V_f can be written as:

$$\vec{F}_f = \vec{F}_g + \vec{F}_m = \rho_f V_f \vec{g} + \frac{1}{\mu_0} \kappa_f V_f B \nabla B \tag{3.140}$$

where ρ_f and κ_f are the physical density and volume magnetic susceptibility, respectively, of the magnetic fluid.

For a ferromagnetic (*s.l.*) ferrofluid, the magnetic force can be expressed as

$$\vec{F}_m = \frac{1}{\mu_0} J_f V_f \nabla B \tag{3.141}$$

where J_f is the magnetic polarization of the ferrofluid.

When the field gradient is parallel with the gravitational force and of the same sense, eqs. (3.140) and (3.141) can be rearranged, as

$$\vec{F}_f = V_f \vec{g}(\rho_f + \frac{\kappa_f}{\mu_0 g} B \nabla B) \tag{3.142}$$

and

$$\vec{F}_f = V_f \vec{g}(\rho_f + \frac{J_f}{\mu_0 g} \nabla B) \tag{3.143}$$

The expressions in parentheses can be viewed as an apparent density ρ_a of a magnetic medium exposed to an external non-homogeneous magnetic field:

$$\rho_a = \rho_f + \frac{\kappa_f}{\mu_0 g} B \nabla B \tag{3.144}$$

and

$$\rho_a = \rho_f + \frac{J_f}{\mu_0 g} \nabla B \tag{3.145}$$

In a more general case, when the field gradient makes an angle α with gravity, eq. (3.145) becomes

$$\rho_a = \rho_f + \frac{J_f}{\mu_0 g} \nabla B \cos \alpha \tag{3.146}$$

Equations (3.143) and (3.145) form the basis of ferrohydrostatic separation (FHS).

In a typical case, a water-based ferrofluid with polarization of 0.02 T (200 G), placed in a magnetic field of gradient of 2 T/m (200 G/cm) will exhibit an apparent density of approximately 4200 kg/m^3. Figure 3.54 illustrates the apparent density as a function of the field gradient for two types of kerosene-based ferrofluid.

3.8.2 A particle suspended in a magnetic fluid

A particle suspended in a magnetic fluid is acted upon by several forces illustrated in Fig. 3.55. The force of gravity is given by

$$\vec{F}_{pg} = \rho_p V_p \vec{g} \tag{3.147}$$

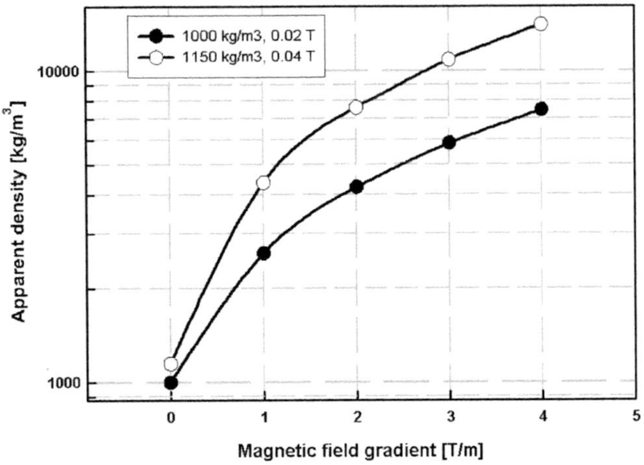

Figure 3.54: The apparent density as a function of the gradient of the magnetic field for two types of kerosene-based ferrofluid, of different densities and magnetic polarizations.

while the magnetic traction force is

$$\vec{F}_{pm} = \frac{1}{\mu_0}\kappa_p V_p B \nabla B \tag{3.148}$$

where κ_p is the volume magnetic susceptibility of the particle.

Such a particle experiences a loss of its weight as a result of two buoyancy forces acting on the particle. The first buoyancy force is the classical Archimedes gravity-related force:

$$\vec{F}_{pgb} = \rho_f V_p \vec{g} \tag{3.149}$$

The other force is the magnetically induced buoyancy force due to the magnetic traction force acting on the ferrofluid:

$$\vec{F}_{pmb} = \frac{1}{\mu_0} V_p J_f \nabla B \tag{3.150}$$

The net vertical force on a particle suspended in a ferrofluid and acted upon by a non-homogeneous magnetic field with vertical gradient can thus be written:

$$\vec{F}_p = \vec{F}_{pg} + \vec{F}_{pm} + \vec{F}_{pgb} + \vec{F}_{pmb} \tag{3.151}$$

or, in an explicit form:

$$\vec{F}_p = V_p(\rho_p - \rho_f)\vec{g} + \frac{\nabla B}{\mu_0} V_p(\kappa_p B - J_f) \tag{3.152}$$

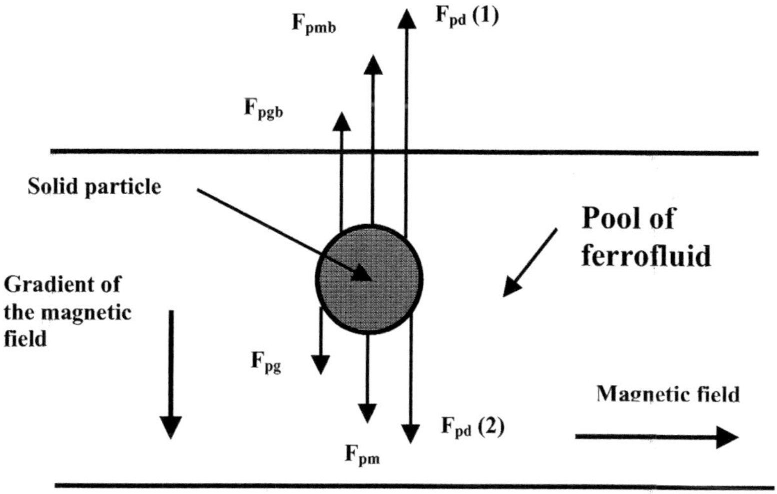

Figure 3.55: Forces acting on a particle in a stationary ferrofluid, placed in a non-homogeneous magnetic field.

Defining the effective cut-point density ρ_{cp} for separation as that particle density ρ_p for which $F_p = 0$ (i.e. equilibrium of the forces acting on the particle), eq. (3.152) yields:

$$\rho_{cp} = \rho_f + \frac{\nabla B}{g}(J_f - \kappa_p B)\cos\alpha \qquad (3.153)$$

For non-magnetic particles $\kappa_p = 0$ and eq. (3.153) reduces to

$$\rho_{cp} = \rho_f + \frac{J_f}{\mu_0 g}\nabla B\cos\alpha \qquad (3.154)$$

where α is the angle between the vectors of the gravitational force and the field gradient. When the force of gravity and the gradient of the magnetic field are parallel, eq. (3.154) becomes

$$\rho_{cp} = \rho_f + \frac{J_f}{\mu_0 g}\nabla B \qquad (3.155)$$

which is equivalent to eq. (3.145).

It can be seen that for non-magnetic particles, the density cut-point is equal to the apparent density of the ferrofluid. Particles whose densities ρ_p are smaller than the apparent density of the ferrofluid ($\rho_p < \rho_a$) will float in ferrofluid, while particles with density greater than the apparent density of the ferrofluid ($\rho_p > \rho_a$) will sink. Trajectories of sink and float particles in a stationary ferrofluid placed in a non-homogeneous magnetic field are shown in Fig. 3.56.

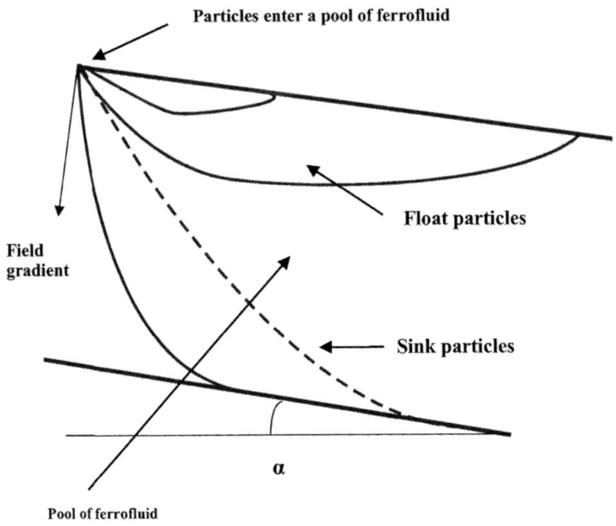

Figure 3.56: Trajectories of float and sink particles in a stationary ferrofluid placed in a non-homogeneous magnetic field.

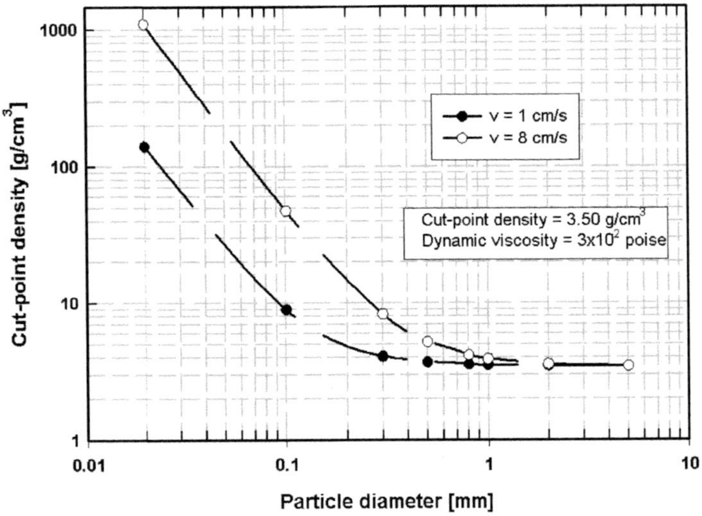

Figure 3.57: The effect of hydrodynamic drag on cut-point density, for two particle velocities. The nominal "large-particle" (eq. 3.155) cut-point density $\rho_{cp} = 3500$ kg/m^3.

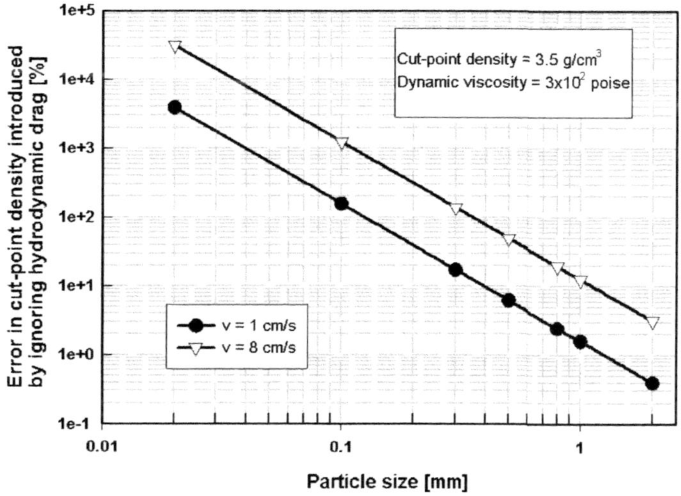

Figure 3.58: The inaccuracy in cut-point density introduced by ignoring hydrodynamic drag, for two different particle velocities. The nominal cut-point density $\rho_{cp} = 3500$ kg/m^3.

3.8.3 The effect of the hydrodynamic drag

In many applications of ferrohydrostatic separation it is legitimate to ignore the effect of the hydrodynamic drag. For small particles, however, the influence of the drag on the cut-point density can be significant. If we assume that the Stokes law applies, the hydrodynamic drag is given by eq. (1.16). By including this drag in the expression of the total force (eq. (3.151)), the equilibrium of forces on a non-magnetic particle yields

$$\rho_{cp} = \rho_f + \frac{J_f}{\mu_0 g}\nabla B + \frac{9\eta v}{2gb^2} \tag{3.156}$$

Equation (3.156) applies to a particle that is moving vertically downwards, i.e. the hydrodynamic drag $F_{pd}(1)$ acts on the particle. Such a particle is either a sink particle, i.e. a particle whose density is greater than ρ_a, or a float particle that is temporarily moving downwards because of the non-zero initial velocity with which it entered the fluid, as shown in Fig. 3.56.

Once a float particle reaches its terminal velocity (equal to zero), it will reverse the direction of its motion and will start moving upwards. At that point the opposite hydrodynamic drag $F_{pd}(2)$ will be acting on the particle and the particle will experience an apparent density given by

$$\rho_a = \rho_f + \frac{J_f}{\mu_0 g}\nabla B - \frac{9\eta v}{2gb^2} \tag{3.157}$$

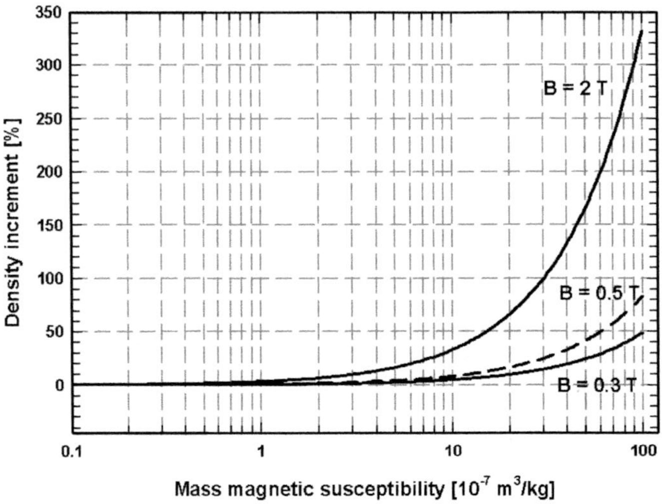

Figure 3.59: Density increment, for paramagnetic particles, as a function of magnetic susceptibility, at three different values of the applied magnetic field. $\nabla B = 2$ T/m.

Therefore, it becomes clear that, in general terms, the presence of the hydrodynamic drag increases the cut-point density for sinking particles, i.e. sink particles or float particles in their initial phase of motion within the ferrofluid. For particles reporting to the float fraction, however, the hydrodynamic drag decreases the cut-point density.

It can be found from eq. (3.156) that for particles greater than, for example, 1 mm the hydrodynamic drag can be neglected and the cut-point density is independent of particle size, and is given by eq. (3.155). However, by inclusion of the hydrodynamic drag, the cut-point density begins to deviate, for particles smaller than 1 mm, from its "large particle value", as is illustrated in Fig. 3.57. Figure 3.58 shows the inaccuracy in the cut-point density introduced by ignoring the hydrodynamic drag.

3.8.4 The effect of particle magnetic susceptibility

A magnetic particle ($\kappa_p \neq 0$) will exhibit an effective density ρ_{eff} different from its natural (true) density ρ_p, as a result of its attraction (or repulsion) in the direction of the field gradient. This effective density can be written:

$$\rho_{eff} = \rho_p + \frac{\kappa_p B}{\mu_0 g} \nabla B \cos \alpha \tag{3.158}$$

It can be seen that the difference between the effective and true densities $\Delta\rho_p = \rho_{eff} - \rho_p$ increases, for paramagnetic particles, with increasing magnetic susceptibility, magnetic field strength and the field gradient. A relative density increment as a result of non-zero magnetic susceptibility of a particle, follows from eq. (3.158)

$$\frac{\rho_{eff} - \rho_p}{\rho_p} = \frac{\chi_p B}{\mu_0 g} \nabla B \tag{3.159}$$

where χ_p is the mass magnetic susceptibility of the particle. A typical dependence of the density increment as a function of magnetic susceptibility of paramagnetic particles is shown in Fig. 3.59. It is evident that the density increment increases with increasing magnetic susceptibility of the particle and the field strength. This is a very important implication that affects the accuracy of separation of materials of non-zero magnetic susceptibility.

On the other hand, the effective density of diamagnetic particles decreases, as a result of their negative magnetic susceptibility. In view of the low values of diamagnetic susceptibility, the density decrease is very small; at the magnetic induction of 2 T, the decrease is approximately 0.2%. For example, the effective density of quartz (χ = - 6.0$\times 10^{-9}$ m^3/kg) is reduced, at 2 T, from its physical density of $\rho = 2700$ kg/m^3 to approximately 2695 kg/m^3.

The validity of eq. (3.159) has been tested using magnetic tracers, particles of constant density and of different values of magnetic susceptibility [R31]. Magnetic tracers, 5 mm in size, and of density 3550 kg/m^3, with mass magnetic susceptibility ranging from 2.5$\times 10^{-7}$ m^3/kg (20$\times 10^{-6}$ cm^3/g) to 1.25$\times 10^{-5}$ m^3/kg (1000$\times 10^{-6}$ cm^3/g), were separated in a ferrofluid the apparent density of which was varied from 3600 kg/m^3 to 4200 kg/m^3. The magnetic induction to which the ferrofluid was exposed in the separation region was 0.2 T and the field gradient was 1.5 T/m. Partition curves thus obtained are shown in Fig. 3.60.

It can be seen that 50% of tracers with density of 3550 kg/m^3 sank in a ferrofluid of apparent density of 3700 kg/m^3 only when their susceptibility was greater than 1.25$\times 10^{-6}$ m^3/kg. When the density of the ferrofluid was set to 4200 kg/m^3, 50% of magnetic tracers of density 3550 kg/m^3 sank when their magnetic susceptibility was greater than 8.2$\times 10^{-6}$ m^3/kg.

This experimental observation can be compared with the theoretical prediction expressed by eq. (3.159). Using the above values of particle susceptibility, and taking into account that $B = 0.2$ T and $\nabla B = 1.5$ T/m, the right-hand side of eq. (3.159) is equal to 0.19. This value agrees well with the relative density increment (4200 - 3550)/3550, obtained from the left-hand side of eq. (3.159).

3.8.5 Motion of particles in a ferrofluid

The equation of motion of non-magnetic particles

The equations of motion of a non-magnetic particle suspended in a stationary ferrofluid, as depicted in Fig. 3.61, can be expressed [M18] as:

$$\frac{d^2x}{dt^2} = \frac{dv_x}{dt} = \frac{1}{\rho_p V_p}(F_{pb}^x + F_{pm}^x + F_{pd}^x) \tag{3.160}$$

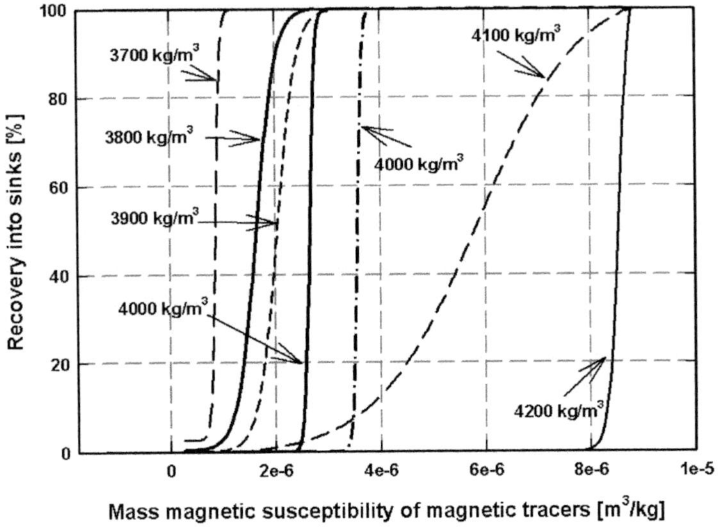

Figure 3.60: The partition curves for magnetic tracers of density 3550 kg/m^3, as a function of their mass magnetic susceptibility, separated in a ferrofluid of different apparent densities [R31].

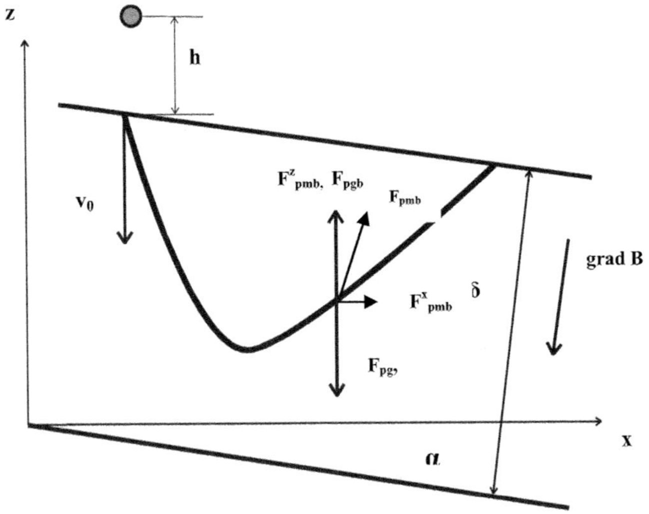

Figure 3.61: Section through a pool of ferrofluid of depth δ, inclined at an angle α.

$$\frac{d^2z}{dt^2} = \frac{dv_z}{dt} = \frac{1}{\rho_p V_p}(F^z_{pb} + F^z_{pd} + F^z_{pm} - F_g) \tag{3.161}$$

These equations can be re-written

$$\frac{dv_x}{dt} + Kv_x = W \tag{3.162}$$

$$\frac{dv_z}{dt} + Kv_z = W' \tag{3.163}$$

where parameters K, W and W' have the following forms:

$$K = \frac{6\pi\eta b}{\rho_p V_p}, W = g\frac{\rho_a - \rho_f}{\rho_p}\sin\alpha, W' = g\frac{\rho_a\cos\alpha - \rho_p + 2\rho_f\sin^2(\alpha/2)}{\rho_p} \tag{3.164}$$

The initial conditions for eqs. (3.162) and (3.163) are:

$$\text{for } t = 0 \;\; v_x = 0 \;\text{ and }\; v_z = v_0 \tag{3.165}$$

The initial velocity of a particle that is being dropped into the ferrofluid from height h is $v_0 = -\sqrt{2gh}$.

Differential equations (3.162) and (3.163) can be solved analytically and the solutions are [M18]:

$$\frac{dx}{dt} = v_x = \frac{W}{K}(1 - e^{-Kt}) \tag{3.166}$$

$$\frac{dz}{dt} = v_z = v_0 e^{-Kt} + \frac{W"}{K}(1 - e^{-Kt}) \tag{3.167}$$

The initial conditions for eqs. (3.166) and (3.167) are

$$\text{for } \; t = 0 \;\; x = 0 \text{ and } z = \delta \tag{3.168}$$

Equations (3.166) and (3.167), with these initial conditions, yield the following solutions:

$$x = \frac{W}{K}t - \frac{W}{K^2}(1 - e^{-Kt}) \tag{3.169}$$

$$z = \delta + (\frac{v_0}{K} - \frac{W"}{K^2})(1 - e^{-Kt}) + \frac{W'}{K}t \tag{3.170}$$

Particle trajectories can be obtained from these parametric equations by calculating pairs of coordinates (x, z). Typical trajectories of non-magnetic particles of different densities are shown in Fig. 3.62.

It can be seen that when $h = 0$ then $v_0 = 0$ and the trajectory of a particle is a straight line.

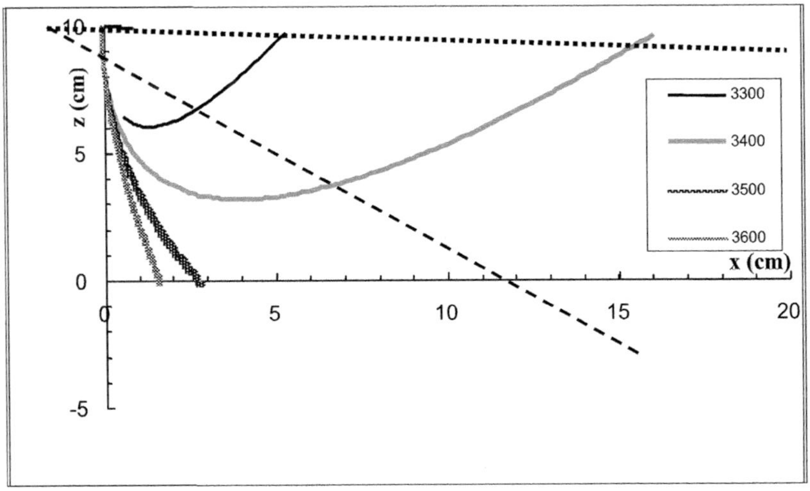

Figure 3.62: Trajectories of non-magnetic particles of various densities: $\rho_a = 3500\ \mathrm{kg/m^3}$, $h = 3$ mm, $\rho_f = 990\ \mathrm{kg/m^3}$, $\eta = 1.65\times10^{-3}$ Pa×s, $\alpha = 2^0$. The straight dashed lines are trajectories of particles of densities of 3500 kg/m^3 and 3550 kg/m^3 for $h = 0$ (adapted from [M18]).

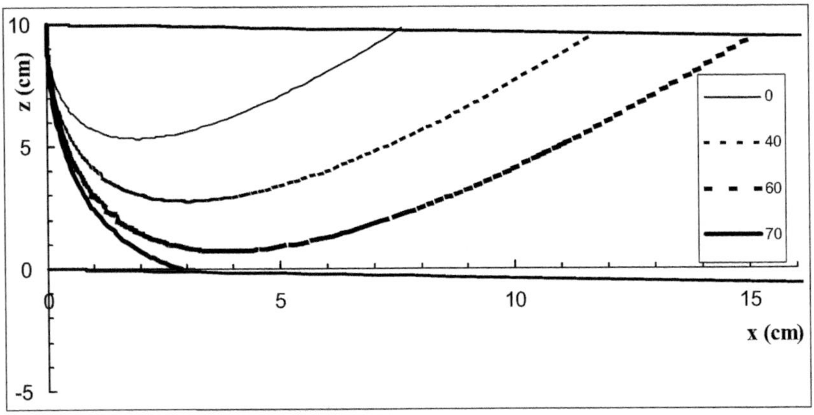

Figure 3.63: Trajectories of magnetic particles of various mass magnetic susceptibilities ($\times10^{-8}\ \mathrm{m^3/kg}$): $\rho_a = 3500\ \mathrm{kg/m^3}$, $\rho_p = 3300\ \mathrm{kg/m^3}$ (adapted from [M18]).

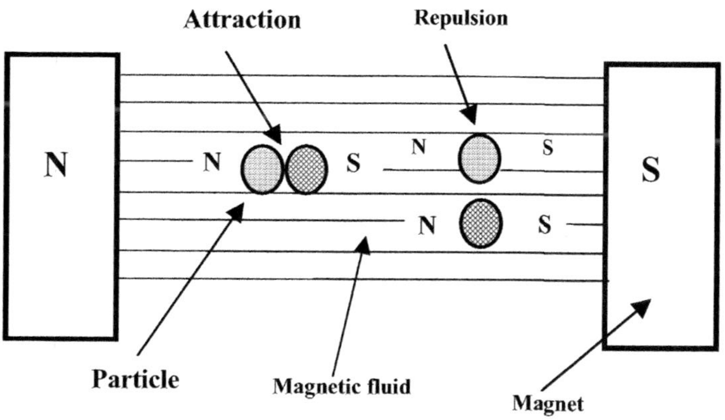

Figure 3.64: Interaction between non-magnetic particles in a magnetized magnetic fluid (adapted from Fujita et al. [F14]).

The equation of motion of a magnetic particle

As has been discussed in Section 3.8.4, a magnetic particle suspended in a ferrofluid placed in a non-homogeneous magnetic field will exhibit an effective density ρ_{eff} (given by eq. (3.158)) different from its natural density. Trajectories of magnetic particles can be obtained by replacing the natural density ρ_p of the particles by effective density. As a result of the non-uniformity of the magnetic field, the effective density is different at different positions of the particle along the vertical axis, being the greatest where the pole tips of the magnet are closest to one another, as shown in Fig. 3.52. The mean effective density can be calculated, for example, by replacing B in eq. (3.158) with the mean magnetic induction $B_{\text{mean}} = B_{\text{min}} + B_{\text{max}}$. In this situation the equations of motion remain the same as parametric equations (3.169) and (3.170). Typical trajectories of magnetic particles are shown in Fig. 3.63.

Interaction of particles during separation

The analysis of the particle motion in a ferrofluid assumed that particles were non-interacting. This simplification ignores, however, the fact that non-magnetic particles suspended in a magnetic fluid possess surface magnetic charges as a result of magnetization of the surrounding fluid. Fujita and Mamiya [F14] determined, theoretically and experimentally, that non-magnetic particles, placed in a magnetized ferrofluid, attracted one another in the direction parallel to the direction of magnetic field and repelled one another in the perpendicular direction. The situation is depicted in Fig. 3.64. The attractive force increases with increasing density of the ferrofluid and magnetic field strength. The attractive interaction was found to reduce the selectivity of separation of non-magnetic particles in a magnetic fluid.

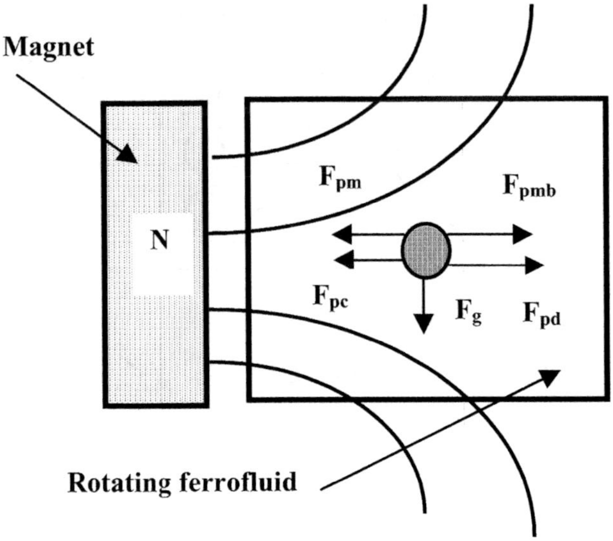

Figure 3.65: Forces acting on a particle in a rotating ferrofluid in a non-homogeneous external magnetic field.

3.8.6 Separation in a rotating ferrofluid

In order to overcome certain practical limitations of gravity-based ferrohydrostatic separation, particularly treatment of fine particles, several proposals were made [R11, A14, W7] to replace the force of gravity by the centrifugal force. Work carried out at Intermagnetic Corp.(USA) in early nineteen eighties resulted in a new rotation-based separation process called Magstream.

The theory and the practice of separation in a rotating ferrofluid have been described in several publications [W20, W21, W22, B22, B23, S44]. Forces acting on a particle suspended in a rotating ferrofluid exposed to a non-homogeneous magnetic field are illustrated in Fig. 3.65. Walker and Devernoe [W22] assumed that the axis of the separator was vertical and ignored the force of gravity and the hydrodynamic drag acting on the particle. In this scenario only the radially outward centrifugal force F_{pc}, magnetic traction force F_{pm} and the inward buoyancy force F_{pmb} resulting from the outward attraction of the magnetic fluid by the magnetic field were included. The centrifugal force can be written

$$F_{pc} = V_p(\rho_p - \rho_f)\omega^2 r \tag{3.171}$$

where r is the radial position of the particle and ω is the rotational velocity of the ferrofluid. The magnetic buoyancy force is given by eq. (3.150) while the magnetic traction force is given by eq. (3.148).

The total force F_{pr} acting on a particle in the horizontal direction, along the radius of the rotating device, can be expressed as:

$$F_{pr} = V_p(\rho_p - \rho_f)\omega^2 r + \frac{1}{\mu_0}\kappa_p V_p B \nabla B - \frac{1}{\mu_0} V_p J_f \nabla B \tag{3.172}$$

In equilibrium $F_{pr} = 0$ and the cut-point density for non-magnetic particles ($\kappa_p = 0$) can be determined from the above equation as:

$$\rho_{cp} = \rho_f + \left(\frac{K}{N}\right)^2 J_f \nabla B \tag{3.173}$$

where N is the number of revolutions of the rotating device per minute ($N = 60\omega/2\pi$), and K is a parameter determined by the design of the magnetic system of the rotating separator:

$$K = 15\left(\frac{1}{\pi^3 r}10^7\right)^{1/2} \tag{3.174}$$

Equation (3.173) is used to calculate the operating conditions in the Magstream separator [W22]. It is apparent from eqs. (3.173) and (3.174) that if the pattern of the magnetic field is designed in such a way that the field gradient is a linear function of the radial position r, then K = const. and the cut-point density is determined solely by the polarization and the rotational velocity of the ferrofluid. This condition is satisfied, for instance, for the magnetic field generated by a sextupole magnet, such that the field magnitude is given by $B = B_0(r/r_0)^{n-1}$, where $2n = 6$, and B_0 is the magnetic induction at r_0, the outside wall of the separation annulus [W20].

Equation (3.173) is equivalent to eq. (3.154) derived for ferrohydrostatic separation. It can, therefore, be seen that the cut-point density in a rotating ferrofluid can be controlled, in a wide range, by varying the rotational velocity and the field gradient. The operating density range of the fluid in a ferrohydrostatic separator is, in contrast, controlled by adjusting the field gradient only. However, in most Magstream models the magnetic field is generated by permanent magnets and the field gradient is given by the design of the magnetic system. For instance, in Magstream Model 200 the field gradient is equal to $\nabla B = 13.2r$ T/m ($1320r$ G/cm).

Simplifying assumptions introduced by Walker and Devernoe [W22], when deriving eq. (3.173), reflect themselves in limited accuracy of determination of the cut-point density and thus in reduced efficiency of separation. Bunge and Fuerstenau [B22], in order to account for the dispersion component of particle behaviour, developed a diffusion model of separation. They concluded that the primary cause of non-ideal separation in a rotating ferrofluid is the dispersion of the settling particles. This fuzzy behaviour was ascribed to the effect of turbulence, particle shape and interparticle interactions. The model of Bunge and Fuerstenau, however, was not able to explain the dependence of the separation efficiency on particle size and fluid viscosity [B22, B23].

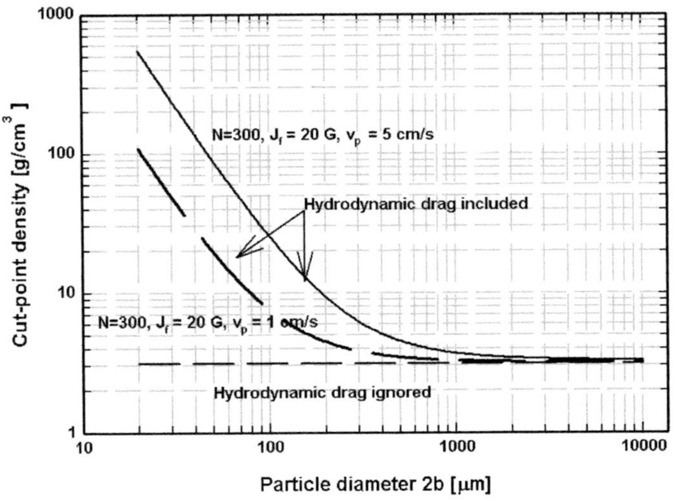

Figure 3.66: The cut-point density of non-magnetic particles in Magstream, as a function of density. $r = 2.1$ cm, $\eta = 5{\times}10^{-3}$ Pa×s, ρ_{cp}(no drag) = 3130 kg/m^3 (adapted from [S44]).

Svoboda [S44] included the horizontal hydrodynamic drag (eq. (1.16)) into the balance of forces and obtained, for the cut-point density,

$$\rho_{cp} = \rho_f + \frac{1}{\omega^2 r}\left(\frac{J_f}{\mu_0} - \frac{\kappa_p}{\mu_0}B\right)\nabla B + \frac{9\eta v_p}{2\omega^2 b^2 r} \tag{3.175}$$

This more general expression shows that the hydrodynamic interaction of particles with the ferrofluid contributes to the determination of the cut-point density. The most significant consequence of the hydrodynamic contribution is that the cut-point density depends on particle size, in contrast to eq. (3.173). It transpires from eq. (3.175) that the correction hydrodynamic term gains importance for weakly magnetic and/or highly viscous ferrofluids (low J_f and high η), and for small particles. In addition, the hydrodynamic term depends on the position of a particle within the separation space.

The effect of particle size on the cut-point density of non-magnetic particles is illustrated in Fig. 3.66. It can be seen, for example, that a cut-point density of 3130 kg/m^3, determined by ignoring the hydrodynamic drag, is accurate down to particle size of 2 mm. For particle diameters smaller than 2 mm the relative importance of the hydrodynamic interaction steeply increases and separation becomes inaccurate.

3.9 Eddy-current separation

It has been stated in Section 2.7 that when conducting particles move through a region of spatially varying magnetic field, then an electromagnetic force, and hence eddy currents, are induced in such particles. The eddy currents are caused by time-dependent changes of the magnetic induction acting on a particle and are given by the Faraday induction law:

$$\vec{\nabla} \times \vec{j} = -\sigma_p \frac{\partial \vec{B}}{\partial t} \tag{3.176}$$

where j is the current density and σ_p is the specific electrical conductivity of the particle.

These eddy currents generate a magnetic field opposed to the external magnetic field. This opposite field then produces a repulsive Lorentz force on a linear element ds of the conductor [S47]

$$\vec{F}_L = I d\vec{s} \times \vec{B} \tag{3.177}$$

where I is the current carried by the conductor.

This force can be made lateral to the motion of the particles and so to deflect them. Negligible eddy currents develop in non-conducting elements, which, therefore, experience no deflection. In this way it is possible to separate conducting particles from non-conducting and this method has been developed into an efficient technique for the recovery of non-ferrous metals from streams of industrial or municipal waste.

The varying magnetic fields can be generated in several ways. In linear motors, the magnetic fields are generated by electromagnets with alternating currents. The repulsive force can also be generated by moving particles through a non-homogeneous static or rotating magnetic fields. These fields can be conveniently generated by permanent magnets.

3.9.1 Static ramp eddy-current separator.

The simplest design of an eddy-current separator (ECS) proposed by Schloemann [S45] comprises a long ramp, as shown in Fig. 3.67. The ramp consists of permanent magnet strips of alternating polarity, mounted on a mild steel base plate inclined at an angle of about 45^0 to the ramp axis. A stream of particles is allowed to slide down the ramp. The non-conducting particles are not affected by the magnetic field and travel straight down the ramp. The conducting particles, under the influence of the eddy-current-induced repulsive Lorentz force perpendicular to the magnet strips, are deflected from the stream of non-conducting particles, as shown in Fig. 3.67. The magnitude of the lateral deflection depends, among other things, on the electrical conductivity, density, size and shape of the particles and the strength and the periodicity of the magnetic field.

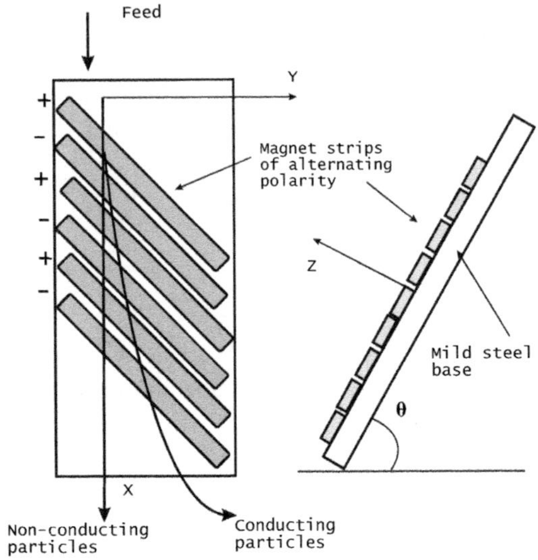

Figure 3.67: Schematic diagram of a ramp eddy-current separator (adapted from Schloemann [S45]).

Schloemann [S45] found that the component of the repulsive force in y-direction acting on a conducting particle is given by:

$$F_{Lp}^{y} = \alpha m_p v_x \tag{3.178}$$

where m_p and v_x are the mass and velocity in the x - direction of the particle, respectively. The coefficient α is a function of particle shape, and for a thin disc of diameter d_p, it is given by [S45]:

$$\alpha = \frac{1}{32} d_p^2 \frac{\sigma_p}{\rho_p} (\frac{\partial B_z}{\partial x})^2 \tag{3.179}$$

and approximate expression for the repulsive force (eq. (2.6)) is quoted by Schubert [S22]:

$$F_{Lp}^{y} \simeq m_p \frac{\sigma_p}{\rho_p} \frac{d_p}{2f} B^2 v_x \tag{3.180}$$

where f is the period of the magnetic field. This equation, supported by additional calculations of the repulsive force on particles of various shapes [S46, V3, V4] shows that the repulsion force is a function of the material parameter σ_p/ρ_p, given in Table 2.1, and of the square of the magnetic induction.

Schloemann [S45] calculated trajectories of disc-shaped particles in a ramp eddy-current separator from the following equations of motion:

$$\frac{dv_x}{dt} = g \sin\theta \text{ and } \frac{dv_y}{dt} = \alpha v_x \tag{3.181}$$

It follows from eqs. (3.181) that

$$x(t) = \frac{1}{2}t^2 g \sin\theta + v_0 t \tag{3.182}$$

$$y(t) = \alpha(\frac{1}{6}t^3 g \sin\theta + \frac{1}{2}v_0 t^2) \tag{3.183}$$

where $v_x(0) = v_0$.

The particle trajectory, in a frictionless situation, is then:

$$y(x) = \frac{\sqrt{2}}{3}\alpha\frac{x^{3/2}}{(g\sin\theta)^{1/2}}\left[\left(1-2\frac{x_0}{x}\right)\left(1+\frac{x_0}{x}\right)^{1/2}+2\left(\frac{x_0}{x}\right)^{3/2}\right] \tag{3.184}$$

where x_0 is the distance along the ramp after which a particle starting at rest has attained the velocity v_0. Therefore, $x_0 = v_0^2/2g\sin\theta$. A more detailed analysis of the effect of friction on particle motion can be found in [S45].

Although ramp eddy-current separators were used in several industrial installations [S46], their commercial success was limited. Poor selectivity, particularly for small particles, and low throughput were responsible for the failure of ramp separators to find a long-term production application.

Further development of the ramp concept resulted in the design of a vertical eddy-current separator [V4, V3], in which the deflection occurred in a vertical channel between two arrays of magnet strips mounted on the inside of two vertical parallel mild steel plates. The strips of alternating polarity were also oriented at an angle to the base plate and, in addition, the magnets on the opposite plate were also arranged with opposite polarity. These static gravity-based separators also failed to gain industrial acceptance and were replaced by dynamic rotating drum eddy-current separators.

A detailed study of the separating Lorentz force, by eliminating the competing forces such as gravity and frictional forces, was conducted theoretically by Fletcher et al. [F15, F16, F17]. These idealized physical models of separation of conducting and non-conducting elements moving in a horizontal plane and incident on a single boundary of a semi-infinite magnetic field were compared with the experimental results. Theoretical trends of the metal recovery with particle size, splitter position and boundary field profile were in good agreement with experimental observations. Although the field geometry and the separation arrangement in these studies differed from those used in practical applications, the methodology used can serve as a starting point for more fundamental research into the mechanisms of eddy-current separation.

3.9.2 Rotating drum eddy-current separators

Presently, the eddy-current separators most frequently used in industrial practice are based on a rotating drum concept. The active part of such a separator is a drum containing rapidly rotating magnetic system. Permanent magnets are arranged, with alternating polarity, along the circumference of the magnetic system. This concept is schematically described in Figs. 2.100 and 3.68.

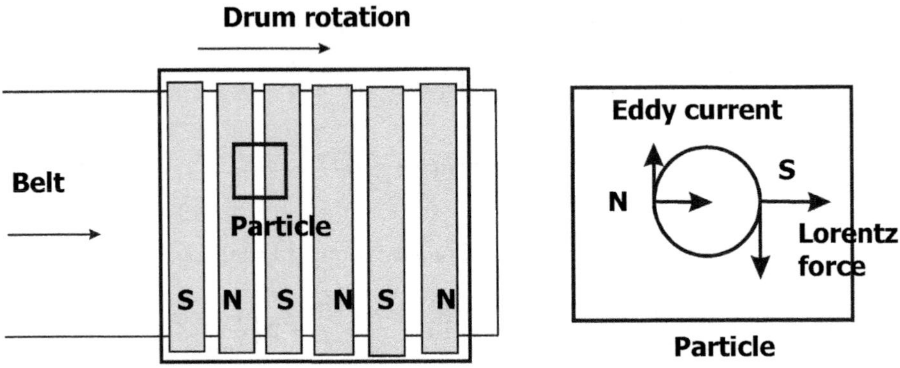

Figure 3.68: A top view of an eddy-current separator based on a rotating drum. The particles are expelled by the Lorentz force in the direction of the direction of drum rotation (adapted from [R17]).

In order to calculate the forces acting on particles due to eddy current, a dipole model was developed by Rem et al. [R17, R18]. It was assumed that the magnetic field induces eddy currents within a particle, which in turn generate a magnetic moment in that particle. The interaction between the magnetic moment μ_M and the external magnetic field B results in eddy current forces. The force can then be expressed as:

$$\vec{F}_{Lp} = \mu_M \nabla B \tag{3.185}$$

where $\mu_M = M_p V_p$. The torque on the particle is given by eq. (3.48).

In order to determine the force acting on a particle, given by eq. (3.185), a magnetic field generated by the magnetic system must be calculated. Rem et al. [R17] have shown that the components of the magnetic induction in the cylindrical coordinate system (r, ϕ, z) relative to the drum axis can be written:

$$B_r = \sum_{n=0}^{\infty} b_n (r/R_{drum})^{-(2n+1)k-1} \sin[(2n+1)k(\phi - \omega_{drum} t)] \tag{3.186}$$

$$B_\phi = \sum_{n=0}^{\infty} -b_n (r/R_{drum})^{-(2n+1)k-1} \cos[(2n+1)k(\phi - \omega_{drum} t)] \tag{3.187}$$

where b_n are the Fourier coefficients, R_{drum} the radius of the drum, k the number of pairs of the magnets used in the drum and ω_{drum} the angular velocity of the drum. The Fourier coefficients can be determined by measuring the magnetic field strength as a function of distance from the magnetic pole surface.

Based on eqs. (3.186) and (3.187), Zhang et al. [Z4] obtained the following expressions for the tangential F_{pt} and radial F_{pr} eddy-current forces acting on the particle:

$$F_{Lpt} = \frac{2\pi s V_p}{\mu_0 w} \frac{(k\omega_{drum} + \Omega)\tau}{1 + (k\omega_{drum} + \Omega)^2 \tau^2} B^2 \tag{3.188}$$

$$F_{Lpr} = \frac{2\pi s V_p}{\mu_0 w} \frac{(k\omega_{drum} + \Omega)^2 \tau^2}{1 + (k\omega_{drum} + \Omega)^2 \tau^2} B^2 \tag{3.189}$$

where w is the width of one pair of magnets, s is the shape factor of the particle and Ω is the angular velocity of particle rotation. τ is a characteristic time with which the induced magnetic field decays in the particle given by [Z4]:

$$\tau = \mu_0 \sigma_p s b^2 \tag{3.190}$$

where b is the particle radius.

The torque on the particle was found to be [Z4]:

$$T = \frac{s V_p}{\mu_0} \frac{(k\omega_{drum} + \Omega)\tau}{1 + (k\omega_{drum} + \Omega)^2 \tau^2} B^2 \tag{3.191}$$

It can be seen that the eddy-current forces and the torque depend on the square of the magnetic field strength. By replacing ceramic ferrite permanent magnets by rare-earth magnets, a considerable increase in the separating force can be, therefore, achieved.

Competing forces

Gravitational, centrifugal, frictional and aerodynamic forces compete with eddy-current forces and their relationship determines the particle trajectories in an eddy-current separator. Eddy current forces and a wide spectrum of competing forces and their various dependencies on particle size and shape, electrical conductivity, magnetic field strength and its frequency (or the angular velocity of the rotating drum) make the determination of particle trajectories and thus the efficiency of separation a problem of considerable complexity. Only rudimentary efforts have been made the develop the balance of forces in a rotating drum eddy-current separator [Z4, R17, R18].

Rem et al. [R18] developed a model for simulation of such a rotating drum ECS. The eddy-current forces and some of the competing forces were included in the model and particle trajectories were calculated. Typical trajectories, shown in Figs. 3.69 and 3.70, comprise the rolling, sliding and finally flying stages. It can be seen that at low belt velocities (Fig. 3.70) light particles tend to lose contact with the belt at an early stage and they collide with the belt at a later stage.

As a result of the dependence of the eddy-current forces and other forces on the cube of the particle size, the particle size is the most significant variable affecting the performance of a separator. It transpires from the above equations that also the particle shape plays an important role in the separation process. For instance, it was observed that flat particles exhibit much greater displacement by eddy-current forces than spherical particles. In spite of considerable complexity of the process, some general rules that govern the efficiency of separation in a drum ECS can be formulated. In order to achieve a high recovery of conducting particles, the eddy-current separating force must be greater than

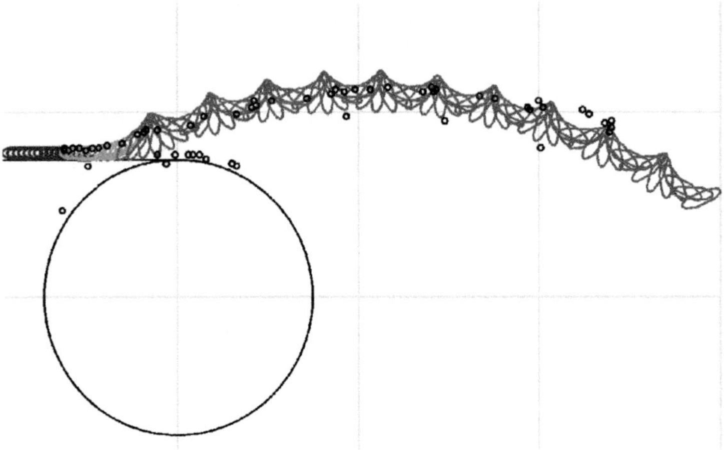

Figure 3.69: Particle trajectories in a rotating drum ECS. Three stages of particle motion, namely rolling, sliding and flying are apparent. Operating conditions: belt speed: 1.5 m/s, drum angular velocity: 35 s^{-1}, material: aluminium particles 15 mm in diameter, 45 mm long (courtesy of P.C. Rem).

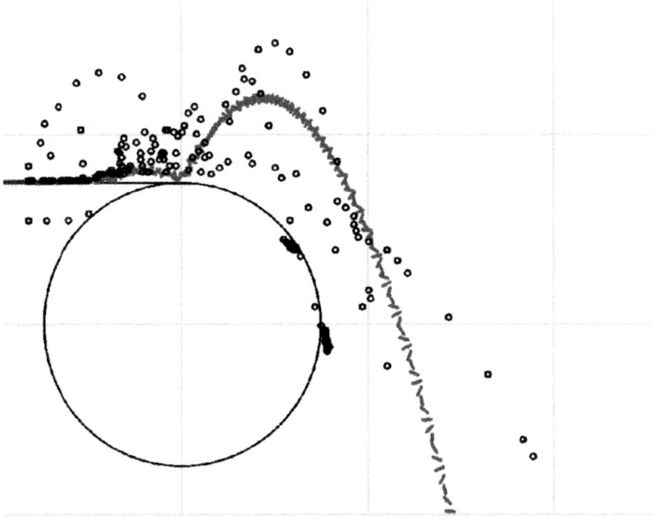

Figure 3.70: Particle trajectories in a rotating drum ECS, at low belt velocity. Belt speed: 0.5 m/s, drum angular velocity: 35 s^{-1}, material: aluminium particles of diameter of 10 mm, 2 mm thick (courtesy of P.C. Rem).

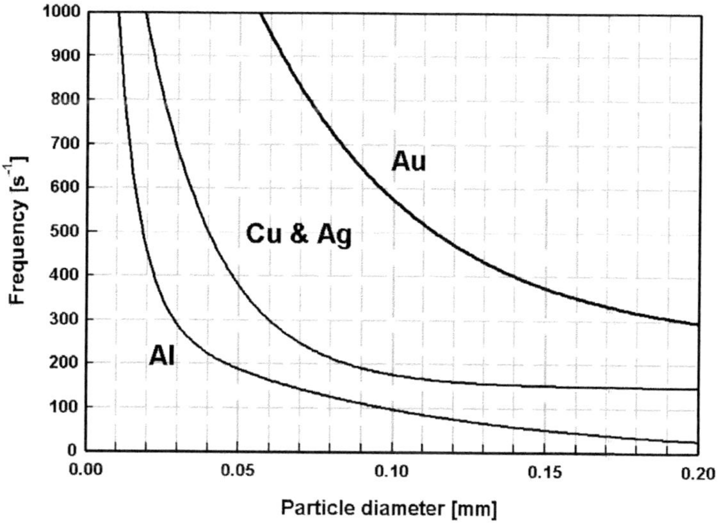

Figure 3.71: Frequencies required to rotate metal particles of various diameters (adapted from Allen [A21]).

the sum of the competing forces:

$$F_{separ} > F_{comp} \tag{3.192}$$

If, however, $F_{separ} >> F_{com}$, selectivity of separation will be low, as no distinction will be made between various classes of conducting particles. Selective separation of one metal from another is determined by the interplay between corresponding eddy-current and competing forces. Selective separation of metal (1) from metal (2) will be achieved when the following relationship is met:

$$F_{separ}^{(1)} > F_{comp} > F_{separ}^{(2)} \tag{3.193}$$

However, when $F_{separ} < F_{comp}$, no separation based on the characteristic property, such as electrical conductivity in ECS, will take place.

3.9.3 Eddy-current separation by particle rotation

With decreasing particle size, the torque on the particle become an important separating force. Also, when the particle size becomes much smaller than the width of the magnet strips used in the rotating drum eddy-current separators, the particles increasingly respond to the field changes by rotation rather than by repulsion [A21]. As a result of decrease in the particle throw and increase of particle rotation as particle size decreases, Allen [A21, A22] proposed to use particle rotation as the separating mechanism of conducting particles.

Allen found that the magnitude of the torque acting on a cubic conducting particle by a rotating field is

$$T_c = \frac{\pi B^2 L^5 w}{8\varrho_p} \tag{3.194}$$

where L is the length of the cube edge, w is the magnetic field rotation frequency and ϱ_p is the resistivity of the material. For a sphere of diameter d_p, the torque is given by

$$T_c = \frac{\pi B^2 d_p^5 w}{60\varrho_p} \tag{3.195}$$

If such a torque is exerted on a particle by a rotating magnetic field, and the particle is resting on a surface, the particle will be rolled when the torque becomes greater than the gravitational force holding the particle against the surface. The gravitational torque preventing a cubic-shaped particle from rolling can be written:

$$T_g = \frac{L^4}{2}\rho_p g \tag{3.196}$$

Combining eqs. (3.195) and (3.196) an estimate of the minimum size of a cubic particle that could be rolled at a given magnetic field strength is obtained [A21]:

$$L_{\min} = \frac{4\rho_p \varrho_p g}{\pi B^2 w} \tag{3.197}$$

Equation (3.197) gives a reasonable estimate of the minimum size of a "rollable" cubic particle when the rotating magnetic field has its rotation axis at the centre of the particle. In a more practical situation of a rotating drum eddy-current separator, the magnetic field not only rotates the particle, but it is also moving through the particle. This moving field produces a field which rotates in the opposite direction, at half the field rotation frequency [A21]. The minimum "rollable" cubic particle size, ignoring inter-particle interactions, can, therefore, be estimated as

$$L_{\min} = \frac{8\rho_p \varrho_p g}{\pi B^2 w} \tag{3.198}$$

The rotating frequencies of the magnetic field, required to rotate and thus to separate particles of various metals and various sizes in a magnetic field of 0.45 T, are shown in Fig. 3.71 [A21]. It can be seen, for example, that particles of gold, as small as 60 μm, could be separated by rotating them at the frequency of 1000 s^{-1}. The concept has been demonstrated by separating aluminium particles from their mixture with sand, with the recovery of 86% and the concentrate grade of 95 % Al, at the feed rate of 5 t/h/m [A17].

Chapter 4

Elements of Design of Magnetic Separation Equipment

4.1 Introduction

The design of a magnetic circuit for a magnetic separator is based on the need to optimize the magnetic force required to separate efficiently, and selectively, particles of different magnetic susceptibilities. As transpires from eq. (1.8), the magnetic force is the product of the magnetic field strength and of the field gradient. The desired magnetic field strength, the field gradient and their combination are a function of the type of material to be separated, mode of separation, required metallurgical performance and economic criteria. In this chapter the basic design criteria and techniques of the generation of the magnetic field and its gradient, as used in magnetic separation, will be reviewed.

Magnetic circuit design has been traditionally difficult because of the usual non-analytic nature of the solutions for fields emanating from arbitrary sources. Further, unlike currents in electric circuits, magnetic flux is not confined to the wires but passes through various components of the circuit, including the air. Over the years, several analytical techniques have been used to give approximate solutions to magnetic design problems. In the simplest cases, the analytical solution can be exact, but in general for any practical problem, analytical solutions lead to large errors. Most designs are optimized by measurement and extrapolation, and experience of working with a particular type of a magnetic circuit is important.

The advent of powerful computational techniques, such as finite element analysis, allows us to find accurate solutions for two- or three-dimensional field distributions in complex geometries, which in turn may be used to predict device performance with similar precision. However, these modelling techniques require a detailed definition of the geometry and boundary conditions to be solved, which assumes that an initial design already exists. While providing an accurate field solution for a given geometry, these techniques will not optimize it. Suggestions for changes in dimensions, materials etc. must come from the designer. Consequently, while computer-based field analysis is an effective tool for simulating a known device, it is too cumbersome for design optimization [C16].

Field equations

Calculation of the flux in a magnetic circuit is based on two Maxwell's equations. The first, in its differential form is:

$$\vec{\nabla} \times \vec{H} = \vec{j} \tag{4.1}$$

where j is the current density. After integrating over an arbitrary surface S we obtain the above equation in the integral form:

$$\int_S (\vec{\nabla} \times \vec{H}) \cdot d\vec{a} = \int_S \vec{j} \cdot d\vec{a} \tag{4.2}$$

Then, using Stokes's theorem, the left-hand side of this equation can be transformed into a line integral around a closed path C, which bounds the surface S, and

$$\oint_C \vec{H} \cdot d\vec{l} = \int_S \vec{j} \cdot d\vec{a} = I \tag{4.3}$$

where I is the total current. This is Ampere's circuital law: the line integral of H around a closed path is equal to the total current crossing any surface bounded by the line integral path. The left-hand side of eq. (4.3) is known as magnetomotive force or mmf.

The second fundamental relationship required for a magnetic circuit design expresses the fact that flux must be conserved through any closed volume, or, that the total flux flowing out of any volume must be zero. Therefore

$$\int_S \vec{B} \cdot d\vec{a} = 0 \quad \text{or} \quad \vec{\nabla} \cdot \vec{B} = 0 \tag{4.4}$$

Equations (4.3) and (4.4) are two basic equations used for magnetic circuit design.

4.2 Design of circuits with permanent magnets

4.2.1 Basic concepts and figures of merit

The general relationship between magnetic induction and intrinsic material magnetization or polarization is given by

$$\vec{B} = \mu_0(\vec{H} + \vec{M}) = \mu_0\vec{H} + \vec{J} \tag{4.5}$$

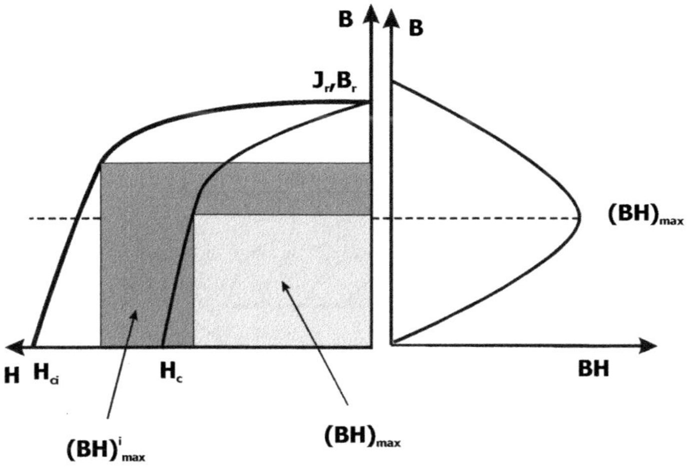

Figure 4.1: The maximum energy products of permanent magnets.

The relationship between B and H is illustrated by the hysteresis curve shown in Fig. 1.7. The second (demagnetization) quadrant of such a hysteresis curve is important for description of the quality of a permanent magnet material. Relative contributions of J and H greatly influence the magnetizing, demagnetizing and stability properties of permanent magnets. For instance, the coercive force H_c can never exceed the remanent magnetic induction B_r:

$$\mu_0 H_c \leq B_r \tag{4.6}$$

while the intrinsic coercive force H_{ci} is not under such a restraint.

There is clearly one point between B_r and H_c for which the product of B and H has a maximum value. The optimum operation of a magnet corresponds to conditions under which this BH product is maximum. This maximum value of the energy product is given by

$$(BH)_{\max}^{\lim} = \frac{J_s^2}{4\mu_o} = \frac{\mu_0 M_s^2}{4} \tag{4.7}$$

where J_s (or M_s) is the saturation polarization (or magnetization) of the magnet material. It transpires from this equation that, in order to establish the maximum capability of a permanent magnet, one has to know only the value of the saturation polarization. The maximum value of the energy product corresponds to the largest rectangle which can be drawn under the demagnetization curve corresponding to the magnetic induction, as shown in Fig. 4.1. Another important figure of merit is the intrinsic energy product $(BH)_{\max}^{i}$. This quantity is of importance when the magnet is subjected to large demagnetizing influences in its operating environment. The maximum intrinsic energy product corresponds to the largest rectangle that can be drawn under the demagnetization curve

Table 4.1: Relative permeabilities of permanent magnet materials.

Material	Permeability
Alnico 6	4.2
Ferrimag 5	1.06
NdFeB	1.08 - 1.1

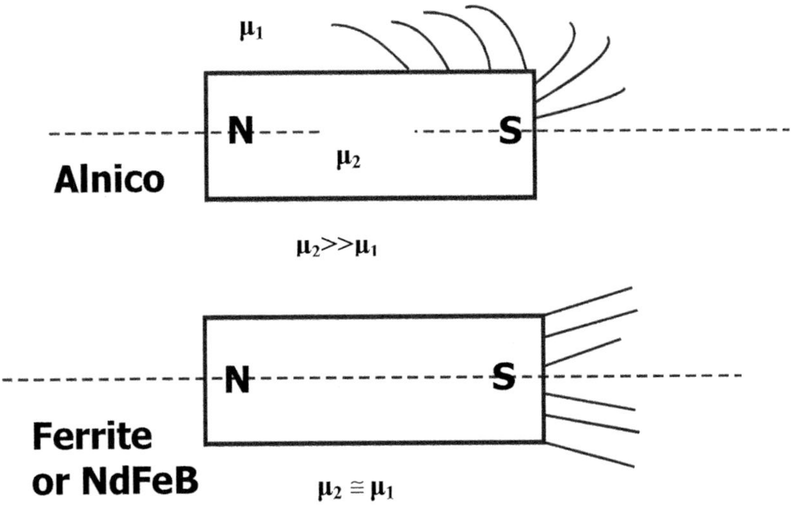

Figure 4.2: Flux leakage in two different types of permanent magnets.

corresponding to the material polarization. The situation is depicted again in Fig. 4.1.

Different types of permanent magnets can differ considerably in their magnetic behaviour. For instance, the two energy products of an Alnico magnet are very similar, namely $(BH)^i_{\max} \cong (BH)_{\max}$, while for ferrite and rare-earth magnets $(BH)^i_{\max} >> (BH)_{\max}$. The permeability of Alnico magnets differs considerably from that of the air, namely $\mu_{material} > \mu_{air}$, while for ferrite and rare earth magnets these permeabilities are similar, so that $\mu_{material} \cong \mu_{air}$. Typical values of magnetic permeabilities are shown in Table 4.1.

As a result of the above differences, the patterns of the magnetic flux around a magnet also differ. As is shown in Fig. 4.2, the flux lines in Alnico magnets leave the magnet limb perpendicularly, since the poles of the magnet are not at the terminals, but appear to be spaced at about 0.7 of the length of the bar [P4]. On the other hand, in ferrite and rare-earth magnets the flux lines tend to reach the terminal and the pole spacing is, therefore, the full length of the bar. In a long bar of Alnico 5, only 45% of the lines in the neutral section of the magnet reach the end of the bar. Therefore, a high percentage of the total energy is

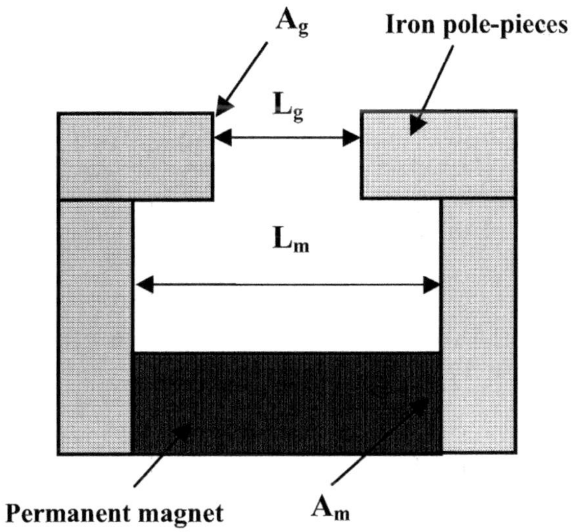

Figure 4.3: Magnetic circuit with a permanent magnet.

stored in leakage fields. On the other hand, in a ferrite bar approximately 95% of the flux at the neutral section reaches the end of the magnet. These magnets, therefore, store most of their energy within their volume and their useful energy approaches $(BH)_{\max}$ [P4].

4.2.2 The static gap design

In a circuit with permanent magnets there are no currents, $I = 0$, and eq. (4.3) reduces to

$$\oint \vec{H} \cdot d\vec{l} = 0 \tag{4.8}$$

In a simple magnetic circuit of Fig. 4.3, with a permanent magnet and an air gap, under static conditions, in which the dimensions of the air gap are constant, eq. (4.8) can be re–written

$$\int_{gap} \vec{H} \cdot d\vec{l} + \int_{magnet} \vec{H} \cdot d\vec{l} = 0 \tag{4.9}$$

Therefore, the sum of the magnetomotive force (mmf) in the magnet and the magnetomotive force in the air gap equals zero:

$$H_g L_g + H_m L_m = 0 \text{ or } H_g L_g = -H_m L_m \tag{4.10}$$

assuming that the soft-iron parts have infinite magnetic permeability. Also, because the flux is continuous

$$B_g A_g = B_m A_m \tag{4.11}$$

In the above equations L_g, A_g and B_g are the length, cross-sectional area and the magnetic induction of the air gap, respectively, while L_m, A_m and B_m are the length, cross-sectional area and magnetic induction of the magnet. It can be seen from eq. (4.9) that the magnetomotive force inside the magnet is in the opposite direction to the magnetomotive force in the air gap. Permanent magnets thus operate in the top left-hand quadrant (demagnetization quadrant) of the hysteresis curve.

By multiplying eq. (4.10) by eq. (4.11) we get

$$H_g B_g V_g = -H_m B_m V_m \tag{4.12}$$

where V_g and V_m are the volumes of the gap and magnet, respectively. This equation can be re-written as

$$\frac{1}{\mu_0} B_g^2 V_g = -H_m B_m V_m \tag{4.13}$$

These equations illustrate that the larger the product $H_m B_m$ and/or the volume of the magnet, the greater the magnetic induction in the air gap. It is also possible to determine, from the above equation, the volume of the magnet required to generate a desired flux density in a given air gap:

$$V_m = \frac{B_g^2 V_g}{\mu_0 H_m B_m} \tag{4.14}$$

By combining eqs. (4.12) and (4.13) we can also write

$$\frac{B_m}{H_m} = -\mu_0 \frac{A_g L_m}{A_m L_g} = -P_g \frac{L_m}{A_m} \tag{4.15}$$

where $P_g = \mu_0 A_g / L_g$ is the so-called permeance of the air gap, while the permeance of the magnet can be defined as $P_m = \mu_0 A_m / L_m$. Permeance is a measure of magnetic conductance which, if known, enables an accurate prediction to be made concerning how much of the total flux of a magnet will reach the air gap. The reciprocal of permeance is called reluctance, $\Lambda = 1/P$.

It can be seen from eq. (4.15) that by increasing the air-gap area and/or reducing the air gap length the ease with which flux travels across this gap will increase.

Equation (4.15) can be plotted as a straight line with a negative slope on a demagnetization curve. The intersection of this so-called load line with the demagnetization curve represents the operating point of the magnet, as shown in Fig. 4.4. The slope of the load line is equal to $-\mu_0 A_g L_m / A_m L_g$. For a very small air gap, in the limit $L_g \to 0$, the magnet is short-circuited and $B_m / \mu_0 H_m \to -\infty$, as shown in Fig. 4.4. In the opposite case, if the soft iron pole pieces are removed and the magnet stands alone, the air gap becomes large and the magnet operates in the open circuit regime. The optimum operation of the magnet is obtained when the working point on the demagnetization curve corresponds to the maximum value of the product (BH), as illustrated in Fig. 4.1.

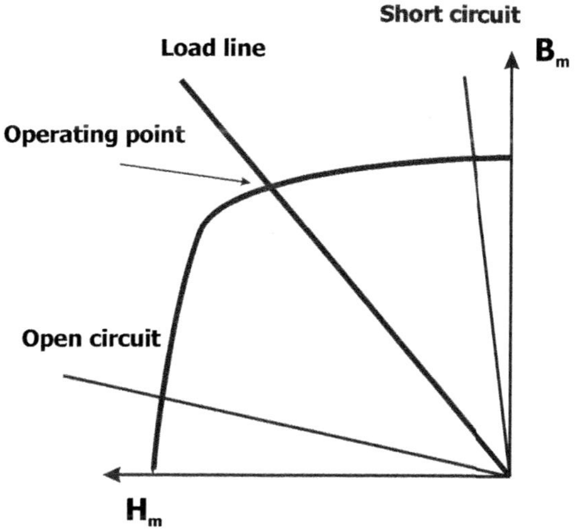

Figure 4.4: The load lines of permanent magnets operating in different regimes.

4.2.3 Leakage factors

In all practical situations the pole pieces will have finite magnetic permeability and even more importantly, there will be a considerable leakage of the magnetic flux. Allowance for these imperfections can be made by introducing correction factors K_1(leakage coefficient) and K_2 (loss factor) [P4]:

$$K_1 = \frac{\text{Total flux in circuit}}{\text{Useful flux}} = \frac{\text{Useful flux } + \text{ Leakage flux}}{\text{Useful flux}} \tag{4.16}$$

$$K_2 = \frac{\text{magnet mmf}}{\text{useful mmf}} \tag{4.17}$$

The equation for the load line then becomes

$$\frac{B_m}{H_m} = -\mu_0 \frac{K_1}{K_2} \frac{A_g L_m}{A_m L_g} \tag{4.18}$$

It is of interest to note that the B_m/H_m ratio is entirely determined by the physical relationships between the permanent magnet and the entire magnetic circuit within which it functions. This ratio is independent of the material from which the magnet is made.

Although the correction factor K_1 is rarely smaller than 2 and has no upper limit, for reasonably efficient permanent magnets it is in the range from 2 to 3. The loss factor K_2 is generally much smaller than K_1, usually within the range 1.05 to 1.45 [H18].

Dynamic operation

If the magnetic circuit geometry does not remain unaltered, then the load line slope in eq. (4.18) will change through any of the parameters L_g, A_g, K_1 and K_2. The operating point of the magnet may undergo an excursion within the short and open-circuit limits shown in Fig. 4.4. A portion of the flux changes from the leakage paths to the useful paths and back again, as the useful permeance is changed. The operation of a lifting magnet is a typical situation in which the magnetic system undergoes dynamic changes.

4.2.4 Determination of permeance and leakage factors

The primary requirement for magnetic circuit design, and therefore the design of the basic magnet used in the circuit, is to be able to calculate the relationships between flux paths. As pointed out in the previous section, the easiest approach to such calculations is by means of permeances. In general, the permeance of the flux paths varies from a minimum near the neutral section of the magnet to a maximum for zones at the end of the magnet or close to the air gap. Relatively simple formulae for various circuit geometries are available in the literature [T6, P4, M19, H18, M20]. Adding the gap and leakage permeances using these relationships is one of the methods of estimating the total permeance.

For instance, permeance of the air gap between rectangular surfaces of width a and thickness b (Fig. 4.5) is given by

$$P = \frac{\mu_0 ab}{L_g} \tag{4.19}$$

This equation is a special case of a more general expression

$$P = \frac{\mu_0 A_g}{L_g} \tag{4.20}$$

which is valid for any shape of the air gap. The leakage permeance of the geometry shown in Fig. 4.5 is given by an approximate formula [H18]

$$P = \frac{\mu_0 a}{\pi} \ln(\frac{2b}{L_g} + 1) \tag{4.21}$$

In order to demonstrate the difficulties facing a designer, Parker and Studders [P12] quote, for the same geometry, another formula that leads to values of P about 20% to 30% higher.

For a bar magnet that operates in the open-circuit mode, the permeance is the permeance from one limb of the magnet to the other, since there are no pole pieces or a well-defined air gap. Permeance of such magnets, of various shapes such as bars, rods and rings, is

$$P = 1.77\sqrt{A} \tag{4.22}$$

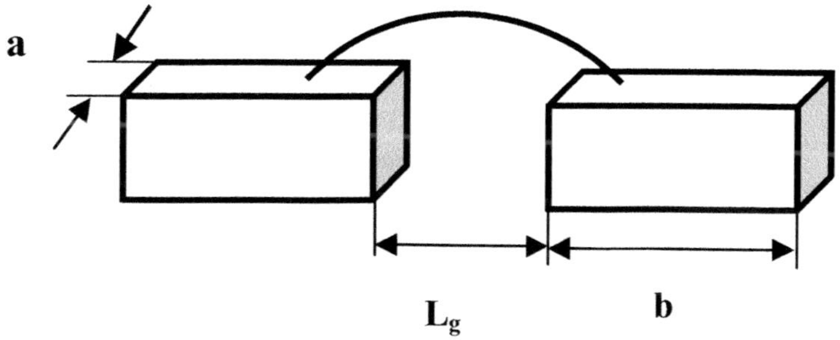

Figure 4.5: Geometry of a magnetic circuit for calculation of the leakage permeance.

where A is the cross-sectional area of the pole. This relationship is applicable to ferrite and rare-earth magnets, while for Alnico magnets the relation is

$$P = 1.77 \times 0.7\sqrt{A} \tag{4.23}$$

This equation takes into account the fact that for Alnicos the effective spacing of the poles is only 0.7 of the total length, as shown in Section 4.2.1.

It is clear that the calculation of individual permeances is very laborious and various empirical formulae have been developed for the determination of the total leakage. A very simple example of such a formula [H18] is an expression for the leakage factor K_1 for a gap of length L_g between circular pole-pieces of diameter d :

$$K_1 = 1 + \frac{7L_g}{d} \tag{4.24}$$

Comprehensive sets of formulae, usually attributed to Tenzer [e.g. T6], are available in the literature [e.g. H18]. For instance, in a magnetic circuit shown in Fig. 4.6, in which the permanent magnet material is adjacent to the gap, leakage from the sides and base, which are at the same potential, can be ignored. The correction factor K_1 is then given by [H18, M19]:

$$K_1 = 1 + \frac{L_g}{A_g}(1.1U_a\frac{0.67a}{0.67a + L_g})(1 + \frac{L_g}{a}) \tag{4.25}$$

where U_a is the perimeter of a part shown as length a. Similar equations are available for other geometries [H18, M19].

4.2.5 Magnetic field around a permanent magnet

The distribution of the magnetic induction at various points of magnets operating in the open-circuit regime is shown in Fig. 4.7 and Table 4.2. The long thin bar is magnetized along its long axis while the thin disc is magnetized along its short axis. The sphere is magnetized along its diameter. It is of interest to

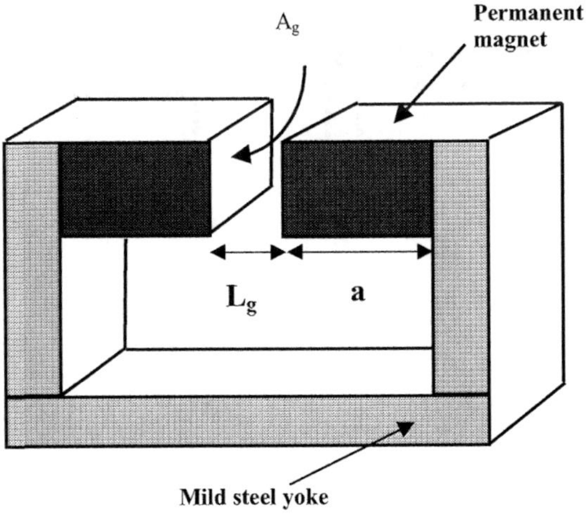

Figure 4.6: A permanent magnet circuit for the calculation of the leakage coefficient.

note that a thin disc magnetized through its thickness produces no magnetic flux density although the magnetic field strength H at point C is equal to J/μ_0.

Calculation of the magnetic field around a permanent magnet can be performed either using the pole concept or the surface current models. In the former approach a magnet is treated as having a uniform distribution of poles on the ends of the magnet. Using this assumption, magnetic induction on the z axis of a circular infinitely long magnet, as a function of the distance from the

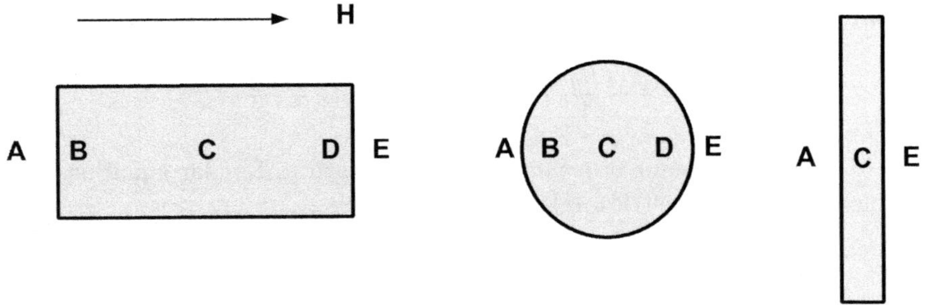

Figure 4.7: Distribution of magnetic induction inside and outside magnets of various shapes.

Table 4.2: Distribution of magnetic induction B inside and outside magnets of various shapes [H18].

Point	Bar	Sphere	Disc
A	J/2	2J/3	0
B	J/2	2J/3	
C	J	2J/3	0
D	J/2	2J/3	
E	J/2	2J/3	0

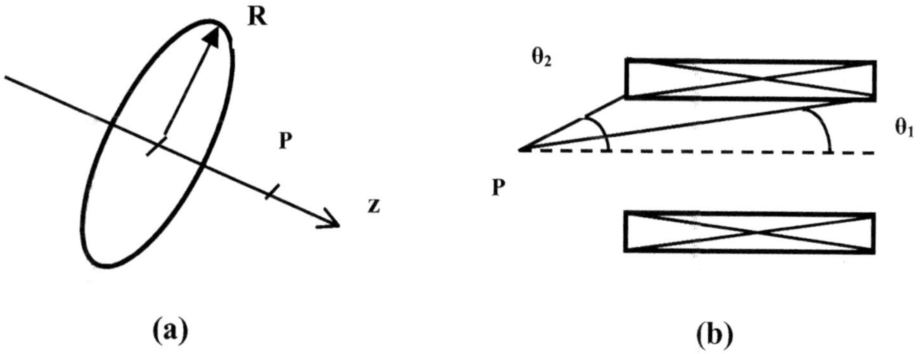

Figure 4.8: Calculation of magnetic induction on the axis of a cylindrical magnet.

magnet (Fig. 4.8), is given by

$$B_z = \frac{J}{2}\left[1 - \frac{z/R}{\left(1 + \frac{z^2}{R^2}\right)^{1/2}}\right] \tag{4.26}$$

where R is the radius of the magnetic cylinder. By symmetry, B has no components perpendicular to z. It can be seen that on the surface of the magnet ($z = 0$) the magnetic induction $B = J/2$, as shown in Table 4.2. Numerical computations are required to determine the magnetic induction off the axis. Expressions similar eq. (4.26) can also be obtained for the components of magnetic induction generated by rectangular surfaces [M19].

A uniform solenoid of rectangular or circular cross-section with surface current density j_s (A/m) is equivalent, for the purpose of calculating magnetic induction, to a bar with uniform magnetization $M = j_s = J/\mu_0$. If a solenoid has a circular cross-section, magnetic induction B, at point P on its axis, is

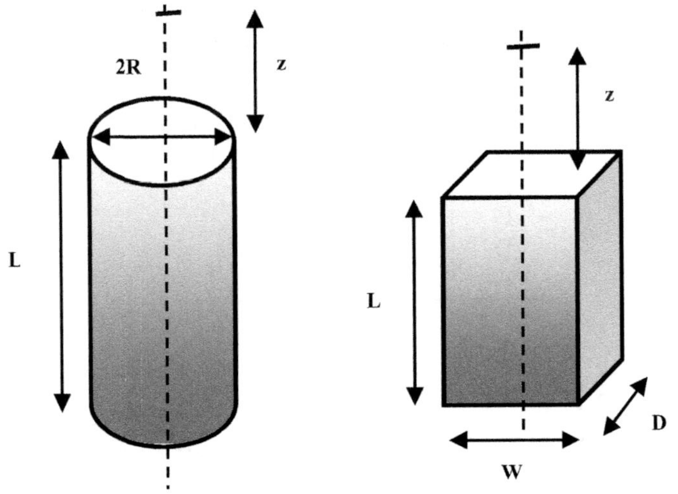

Figure 4.9: Definition of geometrical parameters of a cylindrical and a rectangular permanent magnets for calculation of the magnetic field.

given by

$$B = \mu_0 \frac{j_s}{2}(\cos\theta_1 - \cos\theta_2) \tag{4.27}$$

It can be seen that on the surface of an infinitely long solenoid ($\theta_1 = 0$, $\theta_2 = 90^0$) $B = \mu_0 M/2$, as could have been expected.

Selvaggi and Selvaggi [S48] developed mathematical equations that allow the external and internal magnetic fields of cylindrical and rectangular permanent magnets of finite dimensions to be determined. These magnets are shown schematically in Fig. 4.9. The authors assumed that the surface charge density at the poles is equal to J_r and that only two parallel faces contribute to the field.

The equation for the external field at distance z from the face of a circular magnet of radius R and length L is

$$B_{ext} = \frac{J_r}{2}\left\{\frac{z+L}{[R^2+(z+L)^2]^{1/2}} - \frac{z}{[R^2+z^2]^{1/2}}\right\} \tag{4.28}$$

The equation of the internal field, at the centre of the magnet is

$$B_{int} = \{\frac{J_r}{2}\frac{L}{(R^2+\frac{L^2}{4})}\} \tag{4.29}$$

Clearly, for $L \to \infty$, $B_{int} \to J_r$.

It can be seen that both terms in parentheses in eqs. (4.28) and (4.29) are dimensionless and thus any system of units can be used.

Similar expressions were derived for a rectangular permanent magnet of finite dimension, as defined in Fig. 4.9. The equation for the external field is

$$B_{ext} = \frac{J_r}{\pi} \left\{ \begin{array}{c} \arcsin \dfrac{WD}{4\left[(\dfrac{W^2}{4} + z^2)(\dfrac{D^2}{4} + z^2)\right]^{1/2}} - \\ \arcsin \dfrac{WD}{4\left[(\dfrac{W^2}{4} + (z+L)^2)(\dfrac{D^2}{4} + (z+L)^2)\right]^{1/2}} \end{array} \right\} \tag{4.30}$$

while the expression for the internal field is given by

$$B_{int} = \frac{J_r}{\pi} \left\{ \pi - 2\arccos \left[\frac{WD}{4[(\frac{W^2}{4} + \frac{L^2}{4})(\frac{D^2}{4} + \frac{L^2}{4})]^{1/2}} \right] \right\} \tag{4.31}$$

Considerably more complicated expressions for the off-centre magnetic field can also be found in [S48]. Good agreement between the measured values and those determined from eqs. (4.28) to (4.31) was found.

4.2.6 Magnetic forces

Permanent magnet circuits can be designed either for holding magnetic bodies or for attracting them. The distinction is that the holding magnet has to exert a magnetic force on a material with which it makes a good contact, while the attracting magnet has to exert a force on a magnetic body at a considerable distance. The distinction is important, because the optimum design for the two purposes differs considerably.

While the attraction force is essential in many forms of magnetic separators, such as suspended magnets or wet low-intensity drum magnetic separators, in other types of separators it is the holding force that plays a decisive role. The correct compromise between holding and attracting particles from a distance must be achieved.

A force between two magnetic charges Q_1 and Q_2 separated by a distance r is given by Coulomb's magnetostatic law:

$$\vec{F} = \frac{Q_1 Q_2}{4\pi\mu_0 r^2}\vec{r_0} \tag{4.32}$$

where r_0 is the unit vector in the direction of r. For a thin cylindrical magnet $Q = JA$ where J is the magnetic polarization and A is the cross-sectional area of the magnet. We can then write, for semi-finite magnetic bars of the same cross-sectional area A

$$\vec{F} = \frac{1}{4\pi\mu_0 r^2} J_a J_b A^2 \vec{r_0} \tag{4.33}$$

where J_a and J_b are magnetic polarizations of bars a and b, respectively. When these bars are in contact, eq. (4.33) reduces to:

$$F = \frac{1}{2\mu_0} J_a J_b A \tag{4.34}$$

As an example, magnetic polarization of 1 T produces a pull of 40 N/cm^2, or approximately 4 kg/cm^2, while 0.2 T creates a pull of a mere 0.16 kg/cm^2.

Equation (4.34) is often simplified even further, under the assumption $J_a = J_b = B$, into the form

$$F = \frac{1}{2\mu_0} B^2 A \tag{4.35}$$

which is incorrect in many instances [M19]. For instance, if $J_b = 0$, then correctly $F = 0$, according to eq. (4.34). The simplified formula would incorrectly yield $F = J_a^2 A/8\mu_0$. Generally, eq. (4.35) is valid only when bars a and b are of the same material and when there is no source of external magnetic field.

It is obvious from eqs. (4.34) and (4.35) that the holding force is proportional to the square of flux density and only to the first power of the magnet area. By using tapered soft-iron pole-tips it is possible to concentrate the flux to a higher density than the magnets themselves can provide. Although the cross-sectional area of the magnet would decrease somewhat, the holding force will increase nevertheless.

Although by increasing the magnetic flux it is possible to increase the holding force for zero air gap, the reach of the magnetic force would decrease once the air gap is introduced. The force of attraction would thus decrease. Therefore, in those magnetic separators in which a pull at a distance (traction force) is important, wide pole-pieces working at a low induction and allowing high force penetration into the air gap, are required. On the other hand, closely spaced poles that produce high holding force are incorporated into those magnetic separators that rely on the holding force rather than traction force. Dry drum and roll magnetic separators are an example of such machines.

4.2.7 Soft iron components of the magnetic circuits

Pole-pieces of a solid or laminated soft iron or mild steel , or occasionally cobalt-iron, are an essential part of magnet design. These flux-conducting members are used to complete a return path for the flux and to concentrate or change the flux density in a circuit. Cost is usually an important factor in the choice and majority of magnetic circuits use ordinary cold-rolled steel for these components. Cobalt-iron alloy is used in some high-performance circuits. It is a good practice to operate these materials near their maximum permeability, as shown in Fig. 1.9. This will ensure the maximum flux-carrying capacity of the circuit. Magnetization curves for some high-permeability materials is shown in Fig. 4.10.

Relative position of permanent magnets and mild steel components is also of significant importance. Figure 4.11 illustrates three possible arrangements

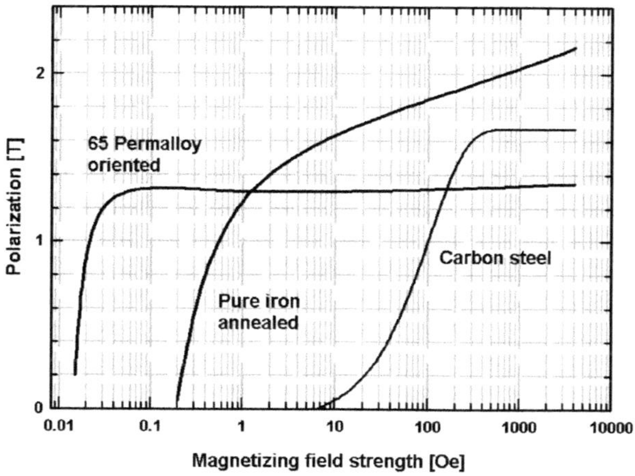

Figure 4.10: Magnetization curves for several high-permeability materials.

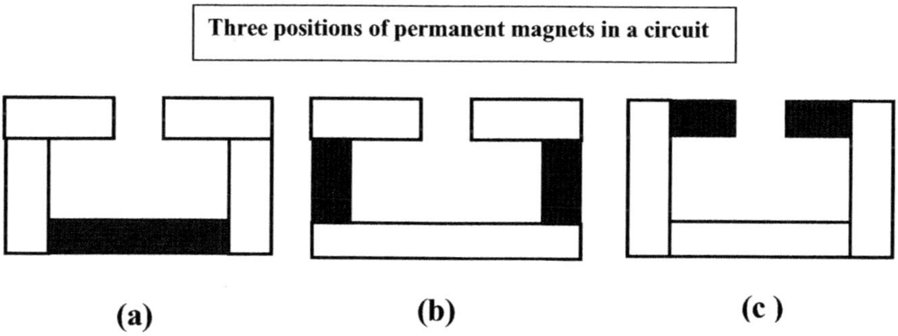

Figure 4.11: Three arrangements of a circuit with permanent magnets.

Table 4.3: Reversible temperature coefficients of permanent magnetic materials.

Material	$\alpha(\mathbf{B}_r)$ [%/^{0}C]	$\alpha(\mathbf{H}_{ci})$ [%/^{0}C]
Ferrite	- 0.20	+ 0.40
$SmCo_5$	- 0.04	- 0.30
NdFeB	- 0.12	- 0.60

of a circuit with permanent magnets. Arrangement (a) is the least efficient as the leakage between the circuit elements is very high. On the other hand, arrangement (c) is the most efficient as a result of low leakage. Analysis of these circuits with an Alnico 5 magnet showed that the ratio of the gap flux to the total flux in the arrangement (c) is three times higher than in arrangement (a) [P12]. It is clear that permanent magnets should be placed as close to the air gap as possible. A valuable advantage can be obtained by placing a magnet, magnetized through its thickness, on a ferromagnetic plate. Then the magnet behaves like one of twice its thickness.

4.2.8 Stability of permanent magnets

Permanent magnets are expected to maintain a constant flux output over a long period of time, although any magnet, after being fully magnetized, is likely to lose strength slowly with time. In nearly all applications the magnet is subject to influences that tend to alter the magnet flux. However, in contrast to early permanent magnets, the modern materials exhibit remarkable stability, sometimes as high as 10^{-6}.

Reversible changes

The magnetization changes fall into two classes, namely reversible changes and irreversible changes. The reversible changes in the magnetic properties, as a function of temperature, originate in the change in spontaneous magnetization. Reversible changes are functions of temperature and circuit loading and are time-independent. They disappear completely without need for remagnetization when the permanent magnet is returned to its initial temperature. Figure 4.12 illustrates the dependence of the remanent magnetization on temperature for three types of permanent magnet materials.

The reversible changes resulting from temperature changes can be represented by temperature coefficients, expressed as a percentage change in J or B per ^{0}C. A summary of the temperature coefficients α of B_r and H_{ci} for several permanent magnet materials is given in Table 4.3. In addition, Table 1.18 summarizes the Curie temperatures and maximum operating temperatures of various permanent magnet materials.

In many instances the temperature coefficients of magnetization and coercive force do not give enough information about how a magnet will respond to temperature changes. Changes in the shape of a demagnetization curve and the

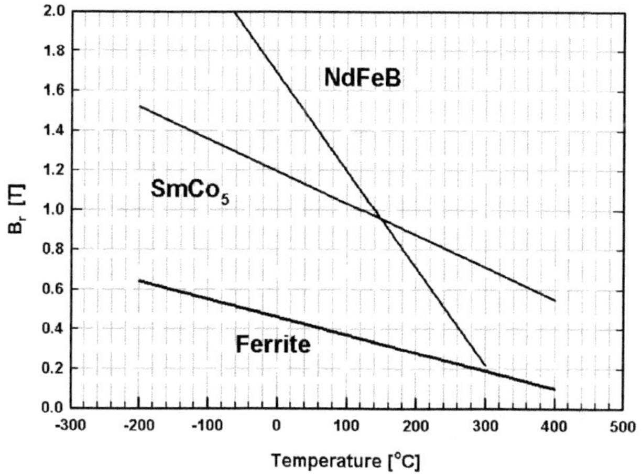

Figure 4.12: Temperature dependence of remanent magnetization of various permanent magnetic materials (adapted from [P4]).

position of the load line and the working point can only be seen from a complete set of demagnetization curves, measured at different temperatures. Figures 4.13 and 4.14 illustrate families of demagnetization curves at various temperatures for ferrite and NdFeB permanent magnets. Such curves are useful for graphical design of magnets operating under various temperature conditions.

Complete sets of such demagnetization curves are available either in the literature (e.g. [P4]) or the most recent data can be obtained directly from magnet manufacturers.

Irreversible changes

In the case of irreversible changes, after the removal of the disturbing influence, the magnetization does not return to its original value. In certain types of changes however, the magnetization can be fully restored by re-magnetization. Such changes are, for instance, ambient temperature changes and exposure to an external magnetic field. Upon exposure to an external field, the operating point moves down the demagnetization curve and the intrinsic magnetization of the magnet shows a corresponding change. When the external field is removed, the magnetization recovers only partially. This loss is recoverable by re-magnetization. The loss will be greater when a magnet is operating in the steepest part of the demagnetization curve.

Although mechanical shock and vibration contribute energy to a permanent magnet, changing its magnetization, modern materials with high coercive forces seem to be resistant to changes by mechanical effects. At the same time, however, field exposure is not able to completely eliminate changes that might have occurred.

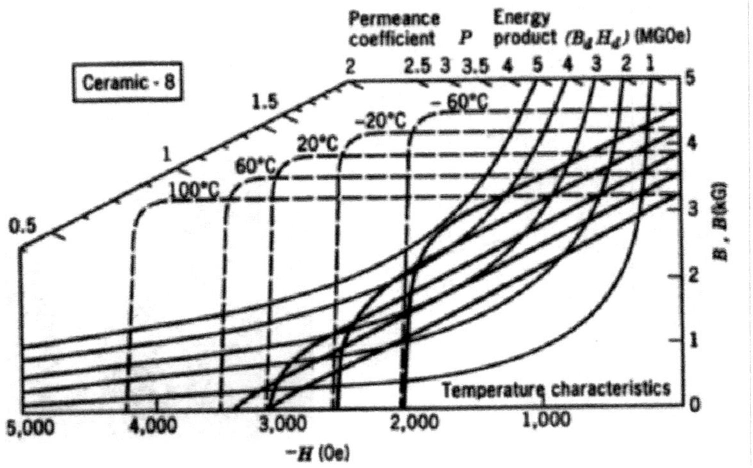

Figure 4.13: Demagnetization curves of a ferrite magnet at different temperatures.

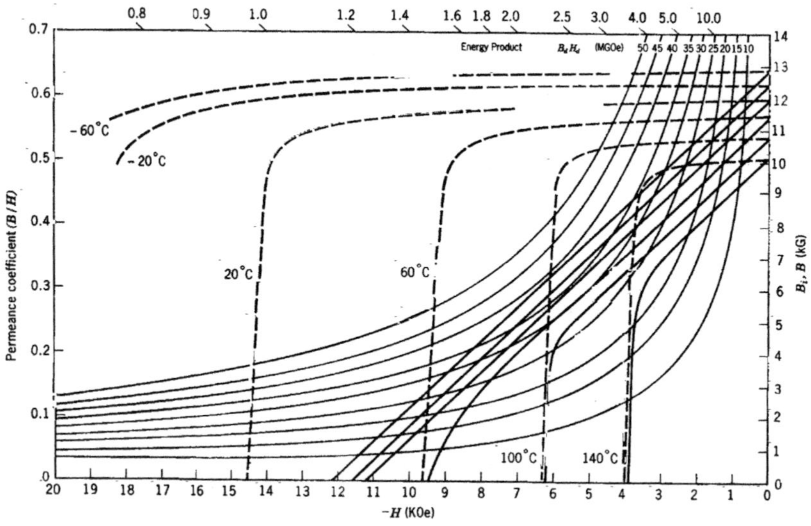

Figure 4.14: Demagnetization curves of an NdFeB magnet at various temperatures.

Table 4.4: Magnetic induction required to magnetize permanent magnet materials.

Material	B_{sat} [T]
Alnico 5	0.3
Alnico 8	0.8
Ferrite	1.0
$RECo_5$	2.0 to 3.0
NdFeB	2.5 to 3.0

While Alnico and ferrite magnets are stable and not subject to corrosion, rare-earth magnets have oxidation problems that limit their maximum operating temperature. If no coating is provided for protection, oxygen diffuses into such a magnet, causing a metallurgical change in a surface layer. The surface layer will have a lower H_{ci}, which will allows this region of the magnet to be demagnetized more easily. Surface oxidation thus leads irreversibly to a reduction in the magnetic properties.

Irreversible changes, resulting from permanent changes in the structural or metallurgical state, are usually time-dependent and re-magnetization does not restore the original state of magnetization. The temperature at which changes in properties first become noticeable corresponds to the maximum operating temperature. Temperature-induced irreversible losses in permanent magnets are shown in Table 1.17.

4.2.9 Magnetization of permanent magnets

It is generally accepted that to magnetize a permanent magnet to saturation, an external magnetic field three to five times its intrinsic coercivity H_{ci} must be applied. Typical values of the magnetic induction required to magnetize various permanent magnet materials are summarized in Table 4.4.

Simple bar or ring-shaped magnets may be magnetized by a larger permanent magnet, a solenoid or an electromagnet. To magnetize rare-earth magnets, very high fields are required and magnetization is often performed with small coils or solenoids energized by large current pulses of short duration. Large magnets cannot be magnetized by a pulse of short duration, because eddy currents prevent the flux reaching the centre of the magnet. In general, the frequency f of the pulse must be chosen so that the magnetizing pulse lasts longer than the eddy current. The depth of penetration, or the skin depth δ, is given by

$$\delta = \left(\frac{\varrho}{\pi\mu_0\mu f}\right)^{1/2} \tag{4.36}$$

where ϱ is the resistivity of the material.

Resistivity plays a very important role since most permanent magnets have very low permeabilities μ. Ceramic ferrite magnets have high resistivity $\varrho > 10^{-4}$ Ωm and full penetration of a pulse is easy. On the other hand, Alnico and

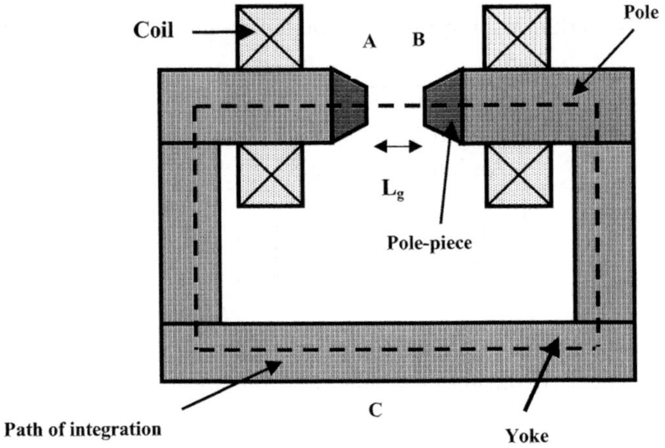

Figure 4.15: Schematic diagram of an electromagnet with iron yoke.

rare-earth magnets, with $\varrho \approx 100\times10^{-8}$ Ωm, pose a problem and a pulse width of at least 10 ms must be used [P4].

4.3 Design of iron-core electromagnets

In section 1.7.2 the position of electromagnets with an iron yoke as a source of magnetic field in magnetic separation was outlined. Although the iron-core electromagnets have been recently replaced by rare-earth permanent magnets in several types of magnetic separators, they still play an important role in various designs of magnetic separators. In those applications where variable magnetic field of moderate strength in a relatively large volume is required, electromagnets with an iron yoke are the field source of choice.

In spite of availability of powerful modelling techniques for optimization of electromagnet design, the initial design ought to be created based on experience and, to some degree, on guesswork. In this section general principles of design of electromagnetic circuits with an iron core will be described. The design of such a circuit, schematically shown in Fig. 4.15, is governed by Ampere's law given by eq. (4.3). If we suppose that coils with a total number of turns N are wound onto the circuit, then Ampere's law can be re-written as

$$\oint_C \vec{H} \cdot d\vec{l} = NI \tag{4.37}$$

The path of integration extends around the magnetic circuit as shown in Fig. 4.15. The magnetic induction B_g in the air gap of length L_g is, therefore, given by

$$B_g = \frac{\mu_0}{L_g}(NI - \int_{BCA} H dl) \tag{4.38}$$

By calculating the actual field B inside the iron part of the circuit, we can then determine the magnetic induction in the air gap from the value of the magnetomotive force (or number of ampere-turns) NI.

We can define reluctances of the air gap Λ_g and of the iron part of the circuit Λ_m as

$$\Lambda_g = \frac{L_g}{\mu_0 A_g} \tag{4.39}$$

$$\Lambda_m = \frac{1}{\mu}\int_{BCA} \frac{dl}{A_m} \tag{4.40}$$

where A_g and A_m are cross-sectional areas of the air gap and iron and μ is the permeability of iron.

If the magnetization of the iron is much less than saturation, then $\mu >> \mu_0$ and $\Lambda_g >> \Lambda_m$. Then we have

$$B_0 = \mu_0 \frac{NI}{L_g} \tag{4.41}$$

This equation expresses a linear relationship between B_0 and I, which is observed for any type of electromagnet at low magnetic fields. If, however, the magnetization of the iron approaches saturation, the permeability μ of the iron is greatly reduced. Under these conditions eq. (4.41) is not valid and we must use the following, more general equation to calculate B_g

$$B_g = \frac{NI}{(\Lambda_g + \Lambda_m)A_g} \tag{4.42}$$

For this purpose we must calculate Λ_m from eq. (4.40), which is a difficult task as we have no straightforward way of determining the value of μ as a function of position l. Furthermore, the leakage of the flux will increase as μ becomes small [I1].

4.3.1 Calculation of the magnetic field

The magnetic field produced in an air gap of an electromagnet can be calculated by integrating contributions of the magnetic charges on the pole surface. For cylindrical flat poles, as shown in Fig. 4.16, the magnetic induction in the gap is given by [M21]

$$B_g = J_s\gamma_c(1-\cos\theta) = J_s\gamma_c\left[1 - \frac{x}{(r_2^2+x^2)^{1/2}}\right] \tag{4.43}$$

where γ_c is a correction factor.

For tapered poles with coinciding apexes (Fig. 4.16) the magnetic induction becomes

$$B_g = J_s\gamma_c(1-\cos\theta+\cos\theta\sin^2\theta\ln\frac{r_2}{r_1}) = \tag{4.44}$$

$$J_s\gamma_c\left[1 - \frac{x}{(r_1^2+x^2)^{1/2}} + \frac{xr_1^2}{(r_1^2+x^2)^{2/3}}\ln\frac{r_2}{r_1}\right] \tag{4.45}$$

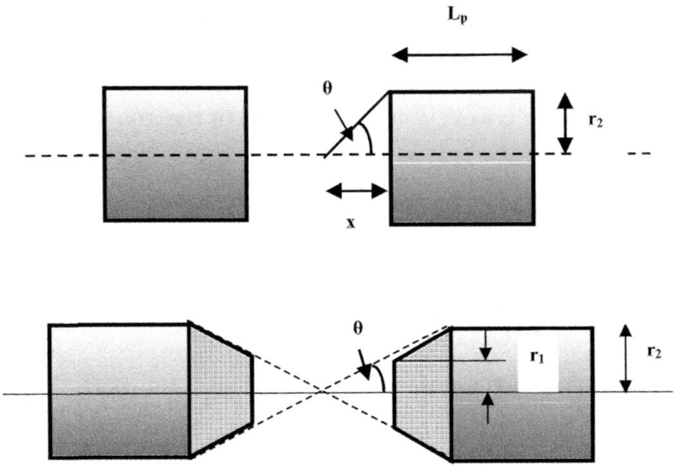

Figure 4.16: Cylindrical flat poles and tapered poles with coinciding apexes.

It can be seen that in this case the magnetic induction consists of two terms, the first term being the magnetic induction for cylindrical flat poles. The second term is greater than zero (provided that $\theta < 90^0$ and $r_2 > r_1$), and magnetic induction generated by tapered poles, therefore, generates a greater magnetic field than flat poles. This second term achieves its maximum for $\theta = 54.7^0$.

It was shown by Montgomery [M21] that eq. (4.43) gives good results if $\gamma_c \cong 0.7$ for $\beta \geq 1$, while for $\beta \cong 0.5$ the correction factor $\gamma_c = 1$ gives a good agreement between calculated and measured values of magnetic induction. β is a geometrical factor defined as $\beta = L_p/2r_2$ and L_p is defined in Fig. 4.16. In magnets with tapered pole-pieces eq. (4.45) has very good predictive capability for $\gamma_c \cong 1$.

4.3.2 Shape of iron parts of the circuit

The yoke, as shown in Fig. 4.15, serves to transport the magnetic flux from the end of one pole to the end of the other pole. The presence of iron permits useful values of the magnetic induction (for example, up to 2 T) to be reached with modest investments in ampere-turns. The iron yoke also confines the stray flux to the immediate vicinity of the magnet gap. The yoke should, therefore, have reasonably high permeability, even when the poles and pole pieces are saturated. Since the magnetization curve of soft iron has a knee at $J/J_s = 70\%$ to 80%, it is recommended to make the cross section of the yoke about 1.2 to 1.4 times that of the poles [I1]. In modern magnets the yokes and the poles are made of laminations.

In order to attain the strongest possible magnetic field, the shape of the poles and pole pieces must be determined to satisfy the following two requirements:

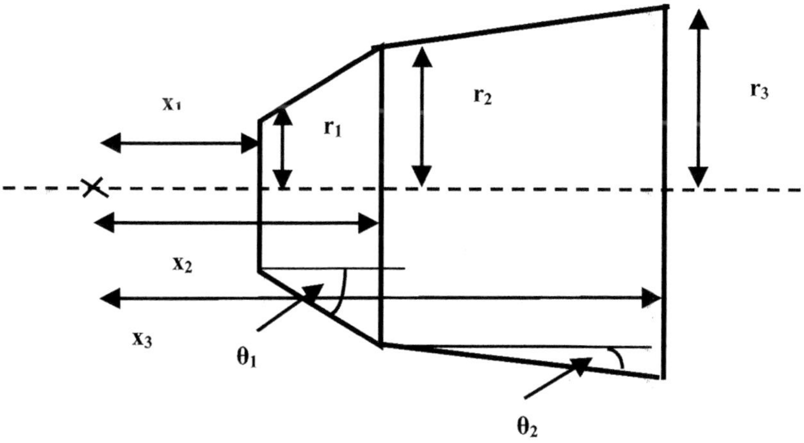

Figure 4.17: Shape of a pole with two conical pole-pieces.

(a) The magnetization of the poles and pole pieces must produce the maximum possible field when they are saturated.

(b) Saturation should be achieved simultaneously throughout the volume of the poles and the pole pieces.

In view of increasing leakage from the poles, as their magnetization approaches saturation, the cross-section of the poles must be reduced as the tip of the pole is approached. The flux density will thus be maintained constant and the requirement (b) will be met.

In order to satisfy the requirement (a), Ewing [E2] showed, as early as the end of the nineteenth century, that the shapes of the pole pieces must be conical, preferably with the vortex angle of 54.7^0. In an actual electromagnet, however, this conical shape must be modified, in order to secure homogeneity of the field in the air gap and to ensure that permeability is constant throughout the pole and the pole piece [I2].

Analyzing eqs. (4.43) and (4.45), Ishikawa and Chikazumi [I2] showed that the pole shape that satisfies the above requirements can be approximated by the shape shown in Fig. 4.17. This shape is composed of a pole-tip of radius r_1, a conical part of radius r_2 and another conical part of radius r_3. Analysis of this shape indicates that that the maximum field in the air gap depends on the value of r_2/x_1 and not on r_1/x_1, while the homogeneity of the field is determined by the value of r_1/x_1. The issue of the field homogeneity is not critically important in magnetic separation and the parameter r_1/x_1 is, therefore, of limited relevance for the design of a circuit of a magnetic separator.

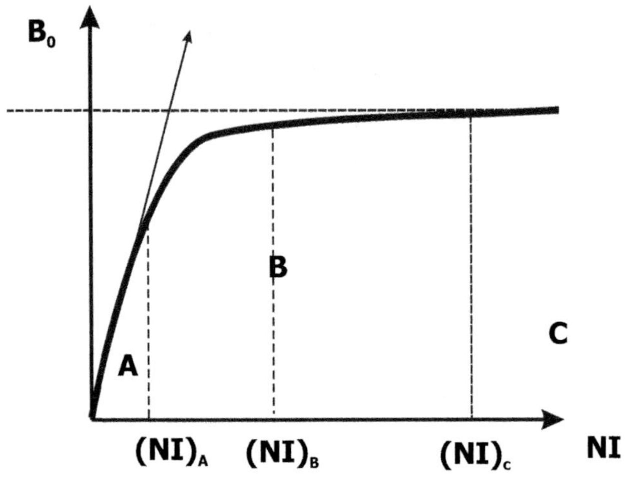

Figure 4.18: The excitation curve of an electromagnet showing the unsaturated zone (A), transition zone (B) and saturated zone (C).

4.3.3 Practical procedure for design of the poles

The design of an electromagnet for a magnetic separator is based on two fundamental specifications, namely the required magnetic field strength B_g and the length $(2x_1)$ and the depth $(2r_1)$ of the air gap. The former parameter is determined by the required separation efficiency while the latter mainly by the desired throughput of the separator. In wet high intensity magnetic separators the depth of the air gap determines the depth of the matrix bed, which in turn also affects the effectiveness of separation.

From an economic point of view, the size and thus the price of the magnet depends strongly on the length of the air gap $L_g = 2x_1$. The volume of the magnet is approximately proportional to $(2x_1)^3$.

Contribution of iron to the magnetic field

The contribution of the iron circuit to the magnetic induction in the air gap can be divided into three regions, as illustrated by the excitation curve shown in Fig. 4.18. In the unsaturated zone A the reluctance of the air gap is much greater than the reluctance of the rest of the circuit so that almost the entire magnetomotive force appears across the gap. Once the transition zone sets in, the flux that reaches the air gap is reduced. In order to achieved the desired magnetic induction, the number of ampere-turns must be increased, as shown in Fig. 4.18.

It was reported by Montgomery [M21] that usually $(NI)_B/(NI)_A \approx 1.1$ at 60% of the maximum magnetic field, while a typical value of this ratio at 85% of the maximum field is about 2. A plot of this ratio against the percentage

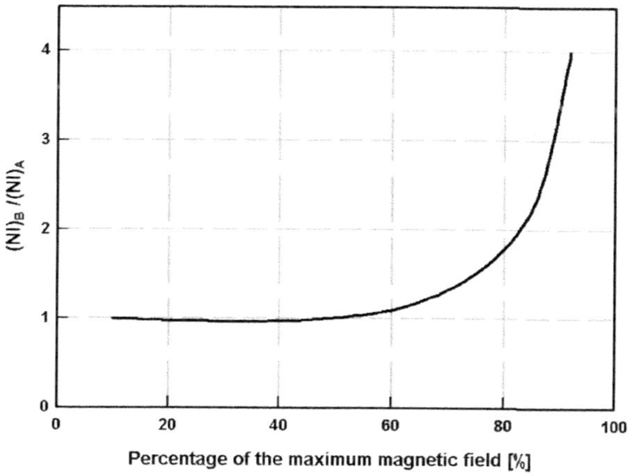

Figure 4.19: Ampere-turns needed to achieve a required magnetic field.

of the maximum induction, for a typical magnet, is shown in Fig. 4.19. With the number of ampere-turns known at 60% and 85% of the maximum field, a satisfactory curve can be drawn representing the magnetization of iron from 0 to 100% [M21]. Figure 4.20 shows the number of ampere-turns required to obtain 85% of the maximum field, as a function of the geometry of the magnet.

Design of coils

Coils in an iron-core electromagnet serve two functions: they provide the electromagnetic force to magnetize the iron, and they contribute a certain magnetic field in the air gap. The magnetic induction generated by a coil, the geometry of which is shown in Fig. 4.21, at the centre of the coil, is given by [M22]:

$$B_g = \frac{NI}{2\mu_0 a_1 \beta(\alpha - 1)} F(\alpha, \beta) \tag{4.46}$$

where $\alpha = a_2/a_1$, $\beta = b/a_1$ and $F(\alpha, \beta)$ is the field factor defined (in SI units) as

$$F(\alpha, \beta) = \beta \ln \frac{\alpha + (\alpha^2 + \beta^2)^{1/2}}{1 + (1 + \beta^2)^{1/2}} \tag{4.47}$$

The field along the axis z of a coil of axially uniform excitation can be written [M22]:

$$B_z(\frac{z}{a_1}) = B_z(0) \left[\frac{F(\alpha, \beta + z/a_1) + F(\alpha, \beta - z/a_1)}{2F(\alpha, \beta)} \right] \tag{4.48}$$

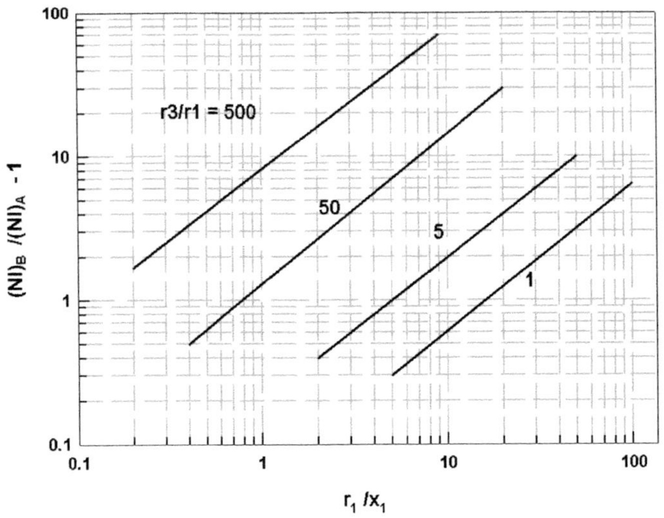

Figure 4.20: Ampere-turns required to reach 85% of the maximum magnetic field in the air gap, as a function of the ratio of the pole face diameter and the gap length, r_1/x_1, for several r_3/r_1 ratios (adapted from [M21]).

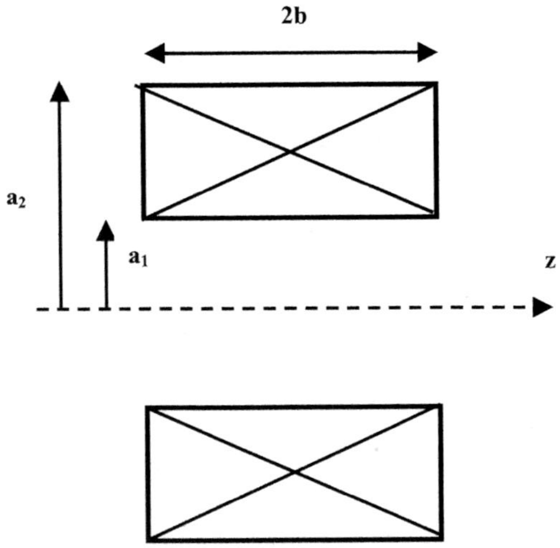

Figure 4.21: Definition of geometrical parameters of a solenoid coil.

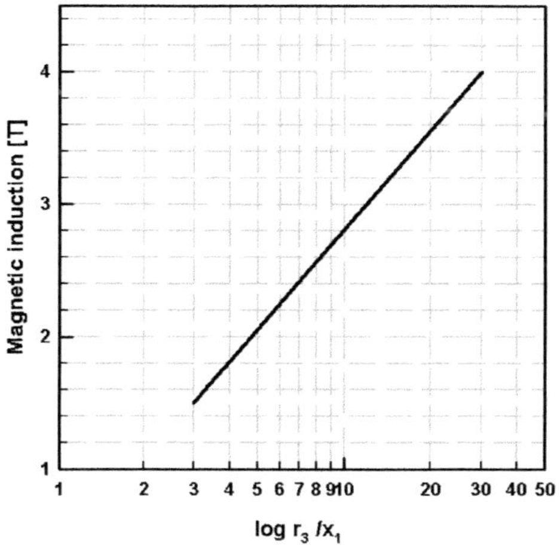

Figure 4.22: Empirically obtained curve expressing the maximum magnetic field as a function of ratio of the pole diameter to air gap length (adapted from [I1]).

If the point at which this field is measured lies outside the coil (e.g. in the air gap of an electromagnet) the following relationship should be used in eq. (4.48):

$$F(\alpha, \beta - z/a_1) = -F(\alpha, z/a_1 - \beta) \tag{4.49}$$

Practical rules for the design

Taking into account the above analysis, the following rules should be followed in the design of an electromagnet:

(1) Although the field homogeneity is not critical in magnetic separation, a suitable choice of the pole radius r_1, in such a way that $r_1 = 2x_1$, usually ensures the field homogeneity within 10^{-3}.

(2) The maximum magnetic field is a function of the maximum pole diameter $2r_3$, irrespective of differences in pole shapes [I1]. Using Figs. 4.20 and 4.22, approximate value of the maximum diameter of the pole from the selected values of B_g and x_1 can be determined.

(3) The optimum value of angle θ_1 depends on the value of r_1/x_1 which again is not of major importance in magnetic separators. θ_1 usually ranges from 50^0 to 60^0.

(4) A generally accepted rule is that $r_3 = 1.2r_2$ to $1.4r_2$.

(5) The number of ampere-turns should be selected in such a way that $(NI)_B = 1.1(NI)_A$ or $2(NI)_A$, if 60% or 85% of the maximum field is to be achieved, respectively.

Figure 4.23 illustrates modelling of the magnetic flux in a iron-core electromagnet.

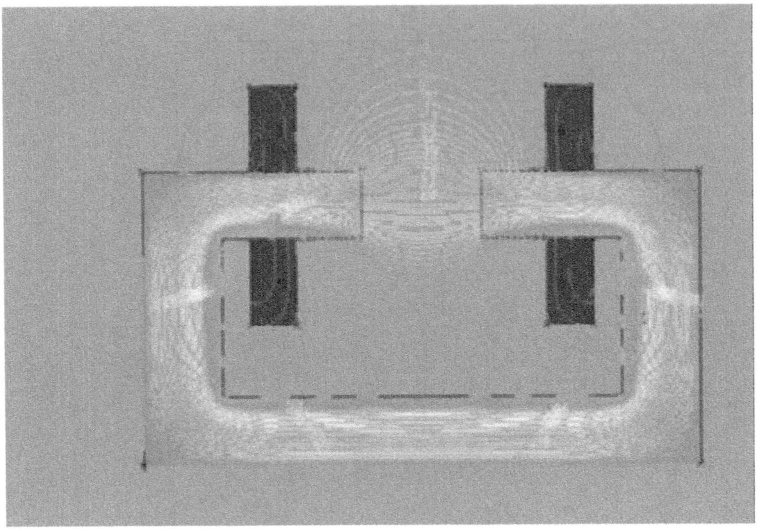

Figure 4.23: Modelling of an iron-core electromagnet.

4.4 Design of solenoid magnets

As has already been discussed in Section 1.7.3, generation of magnetic induction by iron-core electromagnets is limited by the saturation of the iron. In order to overcome this limitation, Bitter [B24] proposed the use of a magnetizing solenoid surrounded by steel cladding. The magnetic induction in the working volume of such a magnet is given by the sum of the contribution from the coil and from the magnetized iron, as expressed by eq. (1.50). Figure 1.40 illustrates the relative contributions from these two components of the magnet.

The idealized circuit of the iron-clad solenoid is shown in Fig. 4.24. The magnetic induction in the air gap is given by Ampere's law (eq. (4.41)). If the magnetic reluctance of the iron frame cannot be neglected, eq. (4.42) must be used. In a more practical form, a leakage factor K can be included in eq. (4.42), and we can write:

$$B_g = \mu_0 \frac{NI}{L_g(1+K)} \tag{4.50}$$

The leakage factor K is often taken equal to 0.2 [T6, M23].

It transpires from eq. (4.50) that the number of ampere-turns required to generate a given magnetic induction is determined solely by the length of the field (or the length of the air gap) L_g and is independent of the radius a_1 of the working volume. Solenoids shown in Fig. 4.25 have the same length of the air gap, while their radii differ, and the number of ampere-turns required to generate a required magnetic induction is the same for both solenoids. This feature is extremely useful in the scale-up of magnetic separators. In iron-core electromagnets a higher throughput of separators is achieved by increasing the

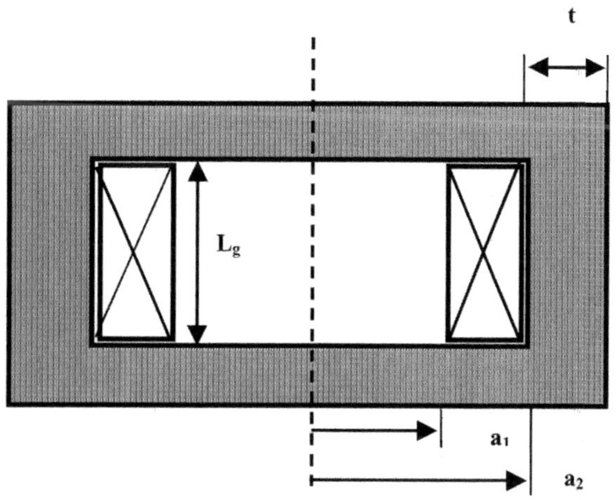

Figure 4.24: Schematic diagram of an iron-clad solenoid.

length of the air gap. This requires an increased number of ampere-turns, as follows from eq. (4.41). This, in turn, results in an increase in the mass, size and cost the magnet.

On the other hand, the scale-up of a magnetic separator based on an iron-clad solenoid is achieved by increasing the diameter of the solenoid, keeping the length of the air gap constant. The number of ampere-turns will be constant, only the length of the conductor, and thus its resistance will increase.

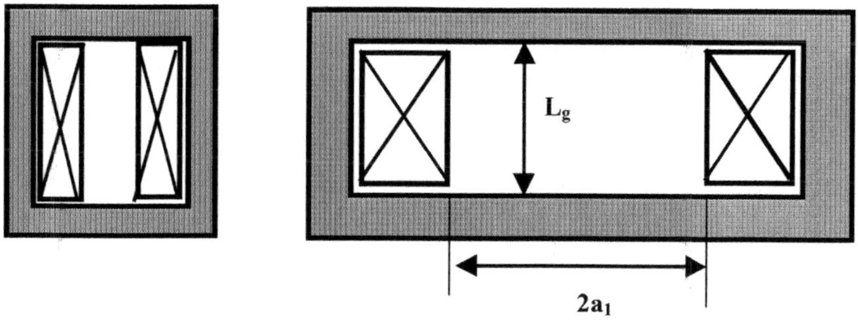

Figure 4.25: Two iron-clad solenoids having the same length of the air gap and different radii.

4.4.1 The power requirements

The magnetic induction generated directly by the coil itself can be calculated using eqs. (4.46) and (4.48). The power W (in watts), required to generate a given magnetic induction, is then given by:

$$W = \frac{\pi \varrho a_1^2}{2b\lambda(a_2 - a_1)^2}(\alpha^2 - 1)(NI)^2 \tag{4.51}$$

where ϱ is resistivity of the conductor and λ is the filling factor of the winding defined as

$$\lambda = \frac{\text{active section of the winding}}{\text{total section of the winding}} \tag{4.52}$$

Combining eqs. (4.41) and (4.51) it can be seen that the power is proportional to the square of the number of ampere-turns and thus of magnetic induction, $W \sim B_g^2$.

While the electric current I and the number of turns N are related by eq. (4.41), the necessary voltage can be found from $V = IR$, where the resistance of the winding can be calculated from:

$$R = \frac{2\pi\varrho N}{\lambda A} D_{mean} = \frac{\pi\varrho(\alpha + 1)}{2\lambda\beta(\alpha - 1)a_1} N^2 \tag{4.53}$$

where A is the cross-section of the winding and D_{mean} is the mean diameter of the winding.

4.4.2 The choice of the conductor

The choice of the conductor, between copper and aluminium, determines the maximum current density, the mass and the volume of the coil and the cost of the magnet. Copper is 3.3 times denser than aluminium (8960 kg/m^3 vs 2699 kg/m^3) and for equal coil sizes aluminium will make a much lighter coil. On the other hand, resistivity of aluminium is about 1.58 times greater than resistivity of copper (2.45×10^{-8} Ωm vs 1.555×10^{-8} Ωm at 0 ^{0}C), 1.58 time more power is required by the aluminium coil to generate the same field. Detailed analysis of the relative merits of aluminium and copper coils, as a function of coil geometry, mass and cost can be found in [M22].

4.4.3 The cooling of coils

The removal of the Joule heat generated inside the winding can be accomplished in several ways. Indirect air cooling can be applied to magnets operating at low current densities, generally not exceeding 3 A/mm^2 for a copper conductor. Direct immersion of the coil into the circulating oil or water greatly improves heat conduction and transfer and higher current densities can be used.

The operating current density can be increased considerably by using hollow conductors through which the cooling liquid circulates. Water is a very efficient coolant in removing heat at conventional temperatures. The oils, for example, have heat-transfer characteristics about one-fifth that of water [D11]. Coils with low resistance, operating at low voltage, are almost always cooled with deionized or demineralized water, mains water causing electrolysis even at low voltage. The cooling water is circulated through the coil, a heat-exchanger, a filter and a deionizer. In the heat-exchanger the heat is carried off by tap water.

The flowrate of water Q (in cm^3/s or m^3/h) required to restrict the temperature rise to ΔT (in ^{0}C) in a magnet with power input of W (in watts) is given [M22] by:

$$Q = \frac{W}{4.186\Delta T}(cgs) = \frac{W}{1163.7\Delta T}(SI) \tag{4.54}$$

For instance, for a 250 kW magnet, with temperature rise of $\Delta T = 30$ ^{0}C, the required flowrate is approximately 2000 cm^3/s, or 7.2 m^3/h.

The operating temperature of a magnet affects the resistance of the winding and thus the power needed to achieve the required magnetic field. The hotter the coil, the higher the resistance will be and the more power will be needed. The temperature dependence of resistivity of conductors can be written

$$\varrho_T = \varrho_{20}[1 + \alpha(T - 20^0C)] \tag{4.55}$$

where ϱ_T and ϱ_{20} are resistivities at temperature T and 20 ^{0}C, respectively and α is the temperature coefficient. For copper $\varrho_{20} = 1.72{\times}10^{-8}$ Ωm and $\alpha = 0.0041/^0$C, while for aluminium $\varrho_{20} = 2.7{\times}10^{-8}$ Ωm and $\alpha = 0.0045/^0$C. Therefore, a copper conductor operating at 60 ^{0}C has resistivity of $2{\times}10^{-8}$ Ωm, while an aluminium conductor has resistivity of $3.2{\times}10^{-8}$ Ωm.

Whether or not the amount and flow velocity of the coolant, as determined by eq. (4.54), can be forced through the winding of the magnet depends on the pumping power available and on the detailed structure of magnet. The pressure drop across a tube is the sum of the entrance and exit losses and frictional drop in the tube. The total pressure drop ΔP (in Pa) becomes

$$\Delta P = (0.74 + 0.015\frac{L}{d})10^3 v^2 \tag{4.56}$$

where L and d are the length and diameter of the cooling path (in m), and v is the coolant velocity (in m/s). A typical pressure drop for copper hollow conductors is 0.5 to 0.8 MPa.

4.4.4 Thickness of the iron cladding

The basic requirement for the iron cladding is that it must not be saturated. The magnetic flux in the cladding is $J_s A_c$ where A_c is the cross-sectional area of the cladding. In order to avoid oversaturation of the cladding, the following inequality must be satisfied:

$$B_g A_g \leqq J_s A_c \tag{4.57}$$

where B_g is the magnetic induction in the air gap of cross-sectional area of A_g. The thickness of the cladding can thus be found from this equation.

Table 4.5: Properties of decarburized steel for high-field magnets [B1].

Chemical composition	0.05 - 0.02% C, 0.2% Mn, 0.02% S 0.05% Cu, 0.01% P, 0.02% Ni
Coercivity H_c^{max}	60 to 80 Am^{-1}
Permeability	B > 0.04 T at H = 40 Am^{-1} B > 1.5 T at H = 1200 Am^{-1} B > 2.0 T at H = 2400 Am^{-1}

The choice of the value of J_s is important for correct dimensions of the iron cladding. The values of saturation polarization for commercial steels are usually not available and the value of 1.5 T or less should be used. Low-carbon decarburized steel is frequently used in high-field magnets and typical specifications are shown in Table 4.5 [B1].

Several important rules relating to the usage of iron cladding can be outlined:

- Magnets for which the length of the air gap is greater than the inner diameter of the solenoid, i.e. $L_g > 2a_1$, do not require the iron pole above and below the working volume. On the other hand, when $L_g < 2a_1$, the presence of iron cladding in the vicinity of the working space is necessary.

- By replacing the external air path with a magnetic circuit of low reluctance, the most significant increase in the magnetic induction, for a given magnetomotive force, can be obtained for solenoids of small axial length. Therefore, the contribution of the iron is largest if the coil has a considerable stray magnetic field. In the case of a long thin solenoid, in which the stray field is small, the effect of an external iron circuit is negligible.

4.5 Design of drum magnetic separators

4.5.1 The basic magnet arrangements

The basic design of the magnetic circuit of drum magnetic separators is common to most commercially available models. Permanent magnet blocks are stacked on top of each other in a specific geometrical arrangement. The simplest arrangement is shown in Fig. 4.26. Strontium or barium ferrite blocks, usually magnetized through their thickness are frequently arranged radially in low-intensity drums. The frequently used size of the blocks is 150 mm long, 100 mm wide and 25 mm thick. The number of blocks per stack is determined by the diameter of the drum. These radially magnetized stacks are most often arranged in rows parallel with the axis of the drum, separated by an air gap. A radial arrangement is also used, as discussed in Section 2.1.5. and as shown in Fig. 2.12. Figure 4.27 illustrates the axial arrangement of magnet blocks in a magnetic drum.

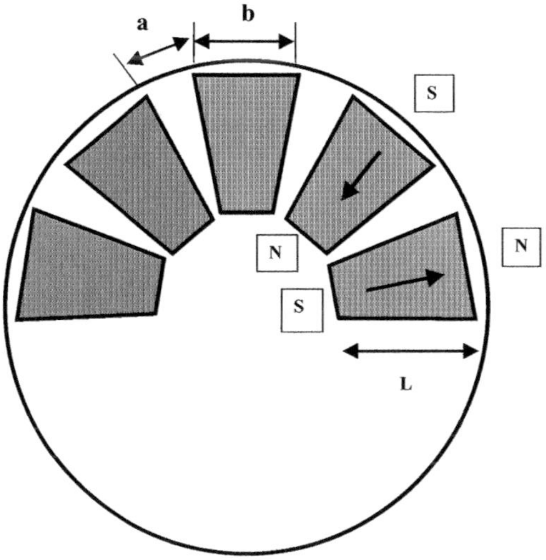

Figure 4.26: Schematic diagram of radially magnetized magnets in a magnetic drum with axial arrangement of the poles.

Figure 4.27: Axial arrangement of magnet blocks in a magnetic drum.

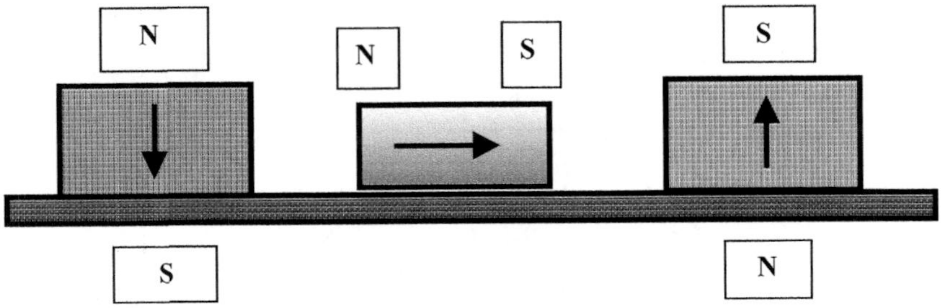

Figure 4.28: Position of a backing magnet (cross pole) in a drum magnetic separator.

High-intensity drum separators employ rare-earth magnets, usually NdFeB, and it is a common practice to simply replace ferrite magnets with rare-earth blocks, with the intent to achieve a higher magnetic field [A13]. The rare-earth blocks are smaller than the ferrite blocks, for example 50×50×15 mm.

Another arrangement of permanent magnets which is frequently employed is shown in Fig. 4.28. A ferrite magnet, magnetized in a transverse direction is inserted between the radially magnetized stacks of permanent magnets. Explanations of the function of this "backing" of "blocking" magnet (or "cross pole") are as numerous as its names. Reduction of the leakage flux between the adjacent main stacks and addition of extra flux [M19] are the most likely explanations of the benefit that this arrangement offers.

4.5.2 Geometry of the magnet assembly

Although drum magnetic separators are one of the oldest types of magnetic separators, and certainly the most widely used, design of the magnetic circuit is still to a great extent empirical. A simple rule for geometrical arrangement of the magnet blocks that was used for many decades states that the ratio of the magnet block width b and the width of the air gap a should be $b/a = 3$ to 5 for ferrite and rare-earth magnets and $b/a = 1.2$ for Alnico magnets and electromagnets [D4]. The design difference is the result of different demagnetization behaviour of the permanent magnet materials, as shown in Fig. 4.2. Using computer modelling, this design rule has been revised, as will be discussed later.

A typical distribution of the field pattern along the circumference of a drum with axial pole arrangement is shown in Fig. 4.29. The maxima in the magnetic induction occur at the centres of the air gaps while the minima occur at the centre of the magnet block. Comparison of the magnetic induction generated by different magnet arrangements in a drum is given in Fig. 4.30.

The relative values of the magnet and air gap widths affect the magnetic field and force distribution around the drum. For instance, by reducing the distance between the magnet blocks, or the so-called pole pitch (defined either

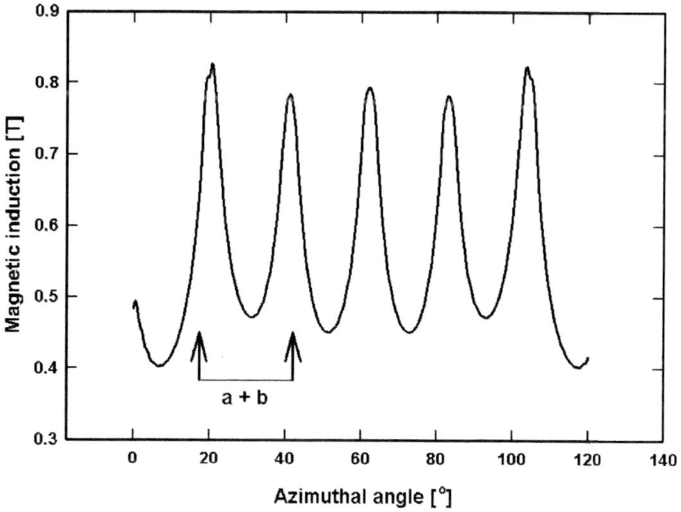

Figure 4.29: Distribution of the magnetic induction along the circumference of an NdFeB drum with axially arranged poles. The width of the radially magnetized magnets b = 100 mm and the air gap width a = 10 mm [M24].

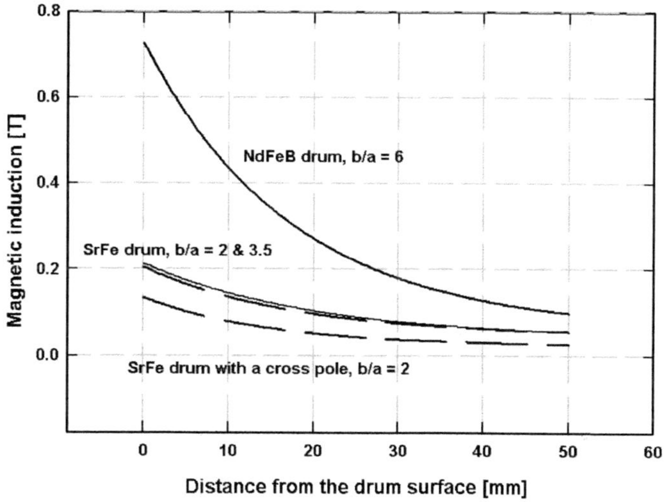

Figure 4.30: The maximum magnetic induction generated by different magnet arrangements in an axial drum with radial magnetization of the magnet blocks (adapted from [M10]).

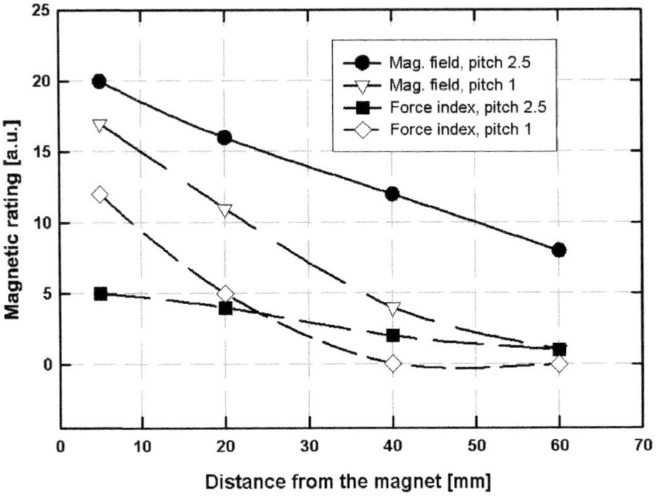

Figure 4.31: Magnetic field and force index (in arbitrary units) as a function of distance from the drum surface, for different distances between the poles (adapted from [S13]).

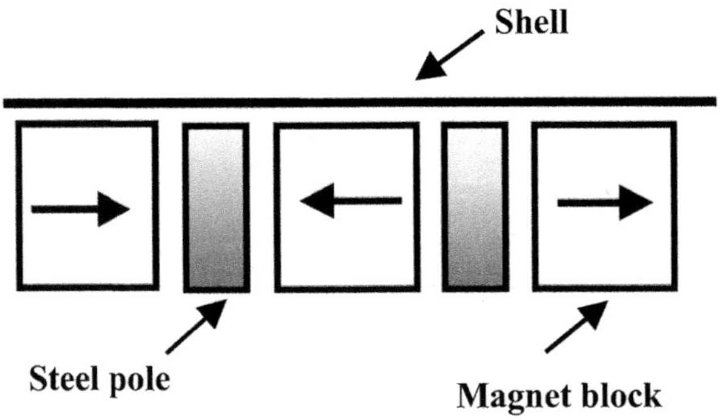

Figure 4.32: Steel poles sandwiched between azimuthally arranged magnet blocks.

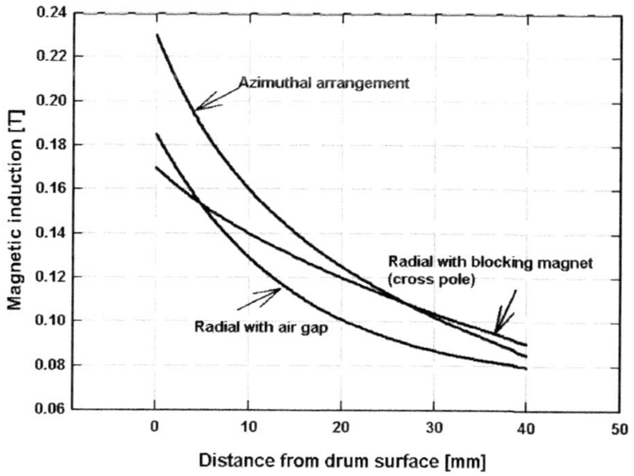

Figure 4.33: Magnetic induction generated by magnetic drums, with three different magnet arrangements, as a function of distance from the drum surface (adapted from [F19]).

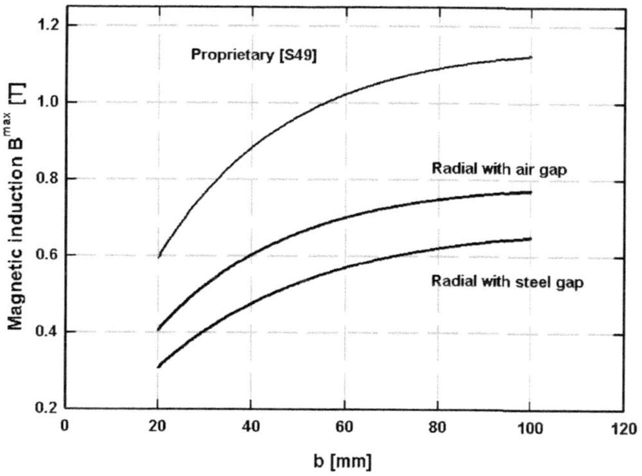

Figure 4.34: Dependence of the magnetic induction on the surface of the drum, on the width b of the NdFeB magnet, for a = 10 mm. Three different configurations are shown [M24].

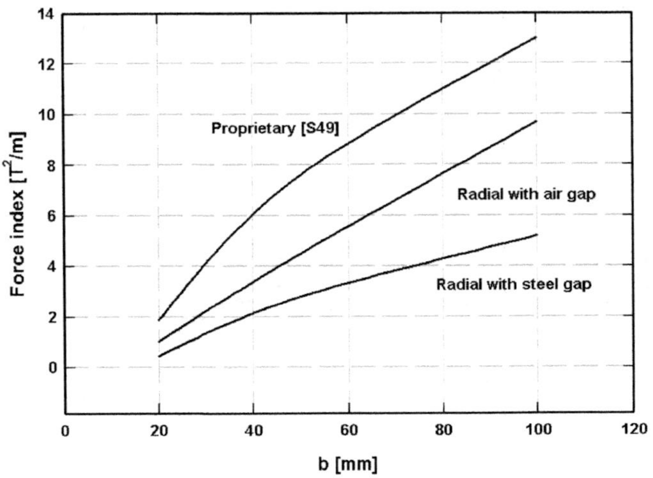

Figure 4.35: Dependence of the force index at 15 mm from the drum surface, on the width b of the NdFeB magnet, for a =10 mm. Three different configurations are shown [M24].

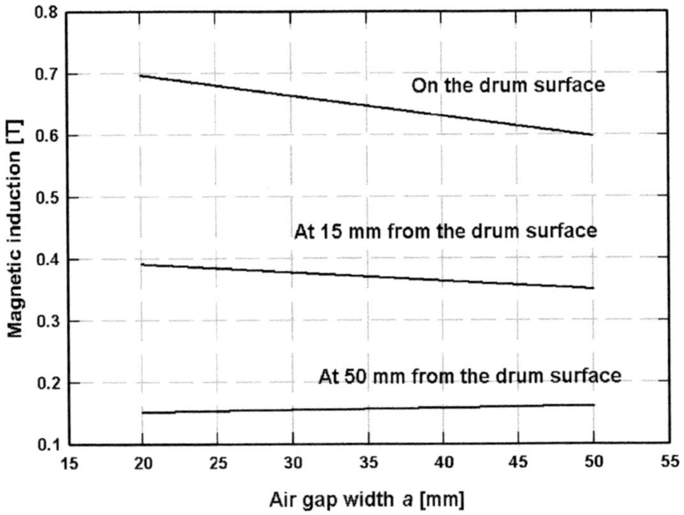

Figure 4.36: Magnetic induction as a function of the width a of the air gap, for an NdFeB magnetic drum, for b = 100 mm [M24].

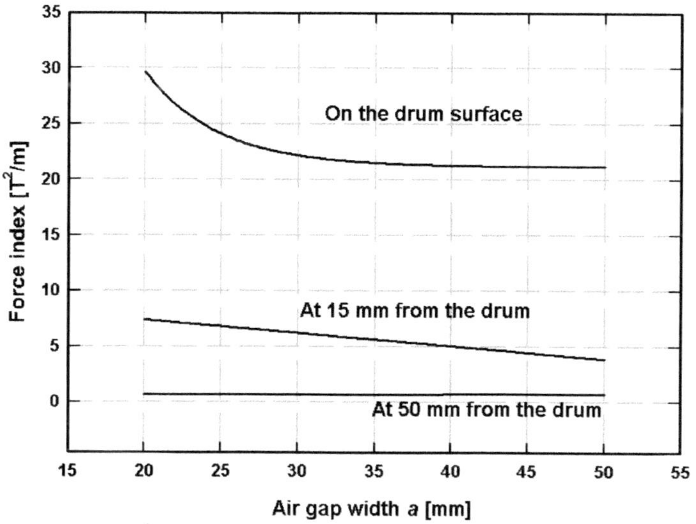

Figure 4.37: Force index for an NdFeB drum, as a function of the width a of the air gap, for $b = 100$ mm [M24].

as a or $a + b$ in Fig. 4.26), it is possible to increase the field gradient. Although the field strength is reduced, the increase in the field gradient is sufficient to result in the increase of the force index. The situation is depicted in Fig. 4.31. It can be seen that the increase in the force index by reducing the pole pitch takes place only in the proximity of the drum surface and that the force index decreases rapidly away from the drum. In other words, the smaller the pitch, the higher the force index on the drum surface and the shorter the reach of the magnetic force.

In the nineteen eighties the Fives-Lille group developed drum magnetic separators with steel poles sandwiched between ferrite magnet blocks with azimuthally arranged direction of magnetization, as shown in Fig. 4.32. As a result of their superior performance, compared to the conventional design, in terms of magnetic force and metallurgical performance, the azimuthal concept was commercially accepted for some applications. Comparison of the magnetic field strength achievable by the conventional and azimuthal designs is shown in Fig. 4.33.

4.5.3 Electromagnetic modelling of the design

Electromagnetic modelling software that has become available in recent years allows for the optimization of the design of conventional magnetic circuits. The modelling allows the investigation of the effect of various geometrical parameters of the drum on the magnetic field and magnetic force. For instance, Fig. 4.34

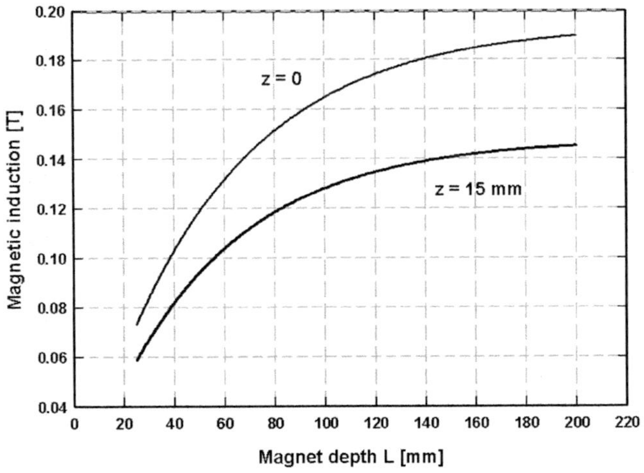

Figure 4.38: The dependence of the magnetic induction generated by a strontium ferrite block 100×150 mm, on its depth L, on the surface of the block and at 15 mm from the surface.

illustrates the variation of the magnetic induction on the surface of a drum with the width b of the magnet block. It can be seen that for a constant width $a =$ 10 mm of the gap, the maximum magnetic induction is achieved for the magnet width of 100 mm. The ratio $b/a = 10$ is thus in disagreement with the empirical rule-of-thumb mentioned earlier.

Figure 4.34 also shows that replacement of the air gap with a steel block results in the reduction of the magnetic field. Similar conclusions can be drawn from Fig. 4.35, which displays the effect of the magnet width on the force index.

The effect of the width a of the air gap on the field at various distances from the drum surface is illustrated in Fig. 4.36. It can be seen that by increasing a, for a constant value of the width of the magnetic block (i.e. with the ratio b/a decreasing), the magnetic induction decreases, at least in the proximity of the drum surface. At larger distances from the drum surface, e.g. at 50 mm, the effect of the value of a (and thus of b/a) is diminishing. Similar trends can be observed for the force index as shown in Fig. 4.37.

The results illustrated in Fig. 4.30 and Figs. 4.34 to 4.37 were obtained for a drum with axial pole arrangement and with magnets magnetized radially. The diameter of the drum was 600 mm and the depth L of the magnet stack was 100 mm.

The effect of the depth of the magnet stack can be determined by using eq. (4.30), which allows for the variation of the depth L , as shown in Fig. 4.9, and for the calculation of the magnetic field generated by such a magnet block, at different distances z. Figure 4.38 illustrates such a dependence for a strontium

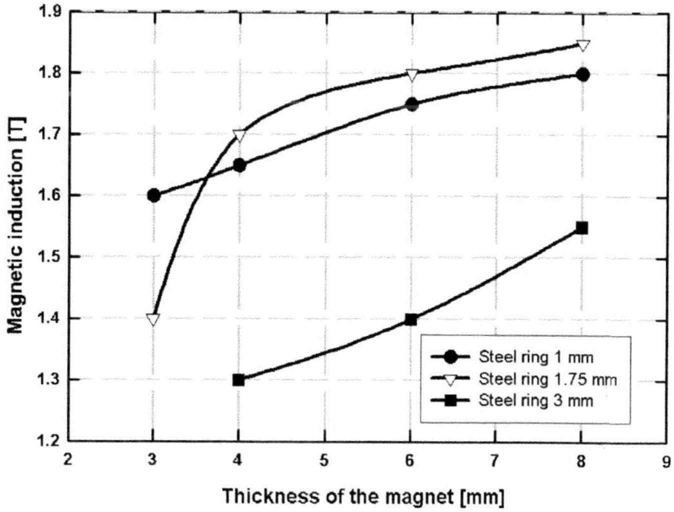

Figure 4.39: Magnetic induction on the roll surface, as a function of the magnet ring thickness. The values of magnetic induction were obtained by computer modelling [F18].

ferrite block, of dimensions 150 mm (W) × 100 mm (D). It can be seen that the benefit of increasing the depth of the magnet stack above 140 mm is limited.

The computer modelling also allowed for the development of new geometries of the magnet arrangement, capable of producing a higher magnetic strength and magnetic force [A13, S8]. Some of these new designs are proprietary and the patents are still pending. Comparison of one such a design, with conventional designs, is shown in Figs. 4.34 and 4.35.

4.6 Design of a magnetic roll

4.6.1 Basic concepts of design

The overall view of a magnetic roll separator is shown in Figs. 2.30 and 2.31. The magnetic roll itself forms a part of a short conveyor fed by a vibratory feeder. The conveyor belt is made of a thin high-tensile material, often with antistatic coating. The belt is supported by second roll (idler). Roll separators are usually of modular design allowing them to be stacked in any number of passes of material. Typical specifications of such a roll separator are shown in Table 4.6.

The efficiency of separation is affected by the roll design and by several operating variables: the number of separation stages, belt speed, splitter position and belt loading. Most of these variables are determined by particle size

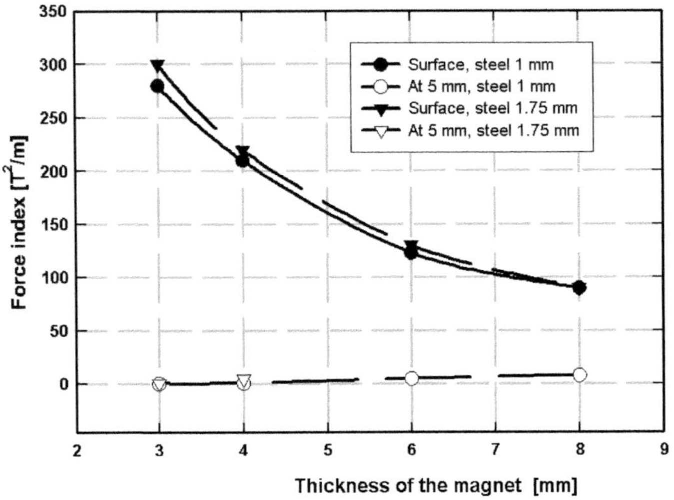

Figure 4.40: Force index as a function of the magnet ring thickness, for thickness of steel rings equal to 1 mm and 1.75 mm, on the roll surface and at 15 mm from the surface.

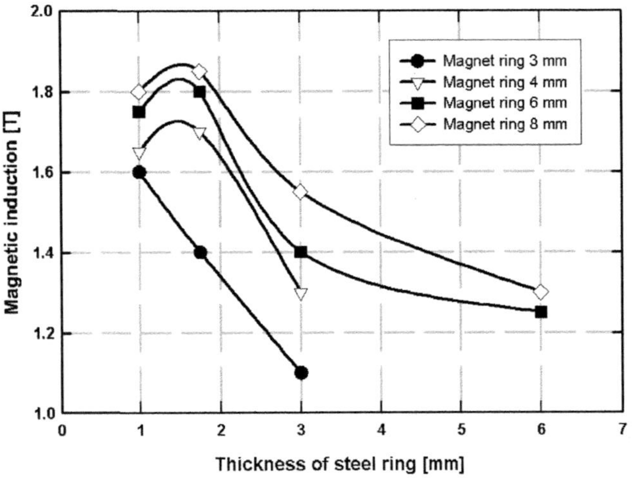

Figure 4.41: Magnetic induction on the surface of a roll, as a function of thickness of steel rings, for several thicknesses of permanent magnet rings.

Table 4.6: Typical specifications of a permanent roll magnetic separator.

Parameter	Specification
Belt thickness	minimum 0.12 mm
Roll diameter	72, 76, 100, 220 & 300 mm
Roll length	250, 500, 1000 and 1500 mm
Max. magnetic induction	NdFeB: 1.9 T, SmCo: 1.5 T, SrFe: 0.5 T

distribution and magnetic properties of the feed and by the recovery vs grade requirement.

4.6.2 Design parameters

Thicknesses of magnet and steel rings

As has been discussed in Section 2.3.3 and shown in Fig. 2.28, the roll consists of discs or rings of permanent magnets, interleaved with mild steel rings. The adjacent permanent magnet rings are arranged with the same polarity facing one another, i.e. rings are arranged in a repulsive mode. The relative thicknesses of the magnets and the steel rings affect the magnetic field strength and its pattern around the roll. This situation is depicted in Figs. 4.39, 4.40 and 4.41. It can be seen that the magnetic induction on the surface of a roll increases with increasing thickness of the magnet ring. At the same time it achieves the maximum value for a certain, rather small, thickness of the steel rings. With further increase in the thickness of the steel rings the magnetic induction decreases. This is a consequence of the fact that the saturation magnetization of steel is much greater than that of the permanent magnet material. If the thickness of the steel exceeds a certain optimum value, the magnet ring does not succeed in saturating the steel, as transpires from eq. (4.12).

As can be seen in Figs. 4.42 and 4.43, for thin steel and magnet rings, which generate a high force index close to the roll surface, the reach of the magnetic force drops dramatically away from the surface. The rate of decrease is much smaller for thicker magnet rings and even more so for an NdFeB drum.

Optimization of the design of the magnetic roll consists not only of maximizing the magnetic induction and force index on the roll surface and at a suitable distance from the roll, but also of matching the dimensions of the magnet and steel rings thicknesses to the particle size distribution of the material to be separated. Distributions of the magnetic induction along the axes of NdFeB and SrFe rolls are shown in Figs. 4.44 and 4.45. The magnetic field (and magnetic force) on the roll surface are the greatest in a relatively narrow region at the interface between the magnet and steel rings. It can also be seen that the field and the force along the circumference of the magnet rings are very feeble, and it is imperative, therefore, that no particles are continuously exposed to this region of the roll.

The thickness of the magnet rings should thus be of magnitude similar to the

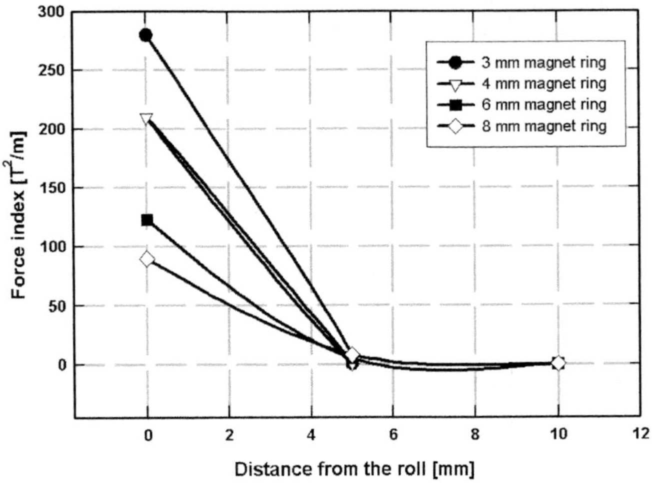

Figure 4.42: Force index as a function of distance from the roll surface, for various thicknesses of magnet rings. Steel ring thickness equals 1 mm.

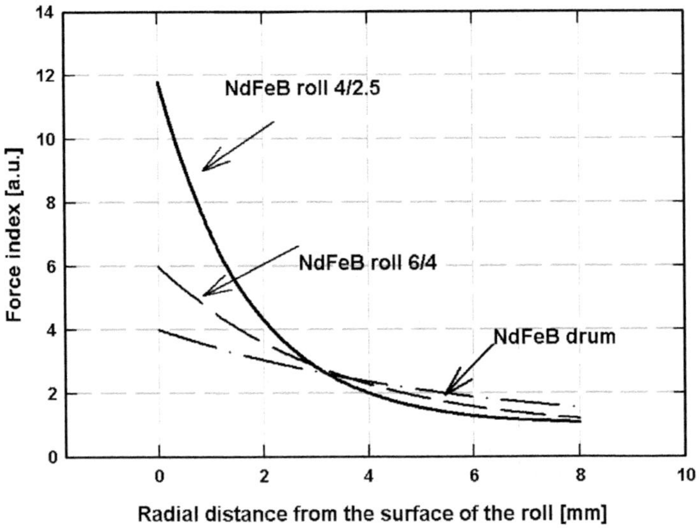

Figure 4.43: Force index (in arbitrary units), for two configurations of NdFeB rolls and an NdFeB drum, as a function of the distance from the surface of the roll [F18].

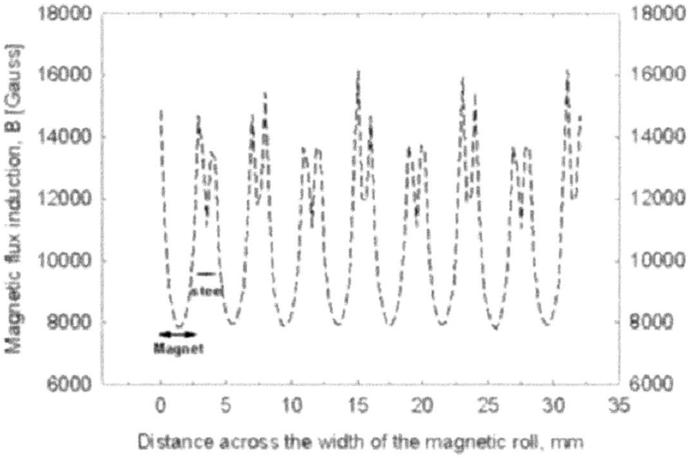

Figure 4.44: Distribution of the magnetic induction on the surface on an Nd-FeB roll with 3/1 configuration (magnet thickness: 3 mm, steel thickness: 1 mm).

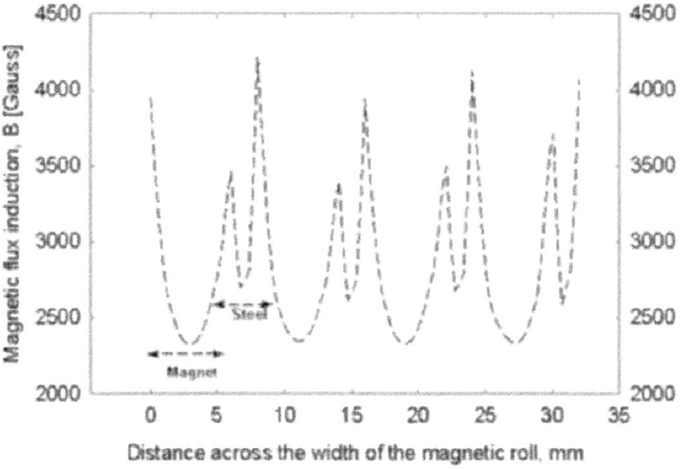

Figure 4.45: Distribution of the magnetic induction on the surface of a SrFe roll with 6/2 configuration (magnet thickness: 6 mm, steel thickness: 2 mm).

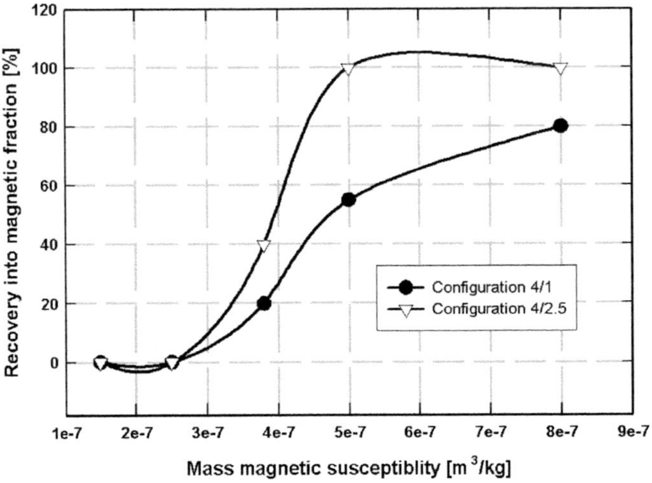

Figure 4.46: The partition curves for the recovery of 4 mm magnetic tracers by NdFeB rolls of different configurations [F20].

mean particle size of the material to be separated. Thin magnet rings should be used to separate fine particles, while for larger particles a thicker magnetic ring offers a larger contact area with the roll, and a longer reach of the magnetic force, of the order of the radius of the particles.

It transpires from the above discussion that the design of the roll, and specifically the thicknesses of the magnetic and steel rings, is dictated by the magnetic properties and size distribution of the feed. The roll design must be optimized in order to maximize the magnetic force and its reach and to match the dimensions of the particles with the magnet thickness. A specific configuration expressed as magnet thickness/steel ring thickness is suitable only for a narrow size fraction and typical configurations are summarized in Table 4.7.

The optimum configuration obtained by electromagnetic modelling can be validated experimentally using magnetic tracers of suitable sizes and accurate values of magnetic susceptibility. Magnetic partition curves can then be constructed, the probable error *Ep* can be calculated and a configuration with the lowest *Ep* can be determined. *Ep* for a magnetic separator can be defined as

$$Ep = \frac{\chi_{75} - \chi_{25}}{2} \tag{4.58}$$

where χ_{75} and χ_{25} are the mass magnetic susceptibilities of particles 75% and 25% of which, respectively, report to the magnetic fraction. Typical partition curves, for 4 mm particles, are shown in Fig. 4.46. It can be seen that the partition curve for the roll configuration 4/2.5 is steeper than that for 4/1

Table 4.7: Various configurations of NdFeB magnetic rolls. Configurations are expressed as the ratio magnet thickness/steel thickness.

Configuration	Size fraction	Comment
3/1, 4/1 or 4.1.5	- 5 + 0.1 mm	Short reach, high field
6/1.5, 6/2, 6/3	- 10 + 5 mm	Longer reach, lower field
8/2, 8/3	+ 10 mm	For large particles

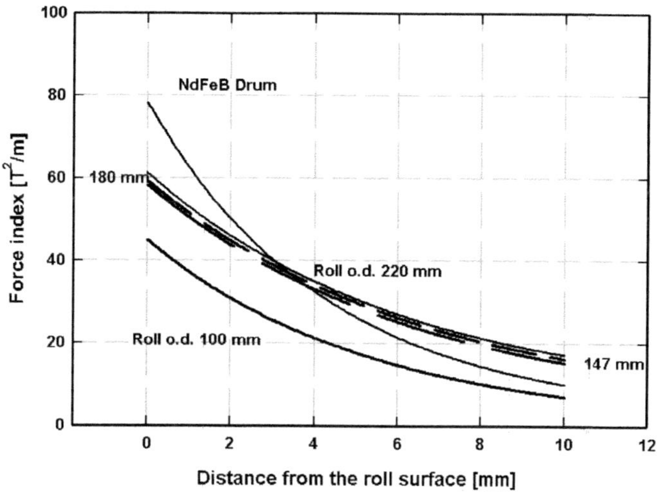

Figure 4.47: Force index for NdFeB rolls of different diameters, as a function of distance from the roll surface. Roll configuration: 12/7.5. The data were obtained by computer modelling (adapted from [F18]).

configuration and Ep is correspondingly lower. Configuration 4/2.5 is, therefore, more selective in separating 4 mm particles of different magnetic susceptibilities.

Diameter of magnet rings

In the past, the diameter of a rare-earth magnetic roll was dictated by availability of the largest magnetic rings. While the first commercial rolls had diameters of 72 and 76 mm, recent progress in manufacture of rare-earth permanent magnet rings and segments allows for the manufacture of rolls with diameter of 100, 220 and 300 mm. The choice is determined by a combination of throughput, performance and cost considerations. The present advantageous cost situation favours larger diameter rolls, particularly when large throughputs are required [A13].

In a simplified assumption of a constant magnetic force and constant resi-

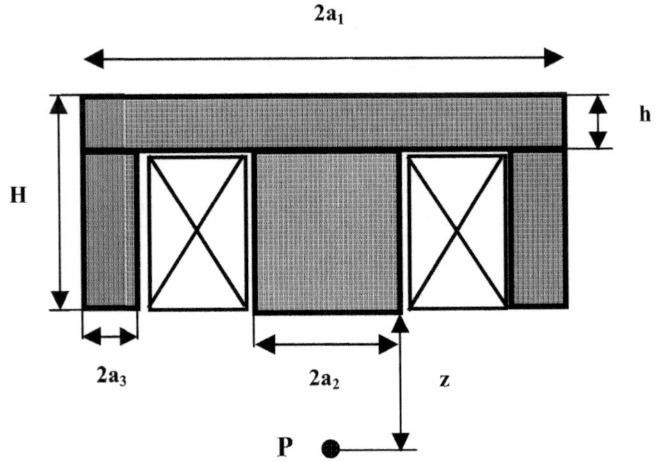

Figure 4.48: Schematic diagram of a suspended magnet.

dence time, a throughput increase by increasing the roll diameter, follows the ratio D_1/D_2, where D_1 and D_2 are the roll diameters, as follows from eq. (5.4). It is sometimes argued that by increasing the magnet diameter the magnetic force also increases and an even higher throughput can be expected. It appears, however, that the increase in the magnet ring diameter above 100 mm brings only negligible increase in the force index, as is illustrated in Fig. 4.47.

On the other hand, a longer retention time of particles in a larger-diameter roll, as a result of a longer arc of the magnet ring, to which the particles are exposed, allows for the increase of speed of rotation of the roll and thus for the increase of the throughput, while maintaining constant residence time. Experience indicates that a throughput increase by factor of two to five, maintaining the same performance, can be obtained with larger diameter rolls [A23, A13].

4.7 Design of a suspended magnet

Suspended magnets are usually installed over conveyor belts, discharge ends of feeders or screens and above chutes or launders. Their design is dictated by the burden depth, the belt speed, the clearance required over the burden and the size, shape and mass of the tramp iron that is being removed. Strength and reach of the magnetic field and of the magnetic force must, therefore, be maximized for high efficiency of the material removal.

A schematic diagram of an electromagnet-based rectangular suspended magnet is shown in Fig. 4.48. The magnet consists of an iron yoke and a coil, which magnetizes the yoke. The magnetic field generated at point P, at distance z from the magnet, is the sum of contributions from the magnetized steel and

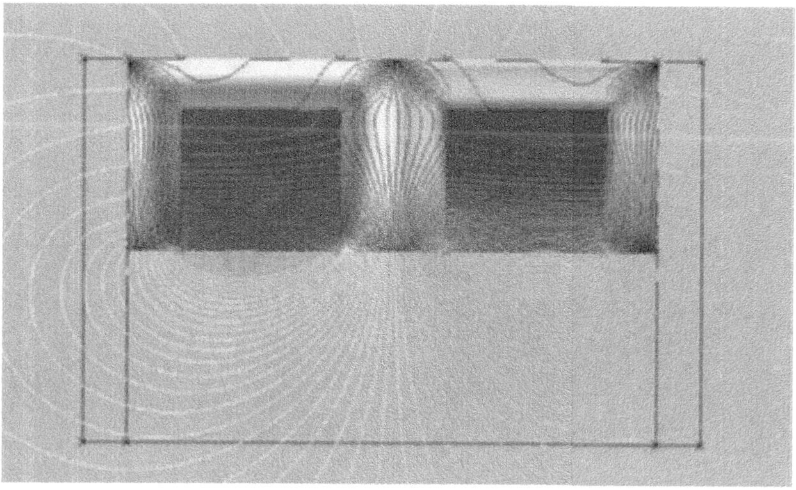

Figure 4.49: Pattern of magnetic flux around a suspended magnet.

the coil. Dimensions of individual parts of the yoke determine the distribution of the magnetic field and magnetic force, their value and reach. Using computer modelling, it is possible to investigate the influence of these dimensions and to determine the optimum design for a given application. The distribution of magnetic flux around a suspended magnet is shown in Fig. 4.49, while a three-dimensional image is presented in Fig. 4.50.

In commercially available suspended magnets, the width $2a_1$ of the magnet typically ranges from 900 mm to 1400 mm, the depth ranges from 700 mm to 1250 mm and the height H is around 500 mm.

Aluminium is the preferred conductor materials, mainly because of its lower density. For the same electric power, the mass of the aluminium coil can be approximately halved, compared with the copper coil [K36]. In addition, the use of the aluminium foil, rather than of the wire, allows for the increase in the filling factor of the winding, by about 40% [K36]. As a result, the number of ampere-turns can be reduced, or, alternatively, at the constant number of ampere-turns, a more powerful magnet can be constructed.

4.8 Design of a ferrohydrostatic separator

Commercial and laboratory ferrohydrostatic separators use either an electromagnet or permanent magnets to generate the magnetic field in the working space. Of fundamental importance in the operation of separators with magnetic fluids are the characteristics of the magnetic levitation force, expressed by the second term on the right-hand side of equation (2.9), namely by $J_f B \nabla B$. These characteristics are determined by the pattern of the gradient of the magnetic

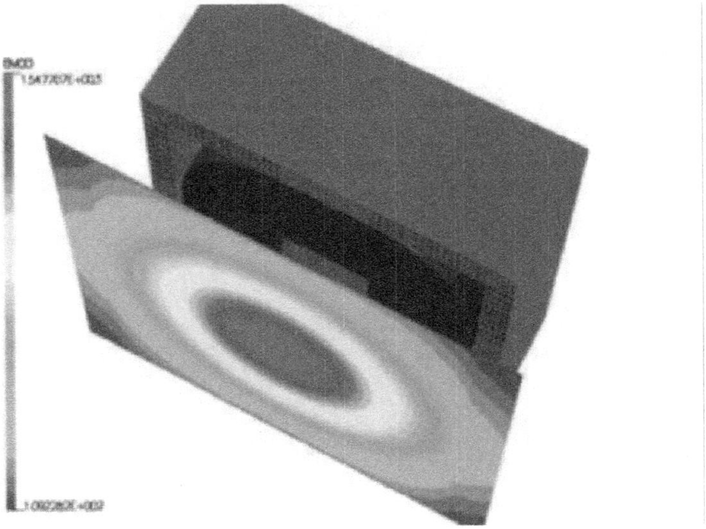

Figure 4.50: A three-dimensional image of the magnetic field around a suspended magnet.

field and, therefore, by the shape of the pole tips.

The shape of the pole tips is determined by several factors, namely by the choice of the magnetic fluid and the number of density fractions that are to be produced. If paramagnetic liquids are to be used, the constant apparent density of separation in the entire volume of the fluid will be obtained if the force acting on a particle is constant, F = const., which corresponds to $B\nabla B$ = const., as transpires from eq. (3.144). This condition can be re-written as

$$\frac{\partial B^2}{\partial y} = const. \tag{4.59}$$

where y is the vertical axis, as shown in Fig. 4.51. This requirement is the result of the linear dependence of the fluid polarization, namely $\kappa_f B$, on the external magnetic field. The field that satisfies condition (4.59) is the isodynamic field.

As can be seen from eq. (3.145), the situation is somewhat different with ferrofluids: if the ferrofluid is magnetically saturated, a constant density of separation will be obtained by maintaining the constancy of the force acting on the particles by satisfying the condition ∇B = const. In other words, the condition

$$\frac{\partial B}{\partial y} = const. \tag{4.60}$$

should be satisfied in a sufficiently large region of the magnetic field.

Two basic shapes of pole tips have been used in ferrohydrostatic separators, namely a simple wedge geometry of the working space, and pole tips with hyperbolic shape.

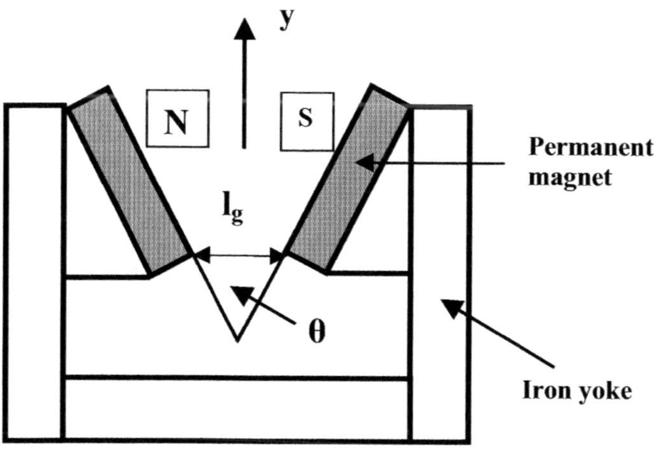

Figure 4.51: Schematic diagram of an FHS magnetic circuit with a wedge-shaped working gap.

4.8.1 FHS separators with a wedge-shaped working gap

Wedge-shaped working space was used in separators with permanent magnets, such as those built by Tohoku University [S50], Nittetsu Mining Co. (Japan) [S51] and the Too Geos (Russia) [S52] machines. A schematic diagram of such a magnetic circuit is shown in Fig. 4.51.

Since the gradient of the magnetic field, in this design, is not constant along the vertical ($\partial B/\partial y \neq$ *const.*), the apparent density of the ferrofluid along the vertical will also vary. A typical distribution of apparent density of the ferrofluid in a separator with wedge-shaped working space, used for recovery of non-ferrous metals, is shown in Fig. 4.52. It can be seen that essentially a continuous spectrum of densities is present in the pool of the ferrofluid. The usefulness of this density distribution is very limited as inter-particle collisions and interactions do not allow accurate and selective separation of individual metals into individual density fractions.

Distributions of the magnetic induction and the apparent density of the ferrofluid along the vertical axis for the Too Geos separators with ferrite permanent magnet-based wedge-shaped pole tips were reported by Solodenko [S52]. The results are summarized in Fig. 4.53. Shimoiizaka et al. [S50] obtained a similar pattern in a separator that used SmCo permanent magnets (see Fig. 4.54). It can be seen again that a magnetic system with wedge-shaped poles creates a plurality of values of the field gradient, and, therefore, of the apparent density of the ferrofluid along the vertical axis of the magnet gap. It is clear that such patterns of the apparent density can be used only for density separation of mixtures of materials with large density differential.

Specifications of the Tohoku University FHS separator, as reported by Shi-

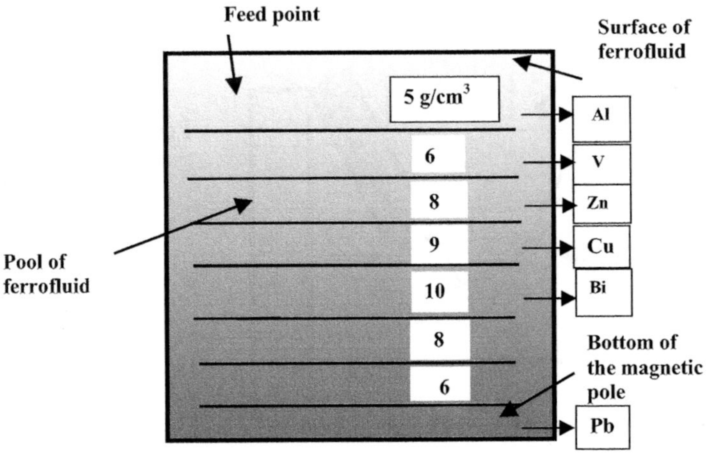

Figure 4.52: Distribution of the apparent density of ferrofluid in an FHS with a wedge-shaped working space. Recovery of various non-ferrous metals at different positions along the vertical axis is indicated.

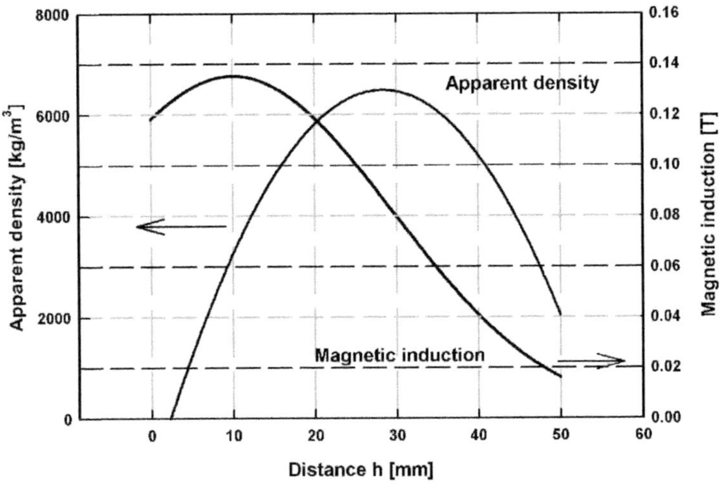

Figure 4.53: The apparent density and magnetic induction in a permanent magnet-based ferrohydrostatic separator with wedge-shaped pole tips, as a function of distance from the base of the pole tips (adapted from [S52]).

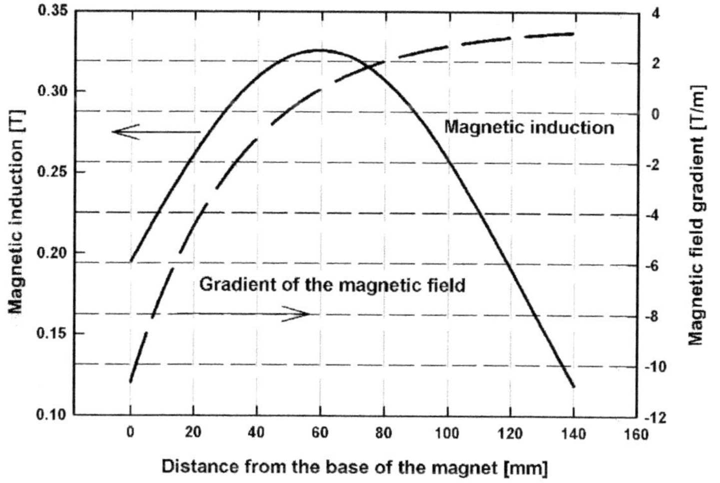

Figure 4.54: Magnetic induction and its gradient in the gap of a SmCo magnet-based ferrohydrostatic separator with wedge-shaped pole tips (adapted from [S50]).

Table 4.8: Specifications of the Tohoku University FHS separator for the recovery of non-ferrous metals, for three different values of the width of the magnet gap [F21].

Parameter	l_g = **60 mm**	l_g = **80 mm**	l_g = **100 mm**
Magnetic field [T]	0.362	0.314	0.265
Field gradient [T/m]	3.5	3.0	2.5
Apparent density [kg/m^3]	9600	8200	6900

moiizaka et al. [S50] and Fujita [F21], are summarized in Table 4.8. SmCo permanent magnets were used to generate the magnetic field in the working space. The height of the ferrofluid pool was 120 mm and the angle θ between the pole tips amounted to 30^0. Water-based ferrofluid of density approximately 1400 kg/m^3 and magnetic polarization of 0.028 T was used.

Another disadvantage of the permanent magnet-based FHS is that the value of the magnetic induction and of the field gradient cannot be easily varied. The magnetic induction can be varied only by changing the distance between the pole tips using a mechanical device that can slide one of the poles against the other. The field gradient can be varied by adjusting the angle θ.

As a result of the absence of constancy of the apparent density of the ferrofluid and of any meaningful control of the operating magnetic field and its gradient, the permanent magnet-based wedge-shaped ferrohydrostatic separators are suitable only for separation of materials that differ in density by at

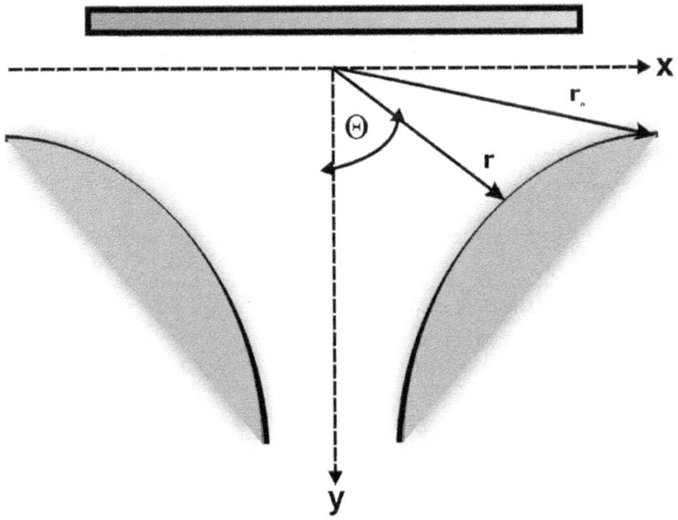

Figure 4.55: Hyperbolic pole profile with a mirror plate.

least 2000 kg/m^3. Recovery of platinum-group metals [S53, S54, G4] and possibly separation of non-ferrous metals [S50] are applications where the limited accuracy of such separators can be tolerated. For more accurate separation of materials with small density differences, hyperbolic pole tips must be used.

4.8.2 Separators with hyperbolic pole tips

A common objective of sink-and-float separation is usually a production of two density fractions selectively and accurately. In order to achieve such a goal in ferrohydrostatic separation, it is necessary for the apparent density of the ferrofluid to be constant throughout the volume of the ferrofluid. As follows from eq. (3.145), this can be achieved by ensuring that the ferrofluid is magnetically saturated and by designing the shape of the pole tips in such a way that the field gradient along the vertical axis is constant, namely $\partial B/\partial y$ = const. An additional constraint on the desired field pattern is that the horizontal gradients, i.e. $\partial B/\partial x$ and $\partial B/\partial z$, are as small as possible. Therefore, for ideal density separation in ferrohydrostatic separators, the following conditions should be met:

$$\frac{\partial B}{\partial y} = const \neq 0, \frac{\partial B}{\partial x} = \frac{\partial B}{\partial z} = 0 \tag{4.61}$$

Such a pattern of the magnetic field can be achieved by a hyperbolic shape of the pole tips [K18], as shown in Fig. 4.55. In deriving the above conditions (eq. (4.61)), it was assumed that the branches of hyperbola forming the pole profile extend to infinity and that their asymptotes (x-axis) are at equal potential [G14]. The latter condition is satisfied in magnetic systems, in which the upper

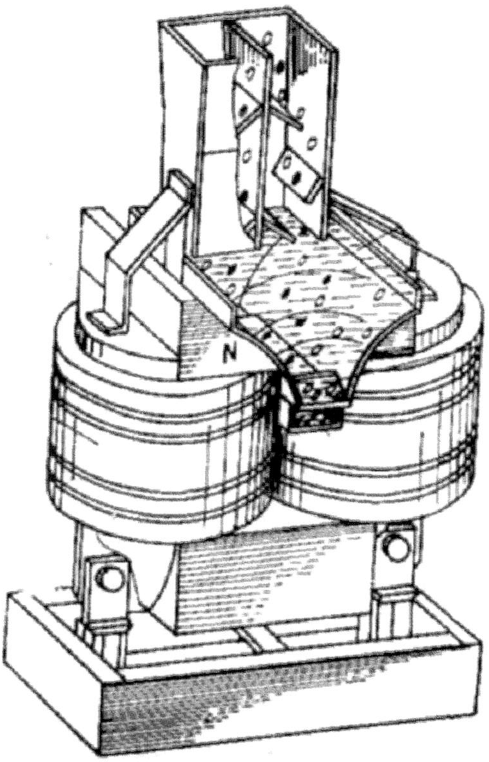

Figure 4.56: GMUO ferrohydrostatic separator with hyperbolic pole tips.

half of the hyperbolic surface is replaced by a ferromagnetic plate, coincident with the x axis, and perpendicular to the y axis, as shown in Fig. 4.55 [K18].

It was shown [K18, G15] that conditions given by eqs. (4.61) cannot be satisfied at all points in the working zone of the magnetic system. These conditions can be met only in the plane of symmetry passing through the centre of the interpolar gap defined by the y axis in Fig. 4.55. The situation is further complicated by the fact that, in practice, the poles have finite dimensions. The hyperbolic surfaces are, therefore, terminated at some finite values of x, y and z. Such a truncation of the ideal hyperbolic shape necessitates corrections to the shape of the pole tips, in order to meet the conditions expressed by eqs. (4.61).

The surfaces of the poles are shaped in such a way that they coincide with one of the elementary solutions of the Laplace equation. A powerful method for the analytical solution of Laplacian fields, namely conformal transformation, was used to calculate the magnetic field generated by poles of finite size and of hyperbolic profile [G14, G15, G16, S55]. This approach is the basis of the design of the GMUO ferrohydrostatic separators, as described in Section 2.8 and

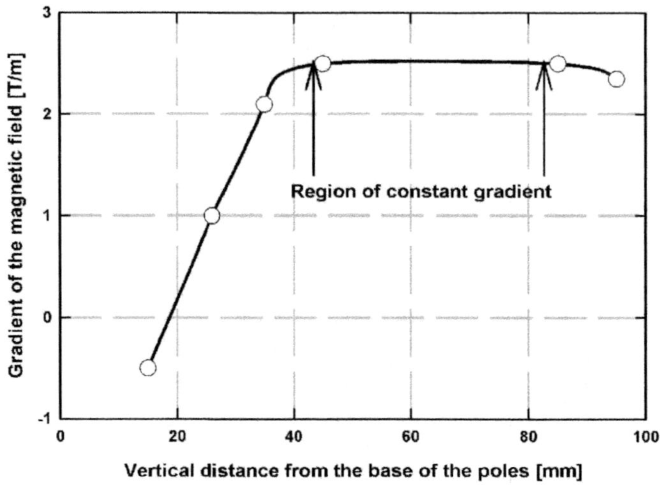

Figure 4.57: The gradient of the magnetic field along the vertical axis of the GMUO FHS-40M separator, as a function of distance from the base of the magnet pole tips.

shown schematically in Fig. 4.56. Similar design of the pole tips was used by the US Bureau of Mines [R9, K17] in their equipment for separation of non-ferrous metals.

The distribution of the magnetic field gradient along the vertical y axis in the GMUO FHS-40M separator is shown in Fig. 4.57. It can be seen that in a 40 mm long region along the vertical the gradient is constant. The apparent density of the ferrofluid is, therefore, also constant and selective and accurate density separation is possible in that region.

It is found, using eq. (3.145), that a field gradient of 2.5 T/m generates an apparent density of approximately 4200 kg/m^3, using, for example, a kerosene-based ferrofluid with magnetic polarization of 0.02 T. The apparent density can be varied by varying the electrical current in the coils and thus varying the magnetic field in the working gap.

The availability of computer electromagnetic modelling software has changed the speed and accuracy of the design of suitable profiles of the pole tips considerably. Although the initial proposal and design of the profiles still has to be worked out from first principles of the magnet design, computer-based field analysis can be used to verify and fine-tune the design. Computer modelling, therefore, allows the pole profiles to be optimized so that the required density range and its accuracy are obtained. An example of the optimized hyperbolic profile of pole tips of FHS is shown in Fig. 4.58 and a general view of a ferro-hydrostatic separator with hyperbolic pole tips is shown in Fig. 4.59.

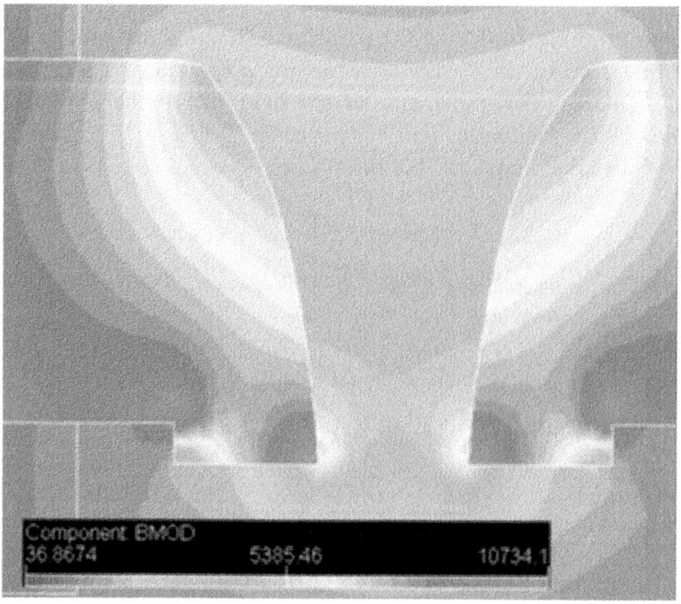

Figure 4.58: Hyperbolic pole tips in FHS, as obtained by computer modelling.

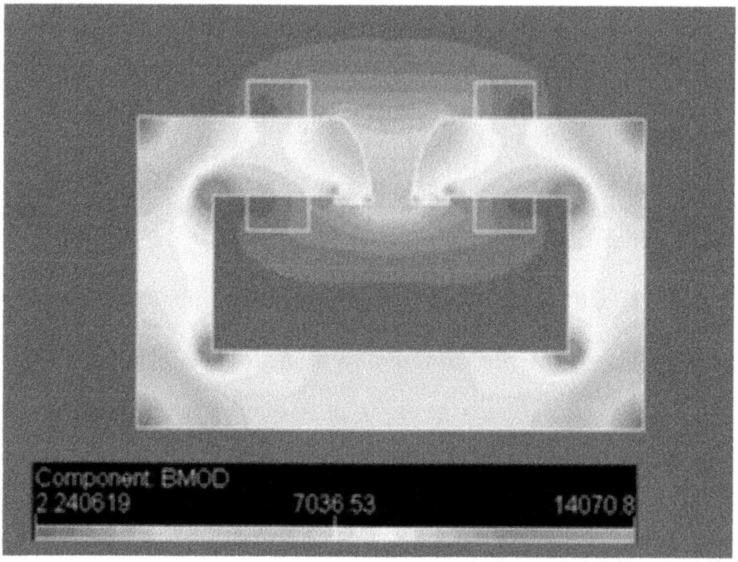

Figure 4.59: Ferrohydrostatic separator with hyperbolic pole tips.

4.8.3 Calculation of profiles of pole tips in FHS

Although computer modelling facilitates optimization and fine-tuning of the pole profiles, the first step in the design of a magnetic system of a ferrohydrostatic separator is, as a rule, the calculation of the field and of the pole profile from first principles. The magnetic field between the poles of the magnet can be represented by a scalar magnetic potential U :

$$H = -\nabla U \tag{4.62}$$

Since the electrical currents are absent in this system, the magnetic potential satisfies the Laplace equation

$$\nabla^2 U = \Delta U = 0 \tag{4.63}$$

which can be written in polar coordinates

$$r^2 \frac{\partial^2 U}{\partial r^2} + r \frac{\partial U}{\partial r} + \frac{\partial^2 U}{\partial \theta^2} = 0 \tag{4.64}$$

A possible solution of the Laplace equation is the potential function

$$U(r, \theta) = R(r) S(\theta) \tag{4.65}$$

where R is a function of r only and S is a function of θ only. It can be shown that these functions have the following form [B25]:

$$R = cr^m + dr^{-m} \tag{4.66}$$

and

$$S = g \cos m\theta + h \sin m\theta \tag{4.67}$$

where c, d, g and h are constants.

A particular solution to the Laplace equation can, therefore, be written in the form of a circular harmonic of order m as

$$U(r, \theta) = (cr^m + dr^{-m})(g \cos m\theta + h \sin m\theta) \tag{4.68}$$

Of practical interest are, for example, solutions of the form [G17, Z7]:

$$U_0 = A_0 \theta \tag{4.69}$$

and

$$U_m = A_m r^m \sin m\theta \tag{4.70}$$

If the surfaces of the poles of the magnet are made to coincide with a given pair of equipotential surfaces (the profile of which is independent of the third dimension), the field between them and the profile of the pole tips are described by eqs. (4.69) and (4.70).

Table 4.9: Properties of pole profiles.

Value of m	U_m/A	Comment
2	$r^2 \sin 2\theta = const.$	constant gradient
3/2	$r^{3/2} \sin \frac{3}{2}\theta = const.$	isodynamic field, $F = const.$
	$\theta = const.$	wedge configuration

Taking into account that $H = -\nabla U$, the modulus of the magnetic field intensity is

$$|H| = mAr^{m-1} \tag{4.71}$$

The gradient of the magnetic field along the vertical (y axis) then has the form

$$\frac{\partial H}{\partial y} = (m-1)mAr^{m-2} \sin m\theta \tag{4.72}$$

while the product of the magnetic field and its gradient (force index) are expressed as

$$H\frac{\partial H}{\partial y} = (m-1)m^2A^2r^{2m-3} \sin m\theta \tag{4.73}$$

These equations can be used to evaluate and design various shapes of the pole profiles. For $m = 1$, $H\partial H/\partial y$, or the magnetic force, is zero. The simplest practical case of a wedge pattern is expressed by eq. (4.69). The wedge configuration is represented by a straight line described by the equation $\theta =$ const.

The isodynamic profile, in which $H\partial H/\partial y =$ const., can be obtained from eq. (4.73) when $m = 3/2$. The isodynamic profile of the pole tips is then given by

$$U_m = Ar^{3/2} \sin \frac{3}{2}\theta = const. \tag{4.74}$$

and

$$H\frac{\partial H}{\partial y} = A^2 \sin \frac{3}{2}\theta \tag{4.75}$$

Pole profiles that generate a magnetic field whose gradient in constant along the vertical axis can be obtained from eq. (4.72) when $m = 2$ and the equations of the profile are as follows:

$$U_m = Ar^2 \sin 2\theta = const. \tag{4.76}$$

and

$$\frac{\partial H}{\partial y} = 2A \sin 2\theta \tag{4.77}$$

Equations (4.74) to (4.77), in which m determines the geometrical features of the field and A describes the magnetic characteristics of the system, provide the full description of the profile of the pole tips. Properties of these configurations are summarized in Table. 4.9.

Design procedure

We will assume that we want to design the pole profiles for a ferrohydrostatic separator operating with a ferrofluid. In order to obtain selective and accurate separation of the feed material into two density fractions, the apparent density of the ferrofluid in the working volume of the separator must be constant. This condition will be met if the field gradient along the vertical axis is constant, i.e. if conditions (4.61) are met. Therefore, the profile pattern must satisfy eq. (4.77).

The practical procedure of designing the pole profile and the magnetic field characteristics is as follows:

1. Determine, from eq. (3.145), what field gradient $\partial H/\partial y$ is needed to obtain the required density of separation, i.e. the apparent density of the ferrofluid.

2. Using $\partial H/\partial y = 2A$ (eq. (4.76)) calculate A.

3. From $H = 2Ar$ (eq. (4.71)) calculate H, for a given value of r. The length of the pole tips and the width of the gap between the pole tips are determined by r. The simplest approach would thus be to take r to represent the radius vector of the point of the closest distance L between the pole tips.

4. Based on the condition $U = Ar^2 \sin 2\theta = const.$, it is possible to calculate the constant from

$$r_0 = \frac{const}{(\sin 2\theta_0)^{1/2}} \tag{4.78}$$

where r_0 is the starting radius vector and θ_0 is the starting azimuthal angle, as shown in Fig. 4.55. Using this constant, the pole profile is then given by

$$r = \frac{const}{(\sin 2\theta)^{1/2}} \tag{4.79}$$

A similar procedure can be used to construct other pole profiles, for instance the isodynamic shape.

The role of the mirror plate

It follows from the pioneering work of Kaiser and Mir [K18] that a ferromagnetic mirror plate simulates the role of the upper half of the hyperbolic surfaces. Computer modelling of the hyperbolic pole profiles has shown [F20] that the presence of the mirror plate negatively affects the constancy of the vertical field gradient for the isodynamic field pattern. On the other hand, the constancy of the field gradient in a magnetic system with constant vertical gradient improved by including the mirror plate.

The position of the plate was found to have an influence on the magnetic characteristics of the circuit, particularly the constancy of the vertical field gradient. For the GMUO FHS-40M separator, shown in Fig. 4.60, the optimum distance of the plate from the top of the pole pieces was about 28 mm. On the other hand, the thickness the plate did not affect the magnetic parameters.

Figure 4.60: GMUO FHS-40M ferrohydrostatic separator with a mirror plate.

4.8.4 Scale-up of ferrohydrostatic separators

The application of ferrohydrostatic separation on production scale, particularly for the recovery of non-ferrous metals from automobile or electronic scrap, requires that separators with throughputs of at least 1 t/h, but preferably more, are available. In existing models of FHS, that are based on the iron-yoke concept with the air gap between the pole pieces, the increase in throughput can be accomplished by increasing the gap between the pole pieces. As has been discussed in Section 4.3, the increase in the length of the air gap results in an increase in the ampere-turns and thus of the mass, size and cost of the magnet.

With ferrohydrostatic separators, the situation is aggravated even further by the fact that in order to maintain the required field gradient, the necessary magnetic field increases linearly with dimensions of the air gap. In other words, since $\dfrac{\partial H}{\partial y} = 2A$ for $m = 2$ (the constant gradient field pattern), the magnetic field strength, given by $H = 2Ar$ (eq. (4.71)), required to maintain this gradient, increases with r. The scale-up of a ferrohydrostatic separator by increasing the width of the air gap thus results not only in an increase in the magnetic field, with attendant uneconomical increase in the volumes of iron and copper, but also in the increase of the height of the pole pieces well beyond the height of the pool of the ferrofluid required for efficient separation.

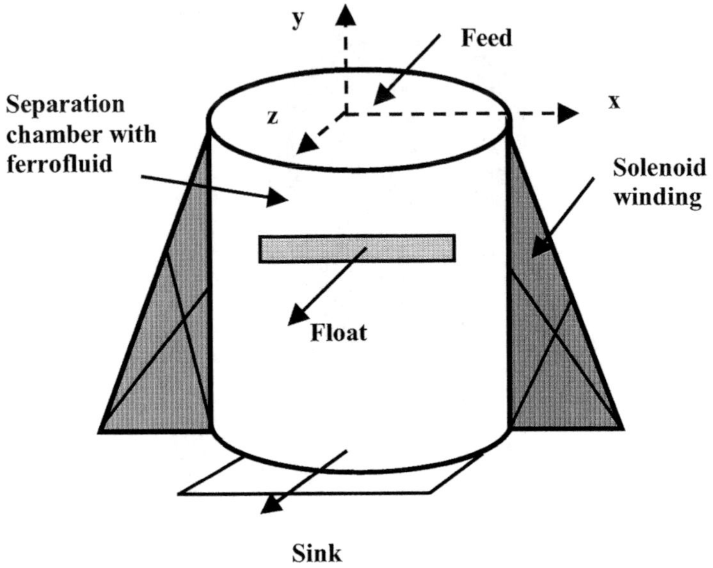

Figure 4.61: Schematic diagram of a ferrohydrostatic separator with a solenoid magnet [S56].

Solenoid-based FHS

It is clear that the design of a large-throughput ferrohydrostatic separator must not be based on a simple increase of the air gap between the pole pieces. A natural approach to the scale-up of FHS would thus be to replace an iron-core magnet by a solenoid magnet as was proposed by Svoboda [S56]. Figure 4.61 illustrates such a solenoid-based FHS separator. A solenoid, and particularly an iron-clad solenoid, has numerous advantages compared to an iron yoke magnet.

While it is necessary to increase the air gap in an iron yoke magnet in order to achieve an increase in throughput of material to be separated, with the attendant drawbacks mentioned above, the throughput of a solenoid-based separator can be increased by merely increasing the relevant transverse dimension (diameter) of the solenoid. The axial length (along the y - axis, as shown in Fig. 4.61) of the air gap and thus the number of ampere-turns required to generate a given magnetic field remain constant.

A solenoid allows the design of the magnetic field pattern to be achieved in a simple and accurate manner. Such an approach facilitates the provision of a magnetic field gradient which is constant, thereby enabling close control to be maintained over the apparent density of the ferrofluid and accordingly over the cut point density. This can be achieved, for example, by precisely designing the winding of the solenoid, or by varying the current density at different positions in the winding, or by using a multiple winding arrangement. This is in contrast

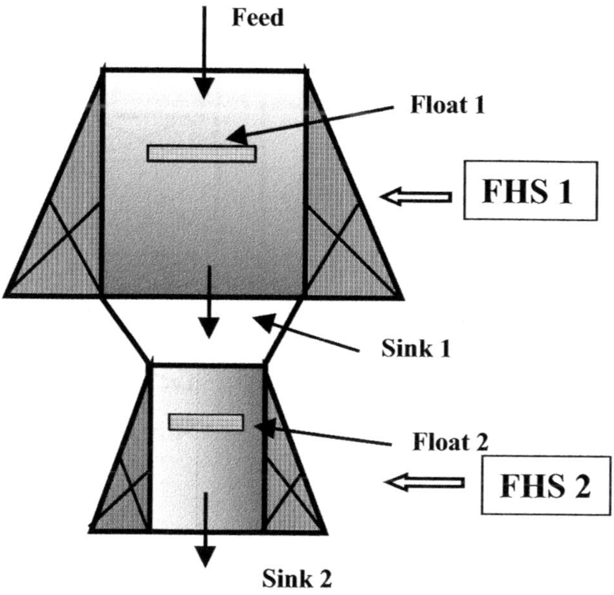

Figure 4.62: A two-stage ferrohydrostatic separator with solenoid windings [S56].

to iron-core magnets, where approximate solutions of the Laplace equation are used to design approximate shape of pole tips, a task that is further complicated by non-linear magnetization characteristics of iron.

The magnetic field across the transverse dimension of a solenoid (along the x - axis) is homogeneous in a large region within the solenoid, which means that the same constant apparent density of ferrofluid can be achieved across the essentially full transverse dimension. Therefore, almost the entire diameter of the solenoid can be used for separation. This is in contrast to iron-core magnets where the constant gradient exists only at the plane defined by the y - axis and considerable deviations from a constant gradient exist in the proximity of the pole tips.

Because of the relatively small mass and size of a solenoid compared to an iron-yoke magnet, it is possible to arrange for two or more FHS units to provide for multi-stage separation, as shown in Fig. 4.62. In this arrangement the top unit, operating, for example, at the apparent density of 8000 kg/m^3, can float zinc and sink copper and lead, while copper could be separated from lead into the floating fraction in the bottom separator operating at the apparent density of 9500 kg/m^3.

Computer modelling of the solenoid-based FHS [S8] showed that it is fairly simple to achieve field gradients of the order of 3 T/m in sufficiently large volumes of ferrofluid. It was also shown that the use of steel cladding considerably

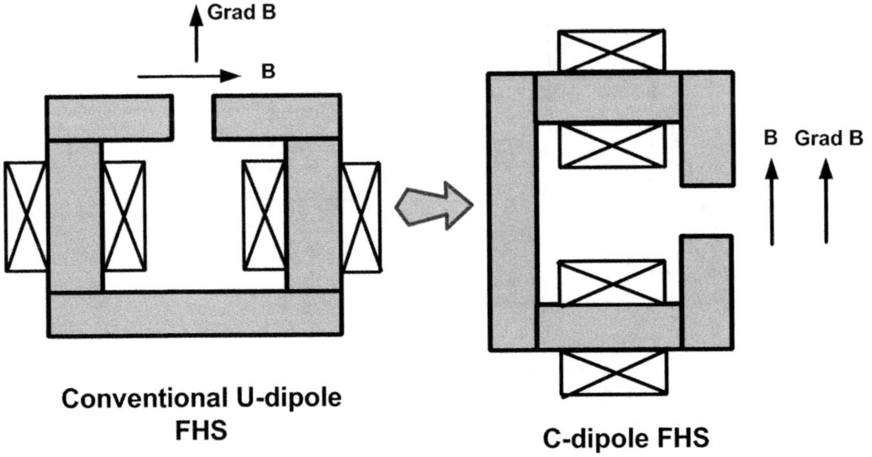

Figure 4.63: U-dipole magnetic circuit versus C-dipole circuit.

increased this volume in both axial and radial directions. The modelling also confirmed the technical and economic advantages of this design by the use of a superconducting solenoid.

C-dipole-based FHS

The scale-up of a conventional iron-yoke-based ferrohydrostatic separator can also be accomplished by re-orienting the U-shaped magnet (Fig. 3.53), to form a C-dipole circuit, as shown in Fig. 4.63. Svoboda [S57] proposed the use of C-dipole magnet to generate a vertical magnetic field, the pattern of which can be controlled by appropriate design of the magnetizing coils on upper and/or lower legs of the C-dipole, and/or by controlling the relative polarity of the electrical current flowing through these coils. Appropriate shaping of the pole-tips of the upper and lower pole pieces can also be employed to create a required field pattern.

The scale-up of such a C-dipole-based separator can be accomplished by merely increasing the length of the magnet L_m, and maintaining a constant length of the air gap L_g between the pole-pieces, as shown in Fig. 4.64. The magnetic field along the length of the C-dipole is homogeneous, and the same magnetic field pattern and apparent density of the ferrofluid, along the vertical axis, can be maintained along the full length of the magnet. One possible embodiment of such a concept of the C-dipole FHS is shown in Fig. 4.65.

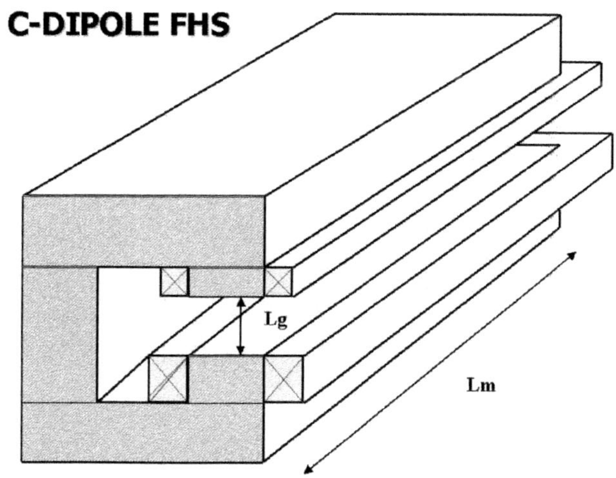

Figure 4.64: Schematic diagram of a ferrohydrostatic separator based on a C-dipole magnet [S57].

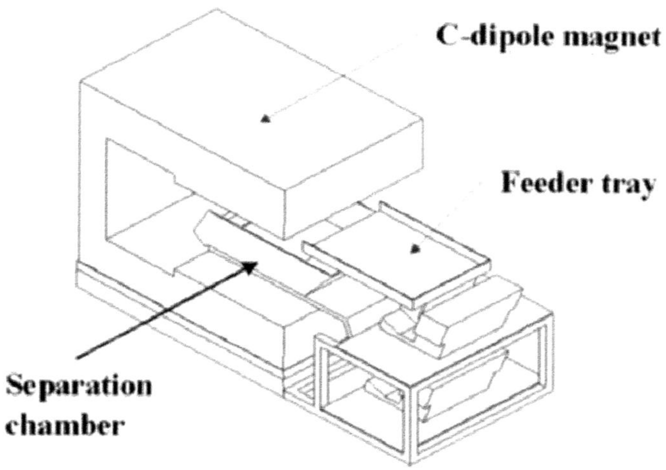

Figure 4.65: Schematic diagram of a ferrohydrostatic separator based on a C-dipole magnet [S57].

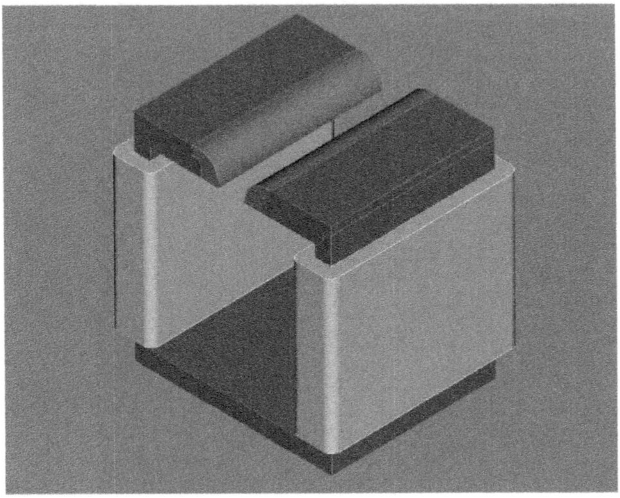

Figure 4.66: Magnetic circuit of an FHS separator with across-the-pole feeding system.

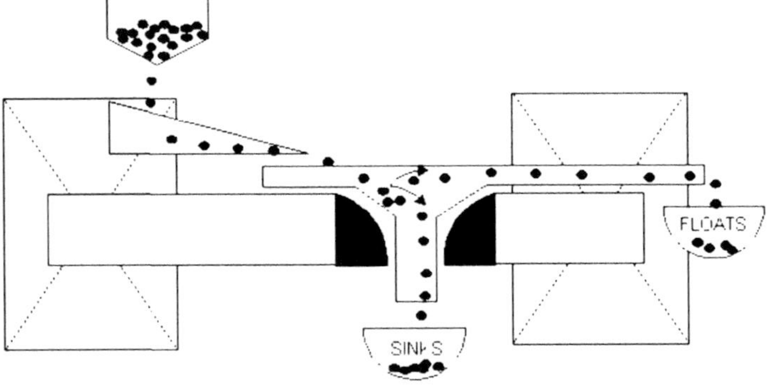

Figure 4.67: Separation process in the across-the-pole FHS concept [K19].

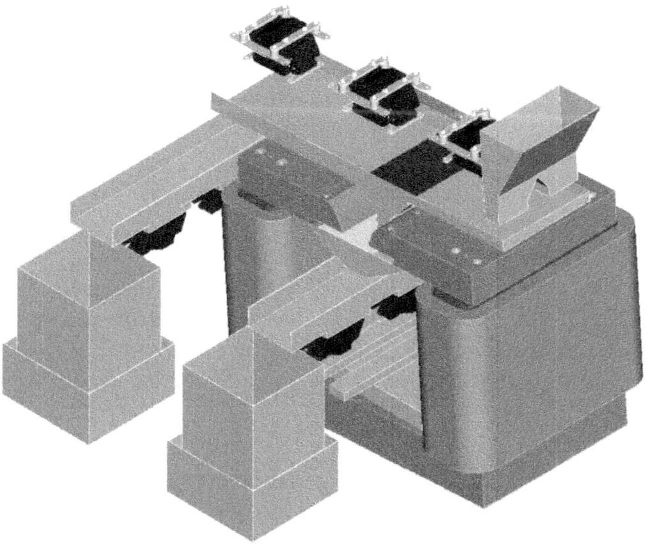

Figure 4.68: Layout of the ATP ferrohydrostatic separator [K19].

Figure 4.69: A pilot-plant ATP ferrohydrostatic separator (De Beers Consolidated Mines (Pty.) Ltd.)[K19].

"Across-the-pole" FHS

An even simpler concept for overcoming the throughput limitation of the conventional U-pole design of FHS was developed at De Beers Consolidated Mines (Pty.) Ltd. [S8, K19]. By rotating the feed direction through 90^0, the throughput increase could be achieved by lengthening the pole pieces rather than increasing the width of the air gap. Figure 4.66 gives a three-dimensional impression of this across-the pole (ATP) concept, while Fig. 4.67 gives a schematic description of the separation process. As can be seen, instead of feeding the material across the width of the air gap of the magnet, the particles are introduced into the ferrofluid across the poles. The float fraction passes over one of the pole pieces, while the sink fraction is collected underneath in the bottomless separation chamber. An illustration of the layout of the separator is shown in Fig. 4.68.

The width of the air gap is determined by the required residence time of material in the ferrofluid and remains the same for any length of the pole-pieces, and, therefore, for any throughput of the separator. The magnetic pattern and the apparent density of the ferrofluid are determined by the profile of the pole tips, in the same way as in conventional FHS units. A view of the pilot-scale ATP separators is shown in Fig. 4.69.

Chapter 5

Practical Aspects of Magnetic Methods for Materials Treatment

Magnetic separation is used in a wide spectrum of areas of material handling whose requirements differ considerably. In general terms, magnetic methods of separation are used in several distinct processes:

- Magnetic filtration
  - recovery of liquid (the magnetizable solids are discarded)
  - recovery of magnetizable solids (liquid is discarded).
- Concentration of a valuable "non-magnetic" component of a mixture by discarding the magnetizable component recovered into the magnetic fraction.
- Concentration of magnetizable component into the magnetic fraction and rejection of the "non-magnetic" waste.
- Concentration of non-ferrous metals by discarding the non-metallic component of a mixture, and separation of individual non-ferrous metals into individual fractions.
- Separation of materials into individual density fractions.

The selection of the most appropriate magnetic technique to be used in a given application is a formidable task, and clear definition of the objective of the process is an important prerequisite for successful choice of the separation technique.

5.1 Selection of magnetic separation technique

There are numerous types of magnetic separators, each one being suitable for only a limited range of applications. The choice is influenced by a wide variety of factors and the successful operation of a magnetic separator depends, to a great degree, on practical experience. The results of separation cannot generally be accurately predicted.

As with most other physical techniques of material treatment, magnetic separation cannot be successfully performed simply on the basis of theoretical models and experience is still the foundation. At present, there is no branch of magnetic separation where theoretical treatment of the subject could be considered entirely adequate.

Each magnetic separation technique has its merits as well as limitations; no single technique is applicable to all applications of magnetic separation. The evaluation of magnetic separation as a means of beneficiation, or generally of recovering solids, liquids or gases, is often based solely on the data collected from a few brief tests, not always carried out in a suitable magnetic separator. The likelihood of disappointing results and even of rejection of the technique is high, in spite of adaptability and large scope of the magnetic methods of materials manipulation.

The choice of magnetic equipment, and the selection of appropriate operating conditions differ widely between these processes. The mode of operation, namely wet or dry, and the throughput requirements impose additional complicating restrictions on the selection of the most suitable magnetic technique.

Since particle size of the material to be treated is the most important variable that determines the selection of the magnetic separation technique, the general guidelines of the process selection are based on particle size distribution, as shown in Fig. 5.1.

If the size of particles to be manipulated by magnetic separation is greater than, for example, 75 μm, we have an option of a dry or wet process. If, however, the particles are finer than 75 μm, the wet technique is more appropriate, although this threshold particle size can be smaller or greater than 75 μm, depending upon the amounts of the fines present in the feed and on the physical properties of the feed.

The subsequent choice of the magnetic separation technique depends on the magnetic properties of the material. If a strongly magnetic component is present in the feed, low-intensity drum magnetic separators can be used, either in wet or dry modes of operation.

Coarse, weakly magnetic materials can be treated by wet roll magnetic separators, rare-earth drum separators and possibly by wet high-intensity high-gradient magnetic separators. In the latter technique, there is an upper limit imposed on the particle size by the type of the matrix and particles greater than 1 mm usually cause clogging, even of a coarse matrix.

In a dry mode, coarse weakly magnetic particles can be recovered by a variety of high-intensity magnetic separators, including induced magnetic roll, cross-

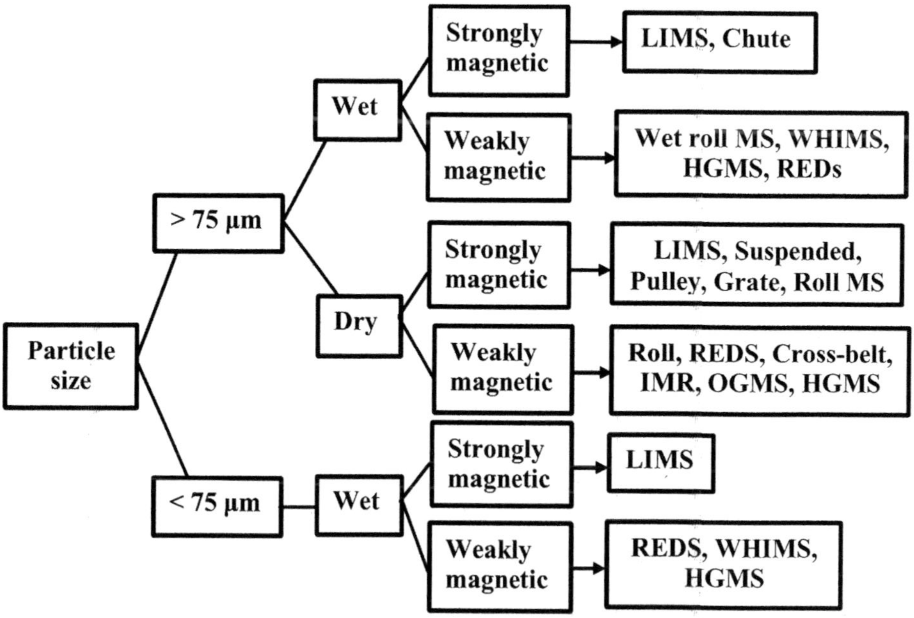

Figure 5.1: Classification of magnetic separation processes (adapted from Svoboda [S1]).

belt separator, permanent magnet roll separator, rare-earth drum separator and open-gradient magnetic separator.

If the particles are finer than 75 μm, the best approach is to use wet magnetic separation. Fine, strongly magnetic particles are usually recovered by wet low-intensity drum separators, while for weakly magnetic materials the wet high-intensity or high-gradient magnetic separation is appropriate.

The decision making process for selection of the magnetic separation technique is schematically shown in Fig. 5.2.

5.2 Dry magnetic separation

Dry magnetic separation is a technique dating back to the beginning of the last century. It has been practiced successfully in the removal of tramp iron from process streams, in beneficiation of strongly magnetic minerals, as well as feebly magnetic minerals such wolframite, cassiterite or ilmenite. Purification of numerous industrial minerals, such as andalusite, glass sand, fluorspar, feldspar, diamonds and others, by removing fine strongly magnetic impurities, is very often carried out in a dry mode. More recently developed eddy-current separators also operate in a dry mode.

For dry magnetic separation to be successful, the material must be com-

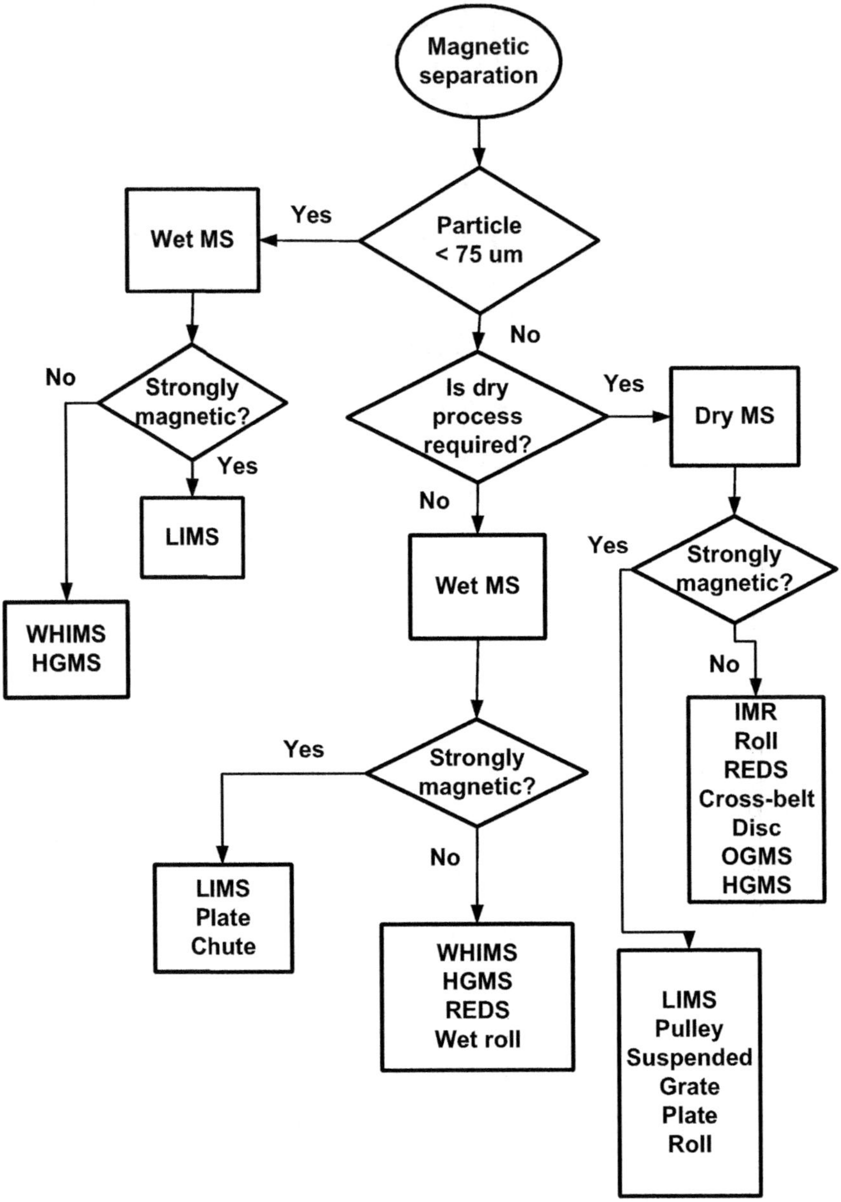

Figure 5.2: Selection of magnetic separation techniques (adapted from Svoboda [S1]).

pletely dry, which sometimes involves an expensive heating stage. For particles smaller than a certain threshold size (for instance 75 μm), the attractive electrostatic and van der Waals forces will cause sticking and the formation of aggregates. Selectivity of separation can thus be severely affected.

Dry processes often require the material to be sized into several rather narrow size fractions, each of which must be spread in a monolayer over the separating surface. It is a well-established fact that when magnetic separation is performed on materials sized into, for example, three or more fractions, the efficiency of separation is usually much improved in comparison with unsized samples.

The loading of the separating surface, such as belt or drum, influences the throughput and the efficiency of the separation, and a compromise must be found between these contradictory requirements. As a rule of thumb, when the particles are introduced onto a belt in a monolayer, a safe belt loading (portion of the belt covered with feed particles) should be approximately 25%, even when the feed contains a high percentage of the magnetic material.

As a rule, a larger proportion of dry fines necessitates higher rotating speeds of a roll or drum. Although increase in speed will improve the grade of the magnetic product and the throughput of the separator, it will also increase the losses of the magnetic component. The recovery can be improved by re-treating the non-magnetic fraction in a second stage. On the other hand, when the valuable product is the non-magnetic fraction, high speed of rotation will result in high recovery and poor selectivity of separation. The grade of the concentrate can be improved by cleaning the non-magnetic fraction in the second pass.

5.2.1 Electrostatic phenomena in dry separation

During dry material handling operations, the dielectric particles invariably become electrostatically charged. The dominant charging mechanisms are triboelectrification (or frictional charging) and contact charging. These mechanisms can happen during any dynamic contact of surfaces, such as sliding, rolling and vibration of surfaces at contact. When two dissimilar dielectric materials make and then break contact, charge transfer from one of the materials to the other takes place. This is not to say that conductors will not charge by contact electrification - they usually charge - but since they are conductors, they often lose their charge quickly.

While physical mechanisms of contact charging are well understood, triboelectrification, the oldest form of charging, is poorly understood and explanations are of phenomenological nature. One of the controversial rules of thumb often used for predicting the sign of the surface charge resulting from contact electrification is Coehn's rule, which states that "when two dielectric materials are contacted and separated, the material with the higher dielectric constant becomes positively charged" [L12]. Coehn's rule was quantitatively formulated as

$$\sigma_s = 15 \times 10^{-6}(\epsilon_{r1} - \epsilon_{r2}) \tag{5.1}$$

where σ_s is the surface charge density (in $[C/\mathrm{m}^2]$) and ϵ_{r1} and ϵ_{r2} are dielectric

Table 5.1: Dielectric constants of selected materials.

Material	**Dielectric constant** ϵ_r
Water	80.4
TiO_2	90
Barium titanate	1200
Carbon	5.7
Ilmenite	33.7 to 81
Chromite	11
Rutile	89 to 173
Zircon	12.6
Glass	4 to 6
Quartz	4.3
Hematite	25
Limestone	6.1
Fluorspar	7.4
Bakelite	4.9
Perspex	3.4

constants (relative permittivities).

Although Coehn's rule is reported to have been verified for more then 400 substances [L12], it is of limited value to the metallurgist because of the difficulty in determining the relative permittivities of the surface layers of two materials that are to make contact. Typical values of dielectric constant of selected materials are shown in Table 5.1.

Sometimes several materials can be arranged into the so-called triboelectrical series such that each material is charged positively when it is rubbed against the previous one. Examples of such series are:

- zinc-glass-cotton-filter-paper-silk-zinc [B26] and
- teflon-metals-resins-wood-mica-quartz-glass [H19].

The triboelectric charging behaviour of materials is generally sensitive to an atmospheric environment. At high values of relative humidity, considerable negative charging takes place. At low values of relative humidity, although fine particles still charge negatively, the coarse particles show a change of charging polarity such that the net charge conveyed by the powder material is positive [B27].

When a particle undergoes charging during handling, the charge transferred at the particle contact point will re-distribute itself over the particle surface by electrostatic forces. The rate at which re-distribution proceeds depends on the relaxation time τ_e (in [s]) given by

$$\tau_e = \epsilon_r \epsilon_0 \varrho \tag{5.2}$$

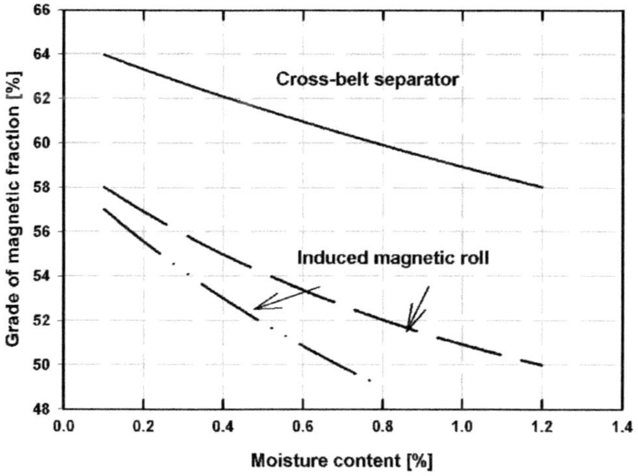

Figure 5.3: The effect of moisture on the grade of the magnetic concentrate obtained from - 2 mm magnetite and martite ores treated by cross-belt and induced magnetic roll separators (adapted from [D4]).

where ϱ is the resistivity of the material (in [Ωm]) and ϵ_r its relative permittivity (dielectric constant) and $\epsilon_0 = 8.85\times10^{-12}$ F/m is permittivity of vacuum.

The time dependence of the surface charge density, after completion of the charge transfer, is given by:

$$\sigma_s = \sigma_0 \exp(-t/\tau_e) \tag{5.3}$$

where σ_0 is the surface charge density at $t = 0$.

As a general rule, if electrostatic phenomena arise in a material handling situation, they are likely to become increasingly significant as particle size decreases. Particles charged triboelectrically, either positively or negatively, may adhere strongly to surfaces. Moisture contributes to the strength of adhesion and Fig. 5.3 illustrates the effect of moisture on selectivity of dry magnetic separation. It can be seen that the grade of the magnetic concentrate is seriously impaired by the presence of moisture and the permissible moisture content decreases with decreasing particle size. For instance, if an ore with size distribution - 20 mm can tolerate 4% of moisture, then the maximum moisture content in a fine ore - 2 mm should not exceed 1% [D4].

The effect of air humidity

The moisture concentration in an ore is also influenced by the relative humidity of the air. This is a consequence of the capillary attraction between particles, which results in adsorption and subsequent capillary condensation of the water vapours in the particle contact area. If the force of capillary attraction F_h

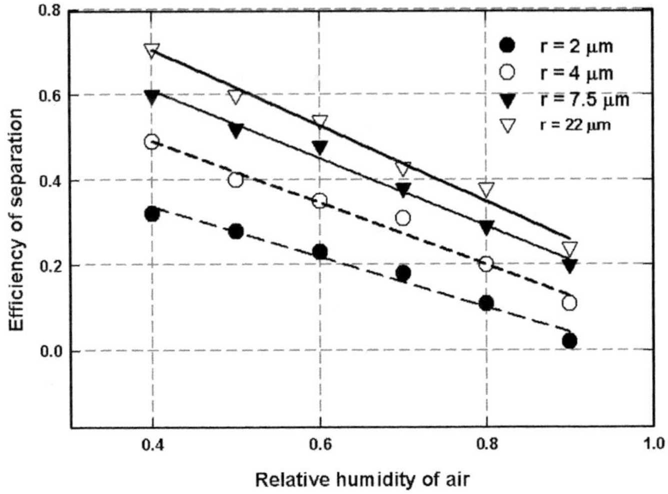

Figure 5.4: The effect of air humidity on dry magnetic separation of magnetite-quartz mixtures, for different values of particle radii (adapted from [V10]).

between a non-magnetic particle and a magnetic particle is greater than its weight G, the non-magnetic particle will report in the magnetic fraction. The force F_h is the sum of of the surface tension and the Laplace pressure force:

$$F_h = 8\pi r_1 \gamma + 4\pi r_2^2 \gamma (\frac{1}{r_2} - \frac{1}{r_1})$$

where r_i ($i = 1$ and 2) are the radii of the curvatures of the meniscus between the contacting particles 1 and 2, and γ is the surface tension (in N/m) of the fluid.

Experimental investigation [V10] of the effect of the air humidity on dry magnetic separation of magnetite-quartz mixtures, as a function of particle size, confirmed that the efficiency of separation varies linearly with the ratio G/F_h. It can be seen from the results illustrated in Fig. 5.4 that with increasing humidity, the separation efficiency decreases steeply. The effect of the air humidity becomes particularly pronounced with decreasing particle size.

5.2.2 Density and magnetic tracers

The use of the Tromp curve, or partition curve, is a well established practice for the evaluation and representation of gravity concentration processes. The partition curve can be conveniently constructed with the aid of density tracers [N15]. The density tracers are manufactured as cubes in sizes ranging from 0.8

to 20 mm. They are colour-coded according to the magnitude of their density and are commercially available, for example from Bateman Engineering (Pty.) Ltd. (South Africa), in the density range from 1300 kg/m^3 to 4000 kg/m^3, with density interval of 50 kg/m^3. Partition Enterprises (Pty.) Ltd. (Australia) supply density tracers that are strongly magnetic for easy recovery, by magnetic means, from density separation products.

The partition number, or the percentage recovery R into the concentrate, for each size and density fraction, is usually determined as

$$R = 100\frac{n_c}{n_c + n_t}$$

where n_c and n_t are the numbers of tracers in the concentrate and in the tailings, respectively.

Magnetic tracers, which have become commercially available only recently, can be used, in a fashion similar to the way density tracers are used in gravity concentration, to determine and optimize the separation efficiency of magnetic separators. A magnetic tracer is a simulant designed to have the same size and magnetic susceptibility as the material which is being recovered or removed in a magnetic separator. Magnetic tracers of mass magnetic susceptibility ranging from 2.5×10^{-7} m^3/kg to 125.5×10^{-7} m^3/kg (20×10^{-6} cm^3/g to 1000×10^{-6} cm^3/g), in steps of 2.5×10^{-7} m^3/kg, for low-susceptibility tracers, and $12.5\times10^{-7} m^3/kg$ for high susceptibility tracers, are available from Bateman Engineering (Pty.) Ltd., South Africa [D12]. These tracers, whose size ranges from 1 mm to 20 mm, are colour- and shape-coded, according to the magnitude of their magnetic susceptibility for easy evaluation of the separation process. Tracers smaller than 1 mm are impractical as they cannot be easily identified among the mineral particles. The magnetic partition curve can be constructed using the partition numbers, or percentage recovery into magnetic or non-magnetic fractions. In terms of the recovery of the tracers into the magnetic fraction, for example, the partition number, for a given size and magnetic susceptibility, is expressed as:

$$R = 100\frac{n_{mag}}{n_{mag} + n_{non-mag}}$$

where n_{mag} and $n_{non-mag}$ are the numbers of tracers in the magnetic and non-magnetic fractions, respectively. The probable error of separation is given by eq. (4.58).

While magnetic tracers are ideally suited for quantification of the efficiency of a magnetic separation process, their practical use is restricted to the separation of materials coarser than 1 mm. Although they can be used to evaluate the efficiency of separation of even finer materials in dry separators such as roll and drum, the particle-size dependence of the magnetic susceptibility cut point can result in considerable inaccuracy of the audit. In wet separators, such as low-intensity drum separators or WHIMS or HGMS, even the smallest 1 mm magnetic tracers would interfere with the separation process

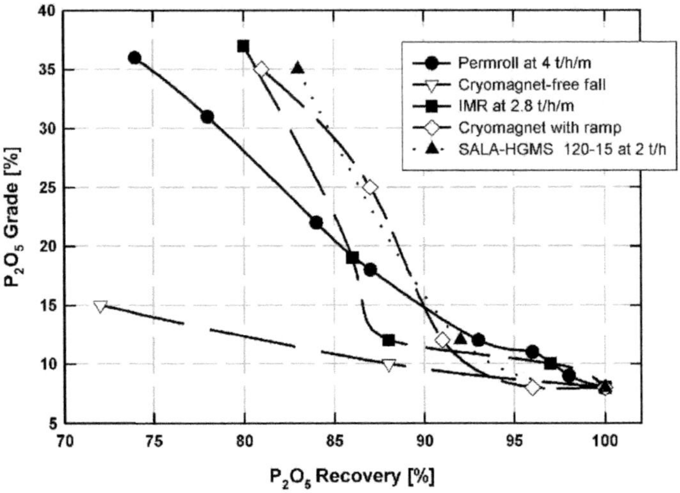

Figure 5.5: Comparative metallurgical efficiencies of various dry magnetic separators as applied to apatite recovery (adapted from Roux et al. [R19]).

5.2.3 Comparison of dry magnetic separation techniques

The minerals industry has always had a great need to beneficiate coarse, weakly magnetic minerals and a wide spectrum of dry magnetic separators has been developed over the years. The range of various types of separators, based on different principles of the generation of the magnetic field, makes a correct choice of the most suitable technique a difficult task. Comparisons of the separation efficiencies of different techniques are usually not published and if they are, caution should be exercised when taking these comparisons at face value. Every comparison is only conditional since the relative merit of each technique is a function of the mineral to be treated and its physical properties. Sometimes comparisons are offered by manufacturers of magnetic separators who might be biased towards their own product.

Comparison of dry separators for beneficiation of the pyroxenite ore

One of the rare comprehensive comparative production-scale exercises was conducted by Roux et al. [R19]. In order to beneficiate pyroxenite ore to yield apatite concentrate, several dry magnetic separators were tested at Foskor (South Africa). Figure 5.5 summarizes the metallurgical efficiencies of different types of separators, while Fig. 5.6 shows the relationship between phosphate recovery and particle size for IMR, HGMS and OGMS machines. A thorough practical experience with these magnetic separators was gained at Foskor and their

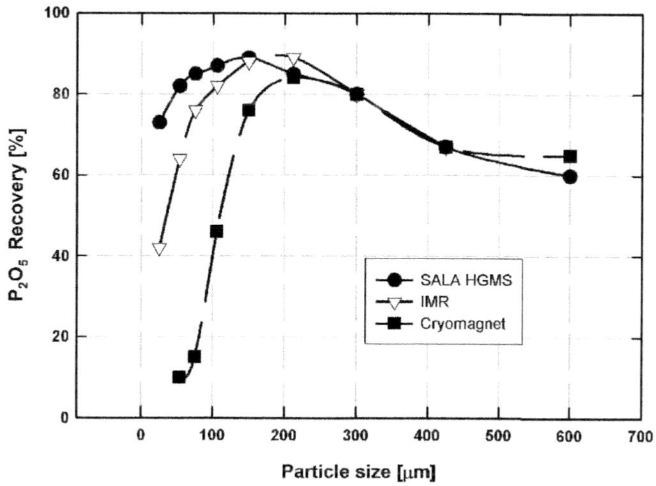

Figure 5.6: Phosphate recovery as a function of particle size for various magnetic separators (adapted from Roux et al. [R19]).

comparison is summarized in Table 5.2.

A gradual deterioration in the results was observed with IMR, caused by abrasive wear on the epoxy filler between the steel discs of the rolls. Furthermore, because the separation distance between the magnetics and non-magnetics is very small, a precision setting of a splitter was required. As the mineralogy of the ore varies, constant attention to the optimization of roll speed and splitter settings was required, which made it difficult to maintain good metallurgy [R19].

A Sala HGMS 120-15 operated reliably and well for 1000 hours and it was found that single-stage separation yielded metallurgical results equivalent to that of two stages of IMR separation. The expanded metal matrix showed no measurable wear, while the metallurgy was shown to be adversely affected by oversize material, which gradually accumulated in the matrix. In order to keep coarse particles out of the feed, and to avoid the onerous task of cleaning the matrix chambers, it became necessary to fit a second screen which could be kept under continuous visual observation.

An extensive series of experiments with SmCo permanent roll separators showed that the feed rate, roll speed and thickness of the magnet and steel rings must be optimized for each separation stage.

The open-gradient, split-coil superconducting magnet separator was tested in two modes, shown in Chapter 2 in Fig. 2.85(a) and (b). The free-fall system gave acceptable metallurgical results only at very low feed rates. A much more satisfactory ore presentation system was found to be the one in which the ore slid down a ramp, as shown in Fig. 2.85(b). The ramp imparts an outward

Table 5.2: Comparison of different types of dry magnetic separators, as applied to phosphate recovery (after Roux et al. [R19]).

Parameter	IMR	RE Roll	HGMS	OGMS
Capacity	Low, 4 t/h/m	Low, 8 t/h/m	High, 400 t/h	High, 60 t/h
Power input	2 kW/t	0.6 kWh/t	5.6 kWh/t	0.5 kWh/t
Ore distribution	Complex	Complex	Simple	Simple
Mass [kg] per t/h of feed	2200	440	4100	41
Sensitivity to ore variation	High	Low	Low	Very high
Recovery of the fines	Poor	Very poor	Good	Very poor

Table 5.3: Comparison of IMR and RE roll in ilmenite-rutile separation (after Arvidson [A26]).

Product	IMR @ 2.5 t/m/h	RE roll @ 4 t/h/m
Magnetics (mass %)	25	30.5
Non-mags in mags (%)	2.5	0.46
Middlings (%)	10	0.8
Non-magnetics (%)	65	68.7
Mags in non-mags (%)	0.5	0.28

velocity to the particles, thereby counteracting, to a considerable extent, the sweeping across of non-magnetic particles by the magnetic ones. This approach made it possible to obtain metallurgical results comparable with those of IMR separators, at feed rates of up to 6 t/h per meter width, or 3.5 times the optimum feed rate of an IMR.

Comparison of dry separators for beneficiation of beach sands and other industrial minerals.

High-intensity magnetic separation has been a standard beneficiation method for heavy mineral sands and various industrial minerals, and a variety of dry magnetic separators has been used. With the advent of affordable rare-earth permanent magnets, and as a result of design improvements, the conventional dry separators, such as induced magnetic roll and cross-belt separators are being replaced by reliable and economical rare-earth roll and drum separators [A24, A25, A26]. Although the plant data are usually confidential, comparative studies indicate superior performance of the rare-earth-based separators. An example is shown in Table 5.3, which compares the separation efficiency of IMR and RE roll separators as applied to ilmenite and rutile separation.

A similar improvement in the efficiency of separation after replacing the induced magnetic roll with a permanent magnetic roll, was obtained with feldspar.

Table 5.4: Comparison of IMR and RE roll magnetic separation as applied to beneficiation of feldspar [A24].

Product	IMR	RE Roll
Magnetic fraction	12% - 14%	4% - 4.5%
Non-magnetic fraction	86% - 88%	95.5% - 96%

Table 5.5: Reduction of the iron content in various size fractions of feldspar, by IMR and roll permanent magnetic separators. IMR was operated at 1.8 T [A27].

Separator	Fraction [μm]	Fe oxides in feed [%]	Yield into non-mags [%]	Fe oxides in non-mags [%]
IMR	- 500 + 212	0.20	95.5	0.04
	- 212 + 106	0.43	93.0	0.06
	- 106 + 74	0.45	82.3	0.07
Roll	- 500 + 212	0.20	86.6	0.08
	- 212 + 106	0.43	85.8	0.10
	- 106 + 74	0.45	79.5	0.11

This is shown in Table 5.4. On the other hand, results of another comparative study [A27], summarized in Table 5.5, indicate a superior performance of an induced magnetic roll, in the entire range of particle sizes.

Dry magnetic separators can also be efficiently used to recover metal from slags. Comparison of several dry separation techniques to recover ferrochrome from coarse slag is shown in Table 5.6 [S58].

A dry ferrite roll separator was also successfully applied to the recovery of the ferrochrome metallics from a fine fraction (- 0.8 mm) of slag. Comparison of its performance with a wet rare-earth drum separator is shown in Table 5.7.

Replacement of conventional electromagnet-based separators by rare-earth permanent magnet units also often results in economic advantages. Improved quality of the ilmenite concentrate and reduced operating costs contribute to improved plant efficiency, as shown in Table 5.8. Additional cost savings result

Table 5.6: Recovery of ferrochrome metal from slag (- 15 + 2 mm) by magnetic separation [S58].

Separator	Head grade [FeCr %]	Mags grade [FeCr %]	Recovery [%]
Dry LIMS	22.7	60.6	85.0
Lifting IMR	20.1	77.6	85.8
Ferrite roll	27.1	45.7	97.5

Table 5.7: Recovery of ferrochrome metal from fine slag (- 0.8 mm) by magnetic separation [S58].

Separator	Head grade [FeCr %]	Mags grade [FeCr %]	Recovery [%]
Ferrite roll	4.4	16.9	65.6
Wet RE drum	4.7	17.3	70.2

Table 5.8: Comparison of the operating costs of various dry magnetic separators applied to ilmenite beneficiation, in relation to a cross-belt separator [A24].

Magnetic separator	Cost
Cross-belt	100
IMR	42 to 53
RE drum separator	17
RE roll (short belt life)	21 to 42
RE roll (long belt life)	11 to 15

from a smaller size of the plant for a given throughput. Since a plant with rare-earth magnetic separators would use between 6% and 20% of a cross-belt plant space [A24], the cost saving can be considerable. Compared to IMR for non-magnetic cleaning applications, the space saving can be 50% to 70% [A24].

Improved performance can be also obtained by implementing multiple-stage separation, as is demonstrated in Fig. 5.7. Combined with higher roll speeds, the selectivity of separation increases in addition to a higher throughput.

In some applications, it was found advantageous to combine rare-earth roll separators with rare-earth drum separators. While roll separators tend to be more efficient, more selective and can be more readily adjusted to feed variations, drum separators have a higher throughput and are less demanding from the maintenance point of view. Comparison of the roll and roll+drum options is shown in Table 5.9.

Table 5.9: Comparison of roll and roll+drum alternatives for treatment of ilmenite gravity concentrate (- 0.9 + 0.1 mm), at 30 t/h, in 1999 prices [A25].

Parameter	Drum + Roll	Roll separator only
Stages required	1 drum + 2 rolls	3 stages
Capacity	7 t/h/m	4 t/h/m
Quantity	3	5
Price per unit [US$]	$114 000.00	$65 000.00
Total price [US$]	$342 000.00	$325 000.00

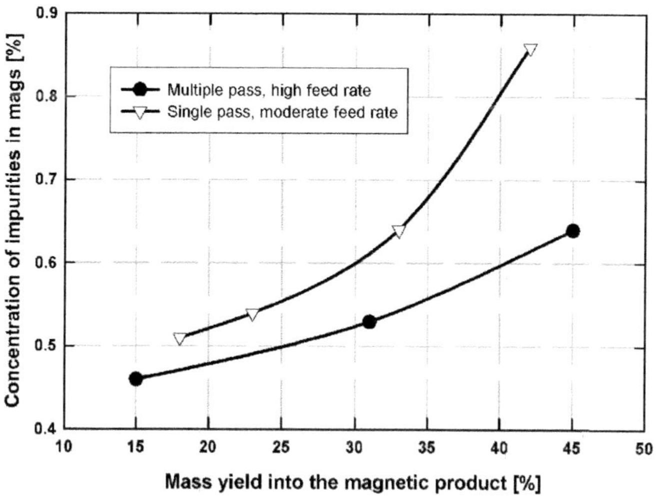

Figure 5.7: Comparison of single and multiple-pass separation of ilmenite, by a roll separator, at different feed rates (adapted from Arvidson [A26]).

Comparison of roll and matrix separators

Jirestig and Forssberg [J7] carried out a detailed comparative study of four dry high-intensity magnetic separators. Three of these separators were roll units: Permroll® (E.L. Bateman) with Kevlar belt, High Force® (Inprosys) with polyurethane belt, induced magnetic roll (Humboldt Wedag) and dry HGMS (Sala). An expanded metal matrix, with and without spacers, was used in the Sala HGMS and particles were propelled through the matrix by a compressed air-driven vacuum pump. Various size fractions of calcite and dolomite samples were used as test materials. Permroll, High force and dry HGMS, operating at 2 T yielded products of similar brightness for fine to medium-size (- 450 + 38 μm) particles, as can be seen in Fig. 5.8. For particles larger than 450 μm, the Sala HGMS machine experienced problems with matrix clogging, while the efficiency of the induced magnetic roll was slightly inferior.

The significant advantage of the matrix separator, compared to belt-type separators, is its high selectivity of separation of fine particles, smaller than 100 μm. The high brightness of products from belt separators was achieved at great cost to the recovery. Only about 5% of 38 μm material reported to the magnetic fraction in the matrix separator, while Permroll and High Force machines experienced significant losses (35% to 80 %) of the valuable material into the magnetic tailings. The different performances of the two belt separators are probably due to different types of belts used and the results indicate the importance of controlling electrostatic properties of the belt material, as discussed in Section 5.2.1.

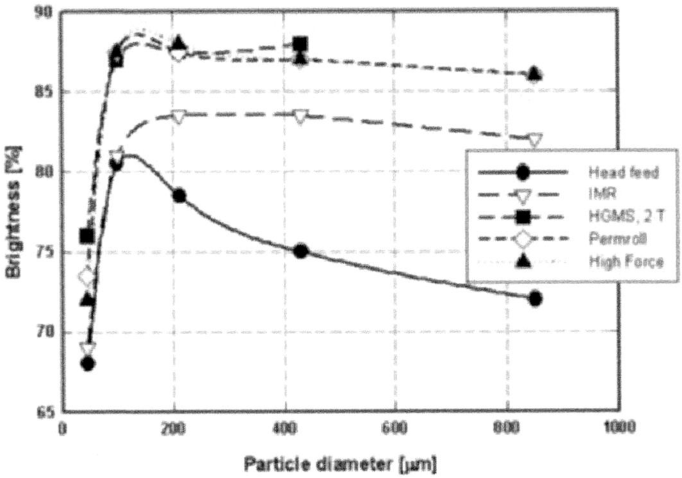

Figure 5.8: Brightness of various size fractions of the calcite non-magnetic product, as obtained in several dry magnetic separators (adapted from Jirestig and Forssberg [J7]).

Comparison of rolls of different diameters

Magnetic rolls of diameter ranging from 72 mm to 300 mm are commercially available and judicious selection of the most suitable unit is essential for optimum metallurgical results and operating costs. Figures 5.9 and 5.10 illustrate that magnetic rolls of larger diameter (300 mm) can operate at higher feed rates than standard 76 and 100 mm diameter rolls. By increasing the speed of rotation the same performance can be maintained. As has been discussed in Section 4.6.2, the magnetic force is practically independent of the magnetic ring diameter. The increase in the rotational speed of the roll, maintaining the same residence time of particles in the magnetic field as in a smaller diameter roll, results in the increase of the centrifugal force on a particle. In turn, this results in a closer balance of the magnetic and competing forces and better selectivity of separation can thus be obtained. This additional benefit of larger ring diameters is depicted in Fig. 5.10.

Comparison of RE roll and RE drum separators

Rare-earth roll separators and rare-earth drum separators are designed to address similar beneficiation applications. The choice between these two types of separators is rarely made on the basis of comprehensive comparative tests. Since the pattern of the magnetic force around the roll and drum differ considerably, correct systematic evaluation of the operation is essential. As has been discussed in Section 4.6.2, magnetic roll separators generate a much higher mag-

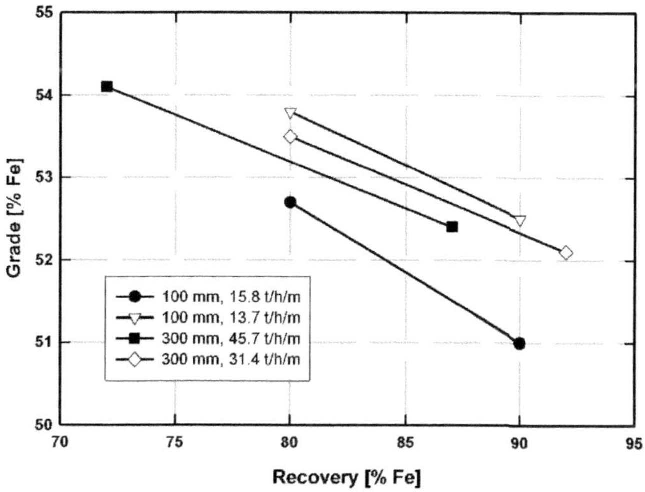

Figure 5.9: Comparison of separation efficiencies of magnetic rolls of diameters 100 and 300 mm, at different throughputs, as applied to beneficiation of - 3 mm iron ore (adapted from Arvidson and Dillé [A23]).

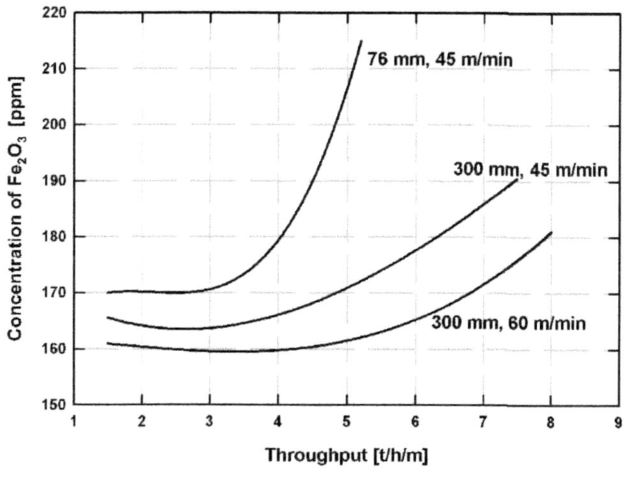

Figure 5.10: Comparison of performances of 76 mm and 300 mm rolls at different speeds of rotation, applied to purification of - 600 + 800 mm silica sand (adapted from Dobney [D8]).

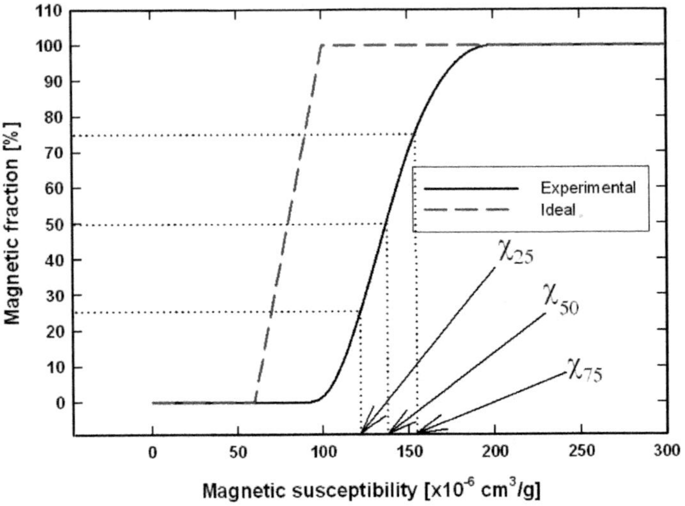

Figure 5.11: Magnetic partition curves. The ideal curve corresponds to a magnetic susceptibility cut-point of $\chi_{50} = 80\times10^{-6}$ cm^3/g (10×10^{-7} m^3/kg) [M24].

netic force, which, however, is of short reach. On the other hand, the magnetic force on the surface of a drum separator is much lower, as a result of a considerably lower magnetic field gradient. The reach of the force is, however, much longer than that in a roll separator. The behaviour of particles in these two types of separator is different and the settings of the optimum operating conditions differ widely.

Detailed comparative tests on rare earth roll and drum separators were conducted by Murariu and Svoboda [M24]. Magnetic tracers of various sizes and mass magnetic susceptibilities were used to set the optimum operating conditions, and magnetic partition curves were generated. The values of Ep, given by eq. (4.58), of these curves were compared. Figure 5.11 illustrates such a partition curve for the cut-point magnetic susceptibility of 80×10^{-6} cm^3/g (10×10^{-7} m^3/kg). The cut-point is defined as the value of magnetic susceptibility (χ_{50}), for which the particles having this susceptibility split equally into the magnetic and non-magnetic fractions.

It was observed that the same cut point could be obtained using different combinations of the speed of rotation of the drum or roll and of the position of the splitter. However, not all combinations of these parameters give the same optimum efficiency of separation, as is demonstrated by Fig. 5.12. It can be seen that, for a drum separator, a lower speed and a closer position of the splitter between the magnetic and non-magnetic fractions, result in lower Ep and thus

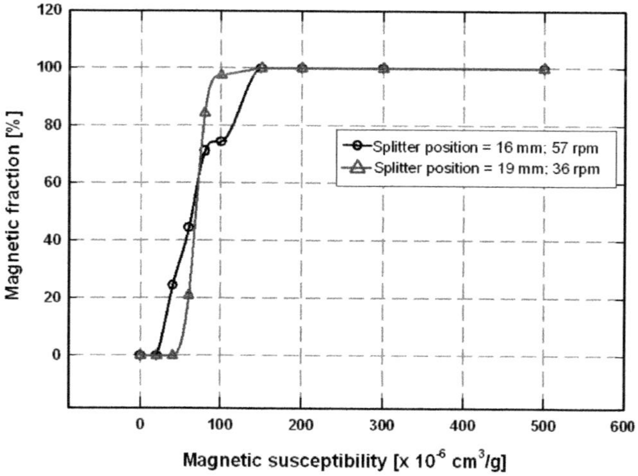

Figure 5.12: Experimental partition curves for an NdFeB drum magnetic separator. The curves were obtained for the same value of the cut point, but for different positions of the splitter and the rotational speed of the drum. Drum diameter: 600 mm, magnetic induction on the drum surface: 0.9 T [M24].

a more efficient separation.

However, with a roll separator (diameter 100 mm), a more selective separation was achieved at a high roll speed and with a splitter far from the roll. This difference in the optimum operations of drum and roll magnetic separators results from the fact that the magnitude of the magnetic force acting on a particle in a drum is much smaller than that in a roll separator. As discussed in Section 1.2, the separation is selective when the magnitudes of the magnetic force and the competing forces are of comparable magnitudes. The magnetic roll, which generates a higher magnetic force than the drum, requires a higher centrifugal force (i.e. a higher roll speed) and a more distant position of the splitter.

Tracer tests showed that the setting of the cut point, that is of the rotation speed of the roll or drum and position of the splitter, is dependent on particle size. The size dependence was found to be approximately linear for both separators, as is illustrated in Fig. 5.13. The slope of the plots indicates sensitivity of a separator to the size of the tracers. It can be seen that roll magnetic separators are considerably more sensitive to particle size and the sensitivity increases with decreasing thicknesses of the magnet and steel rings. A roll separator of a given configuration is, therefore, suitable for accurate separation of only a narrow size range of the feed particles. For example, if the cut point is set to $10\times10^{-7}\mathrm{m}^3/\mathrm{kg}$ (80×10^{-6} cm^3/g) and the required accuracy is $\pm$ 10% (two horizontal lines in Fig. 5.13), a roll separator with configuration 12/7.5 will satisfy the accuracy

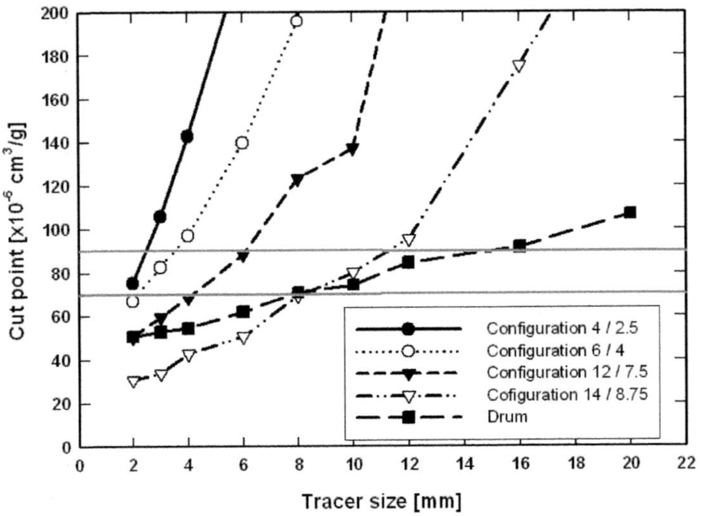

Figure 5.13: Variation of the cut-point magnetic susceptibility with the size of magnetic tracers, for roll and drum magnetic separators of different configurations [M24].

requirement for a size fraction - 6 + 4 mm. A roll separator with configuration 4/2.5 will meet this requirement only for particle size of 2 mm. On the other hand, the drum separator will be able to treat selectively a size fraction of - 16 + 8 mm.

The loading of the belt or the drum The loading of the drum and of the belt of the roll separator, influences the throughput and the efficiency of separation, and a compromise must be found between these two contradictory requirements. As a result of a short residence time of particles on the roll surface, the feed material must be distributed evenly over the belt in a monolayer and a safe belt loading is approximately 25%. Lower belt loadings lead to a low throughput and generally cannot be justified in view of a limited benefit and high cost of the separator.

In drum separators, the residence time is much longer and if the feeding system is optimized and the feeder tray is sufficiently close to the drum, to avoid particle bouncing, the particles usually spread evenly on the drum surface even when particles flow in a thicker layer on the feeder tray. Typical flow of material onto a drum surface is shown in Figs. 5.14 and 5.15. However, adherence of strongly magnetic particles on the drum surface can be a problem, as shown in Fig. 5.16.

Figure 5.14: Flow of the - 2 + 1 mm material on the surface of an NdFeB drum magnetic separator.

Figure 5.15: Flow of the - 16 + 8 mm material on the surface of an NdFeB drum magnetic separator.

Figure 5.16: Attachment of strongly magnetic particles to the surface of a drum magnetic separator.

Throughput of a roll separator The throughput of a roll separator is determined by a variety of factors and only experimental determination can give reliable information on this important specification. The general formula for throughput calculation is

$$C = \frac{V_p \rho_p w v f}{A_p} \tag{5.4}$$

where C is the throughput (in [kg/s]), V_p is the particle volume (in [m^3]), ρ_p is the particle density (in [kg/m^3]), w is the belt width (in [m]), v is the belt speed (in [m/s]), A_p is the cross-sectional area of a particle (in [m^2]) and f is the filling factor of the belt. It is evident from this equation that the shape of the particles must be known before the throughput can be calculated.

Belt tracking As the belts used in roll magnetic separators are frequently very thin, usually between 0.13 mm and 0.25 mm, they are prone to wear and their life time is often short. Accumulation of dust and magnetic particles underneath the belt and misalignment of the belt, result not only in impaired separation performance but also cause damage to the belt. Various manufacturers of roll separators use different mechanisms for the belt tracking, with different levels of resistance to misalignment. Considerable effort has been expended to extend the wear life by developing and applying correct tracking system.

Crown idler rolls have been used successfully for many years, particularly for small diameter rolls and for separators with a large distance between the

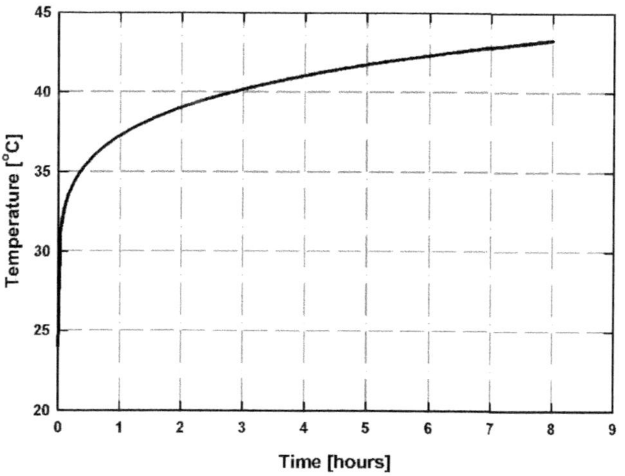

Figure 5.17: The temperature of the shell of an NdFeB drum magnetic separator, as a function of time. Drum diameter: 600 mm, magnetic induction on the shell: 0.9 T, speed of rotation: 50 rpm [M24].

roll and the idler. With the advent of large diameter rolls (e.g. 300 mm), the conventional crowned idler system has been replaced, for rolls longer than 0.5 mm, by a self-adjusting roller system [A23, A13]. Depending on the application, the average lifetime of a thin belt (0.13 mm) ranges from one month to six months, while a thicker belt (0.25 mm) can last as long as 12 months [A24].

The changing of a damaged belt can also be an onerous and time-consuming exercise. It is claimed, however, that the belt changing time for a one meter long roll can be shorter than 5 minutes [D8].

Eddy currents and temperature increase in the drum shell It has frequently been observed that the temperature of the shell of drum magnetic separators increases during operation. This phenomenon is based on the fact that as the metallic shell of a drum separator rotates around a stationary magnetic system, eddy currents are developed in the shell. These eddy currents are generated in such a way that the magnetic field they produce opposes the external magnetic field, as discussed in Section 2.7. This interaction energy is converted into heat which might be detrimental to operation of a drum magnetic separator and it could reduce the lifetime of the permanent magnet material. This is of particular importance for high-field rare-earth drum separators which are often operated at higher rotational speeds than low-intensity magnetic drum separators.

Figure 5.17 illustrates the temperature development on the surface of the shell of an NdFeB drum separator [M24]. The temperature was measured by a thermocouple as a function of time over a period of eight hours. The temperature increased rapidly within the first 30 minutes of operation from the room temperature of 24 ^{0}C to 36 ^{0}C. Afterwards, the temperature kept increasing asymptotically to the temperature of 45 ^{0}C, which remained constant after eight hours. While the temperature within the permanent magnet material inside the drum will be higher than on the shell surface, it is probably still well below the critical operational temperature of the permanent magnet material. Nevertheless, caution should be exercised in this aspect.

The effect of the increased temperature of the magnetic system on the separation efficiency was investigated using 8 mm magnetic tracers and - 4 + 2 mm kimberlite gravel [M24]. It was found that the experimental partition curves determined at room temperature and after six hours of the drum operation, exhibited the same cut-point magnetic susceptibility. Also, the mass recoveries of the gravel into the magnetic fraction were identical, indicating that the separation efficiency was not affected by the increased temperature of the drum.

5.3 Wet magnetic separation

As shown in Fig. 5.1, wet magnetic separation can be applied to coarse as well as fine particles, that are either strongly or weakly magnetic. Strongly magnetic materials can be recovered by wet low-intensity machines, which are by far the most common in use. Weakly magnetic particles, coarse or fine, can be treated by rare-earth drum magnetic separators, while wet high-intensity magnetic separators (WHIMS) and high-gradient magnetic separators (HGMS) are applied, as a rule, to fine weakly magnetic materials only.

5.3.1 Wet low-intensity drum magnetic separation

Wet low-intensity drum magnetic separators (LIMS) are used to recover strongly magnetic solids from a slurry feed. The major areas of application are media recovery in dense media separation plants (DMS) and in magnetite ore beneficiation. Physical construction of wet drum separators is slightly different for the two applications, with the ore concentrator being more rugged in construction and subject to more detailed specifications than the media recovery units. This is particularly true with reference to feed and collection tank design, bearing construction, wear covers on the drums, magnet assembly design and magnetic field strength rating. On the other hand, modern LIMS machines for DMS plant operation have a longer pick-up zone [D13], compared to the conventional separators.

Wet magnetic drum separators for media recovery

In media recovery application the particle size of the feed, particularly the magnetic portion, is quite fine and drum wear is not a serious problem. Maximum magnetic recovery is important, and the highest magnetic purity and solids content in the magnetic concentrate is desired.

Table 5.10: Maximum concentration of solids in the feed into LIMS for dense media recovery [S59].

Separator	Max. solids concentration [% by weight]
Con-current	20
Counter-rotation	25
Double drum	50

Factors that are considered in selecting a wet drum separator for media recovery are:

- Percentage of solids in the feed slurry
- Percentage of magnetics in the feed solids
- Grade of the media used
- Feed volume (m^3/h) to the separator
- Magnetic discharge rate (t/h) from the separator
- Desired magnetic recovery.

Media recovery magnetic drums are available in diameters of 762 mm (30"), 914 mm (36") and 1219 mm (48"), and in drum widths up to 3050 mm (120"). A 1200 mm diameter magnetic separator has approximately twice the capacity of a 900 mm unit. This allows circuits to be designed with fewer magnetic separators and therefore with reduced vulnerability to mechanical failure, uneven feed distribution and poor control [D13].

Both con- and counter-current feed arrangements have been used. The con-current drum arrangement usually gives the highest magnetic content in the magnetic discharge and provides the highest percentage of solids in the discharge. The overwhelming majority of the magnetic separators installed in DMS plants has been of the con-current design. The counter-rotation units have been used successfully as secondary separators [D13].

Solids content in the feed For efficient operation of the DMS plant it is desirable to keep the media as free of fines as possible. The reduced concentration of fines limits the viscosity problem and results in sharper separation. It also improves the magnetic recovery by the magnetic drum and gives a cleaner magnetic concentrate.

Typically, a single LIMS will be used in plants where the feed solids are in the 10% to 15% range. It is significant to note that an extremely dilute feed makes the magnetic recovery more difficult and the feed volume should be reduced by a factor of 0.5 to 0.75 for such a dilute feed [B28, M26].

For solids in the 20% to 25% range, either 900 mm diameter single drum or 760 mm diameter double drum units are recommended. Above 25% solids, 900

Table 5.11: Recovery of ferrosilicon in a magnetic separator at low concentration of the magnetics in the feed [D13].

Mags in the feed [g/ℓ]	Recovery [%]	Mags in effluent [%]
55	98.8	0.83
38	98.0	0.76
5	92.0	0.40
2	67.0	0.67
0.5	56.0	0.22

Table 5.12: Suggested feed volumes into LIMS for dense media recovery [B28, M26].

Separator	Feed rate [$m^3/h/m$]
762 mm single drum	50 - 55
914 mm single drum	55 - 70
762 mm double drum	55 - 70
914 mm double drum	70 - 95

mm diameter double drum are advised. The suggested maximum concentrations of solids in the feed are shown in Table 5.10.

Magnetic content of the feed solids Typically, the magnetic content of solids will be 60% or higher. When the magnetic content of the solids falls below 60%, the feed volume should be reduced and multi-drum arrangement considered, in order to maintain a high magnetic recovery. Table 5.11 illustrates the detrimental effect of low concentrations of the magnetics on their recovery. It is well established that the magnetics-to-ore ratio and the solids content are two parameters that control concentrate purity most effectively [S59].

Recommended feed volume and discharge rate of the magnetics In selecting drum size for media recovery, the feed volume and magnetic solids discharge should be balanced. In Table 5.12. typical feed rates are suggested for various single and double wet low-intensity magnetic separators.

Magnetic discharge rates should not exceed the rates given in Table 5.13, in order to maintain high magnetic media recovery.

The effect of the magnet position Plant experience confirms that the initial settings of a drum magnetic separator, recommended by manufacturers, are only a rough guide. Settings of such parameters as the magnet position and drum-to-tank clearance should be optimized based upon the plant experience. For instance, it was found by Dardis [D13] that the magnet position significantly influenced tailings losses and the return density of the concentrate. Two minima in the loss curve were observed [D13], corresponding to the minima in the profile

Table 5.13: Recommended maximum discharge rates of the magnetics [M26, S59].

Separator	**Max. discharge rate [t/h/m]**
762 mm drum	10
914 mm drum	15
1200 mm drum	24
Double 762 mm	18
Double 914 mm	30
Double 1200 mm	48

of the magnetic force (or force index) shown in Fig. 4.29. Similarly, the critical role of the magnetic position on the recovery of magnetite was reported by Howell et al. [H21]. On the other hand, Rayner and Napier-Munn reported [R20] that the magnet position had no significant effect on the performance of the drum separator.

The distance between the drum and the tank has a two-fold effect on the separation efficiency. With increasing clearance between the drum and the tank, the magnetic field strength and the magnetic force are reduced in the areas furthest from the drum. This negatively affects the efficiency of separation, particularly of fine particles. On the other hand, by increasing the clearance, the velocity of flow through the separation zone decreases and so does the hydrodynamic drag on particles. This results in a more efficient attraction of magnetic particles to the drum and in reduced shear stress acting on particles already captured on the drum. The combined effect of these two contradictory phenomena can lead to improved recoveries when the clearance is increased [D13].

Media types and size The most common medium for separation is a suspension in water of fine magnetite, ferrosilicon (FeSi), or a mixture of the two, depending on the separation density required. Magnetite alone is used when a density is required in the range 1250 to 2200 kg/m^3, a mixture for 2200 to 2900 kg/m^3 and ferrosilicon alone for 2900 to 3700 kg/m^3.

Magnetite, the density of which ranges from 4800 to 5200 kg/m^3, is the most popular choice for medium solids used in coal preparation, while ferrosilicon is invariably used, for instance, for the concentration of diamonds, for the recovery of aluminium in recycling plants, and for beneficiation of iron and manganese ores.

Ferrosilicon containing between 14% to 16% silicon has a density ranging from 6700 to 7100 kg/m^3. It is non-rusting and strongly magnetic. With concentrations of silicon greater than 22% the material is feebly magnetic, while with concentrations of silicon lower than 15%, it is prone to rusting.

Magnetite and ferrosilicon are thus selected as media because they meet a number of essential criteria: they are physically stable and chemically inert, easily removable from the products and are readily recovered for re-use by magnetic

Table 5.14: Particle size distribution of magnetite for DMS [M26].

Size fraction [μm]	Fine grade [%]	Coarse grade [%]
- 300 + 212	3.8	8.2
- 212 + 150	4.4	4.3
- 150 + 106	6.2	14.1
- 106 + 75	7.6	12.0
- 75 + 45	13.5	19.9
- 45	64.5	32.7

Table 5.15: Particle size distributions of selected grades of milled and atomized ferrosilicon (Samancor, South Africa).

Size fraction [μm]	48D	100D	270D	Cyclone 60
+ 212	0 - 2	0 - 1	0	0
+ 150	4 - 8	0 - 1	0	0 - 2
+ 106	12 - 18	1 - 4	0 - 1	0 - 4
+ 75	19 - 27	5 - 10	0 - 3	2 - 6
+ 45	27 - 35	28	7 - 11	18 - 26
- 45	27 - 35	61 - 69	85 - 93	65 - 78

separation.

Media is supplied in several size grades and the grade used will vary with each individual plant. While finer grades improve stability, fine particles, in addition to possible viscosity problems, are more difficult to recover and the feed into LIMS should then be reduced by a factor of 1.5 to 2, to maintain the magnetic recovery. A typical particle size distribution of ground magnetite is shown in Table 5.14, while size distributions of several grades of ferrosilicon are listed in Table 5.15.

There are two basic types of ferrosilicon available to the operators of dense medium plants, namely milled (e.g. 270D) and atomized spherical grades (e.g. Cyclone 60). There is a marked difference in both the shape and nature of ferrosilicon particles produced by grinding and atomizing. Particles produced by grinding are rough and angular in shape, whereas particles produced by atomizing are smooth and largely spherical or rounded. This difference in the particle shape has marked effects on the rheological properties of aqueous suspensions of the two types of ferrosilicon [C7].

The critical value of slurry density for atomized grades is much higher than for the milled grades. At densities above 3000 kg/m^3, the viscosities of suspensions of atomized ferrosilicon are significantly lower than those of the milled grades, owing to the greater friction among the colliding irregularly shaped milled particles [C17]. This superior property of atomized ferrosilicon allows very high suspension densities to be obtained (up to 3800 kg/m^3) and this has made possible the heavy-medium separation of certain ores that were not previously amenable to the treatment. On the other hand, grades of milled ferrosilicon show greater stability than grades of atomized ferrosilicon of equivalent size distribution.

Since the desired conditions of separation are those of minimum viscosity with maximum stability, selection of the grade of ferrosilicon is based on a trade-off between these contradictory requirements and on cost considerations.

Viscosity of dense medium suspensions Magnetite media are characterized by high apparent viscosities, extremely sensitive to rate of shear. Viscosities of the magnetite medium ranging from 700 cP (0.7 Pa$\times$s) at the shear rate of 1000 s^{-1} to 2740 cP (2.74 Pa$\times$s) at 1000 s^{-1} were reported at the operating density of 2800 kg/m^3 [G18]. Viscosities of ferrosilicon under similar conditions ranged from 84 cP (0.084 Pa$\times$s) to 200 cP (0.2 Pa$\times$s). Similar results were obtained for various grades of ferrosilicon [D14]. The modelling data and their verification by experimental measurements show that the apparent viscosity of atomized ferrosilicon (e.g. Cyclone 60), at the operating density of 3800 kg/m^3 ranged from 0.025 Pa$\times$s at the shear rate of 150 s^{-1} to 0.3 Pa$\times$s at 10 s^{-1}.

Media losses Historically, media losses have frequently been the largest single component of the operating costs of dense medium separation plants. Losses of magnetite in coal washing plants range from 0.4 kg to 2 kg per tonne of raw coal. In modern large cyclone plants treating - 45 + 0.5 mm material, the average losses amount to 0.45 kg/t [E3].

In DMS plants that use ferrosilicon, the losses amount to 20% to 40% of the DMS plant operating costs [D13] and may range widely, from 0.1 to 2.8 kg/t [H20], although in modern plants they do not exceed 0.25 kg/t [D13]. In an iron ore cyclone plant, the ferrosilicon losses are reported to be 0.12 kg/t and 0.15 kg/t, for intermediate and fine ore fractions, respectively [K20]. It is generally claimed that the losses in the magnetic circuit usually represent the major component of the total loss. For instance, Hunt et al. [H20] report that, excluding plant housekeeping, the average losses attributable to magnetic separation, adhesion and corrosion occur in the ratio 6:3:1. In another investigation, however, magnetic losses not exceeding 30% of the total loss were observed [N7].

Magnetic properties of ferrosilicon and magnetite The magnetic properties of magnetite and ferrosilicon are important for easy recovery and cleaning of the medium in the circuit. The running costs of a DMS plant are also affected by the magnetic properties. Baran and Svoboda [B29] conducted an extensive study of magnetic properties of both magnetite and ferrosilicon of various grades. A typical hysteresis curve of 270D milled ferrosilicon is shown in Fig. 5.18. Table 5.16 summarizes the magnetic characteristics of ferrosilicon and magnetite, as determined from measurements by a vibrating sample magnetometer.

It can be seen that atomized grades of FeSi exhibit slightly higher values of the saturation polarization (by 5% to 7%) compared to the milled grades. The saturation polarization of magnetite (0.45 T) is lower than that of FeSi by factor

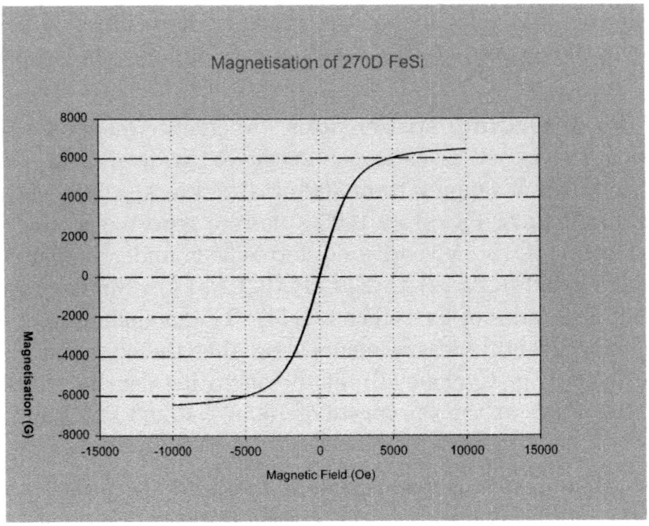

Figure 5.18: A hysteresis curve of 270D ferrosilicon as measured by a vibrating sample magnetometer [B29].

Table 5.16: Saturation polarization Js, remanent polarization Jr and coercive force Hc of various grades of ferrosilicon (Samancor) and magnetite (Amcoal), as measured by a vibrating sample magnetometer [B29].

Material	$\mathbf{J}_s$ **[T]**	$\mathbf{J}_r$ **[T]**	$\mathbf{H}_c$ **[kA/m]**
Cyclone 60	0.93	0.047	9.96
48D	0.83	0.052	10.2
65D	0.87	0.055	10.6
100D	0.87	0.048	9.33
150D	0.86	0.048	9.89
270D	0.83	0.051	9.89
Magnetite	0.45	0.063	14.3

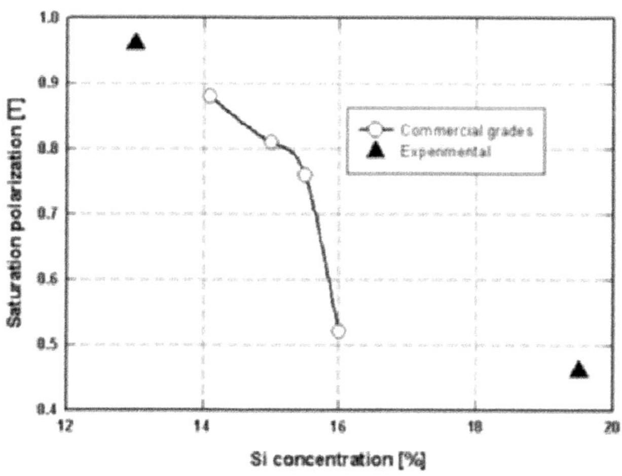

Figure 5.19: Relationship between the Si concentration and the saturation polarization of FeSi (after [B29]).

of two, and it was found to be by about 25% lower than the literature value of 0.6 T, as a result of the presence of weakly magnetic impurities. The coercive force of magnetite is 50% higher than that of ferrosilicon. This might indicate a need for a higher demagnetization field for magnetite.

Although the milled grades of FeSi were found to be much more susceptible to corrosion than the atomized grades, no substantial effect of corrosion on the magnetic properties was observed. Also, no significant differences in the magnetic behaviour of various size fractions of FeSi were observed. Only a slight reduction (less than 5%) in the saturation polarization values for fractions smaller than 106 μm was observed. Losses of fine FeSi in DMS plants can thus be ascribed to the reduced efficiency of magnetic separators to recover fine particles rather than to reduced magnetization.

The initial magnetic susceptibility of all grades of FeSi was approximately 2.8×10^{-4} m^3/kg, while that of magnetite was found to be equal to 3.5×10^{-4} m^3/kg.

It was also observed that a substantial increase in the saturation polarization of FeSi could be achieved by decreasing the concentration of Si as is illustrated in Fig. 5.19. It was found that the saturation polarization changed by 0.16 T for every 1% in the Si content, in the range 14% to 16%.

Demagnetization The passage of heavy medium through magnetic separators in the medium recovery circuit induces remanent magnetization of the medium. This may result in the formation of magnetic flocs with subsequent deterioration of the medium properties, such as stability and viscosity.

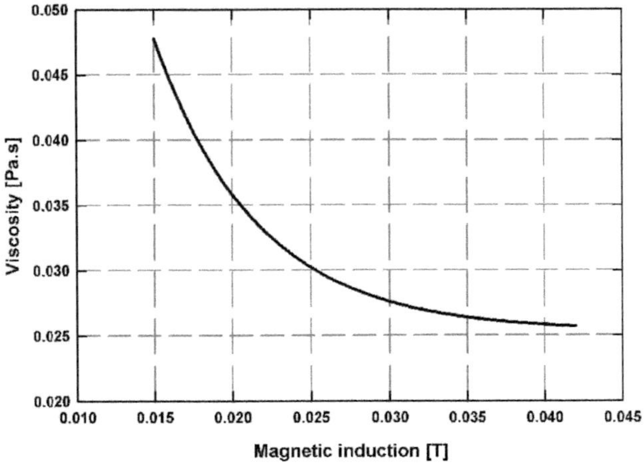

Figure 5.20: The effect of the demagnetizing magnetic induction on FeSi viscosity in a cone circuit at Premier Diamond Mine, South Africa (adapted from [C17]).

It is common practice at DMS plants to install demagnetization coils on the magnetic separator return lines. Although the need to demagnetize media has been recognized in the static-bath processes, the usefulness of demagnetization coils in cyclone plants has been a subject of dispute. The views range from outright rejection to unconditional acceptance. These polarized views are part of the DMS plant folklore, although there is an overwhelming scientific and engineering evidence of the beneficial effect of media demagnetization. In general, the installation of demagnetization coils in DMS cyclone plants is recommended (e.g. [N8]).

The inverse relationship between the magnetic field strength in a demagnetization coil and the viscosity of the circulating medium FeSi medium in the cone circuits is illustrated in Fig. 5.20. A similar beneficial effect of FeSi demagnetization, in a cyclone plant, was reported by Napier-Munn and Scott [N7]. They observed that the coil had a rapid and substantial effect on the viscosity and the results are summarized in Fig. 5.21. The viscosity had started to drop shortly after the coil was activated and to rise again after the coil was de-activated.

Measurements of the magnetic susceptibility of magnetized samples of ferrosilicon showed that the values of magnetic susceptibility were higher than those of unmagnetized samples [B29]. After demagnetization by a coil generating the a.c. magnetic induction of 0.06 T, the magnetic susceptibility returned to its original value.

Practical aspects of demagnetization are discussed in Section 5.5.

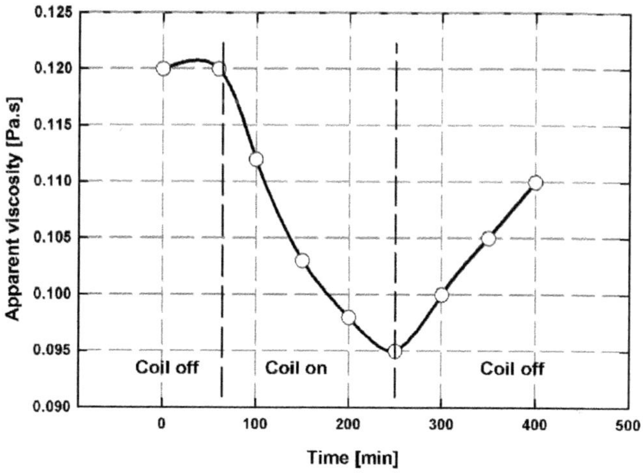

Figure 5.21: The effect of demagnetization on the viscosity of 150D FeSi medium at the cyclone plant of the Mount Isa Mines lead-zinc concentrator, Australia, (adapted from [N8]).

Wet magnetic drum separators for ore concentration

Wet low-intensity drum magnetic separators are also used for the concentration of magnetite ore or other strongly magnetic minerals. Most of the low-intensity magnetic separation technology used today was developed during the 1960s with no significant changes taking place. The most noticeable changes over the years have occurred with the increased diameter and length of the drum. With each increase of the drum diameter, the process capacity of units increased and throughputs as high as 160 t/h per meter length of the drum were recorded. However, tests with 1500 mm diameter drum separators could not reproduce this trend for the capacity increase, and currently not many concentrators of this size have been installed [S13].

The improved design and more efficient permanent magnetic materials allowed the magnetic field rating of the drum magnetic separators, traditionally expressed as the magnetic induction at 50 mm (or 2 inches), to be increased, from 0.05 T to about 0.12 T available today.

In the ore concentration applications, the separators are designed for three basic operations, namely as cobbers, roughers (cleaners) and finishers.

Cobbers Drum separators used in cobbing operations are designed to obtain maximum rejection of a non-magnetic fraction and maximum recovery of the magnetic material. Most iron-ore concentrating drum separators are applied in multi-drum, usually con-current, configurations so that the high-grade magnetic concentrate is obtained. In cobbing operations, two drums are typical, but as many as four drums have been used. The diameter of the drums is either 900 or 1200 mm and the magnetic field rating, at 50 mm, of 0.1 T is standard for cobbers.

Roughers Drum magnetic separators used in roughing or cleaning are applied after the size of the cobber concentrate has been further reduced. Because the cobbing has significantly upgraded the ore, magnetic loading of the drum is usually increased considerably. The objective of roughing is to obtain a high-grade magnetic concentrate, while maintaining high recovery. Double-drum separators, in con- or counter-current modes of operation are typically used in roughers and the required magnetic field at 50 mm is, as a rule, 0.07 T to 0.09 T.

Finishers Finishing drum magnetic separators are designed to produce the highest possible iron content in the concentrate. The feed tank and feed arrangements are usually of semi-counter-current design, with the objective of dispersing the feed particles in order to obtain maximum rejection of non-magnetic particles. Both 760 and 900 mm-diameter drums are used in the finisher applications, with the magnetic field strength at 50 mm lower (e.g. 0.05 T) than in a rougher, to improve selectivity of separation. The feed size is usually reduced to less than 75 μm or 45 μm.

Variables in low-intensity magnetic separation of ores Factors associated with the properties of the feed, to be considered in the selection of a wet drum magnetic separation applied to the ore beneficiation, are [M26]:

- Feed rate (t/h/m)
- Feed particle size (max. 10 mm) and size distribution
- Feed volume (m^3/h)
- Concentration of magnetic particles
- Degree of liberation
- Desired recovery into the magnetic fraction.

As shown in Table 5.17, the maximum feed rate varies with the type of separator considered; maximum and recommended feed volumes are also given. Typical sizes of feed particles for the three modes of operation are:

Cobbers: - 6 mm + 850 μm
Roughers: - 500 μm + 300 μm
Finishers: - 75 μm + 45 μm.

Table 5.17: Feed rates and feed volumes in LIMS, as applied to ore beneficiation [M26].

Application	**Feed rate [t/h/m]**		**Feed volume [m^3/h/m]**	
	Max.	Recommended	Max.	Recommended
Cobber	75	45	100	70
Rougher	60	45	100	70
Finisher	15	15	35	20

Table 5.18: Number of drums and magnetic poles used in LIMS for ore beneficiation [M26].

Separator	**Number of drums**	**Number of poles**
Cobbers	2 to 4	5 or 6
Roughers	1 or 2	5 or 6
Finishers	2 or 3	4 to 10

Magnetic content and degree of liberation The magnetic content of iron ores to be concentrated by LIMS varies over a fairly wide range. Ores with iron concentrations from 10% to 50% or more have been successfully treated. Since liberation is not complete at the cobber stage, a substantial amount of gangue is carried into the cobber concentrate. Typically, the cobber concentrate contains 50% Fe or more and, accordingly, the magnetics content of the rougher feed is quite high. The finisher feed contains 2% to 5% of non-magnetics and thus has a very high concentration of the magnetic material.

Complete liberation of most iron ores that are being treated by LIMS is usually at sizes finer than 150 μm, although some ores have to be reduced to -45 μm.

Magnetic field strength and number of poles Increased magnetic field strength achievable in modern low-intensity drum magnetic separators allows not only more efficient recovery of fine particles, but also the throughput of a separator can be increased without losing recovery. Such an effect of magnetic field strength on throughput and recovery is illustrated in Fig. 5.22.

The number of drums used in various applications varies with the individual ores and the choice is determined by the results of pilot-plant or laboratory tests. In addition, various types of magnetic assemblies and pole combinations have been successfully used. Typical magnetic pole specifications and number of drums are shown in Table 5.18.

Throughput The feed rate in a low-intensity magnetic separator affects its metallurgical performance, as can be seen in Figs. 5.22 and 5.23. With increasing feed rate, the recovery of magnetic particles into the magnetic fraction decreases, the rate of decrease being strongly dependent on the magnetic field

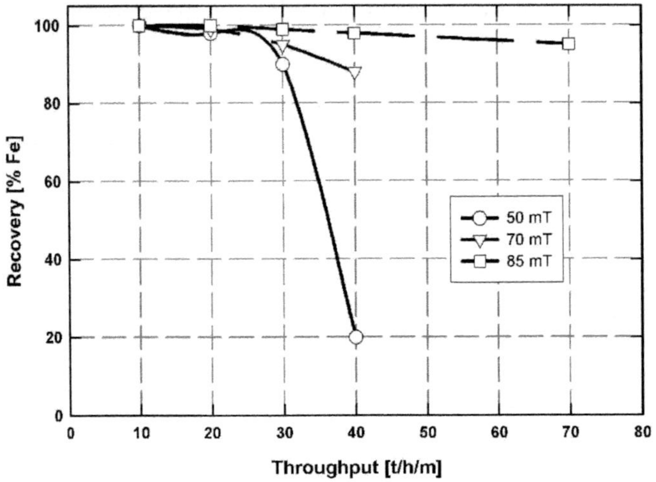

Figure 5.22: Recovery of iron by LIMS, as a function of throughput, for different values of magnetic induction at 50 mm (adapted from Sundberg [S13]).

strength generated by the magnetic drum.

5.3.2 Wet high-intensity magnetic separation

Wet high-intensity magnetic separation is used in three different technological processes, as shown in Fig. 5.24. These separators can thus be used either as a magnetic filter to remove a single solid component from a suspension, or as a magnetic separator to remove an unwanted magnetic component from a mixture of materials, or for a concentration of valuable weakly magnetic materials from a mixture.

The high intensity of the magnetic field in these separators is generated either by rare earth or ferrite permanent magnets (for example in RE drum separators or in a Ferrous Wheel separator), or by resistive electromagnets or superconducting magnets.

Magnetic field gradient, an equally important parameter in magnetic separation, is generated by three different methods. Close-gradient, open-gradient and matrix (polygradient) techniques produce magnetic field gradients of widely different magnitudes, as can be seen in Table 1.20. As a result of these different attributes of high-intensity magnetic separators, a plethora of names and terms is being used to classify them, although the operating principle of many of them are the same or, at worst, they can address similar separation objectives. The resulting chaos is further magnified by the advent of new magnetic materials and advanced magnet designs that allow new approaches to high field and high

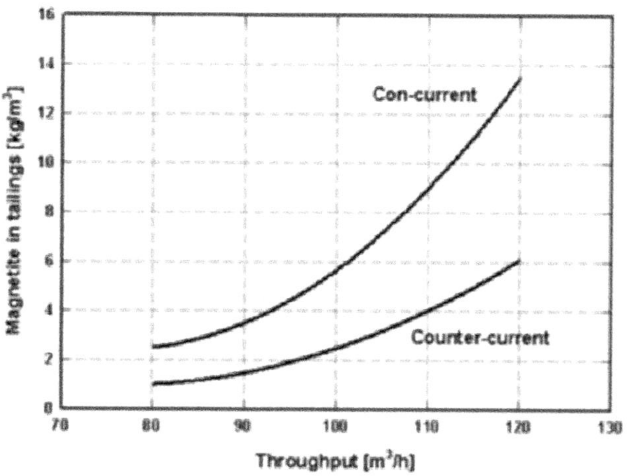

Figure 5.23: The effect of feed rate on the recovery of magnetite, in a low-intensity magnetic drum separator. Drum length: 1.2 m, drum diameter: 900 mm (adapted from [K14]).

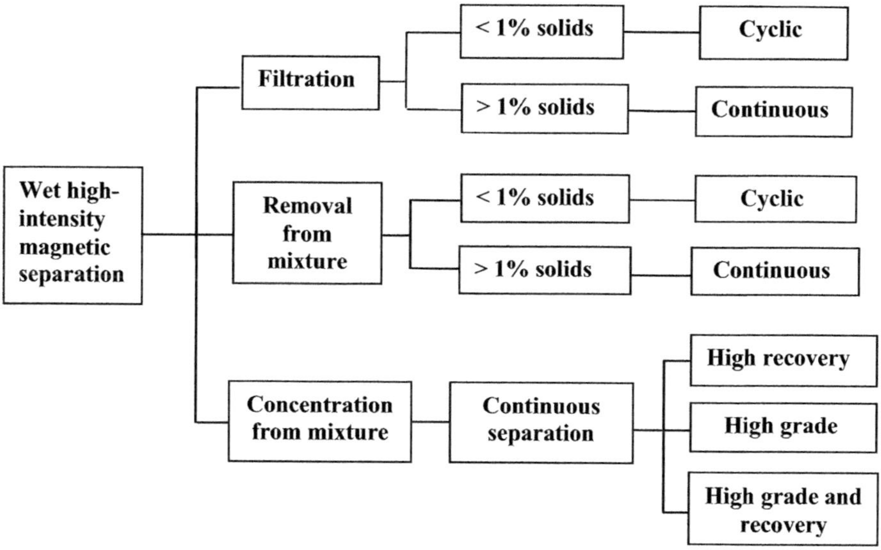

Figure 5.24: The range of applications of wet high-intensity magnetic separation (adapted from Svoboda [S1]).

Table 5.19: Nomenclature for wet high-intensity magnetic separators.

Separator	Acronyms
Rare-earth drum separator	REDS, REDMS, MIMS
Linear multipole open-gradient separator	OGMS
Matrix continuous separator	WHIMS, HGMS, HIMS
Matrix cyclic separator	HGMS
Matrix cyclic magnetic filter	HGMF, HIMF

gradient generation. A summary of the terms most frequently used to describe wet high-intensity magnetic separators is given in Table 5.19.

A schematic diagram of different technological processes, in which wet high-intensity magnetic separators are used, is shown in Fig. 5.24. It is clear that conditions for successful separation in these three categories can differ widely. Their correct selection must be based upon proper identification of the target of the process and the definition of criteria that would distinguish successful separation from unsuccessful. The decision-making process in such a selection is outlined in Fig. 5.25 and its detailed discussion will be presented in subsequent sections.

Wet rare-earth drum magnetic separators

Although dry rare-earth drum magnetic separators have been readily accepted by the mining industry, it has taken somewhat longer to develop and implement wet RE drum separators. One of the reasons for reluctant acceptance of wet RE drums by the industry is the onerous problem with the removal of strongly magnetic material from the drum, even when these particles are present in trace concentrations. Conventional high-pressure spraying or scraping methods are not sufficiently effective in overcoming the strong magnetic force holding the particles on the drum, and new approaches to this problem are being investigated [A23].

Magnetic filtration

Magnetic filtration is used primarily for the removal of fine and quasi-colloidal particulates from dilute fluid suspensions. Removal is carried out by passing suspensions through a magnetized matrix, which leads to the retention of particles throughout the matrix. The fluid stream may be either liquid or gas.

The process of magnetic filtration is analogous to that of deep bed filtration, the difference being that an additional external force, namely the magnetic force, contributes to the filtration efficiency in magnetic filtration. Although the magnetic force does not significantly increase the probability of particle collision with the matrix [S31, S33], the enhanced efficiency of particle attachment caused by the presence of strong, static dipolar magnetic interaction between the matrix and a colliding particle increases significantly the efficiency of the

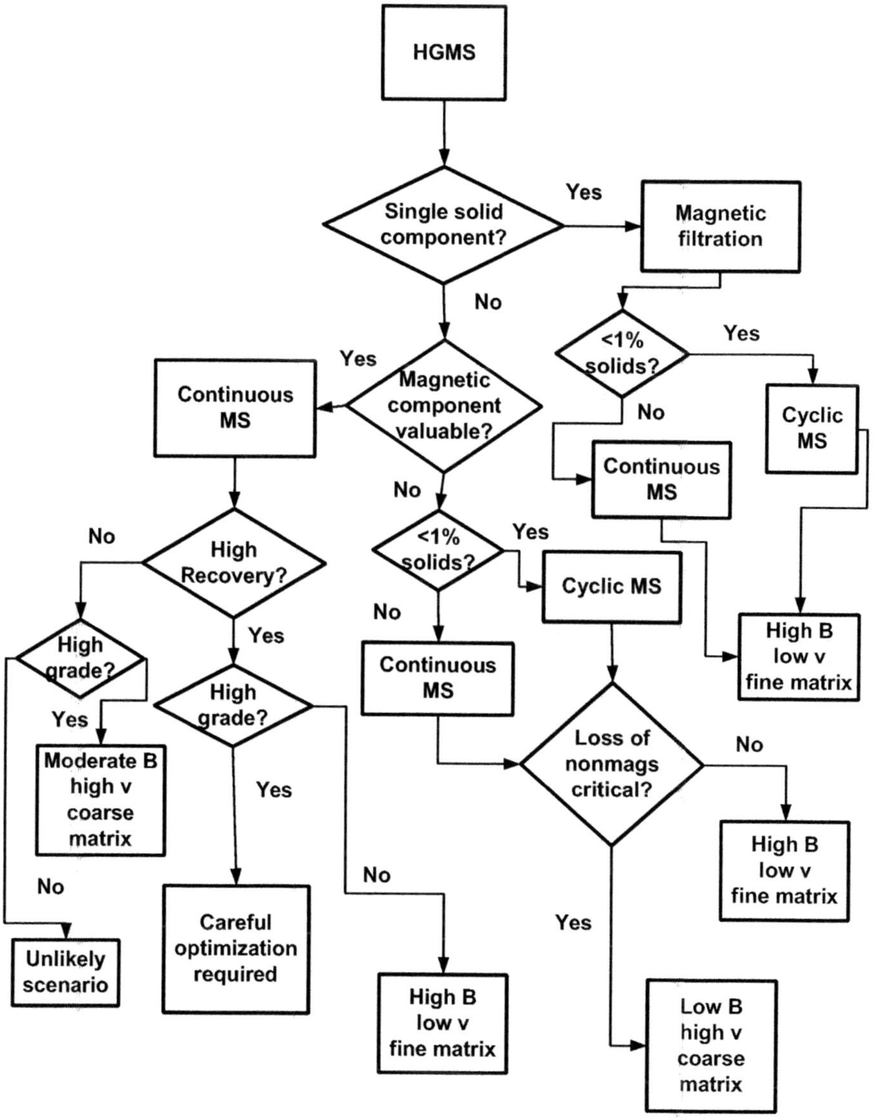

Figure 5.25: Selection of the operating conditions in wet high-gradient magnetic separation (adapted from Svoboda [S1]).

particle capture. The similarity between magnetic filtration and deep bed filtration indicates that the basic rules common to deep bed filtration are applicable to magnetic filtration.

As the aim of magnetic filtration is the removal of a single solid component from a suspension, a brute force approach can be used so that all particles are collected on the matrix. Therefore, the largest possible magnetic force, compatible with the magnetic properties of the material to be removed, should be used. Furthermore, the maximum magnetic field, in combination with the finest possible matrix, compatible with the particle size distribution, should be employed.

Simultaneously, in order to ensure the highest degree of extraction of solids, the fluid shear stress on the deposited particles should be kept as low as possible. At the same time, however, the flow velocity of a suspension through the matrix bed should be sufficiently large to ensure a high throughput.

Matrix Although it is usually desirable to use a fine matrix in magnetic filtration, mainly to increase the probability of particle collisions with the matrix, it is also important to avoid mechanical straining. Straining leads to the formation of a surface mat of deposit on the matrix. This causes a high resistance to the flow and ultimately the clogging of the matrix. In order to avoid straining, the ratio a/b of the matrix element radius a to the particle radius b should be of the order of 50 or larger. Coarser matrices with $a/b = 100$ to 500 often give results comparable to those obtained with very fine matrices.

Flow velocity The choice of the flow rate is usually based on the trade-off between the purity of the effluent and the throughput of the filter. For the removal of ferromagnetic impurities from steel-mill waste water, where high hydrodynamic erosion can be tolerated, interstitial velocities of up to 0.3 m/s are used. In applications where fine weakly magnetic particles are to be collected, the flow rate is often as low as 1×10^{-3} m/s.

Removal of magnetic impurities from mixtures

The removal of unwanted magnetic components from mixtures is a well established application of magnetic separation and a wide spectrum of materials can therefore be upgraded by the application of efficient and cost-effective dry magnetic separators. Although dry separation is an answer in many applications, finely ground feebly magnetic materials can be beneficiated only by wet high-intensity magnetic separation. The brightening of kaolin and dolomite and the purification of glass sand and other industrial minerals by wet high-intensity high-gradient magnetic separation are well proven and established applications.

Matrix The common feature of all methods for the removal of magnetic admixtures is that, as in magnetic filtration, the brute force approach can, in principle, be used. This means that a high magnetic field, fine matrix and low flow velocities should be used to achieve efficient removal of impurities. This approach can lead, however, to matrix blockage and to a loss of the valuable non-magnetic product due to its entrainment in the magnetic fraction. In most applications, the excessive loss of the non-magnetic product into the magnetic waste is not acceptable and careful selection of the operating conditions is required. The matrix should be sufficiently coarse to avoid mechanical straining but it ought to be sufficiently fine to guarantee high recovery of the magnetic impurities.

The concentration of materials

The concentration of valuable, weakly magnetic minerals from finely ground ores is the largest application of wet high-intensity magnetic separation and of high-gradient magnetic separation. The objectives of the concentration of minerals by WHIMS and HGMS are threefold, as shown in Fig. 5.25.

a) High recovery of a valuable metal into the magnetic concentrate. In this case a high magnetic field, in combination with low flow velocity should be used. A fine matrix would be a natural choice, although a coarse matrix often performs equally well or even better.

b) High quality of the magnetic concentrate. If the recovery is of secondary importance, the operating conditions should be chosen in such a way that selectivity of the separation process is high. Only a moderate magnetic field combined with a high flow velocity should be used. An excessively high magnetic field causes the formation of flocs, comprising both magnetic and non-magnetic particles, and also increases the possibility of the formation of the surface mat of deposit. Correct choice of the matrix is the most critical step in achieving selective separation.

c) High recovery and high grade. This rather natural requirement poses a problem of considerable complexity, since often contradictory requirements are imposed on the operating conditions. The selection of the matrix, the strength of the magnetic field and the flow velocity must be carried out in experimental programme, in which other variables that influence the separation process (e.g. pulp density, dispersion etc.) are simultaneously investigated.

Cyclic or continuous separation?

The choice between cyclic and continuous high-gradient magnetic separators is dictated mainly by the concentration of the magnetically recoverable particles in the feed. If the mass yield into the magnetic fraction is smaller than, for example, 1%, a cyclic separator can be suitable. However, in the majority of applications of HGMS in mineral processing, concentration of the magnetic component is larger than 1%; furthermore, high flow rates are usually required to achieve a high grade of the concentrate and a high throughput. Under these conditions the matrix would be loaded within a matter of seconds and the duty

Table 5.20: Comparison of some operating parameters for various applications of a cyclic magnetic separator.

Application	**Mag. fraction in the feed [kg/m^3]**	**Flow velocity [m/s]**	**Feed time [hours]**
Filtration of thermal power plant water	0.0001	0.24	1000
Steel mill waste water	0.1	0.12	1.5
Kaolin purification	1 to 3	< 0.01	0.1
Hematite ore	250	0.16	0.0003

Table 5.21: The comparison of cyclic and continuous high-gradient magnetic separators.

Parameter	**Cyclic**	**Continuous**
Duty cycle	Low to medium	100%
Throughput	Low to medium	High
Specific processing cost	High	Moderate
Versatility	Low	High
Mechanical complexity	Low	High

cycle of the separator would be very short. Table 5.20 illustrates the range of feed times for a cyclic high-gradient magnetic separator in different applications.

It can be seen that the main application of cyclic separators is filtration of cooling and waste waters. For the beneficiation of weakly magnetic ores, the utilization factor of the separator is very low. Although the situation can be improved somewhat by the application of reciprocating technology, the use of continuous magnetic separators is inevitable. A continuous separator which operates with a 100% duty cycle, can treat a wide range of minerals with varying content of the magnetic particles. The comparison of cyclic and continuous high-gradient magnetic separators as applied in mineral processing is given in Table 5.21.

5.3.3 Magnetic field in high-intensity magnetic separators

The magnitude of a magnetic field generated by the magnet of a magnetic separator is one of the most important parameters. It signifies the potential of a separator for the treatment of various classes of magnetizable materials.

Various manufacturers of magnetic separators use different approaches upon how to ascertain and specify the magnetic field strength, and the reported values are often contentious and occasionally misleading. The magnetic field in mag-

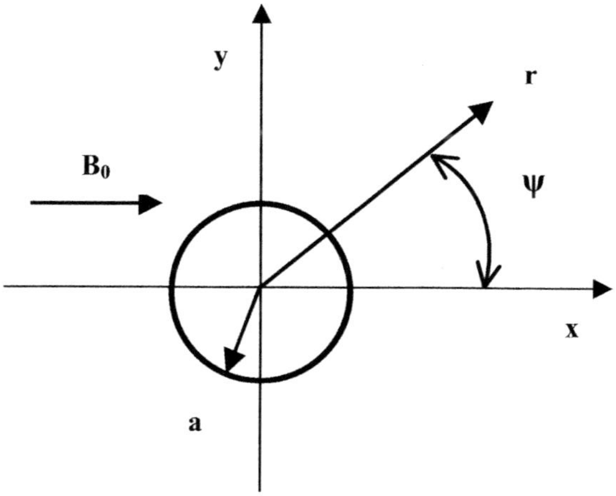

Figure 5.26: The geometrical configuration for calculation of magnetic induction in the vicinity of a matrix element of radius a.

netic separators is generally very non-uniform. Particularly in high-intensity separators, such as rare-earth rolls or WHIMS or HGMS, the value of the magnetic field varies considerably over a scale of several hundred micrometers, as has been discussed in Section 1.6. Commonly available measuring instruments are able to measure only the mean magnetic field over a relatively large cross-sectional area, defined by the size, for example, of the Hall sensing element, and the reported values of the magnetic induction generated, for instance, by a rare-earth roll separators, may range from 1.0 T to 2.0 T, depending upon which measuring (or estimating) technique was employed. In matrix separators, the field is measured either as a background magnetic field or in the vicinity of a matrix element. The importance of the latter value of the magnetic field is insignificant, as it will be very similar for low-field and high-field separators.

The total external magnetic induction in the vicinity of a matrix element is given by the superposition of the background magnetic induction B_0 and the induction due to the matrix element B_m:

$$\vec{B} = \vec{B}_0 + \vec{B}_m \tag{5.5}$$

The induction B_m causes distortion of the essentially uniform background induction B_0 in the vicinity of each matrix element. This matrix material is magnetized to polarization J, which is proportional to the background induction B_0 until the material is saturated. Above its saturation value J_s, the polarization is constant and does not increase with further increase in B_0.

Under the assumption that the matrix material is magnetically saturated, the total magnetic induction at a point r outside the matrix can thus be calcu-

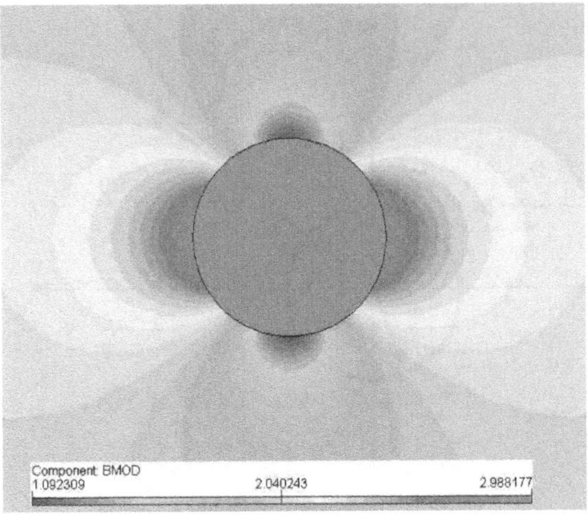

Figure 5.27: Pattern of the magnetic field outside a horizontally magnetized steel sphere.

lated from the expression:

$$B = \left[B_0^2 + \frac{1}{4} J_s^2 \left(\frac{a}{r} \right)^4 + B_0 J_s \left(\frac{a}{r} \right)^2 \cos 2\psi \right]^{1/2} \tag{5.6}$$

where a is the radius of the matrix element and the angle ψ is defined in Fig. 5.26.

The pattern of the magnetic induction and the radial component of the magnetic induction, for $\psi = 0$, in the vicinity of a magnetized steel wire, obtained using the electromagnetic VF modelling software, are shown in Figs. 5.27., 5.28 and 5.29, for two values of the saturation polarization. It can be seen that, in order to generate, for example, the magnetic induction of 2 T on the surface of the matrix, a background magnetic induction of 1 T is needed. If a steel with rather poor magnetic properties (e.g. $J_s = 1$ T) is used for the matrix, the background induction of 1.5 T is required to create 2 T on the matrix surface.

It also transpires from eq. (5.6) and Figs. 5.28 and 5.29, that with increasing distance from the matrix, the role of the saturation magnetization of the matrix material diminishes and, for a distance $r = 3a$, the difference between the induction generated by $J_s = 2$ T or 1 T materials is negligible.

Similarly, it is possible to plot the gradient of the magnetic induction outside a magnetized steel wire. Figures 5.30 and 5.31 depict such patterns of the field gradient. These graphs reveal several interesting facts. It can be seen that the gradient is essentially independent of the values of the background magnetic induction B_0, with the exception of a case where $B_0 = 0.5$ T and $J_s = 2$ T. At the same time, the gradient is independent of the magnetic polarization of the wire material for distances from the wire surface greater than $2a$.

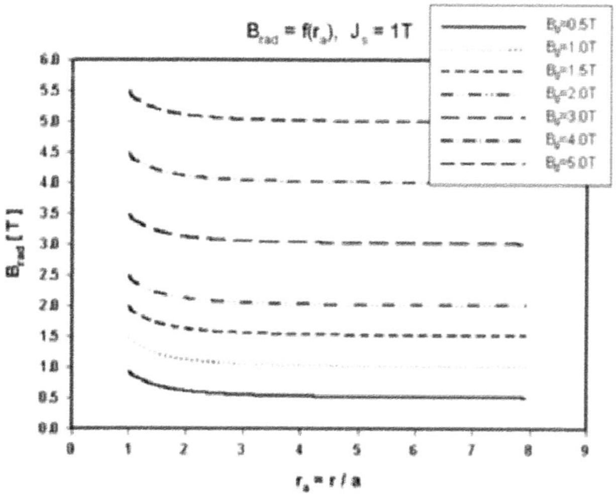

Figure 5.28: Distribution of the radial component of the magnetic induction in the vicinity of a magnetized wire ($a = 2.5$ mm), as a function of the normalized distance r_a from the wire. $J_s = 1$ T.

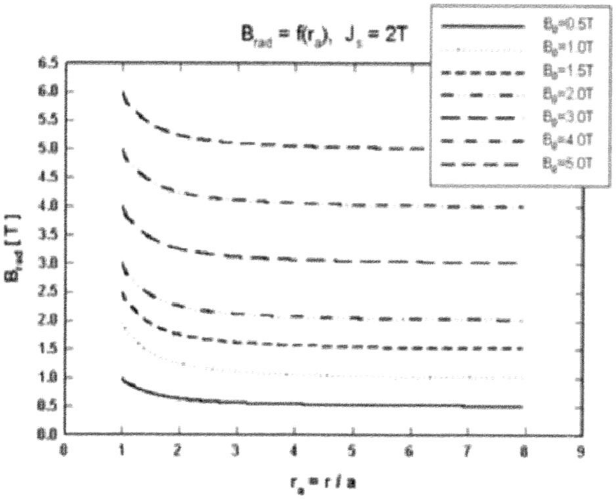

Figure 5.29: Distribution of the radial component of the magnetic induction in the vicinity of a magnetized wire ($a = 2.5$ mm), as a function of the normalized distance from the wire. $J_s = 2$ T.

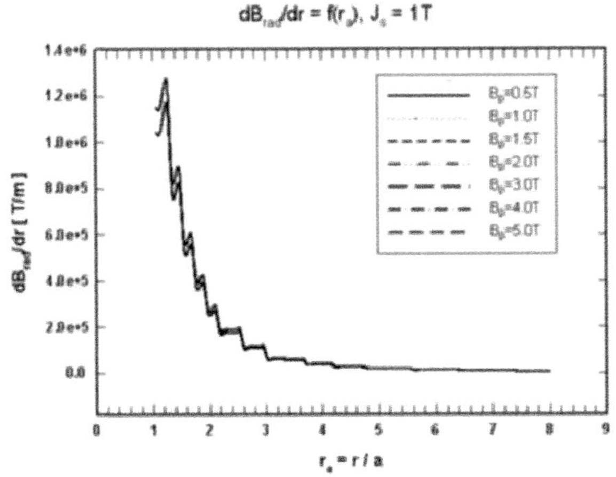

Figure 5.30: Distribution of the gradient of the radial component of the magnetic induction in the vicinity of a magnetized wire ($a = 2.5$ mm). $J_s =$ 1 T.

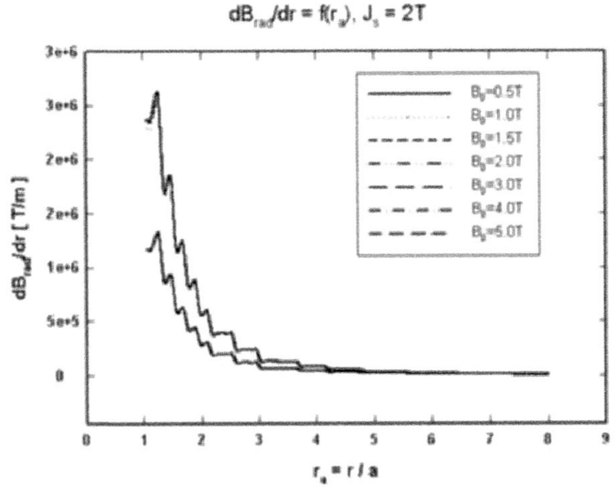

Figure 5.31: Distribution of the gradient of the radial component of the magnetic induction in the vicinity of a magnetized wire ($a = 2.5$ mm). $J_s =$ 2 T.

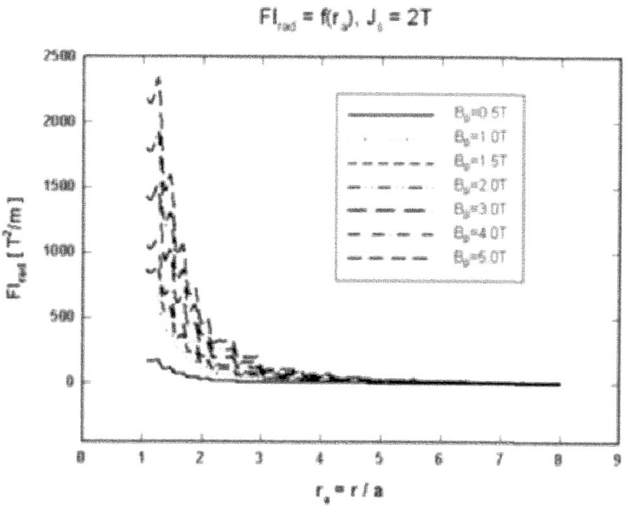

Figure 5.32: Distribution of the radial component of the force index in the vicinity of a magnetized wire ($a = 2.5$ mm), as a function of the normalized distance from the wire. $J_s = 2$ T.

The total magnetic force outside a magnetized steel wire, expressed as the force index, is illustrated in Fig. 5.32. It can be seen that the force index on the surface of the wire decreases with the decreasing background magnetic field. However, with increasing distance from the matrix surface, the influence of the background magnetic field diminishes and at distances greater than $3a$, the force is independent of the external background magnetic field.

A force holding already captured particles on the matrix surface increases, therefore, with increasing background magnetic field. This force decreases for particles captured onto more distant layers of the particle build-up. However, the traction force that is supposed to attract particles from a distance onto the matrix surface is, for practical purposes, independent of the background magnetic field. These observations contradict simple models of particle capture based on single collector-single particle approach, discussed in Section 3.4.1.

The effect of the magnetic field strength on the performance of a magnetic separator

The performance of a magnetic separator is, to a great extent, determined by the magnetic field strength, as discussed in Chapters 1 and 3. Simple theoretical models show that an increase in the magnetic induction increases the efficiency of magnetic separation. A more realistic model of the process [S31, S32, S33], and numerous practical applications of the technique, paint a somewhat different picture and show that an increase in the magnetic field strength can have even

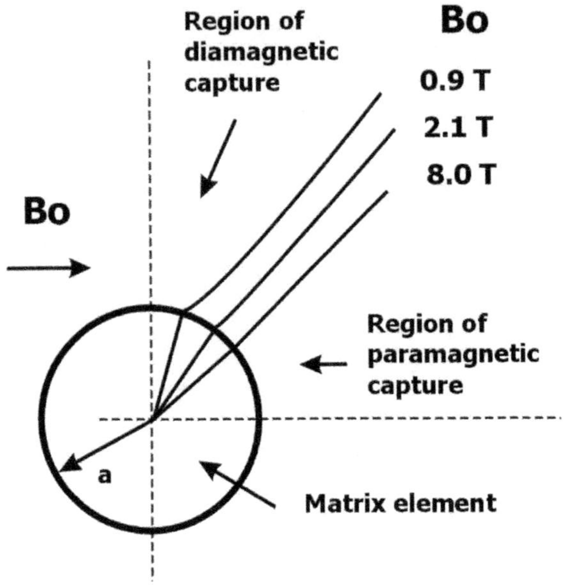

Figure 5.33: The attractive and repulsive regions for paramagnetic particles, around a magnetized matrix collector, as a function of magnetic induction.

a negative effect on the performance of a magnetic separator.

There are several reasons for the failure of the magnetic field to play the role that follows from simple theoretical models. Most materials that are treated by high-intensity magnetic separation are not paramagnetic. They usually possess some degree of magnetic ordering, which results in the field dependence of their magnetic susceptibility, as given by the Owen-Honda equation (1.44): $\kappa = \kappa_\infty + M/H$.

In the first phase, before the onset of magnetic saturation of the ferromagnetic (*s.l.*) components, the magnetic susceptibility remains constant when the field increases, since the magnetization also increases. Once the saturation is reached, which, for most ferromagnetic materials, occurs at quite low fields of the order of a few mT, the susceptibility decreases with increasing applied magnetic field and the drop in the value of magnetic susceptibility can often be quite appreciable. At high magnetic fields, the second term in eq. (1.44) can be neglected and the susceptibility becomes constant and equal to κ_∞. Therefore, the reduction in the magnetic susceptibility that accompanies the increase in the magnetic field strength can often result in stagnation, or even a decrease, in the recovery.

Another factor that adversely affects the recovery into the magnetic fraction is the decrease in the surface area of the matrix elements which is available for the capture of paramagnetic particles, when the applied field is increased.

The situation is depicted in Figs. 3.15, 3.16 and Fig. 5.33 and was discussed in greater detail in Section 3.4.1. It is evident that by increasing the magnetic field, the loading of material on the matrix increases as a consequence of a smaller surface area of the matrix available for paramagnetic capture.

A high magnetic field often induces the formation of clusters of magnetic particles through the process of magnetic flocculation. Since the formation of such flocs is usually non-selective, the grade of the magnetic concentrate deteriorates and the recovery can drop as a consequence of excessive matrix loading and of increased interstitial velocity of the slurry in the matrix.

It can therefore be seen that a high magnetic field does not necessarily mean better results. The optimum operating magnetic field must be determined in such a manner that the recovery/grade relationship satisfies the objective of the application.

In order to illustrate the role of the magnetic field strength in high-gradient magnetic separation, typical examples, as observed in the laboratory or on production scale, will be discussed.

Magnetic filtration The magnitude of the applied magnetic field is dictated mainly by the magnetic susceptibility of the suspended solids. The usual picture in magnetic filtration is that the efficiency of the removal of solids initially increases with increasing field strength up to a certain point, after which the filtration efficiency remains constant. In filtering ferromagnetic particles, a magnetic induction of 0.3 T is usually sufficient. Typical applications are the purification of steel mill waste water and cooling water in power plants, and the removal of suspended solids from municipal waste water seeded with magnetite. A typical example of such a straightforward "all-or-none" relationship between the magnetic field and the separation efficiency is shown in Fig. 5.34.

A similar observation was made in the application of a superconducting magnetic separator to the purification of municipal sewage water seeded with a ferromagnetic material. Isogami et al. [I5] reported that when a sufficient concentration of the seeding material was used, the filtration efficiency was essentially independent of the magnetic field strength in the interval from 0.1 T to 1.5 T. The experimental results are summarized in Fig. 5.35.

In more demanding applications, in which feebly magnetic solids are to be removed, a much higher magnetic induction, sometimes as high as 2 T, is required. However, often little benefit is obtained with fields exceeding 1.5 T, as is demonstrated in Fig. 5.36. It is evident that the efficiency of filtration of very fine, weakly magnetic kimberlite particles from diamond waste water does not improve with increasing magnetic field for a magnetic field strength exceeding 1.5 T.

These two examples typify the situation usually encountered in magnetic filtration, namely that the removal of solids is generally not improved by increasing the field strength above a certain value.

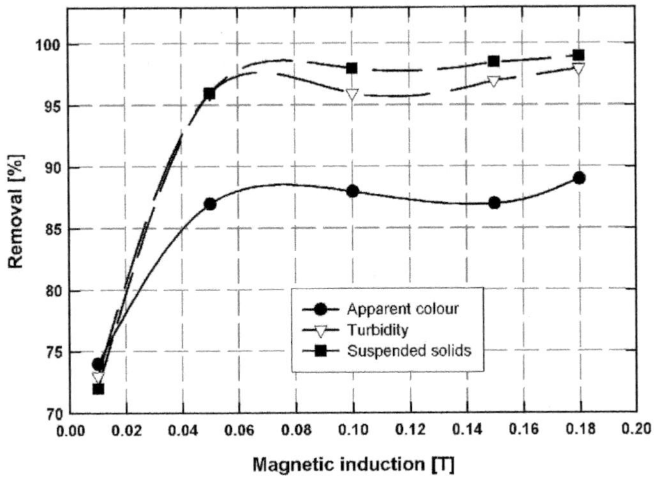

Figure 5.34: The efficiency of the removal of solids from municipal waste water seeded with magnetite (adapted from [A28]).

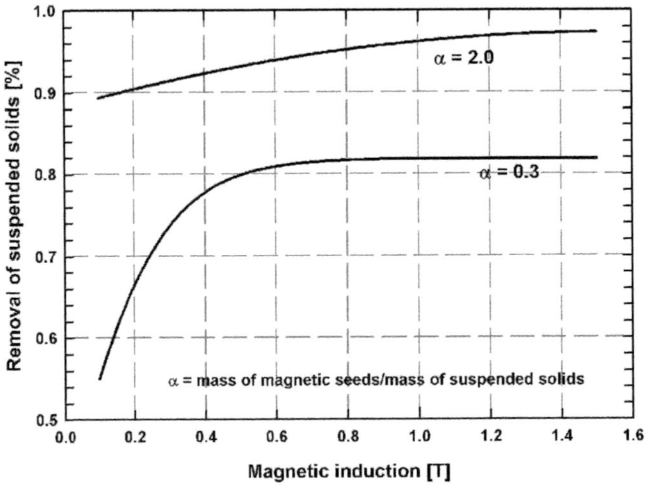

Figure 5.35: Removal of suspended solids from sewage water, seeded with a ferromagnetic material, as a function of the magnetic induction (adapted from Isogami et al. [I5]).

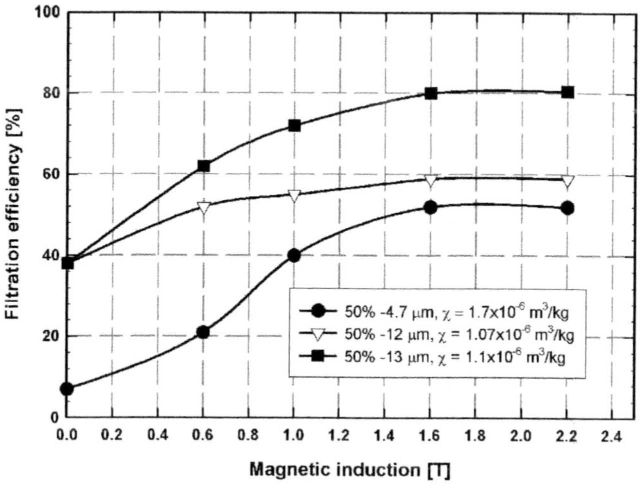

Figure 5.36: The efficiency of magnetic filtration of kimberlite solids from three types of diamond mine waste water (adapted from [S32]).

Removal of magnetic impurities from mixtures Although a high magnetic field strength is often desirable for an efficient removal of magnetic impurities, care must be exercised in the selection of the magnetic induction. In contrast to magnetic filtration, where excessive magnetic field strength usually does not affect the performance of a magnetic filter, results of the removal of magnetic impurities from mineral mixtures can often be impaired by undesirable losses of the non-magnetic product.

As in magnetic filtration, the efficiency of removal of weakly magnetic impurities does not increase indefinitely with increasing magnetic field strength and a leveling-off of the removal efficiency at higher magnetic fields is usually observed.

This effect is illustrated in Figs. 5.37 and 5.38. Figure 5.37 shows the effect of the magnetic field on the concentration of iron oxide in kaolin (South Africa). It can be seen that, with a steel wool matrix, a magnetic induction greater than 0.8 T did not improve the grade of the kaolin product. With the ball matrix, this threshold magnetic field was higher, namely 1.2 T. Moreover, the quality of the product appears to deteriorate in the magnetic field of the order of 2 T. An even more dramatic leveling-off of the grade of the non-magnetic glass sand product, as a function of the magnetic field can be seen in Fig. 5.38.

It can be seen that very often there is a threshold field, which marks the maximum field for which the grade of the non-magnetic product still improves. Magnetic field smaller or equal to the threshold field should be used.

However, in some applications, particularly in kaolin beneficiation, very high magnetic fields are reported to be beneficial. Figure 5.39 illustrates that with

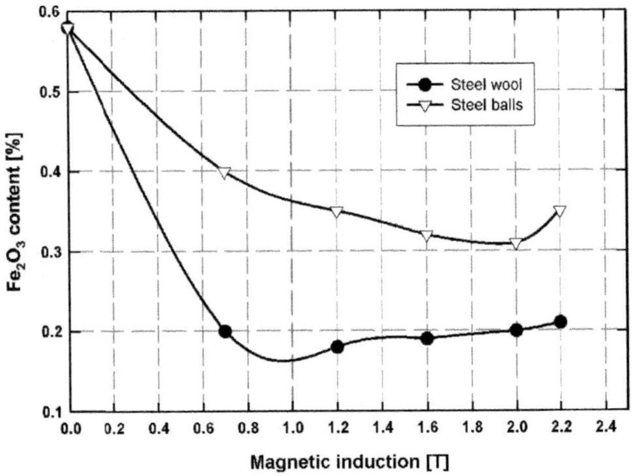

Figure 5.37: The effect of the magnetic induction on the concentration of iron oxide in kaolin (South Africa) (adapted from [S32]).

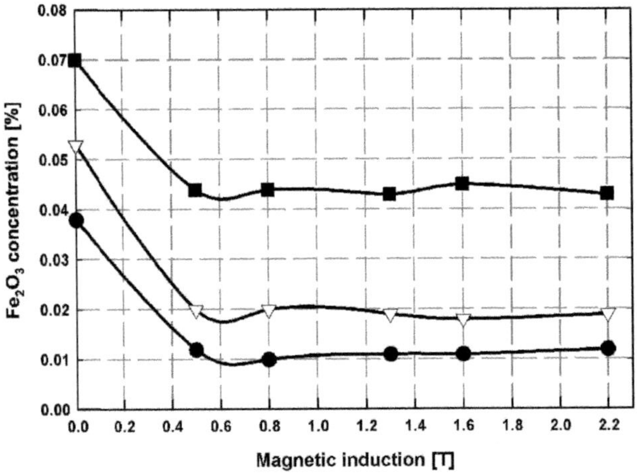

Figure 5.38: The effect of the magnetic induction on the concentration of iron oxide in three glass sand materials (adapted from [S32]).

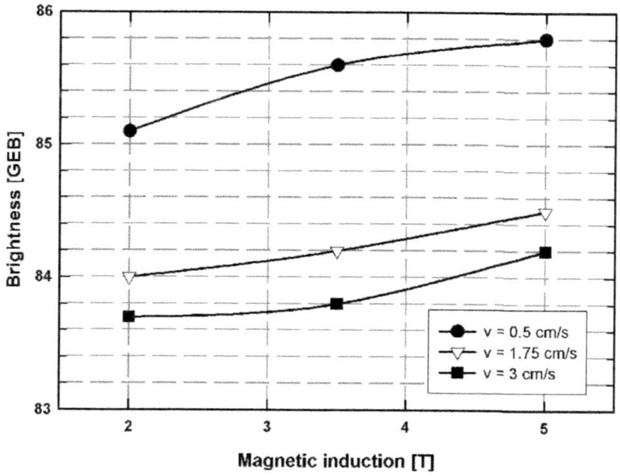

Figure 5.39: The effect of the magnetic induction on the brightness of the Georgia kaolin, for different values of feed velocity. Matrix length: 500 mm, matrix loading: 3.7 bed volumes (adapted from [S60]).

increasing magnetic field, the brightness of the Georgia kaolin clay improved [S60].

In addition to the conflict between grade and recovery, there are valid economic constraints on the selection of the magnetic field strength. In the design of magnetic separation systems, the capital investment and processing costs increase rapidly as magnetic induction is increased. It is, therefore, desirable to use the lowest practical magnetic field strength consistent with acceptable product quality.

The concentration of minerals The selection of the operating magnetic field in the concentration of weakly magnetic minerals is a sensitive task and only a well defined objective of separation will determine the approach taken. The traditional conception is that an increase in the magnetic field strength will result in an increase in recovery of weakly magnetic materials. Recovery of iron from feebly magnetic Anshan (China) iron ore, illustrated in Fig. 5.40 exhibits such a trend.

There are, however, certain factors, already discussed, that can cause the recovery into the magnetic product to decrease with increasing magnetic field, and there is usually an optimum magnetic field strength at which recovery reaches its maximum. With further increase in the field, recovery either remains constant or even decreases. Such trends are illustrated by the beneficiation of gold ore flotation tailings (Fig. 5.41) and manganese ore (Fig. 5.42).

Moreover, in contrast to magnetic filtration, or to many applications of the

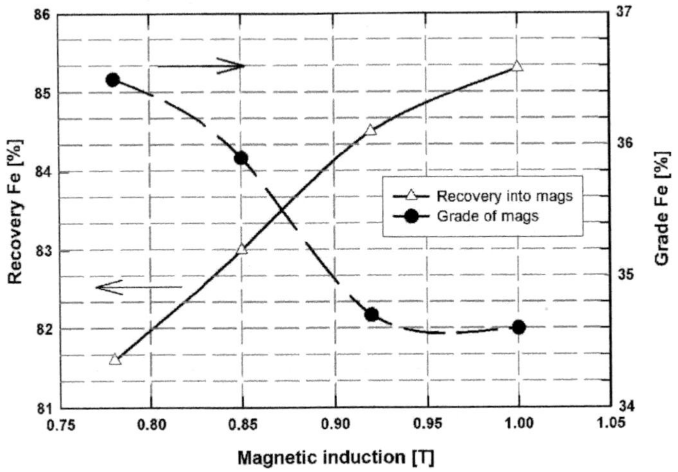

Figure 5.40: Beneficiation of the Anshan (China) iron ore by SLON HGMS (adapted from Xiong [X2]).

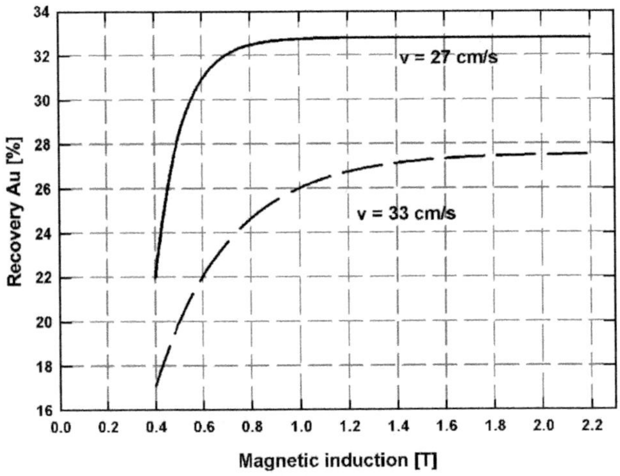

Figure 5.41: Recovery of gold by HGMS from East Rand (South Africa) flotation tailings (- 75 μm), as a function of the magnetic field, for two values of flow velocity (adapted from [S32]).

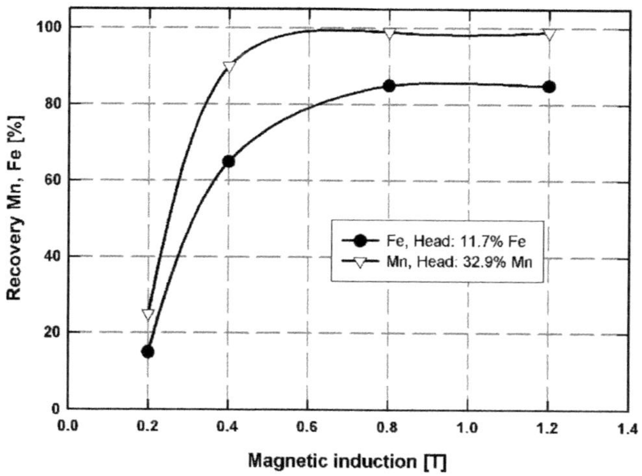

Figure 5.42: Recovery of Fe and Mn from manganese ore (South Africa, - 75 μm) by HGMS (adapted from [S32]).

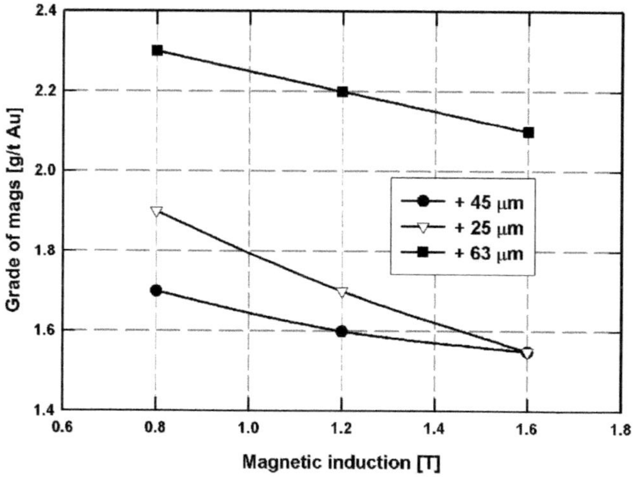

Figure 5.43: The grade of the magnetic concentrate from HGMS of gold flotation tailings (South Africa), as a function of magnetic induction, for three size fractions (adapted from [S32]).

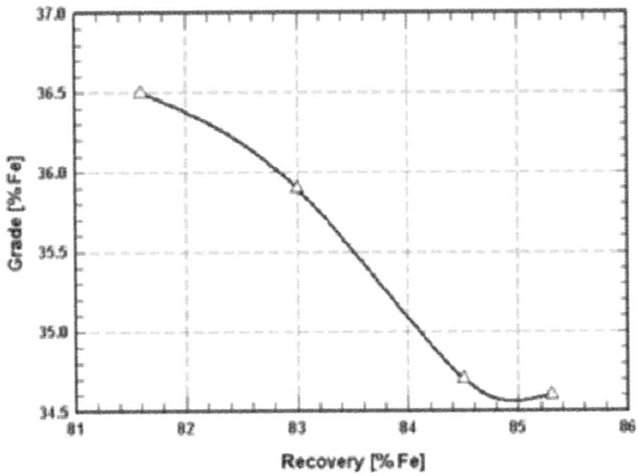

Figure 5.44: A typical relationship between the recovery of iron into the magnetic concentrate and the grade of the magnetics (based on Fig. 5.40).

removal of magnetic impurities from industrial minerals, the recovery of a valuable component is not the only criterion of the performance of a magnetic separator when applied to concentration. The grade of the magnetic product is usually equally important and the magnetic field strength has, therefore, to be selected in such a way that recovery and grade are optimized simultaneously. The grade of the magnetic concentrate usually decreases with increasing magnetic field, as a consequence of non-selective magnetic capture caused by the entrainment of non-magnetic gangue particles in clusters of magnetizable particles, and by the capture of non-liberated grains containing magnetic inclusions.

A decrease in the grade of the magnetic concentrate with increasing magnetic field represents a characteristic feature of a majority of magnetic separation applications to mineral concentration. Figures 5.40 and 5.43 illustrate this trend in the beneficiation of iron ore and gold ore flotation tailings respectively.

Clearly, a selection of the applied magnetic field strength must be based on an evaluation of the interplay between recovery and grade of the concentrate, and on the relative importance ascribed to these two metallurgical factors. The importance of the trade-off between the recovery and the grade is depicted in Fig. 5.44.

It has been shown, therefore, that a higher magnetic field does not necessarily mean better results. The optimum magnetic field has to be determined in each particular case in such a manner that the recovery-grade relationship satisfies the objectives of the application.

5.3.4 Matrix in high-gradient magnetic separators

The importance of a correct choice of matrix for magnetic filtration and material concentration cannot be overemphasized. The shape and size of the matrix play a decisive role in the achievement of optimum recovery and concentrate grade, and determine matrix loading.

In spite of the importance of a correct choice of matrix, little attention has been paid to the theoretical analysis of the optimum matrix size and geometry. The attention was mainly concentrated on maximizing the magnetic force in a single-collector approach, with the aim of applying these rules to multi-collector matrix.

The magnetic force has been shown to reach a maximum for the ratio of the collector radius a to particle radius b equal to approximately three [O5, A29]. Further theoretical [M27] and experimental investigations have indicated that the optimum ratio a/b is a function of a variety of parameters, such as magnetic induction, particle size, type of matrix and matrix loading [S1].

Although the conclusions of the single-collector-single-particle studies are useful for a better understanding of the dynamics of a magnetizable particle in the vicinity of an isolated magnetized collector, their validity for a real matrix is very limited.

Straining and mechanical capture

Attempts to maximize the ratio of traction magnetic force to hydrodynamic drag usually result in particle capture by a straining mechanism. If $a/b < 10$, mechanical straining is dominant and leads to the formation of a surface mat of deposit on the matrix. This layer of deposit causes high resistance to the flow, and increases entrainment of non-magnetic particles in the deposit. This phenomenon is detrimental to the efficiency of separation, irrespective of the type of application. It either increases losses of the non-magnetic concentrate or impairs the grade of the magnetic concentrate.

Therefore, straining should be avoided in the design of magnetic separators. For straining to be minimized, a/b should be larger than 10, but preferably should range from 100 to 500. However, even for high values of a/b, a certain fraction of particles fed into the separator will be mechanically captured. As a rule, mechanical entrapment increases both for coarse particles and for finer matrices.

The presence of straining for fine matrices is demonstrated in Fig. 5.45. It can be seen that straining is absent and particles are deposited almost uniformly throughout the filter for a coarse matrix with $a/b = 1250$, while with finer matrices, particles are predominantly deposited in the vicinity of the filter inlet [E4].

In addition to straining, some particles are retained by the matrix even in the absence of the external magnetic field, as follows from the deep bed model of magnetic separation [S31, S33] discussed in Section 3.4.2. Figures 5.46 and 5.47 illustrate that, for a given type of matrix, mechanical retention increases with

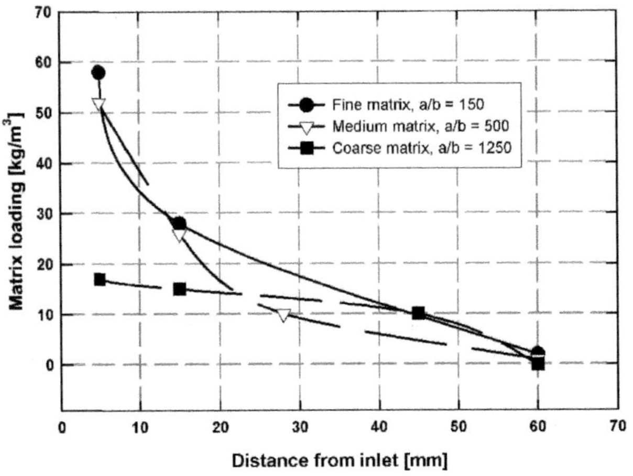

Figure 5.45: Matrix loading as a function of distance from the inlet into a magnetic filter, for three types of matrices (adapted from [E4]).

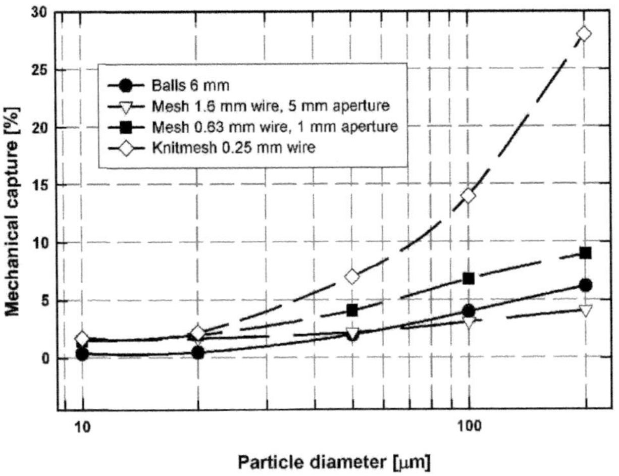

Figure 5.46: Mechanical capture of uraninite particles, as a function of particle size, for various types of matrices, at the superficial velocity of 0.02 m/s (adapted from [S61]).

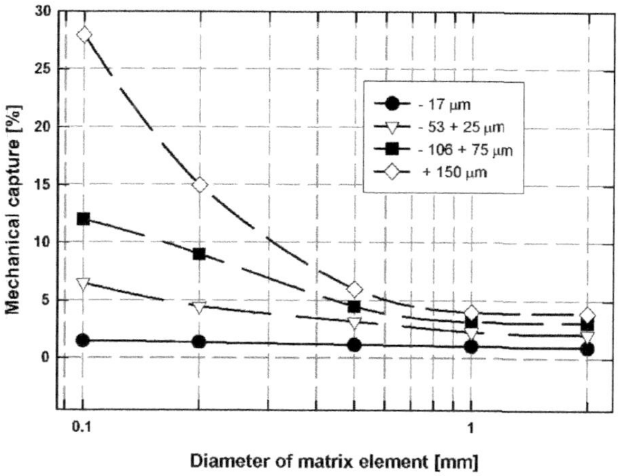

Figure 5.47: Mechanical capture, as a function of the diameter of the matrix element, of various size fractions of uraninite particles, at a flow velocity of 0.02 m/s (adapted from [S61]).

increasing particle size. In addition, it can be seen that mechanical capture increases with increasing fineness of the matrix. Also in agreement with the concept of deep bed filtration, the particle retention decreases with increasing flow velocity of the slurry through the matrix, as depicted in Fig. 5.48.

The optimum size of a matrix

In the application of HGMS to the beneficiation of minerals, the matrix should be matched to the feed properties in such a way that recovery and grade are at their maximum simultaneously. The optimum collector size becomes, therefore, a function of the size distribution of the ore, and of the hydrodynamic conditions in the inter-matrix space. The usual practice is a trial-and-error search for the best size and shape of the matrix.

During this process, several fundamental rules should be kept in mind. With a decreasing collector diameter, the recovery into the magnetic fraction usually increases, whereas the grade of the concentrate decreases. With very fine matrices, the mass yield into the magnetic concentrate becomes high and the grades too low.

Numerous experimental results show that the most satisfactory metallurgical performance is obtained with the ratio of the collector size to particle size ranging from 100 to 300. This is demonstrated in Fig. 5.49, in which the grade of the magnetic concentrate is plotted as a function of the diameter of the matrix elements. It can be seen that, for fine matrices, the grade is very low, higher grades being obtained for fairly coarse matrices, for which a/b is about 350. For

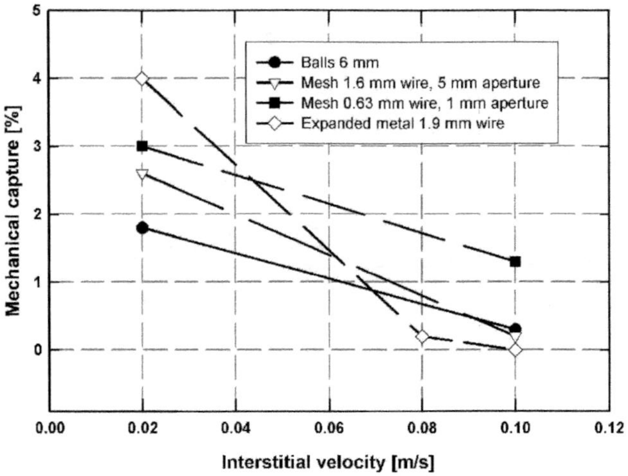

Figure 5.48: Mechanical entrapment, in assorted matrices, of uraninite particles (- 176 μm), as a function of flow velocity (adapted from [S61]).

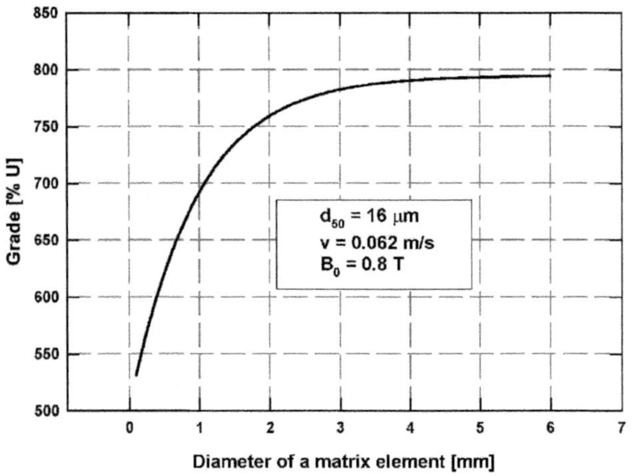

Figure 5.49: The grade of the magnetic uranium concentrate, as a function of the diameter of matrix elements (adapted from [S61]).

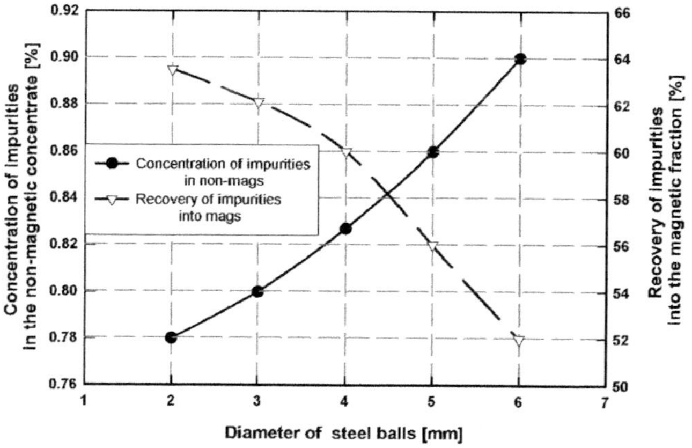

Figure 5.50: The effect of the ball diameter on the efficiency of kaolin purification (the Ukraine). $B_0 = 1$ T, particle size: - 100 μm (adapted from [S62]).

these coarse matrices, the recovery is also sufficiently high.

Similar conclusions may be drawn from Fig. 5.50, which illustrates the effect of the diameter of steel balls on the efficiency of kaolin purification [S62]. For the optimum operation of the separator, the ball diameter is 2 mm to 3 mm, with a/b being approximately 200 to 300.

Several types of production-scale high-gradient separators, such as VMS-100 and SLON, have been using successfully rod matrices. The efficiency of separation is a function of the rod diameter as illustrated in Fig. 5.51. Equally efficient performance of the rod matrix, when applied to beneficiation of fine ilmenite, was reported by Xiong [X3] and the results are summarized in Fig. 5.52.

One of the most widely used matrix magnetic separators is the Jones separator. Vertically arranged grooved plates are used as a matrix, as is shown in Fig. 2.37. The pattern and the magnitude of the magnetic field gradient can be modified by using different types of plates and by altering the gap between the plates. Three different types of plates, with varying tip radii are available and their specifications are given in Table 5.22.

Although grooved plates performed successfully on numerous sufficiently magnetic and coarse ores, their efficiency to concentrate finely ground refractory iron ores has often been found unsatisfactory. The width of the gaps between the plate has frequently been too large to ensure sufficient recovery of weakly magnetic fine particles. At the same time, a decrease in the width of the gap caused rapid clogging of the matrix and deterioration in the performance of the separator.

In order to address this problem, Mekhanobrchermet Institute (the Ukraine)

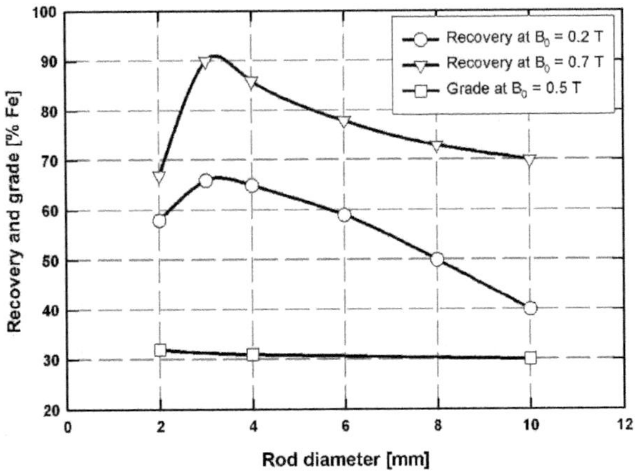

Figure 5.51: The effect of the rod diameter on magnetic separation of siderite ore (adapted from Hencl et al. [H22]).

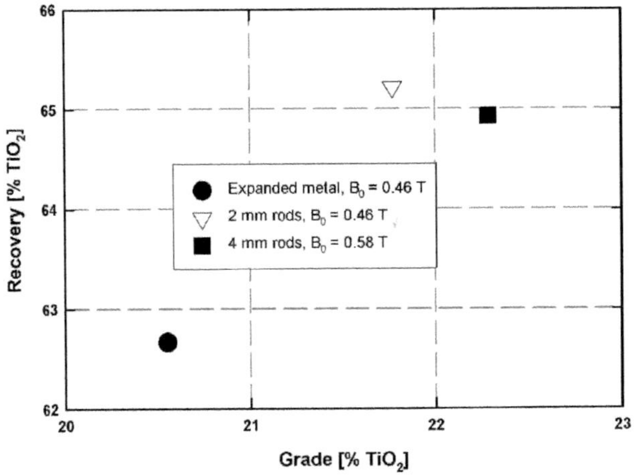

Figure 5.52: Separation of fine ilmenite (- 45 μm) by SLON magnetic separator with rod matrices of different diameters. Comparison with expanded metal matrix is also shown (adapted from Xiong [X3]).

Table 5.22: Grooved plates used as a matrix in Jones magnetic separators.

Plate type	Specification	Applications
4R	4 grooves per inch	For coarse material (- 3 mm), gap width 5 mm
8R	8 grooves per inch	Medium size material (- 0.5 mm), gap 1.2 - 2.5 mm
12R	12 grooves per inch	Very fine material (- 100 μm), gap width 0.5 mm

Table 5.23: Beneficiation of oxidized quartzites (Krivoy Rog) by magnetic separation, using different matrices [M28].

Matrix	Recovery [% Fe]	Grade [% Fe]
Grooved plates, gap 4 mm	59.8	52.1
Grooved plates, gap 2.2 mm	66.4	52.4
Grooved plates + expanded metal, gap 6 mm	67.9	54.0

incorporated expanded metal sheets between the grooved plates. A considerable increase in the recovery of iron was observed when this matrix was applied to the beneficiation of feebly magnetic, finely ground (80% - 45 μm) Krivoy Rog (the Ukraine) and (90% - 74 μm) Baikal (Russia) iron quartzites, as is summarized in Tables 5.23 and 5.24.

Another approach was taken by Turkenich [T2] and Ulubabov and Karmazin [U2] who proposed to use inclined grooved plates so that the slurry passes over the plate surfaces in the form of a film. These plates were designed in such a way that the grooves on opposite sides of the gap were inclined in different directions, as is shown in Fig. 5.53. Central sections of the gap are thus free of the slurry and the efficiency of separation is determined by the thickness of the

Table 5.24: Beneficiation of Baikal oxidized quartzites by magnetic separation, using different matrices [M28].

Matrix	Ore	Recovery [% Fe]	Grade [% Fe]
Grooved plate	Siderite	84.0	33.2
Grooved plate+expanded metal		88.2	32.8
Grooved plate	Limonite	52.6	54.8
Grooved plate+expanded metal		68.2	54.0

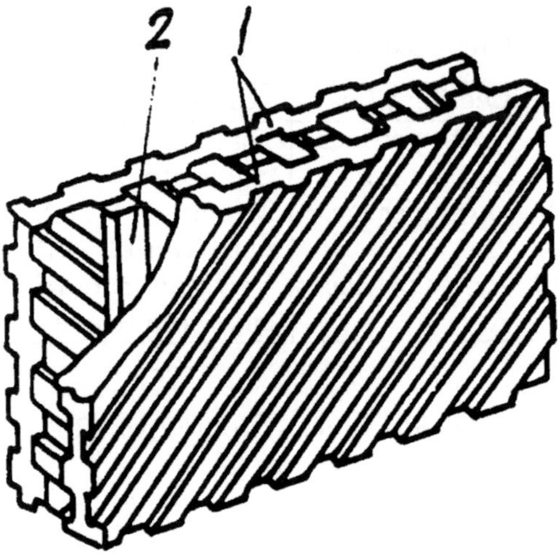

Figure 5.53: Film-flow grooved plates with inclined ridges and grooves. (1) plates, (2) spacers [T2].

slurry film on the surface of the plate. The distance between the plates can be increased to eliminate the matrix blockage.

This approach to the film-flow matrix has been employed in 6-ERM-35/315 magnetic separators, described in Section 2.4.2. A comparison of the performance of conventional vertical plates and inclined plates is shown in Fig. 5.54. Fundamental theoretical issues of the film slurry flow in this matrix were discussed by Turkenich [T2].

The optimum geometry of a matrix

The criteria for selection of the optimum shape and geometry for matrix collectors are even more vaguely defined than those for the size of the collectors. The choice of the matrix depends on the characteristics of the slurry undergoing the separation process and is often dictated by personal beliefs, experience and the availability of a suitable matrix type. Ideally, the matrix should offer maximum collection efficiency, minimum magnetic reluctance and fluid impedance, together with good ability to be cleaned. Various shapes of a matrix are used, such as grooved plates, balls, rods, wire mesh, expanded metal and steel wool.

Besides the size of a collector, a matrix is characterized by its filling factor f, or by porosity (or void fraction) ϵ:

$$f = \frac{\text{volume of matrix}}{\text{volume of separation chamber}} \tag{5.7}$$

and

$$\epsilon = 1 - f \tag{5.8}$$

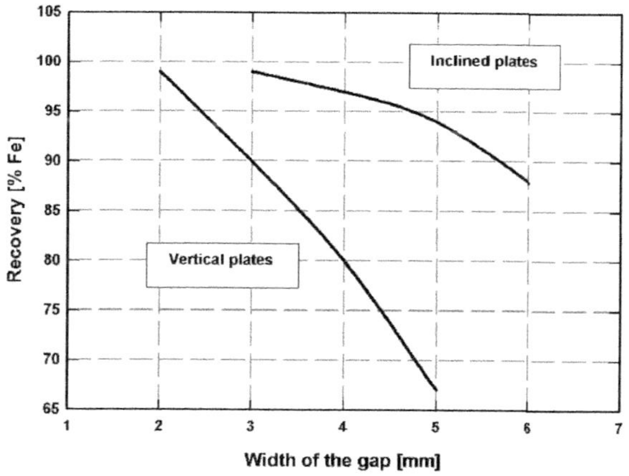

Figure 5.54: Comparison of the performance of vertical and inclined grooved plates, as a function of the width of the gap (adapted from [U2]).

Table 5.25: Filling factors of various matrices.

Matrix	Filling factor f
Spheres	0.50 to 0.64
Steel wool	0.01 to 0.06
Expanded metal	0.12 to 0.28
Wire mesh	0.15 to 0.25
Grooved plates	0.50 to 0.85

The porosity of the matrix bed changes with time, as particle accumulation within the matrix increases:

$$\epsilon = \epsilon_i - \sigma_d(1 - \epsilon_d) \tag{5.9}$$

where σ_d is the specific deposit (volume of deposited matter per unit volume of matrix), ϵ_d is the porosity of a deposit and ϵ_i is the porosity of clean matrix. Typical values of the filling factor for assorted types of matrices are given in Table 5.25, while Figs. 5.55 and 5.56 illustrate some more frequently used matrices.

The performance of various shapes of matrix collectors depends primarily on the particle size distribution and magnetic properties of the solids to be separated. Fine matrices, such as steel wool, are suitable for the removal of very fine and very feebly magnetic particles but in most applications in mineral processing these matrices are not suitable owing to their non-selectivity and susceptibility to clogging.

For the majority of applications in materials treatment, coarse matrices (e.g. balls, expanded metal, grooved plates) are the best choice. There is, quite

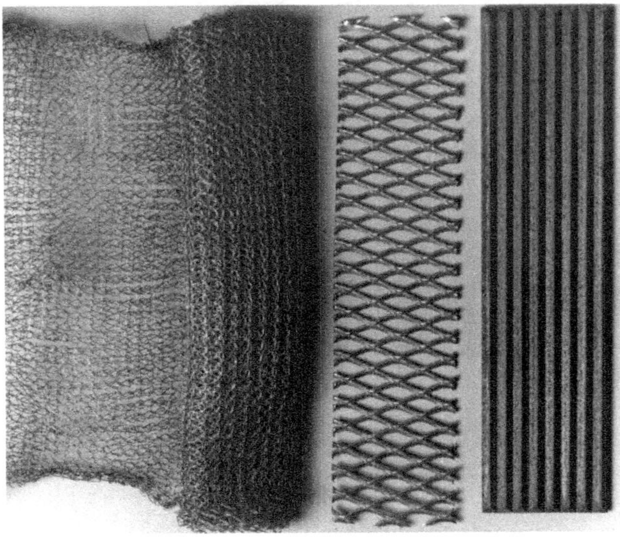

Figure 5.55: Matrices used for magnetic separation: (from right to left) grooved plate, expanded metal 1.9 mm, Knitmesh.

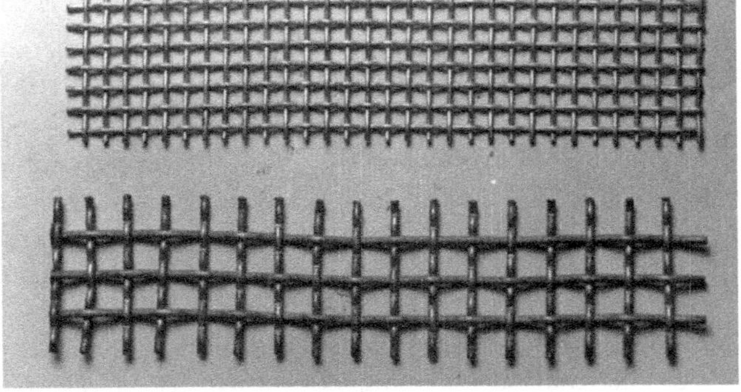

Figure 5.56: Matrices used for magnetic separation: mesh 1.6 × 5 mm (top), mesh 2.5 × 10 mm (bottom).

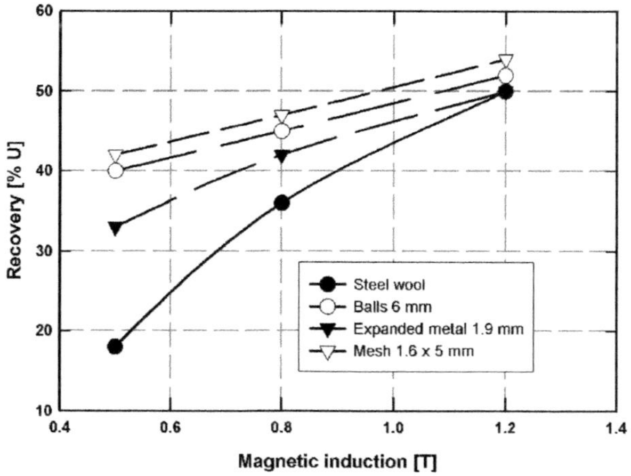

Figure 5.57: Recovery of uranium from leaching residues (70% - 44 μm) by assorted matrices (adapted from [S61]).

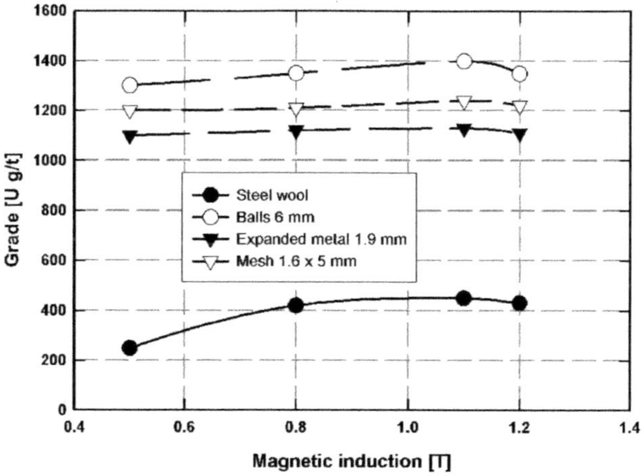

Figure 5.58: The grade of the magnetic concentrate of uranium leaching residues, for different matrices (adapted from [S61]).

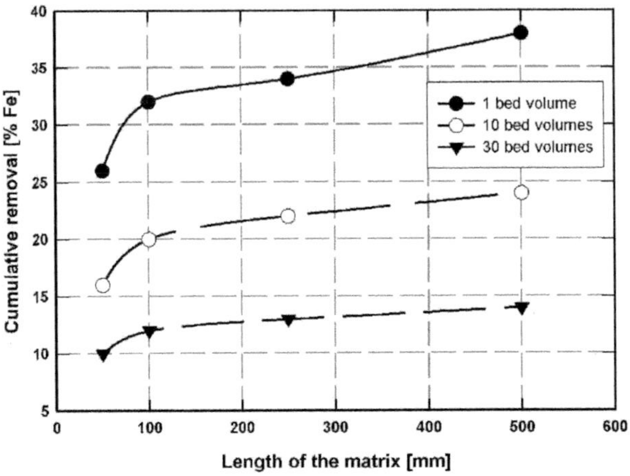

Figure 5.59: The effect of matrix length on the removal of iron from kaolin, for various volumes of the processed slurry (adapted from [N9]).

understandably, no universal type of a matrix that meets all requirements in all applications. Each type of matrix has its advantages and disadvantages and detailed laboratory tests should be carried out to identify the optimum matrix for a given application. Figures 5.57. and 5.58 illustrate the performance of various matrices as applied to recovery of gold and uranium from leaching residues. Although fine matrices, such as steel wool, gave high recoveries, their selectivity was very poor. With frequently used matrices of grooved plates and expanded metal, unsatisfactory recoveries and grades were obtained. On the other hand, steel balls and mesh matrices gave the best results and similar observations were made in other applications of these matrices.

The length of the matrix

The length of the matrix is an important parameter that determines the retention time of particles in the magnetic field, the probability of their collision with the matrix collectors, and thus the probability of their capture. From numerous experimental studies, and from theoretical description of particle capture in a matrix, as outlined in Section 3.4.2, it is evident that the matrix length can have an impact on the efficiency of separation at least as great as the magnetic field strength.

In several early WHIMS or HGMS machines, the length of the matrix was 80 to 100 mm, clearly to the detriment of their efficiency. On the other hand, the length of grooved plates in Jones WHIMS units was 220 mm. The application of such a long matrix is understandable in view of poor filtration properties of the grooved plates. In current cyclic or continuous HGMS separators, the matrix

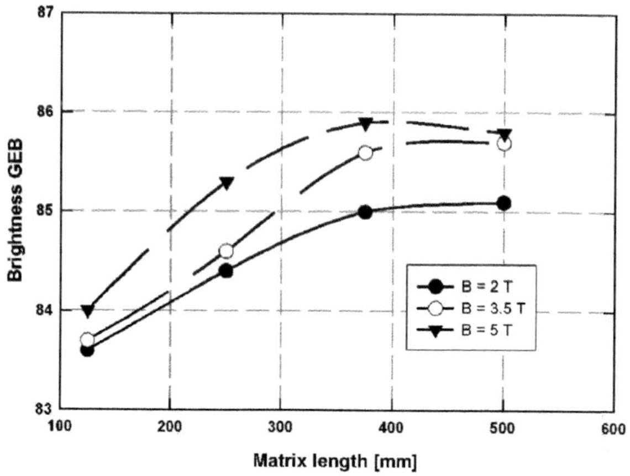

Figure 5.60: The effect of matrix length on the brightness of kaolin at different values of magnetic induction. Slurry velocity $v = 0.5$ cm/s (adapted from [S60]).

length is often about 200 mm, although in very demanding applications, such as kaolin brightening, the length is as much as 500 mm. The beneficial effect of increasing matrix length to 500 mm, on kaolin brightening, is illustrated in Figs 5.59 and 5.60.

For the recovery of moderately weakly magnetic minerals, a matrix length of 200 mm or less is usually sufficient, as shown in Figs. 5.61 and 5.62. Figure 5.61 also illustrates a well established fact that the grade of the magnetic concentrate often does not depend on the matrix length.

Although it is clear than a longer matrix is beneficial for overall metallurgical performance of a matrix separator, it is important that the matrix should be kept as short as possible, consistent with the required metallurgical results. By reducing the matrix length, the size of pole tips of an electromagnet, or the length of a solenoid coil can be reduced. This will result in the reduction of the mass and the costs of the separator, and in lower input power.

Matrix loading

The matrix is able to carry only a limited amount of material particles before its capture efficiency begins to deteriorate. Quantitatively, this capability is usually expressed through the matrix loading. There are several ways to express this quantity, none of them strictly correct, but acceptable if used in a consistent manner.

In kaolin brightening, the matrix loading is often expressed through the number of bed (or canister) volumes processed in the separator. Other definitions

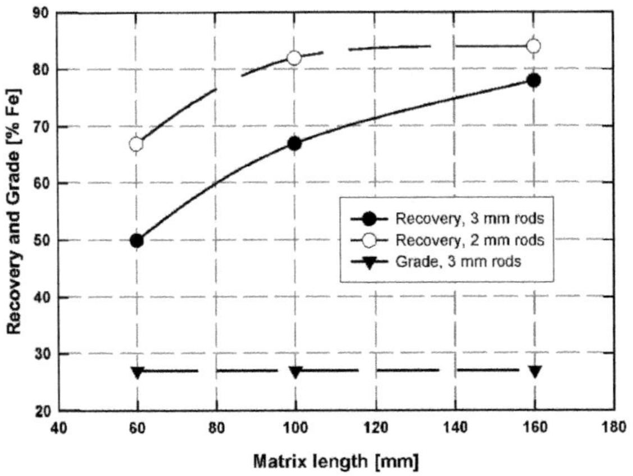

Figure 5.61: The recovery of iron and the grade of the magnetic concentrate, as a function of matrix length, for a matrix of rods of different diameters (adapted from [H22]).

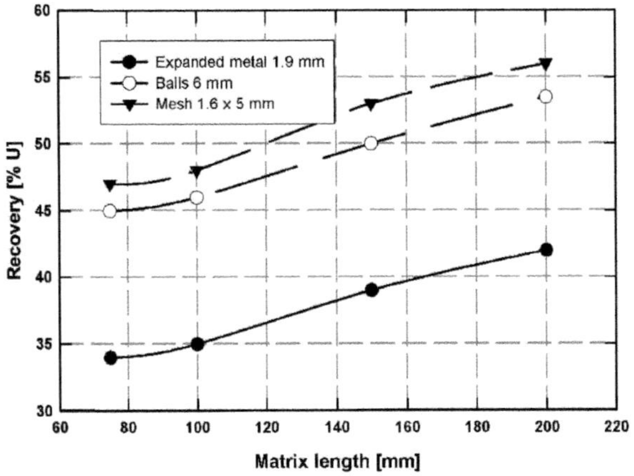

Figure 5.62: The recovery of uranium from uranium-gold leaching residues, as a function of matrix length, for various matrices (adapted from [S61]).

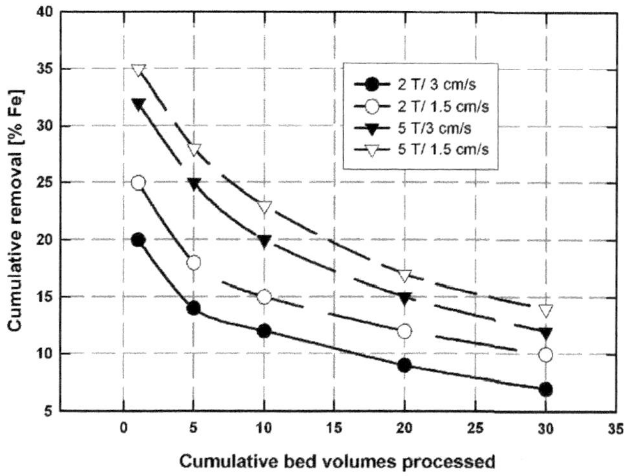

Figure 5.63: The efficiency of iron removal from kaolin, as a function of the number of bed volumes of slurry processed (adapted from [N9]).

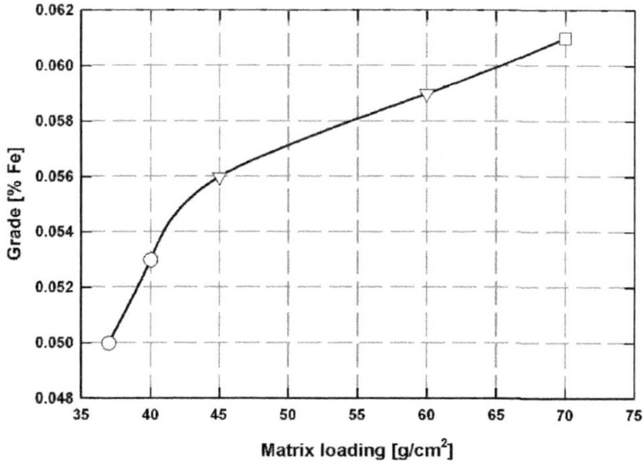

Figure 5.64: The concentration of iron oxide in the non-magnetic concentrate of silica sand, as a function of matrix loading (after [S29]).

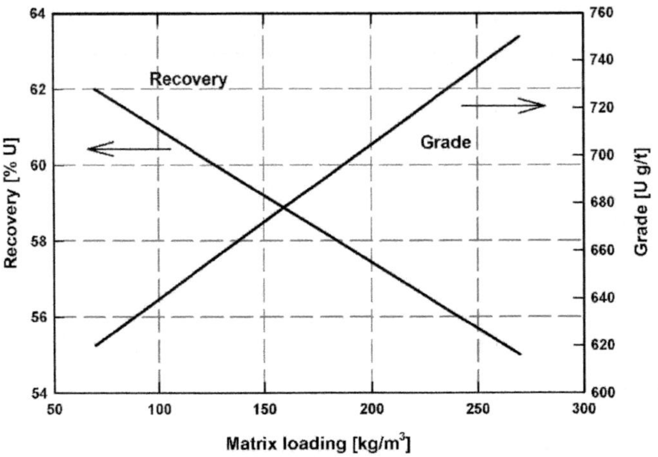

Figure 5.65: The recovery of uranium from cyanidation residues, and the grade of the concentrate, as a function of matrix loading L_3 (adapted from [S61]).

that are occasionally used are expressed by eqs. (5.10) to (5.12).

$$L_1 = \frac{\text{Feed mass}}{\text{Matrix area}} \; [\text{kg/m}^2 \text{ or g/cm}^2] \tag{5.10}$$

$$L_2 = \frac{\text{Feed mass}}{\text{Matrix volume}} \; [\text{kg/m}^3 \text{ or g/cm}^3] \tag{5.11}$$

$$L_3 = \frac{\text{Mags mass}}{\text{Matrix volume}} \; [\text{kg/m}^3 \text{ or g/cm}^3] \tag{5.12}$$

Figure 5.63 illustrates the efficiency of removal of iron from kaolin, as a function of the volume of the kaolin slurry processed. The deterioration of the matrix performance with an increasing number of bed volumes processed is marked. Similarly, the matrix loading curve for beneficiation of silica sand, shown in Fig. 5.64, illustrates a trade-off between the quality of the final product and the throughput per cycle.

In the concentration of minerals, matrix loading affects not only the recovery, but also the grade of the product. As a rule, with increasing loading, the recovery decreases and the grade of the concentrate increases. This trend is demonstrated in Fig. 5.65, where the recovery of uranium from cyanidation residues, and the grade of the magnetic concentrate are shown. Similarly, in beneficiation of siderite iron, illustrated in Fig. 5.66, the same behaviour is observed.

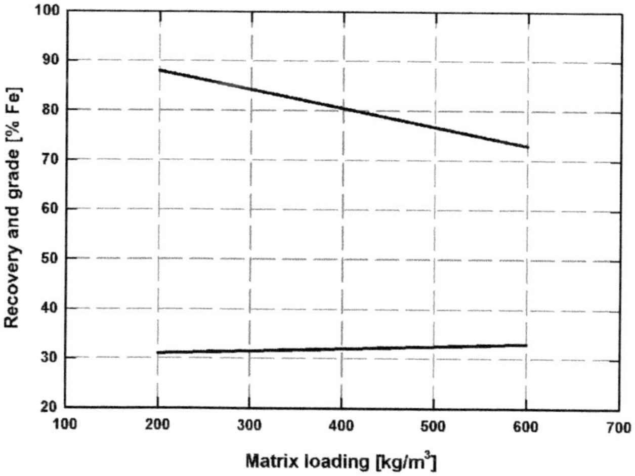

Figure 5.66: The recovery of iron from siderite ore, and the grade of the magnetic concentrate, as a function of matrix loading L_2 (adapted from [H22]).

The trends described above can be explained by realizing that, at a higher matrix loading, the magnetic and entrained "non-magnetic" particles or unliberated grains in higher layers of the material build-up are not held sufficiently strongly and can be stripped off from the matrix by fluid shear stress. With increasing matrix loading, the importance of the erosion wash-off force increases, in comparison with the magnetic attractive force. Although such a process reduces the recovery of magnetic particles, the grade of the magnetic fraction and the selectivity of the process is improved.

We have seen that matrix loading is an important parameter that allows a magnetic separator to be operated in a regime of maximum recovery or maximum grade of a concentrate. Matrix loading also determines the throughput of a separator, the duration of a duty cycle of a cyclic separator, and the rotational velocity of a continuous separator. Typical values of matrix loading for various applications are summarized in Table 5.26.

Magnetic properties of matrix material

There is ample experimental evidence that magnetization of the matrix material plays a minor role in the metallurgical performance of a magnetic separator [S1]. This is in contrast to predictions of simple single-collector single-particle theoretical models, described in Section 3.4.1. On the other hand, these experimental findings do agree well with predictions of theory of HGMS based on deep bed filtration approach [S31, S33]. As has been discussed in Section 5.3.3,

Table 5.26: Matrix loading in different applications.

Application	Matrix	Matrix loading (L_2)
Hematite ore	grooved plates	180 to 300
Siderite ore	rods	200 to 600 (L_1)
Steel mill waste water	steel wool	9000
Power plant cooling water	steel wool	4000
Uranium leach residues	mesh or balls	300 to 3000
Coal desulphurization	steel wool	50 to 250
Silica sand	steel wool	350 to 700 (L_1)
Kaolin purification	steel wool	1 to 20 bed volumes

the modelling of the magnetic field pattern around a matrix collector has shown that the difference between magnetic induction generated by matrix materials with polarization equal to 1 T or 2 T is, for practical purposes, negligible.

Experimental investigation [S63] of the effect of the saturation polarization of the matrix material on the metallurgical performance revealed that, for matrices with magnetic polarizations greater than 0.5 T, the curves of recovery and grade versus magnetic field were essentially identical. For a weakly magnetic steel matrix with a polarization of 0.1 T at the working magnetic field, the recovery was strongly dependent on the external magnetic field. At the background magnetic induction of 1 T, the recovery approached that achieved with strongly magnetic matrices. The results are shown in Fig. 5.67.

Matrices with very feebly magnetic properties (such as austenitic stainless steel or aluminium) can be used, for example, for magnetic scalping of ferromagnetic impurities from a pulp. The aluminium matrix was used by Riley and Watson [R21] for filtration of ferromagnetic particles and the beneficial effect of such matrices on the recovery of uranium was demonstrated by Watson et al. [W23] and by Svoboda [S61].

Clogging of the matrix

Probably the most important problem encountered in pilot or production-scale HGMS separators is a permanent retention of particles in the matrix. Industrial slurries always contain various extraneous materials (wood, plastics, tramp iron), oversize grains and ferromagnetic admixtures. Their presence and frequent shut-downs of the separator for the replacement of a blocked matrix caused by such particles, severely impair the economic and metallurgical performance of the unit.

The basic measure to eliminate matrix clogging is to avoid straining using sufficiently coarse matrices compatible with size distribution of the solids in the feed. Preliminary screening, low-intensity magnetic scalping or desliming to remove fine ferromagnetic particles, are steps that help to minimize retention of particles in the matrix.

However, none of these measures completely eliminates the clogging of the

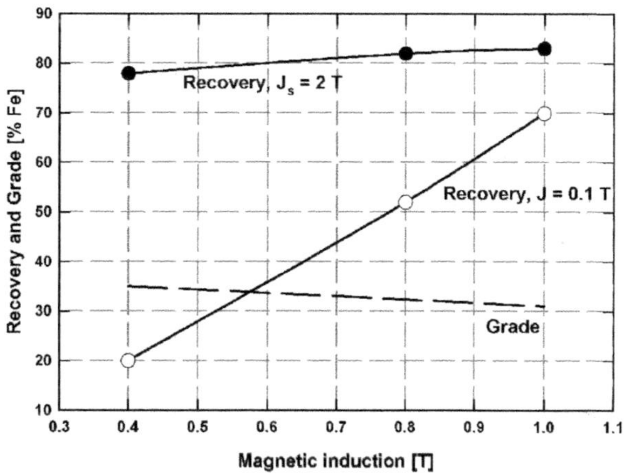

Figure 5.67: The recovery and grade of the magnetic fraction of siderite ore versus magnetic induction, for strongly (J_s = 2 T) and weakly (J = 0.1 T) magnetic matrices (adapted from [S63]).

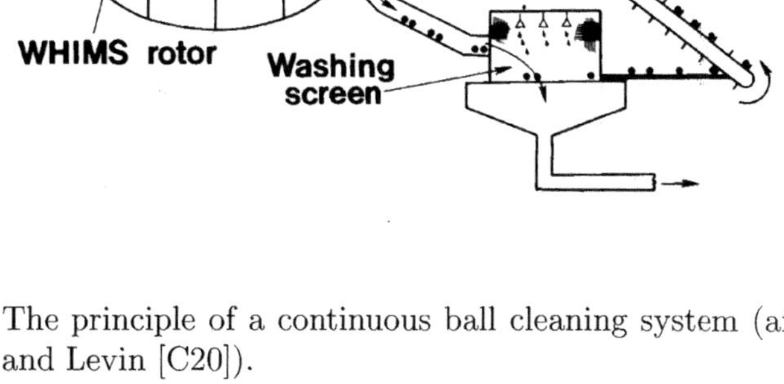

Figure 5.68: The principle of a continuous ball cleaning system (after Corrans and Levin [C20]).

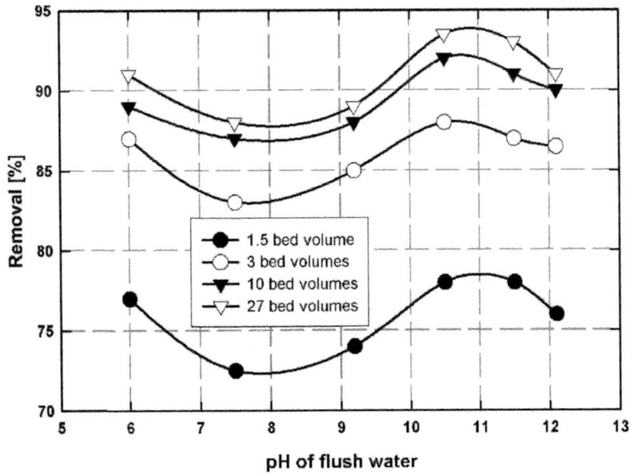

Figure 5.69: Removal of hematite particles from a ball matrix as a function of pH of the flush water, for various volumes of the flush water (adapted from [S64]).

matrix and several efficient long-term solutions to the problem have been introduced over the years. Continuous removal from a separator and external cleaning of steel balls, proposed by Karmazin [K21] and developed by Corrans and Levin [C19, C8], was successfully used in several production-scale applications. The principle of the system is shown in Fig. 5.68.

A back-flushing system [A30] that was introduced into the cyclic and continuous Sala separators was later improved by introduction of a vertical rotor with matrix reversal and back-flush [C7]. Introduction of a flexible self-cleaning matrix with back-flush [C18, S14] and of slurry pulsation in vertical rotor separators [Y2, X1], further reduced susceptibility of matrix to blockage.

Removal of particles from the matrix It was observed by Svoboda and Corrans [S64] that total wash-off of particles from the matrix could not be achieved without affecting the surface interactions between the adsorbed particles and the matrix. It was shown that the removal of hematite particles smaller than 44 μm from the matrix was dependent on the pH value of the flush water, as depicted in Fig. 5.69. It can be seen that the removal is not complete, even under the most favourable conditions. Fractions adhering to the matrix, after ten bed volumes of flush water had been used, amounted to 8% of the feed mass. A substantial increase in volume of flush water resulted in only a marginal improvement.

Flush water is usually pH 7 in practice, and adjustment to pH between 10 and 11 can result in removal improvement by about 7%. Alternatively, a

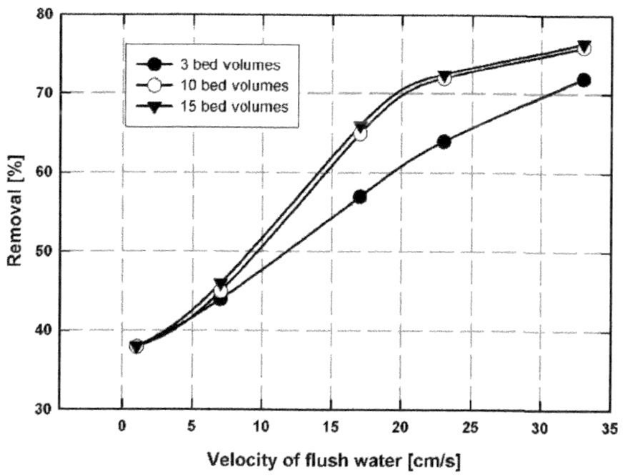

Figure 5.70: Removal of hematite particles (- 44 μm) from a ball matrix, as a function of velocity of flush water (pH 7), for different volumes of flush water (adapted from [S64]).

reduction in the volume of flush water, for a constant removal efficiency, can be achieved by pH adjustment.

Rinse and flush water

The metallurgical performance of a matrix magnetic separator can be significantly controlled by varying the volumes of rinse and flush water.

Rinse water is added at the end of the magnetic head of a continuous separator or at the end of the feed cycle in a cyclic separator. The rinse water is used to wash out mechanically entrained non-magnetic particles and unliberated middling particles.

Flush water is used to remove the magnetic fraction from the matrix. Matrix flushing is performed outside the magnet station in continuous separators, or when the magnetic field is switched off in cyclic machines.

An increase in the volume of rinse water normally results in a decrease in the mass of the magnetic fraction, while the grade of the magnetic fraction increases. These trends are often quite dramatic for small volumes of rinse water, although beyond a certain limit this effect becomes negligible. The optimum volume and flow velocity of the rinse water must be determined experimentally since a wide array of factors affect the influence of rinse water on the metallurgical performance of a magnetic separator.

The function of flush water is to remove captured magnetic material from the matrix. As a result of stray magnetic fields, remanent magnetism of the matrix collectors, and surface interactions between the particles and the matrix, the flow velocity of flush water should be quite high to keep the matrix clean for

Table 5.27: Water consumption (per tonne of feed) in commercial continuous WHIMS machines.

Separator	Rinse water [m^3/t]	Flush water [m^3/t]
Jones	0.5 - 1.0	1.0 - 1.7 [L13]
VMS	0.75	0.80 [C7]
VMS for kaolin		6.0 (total water) [Z3]
Reading	0.54	1.0 - 1.4 [F22], [M29]
6-ERM-35/135		2.5 (total water) [U2]
SALA HGMS	2 - 3	1 - 2.5 [F22]
Ferrous Wheel		1 (total water) [A36]

Table 5.28: Consumption of wash water in cyclic HGMS.

Consumption of wash water expressed in terms of		
Application	Bed volumes	% of processed slurry volume
Power plant cooling water	11	0.01 - 0.02
Kaolin beneficiation	10	125 - 250

a sufficiently long time. Some typical values of rinse and flush water volumes, employed in commercial continuous matrix separators are summarized in Table 5.27, while data for a cyclic HGMS are shown in Table 5.28.

Laboratory-scale investigations [S64] have shown that, while the flow velocity of flush water has, up to a point, a dramatic effect on particle desorption from the matrix, the volume of flush water appears to be of secondary importance. The results are summarized in Fig. 5.70.

5.3.5 Flow velocity of the slurry

The metallurgical performance of a magnetic separator varies with the flow velocity of a slurry to such an extent that this variable is one of the primary factors affecting the efficiency of magnetic separation. Moreover, flow velocity determines the throughput of a separator and thus affects the economics of the process.

The velocity of the slurry flowing through the matrix bed determines the shear stress acting on the particles deposited on the surface of the matrix collectors, and thus controls the recovery and the grade of the products. Simultaneously, flow velocity controls the retention time of the slurry in the magnetic field and thus the efficiency of particle capture onto the matrix.

The superficial velocity v_0, given by eq. (5.13), is defined as the average linear velocity the slurry would have in the separation chamber if no matrix was present.

$$v_0 = \frac{Q_v}{S} \tag{5.13}$$

where Q_v is the volumetric flow rate (in [m^3/s]) and S is the cross-sectional area of the matrix (in [m^2]).

The mean flow velocity of the slurry within the matrix is the interstitial velocity v, which is related to the superficial velocity by

$$v = \frac{v_0}{\epsilon} \tag{5.14}$$

where ϵ is the porosity of the matrix bed.

The effective length of a particle trajectory is greater than the matrix length L, because the slurry travels a tortuous path L_t. The velocity $v^{'}$ along the tortuous path is

$$v^{'} = v\frac{L_t}{L} = \frac{v_0 L_t}{\epsilon L} \tag{5.15}$$

If Karman's assumption $L_t/L = \sqrt{2}$ is used [S65], we obtain

$$v^{'} = \sqrt{2}\frac{v_0}{\epsilon} \tag{5.16}$$

The effect of flow velocity on separation efficiency

An increase in flow velocity results in a decrease in the probability of particle collision with matrix, as illustrated by Fig. 3.35. At the same time, the hydrodynamic drag on particles already retained on the matrix increases. The increase in flow velocity, therefore, results in a deterioration in the efficiency of magnetic filtration as shown in Fig. 5.71, which illustrates filtration of PWR cooling water. Removal of magnetite (- 16 μm) from an industrial waste water, as a function of flow rate, is shown in Fig. 5.72.

A similar trend applies to the removal of impurities from mixtures of materials. The lower the slurry flow velocity, the greater is the removal of magnetizable particles from a mixture. This effect is shown in Fig. 5.73 for coal desulphurization by HGMS. Similarly, in kaolin purification, the concentration of magnetic impurities in the product increases, as is shown in Fig. 5.74, and brightness decreases with increasing flow rate, as shown in Fig. 5.75. In contrast to magnetic filtration, the flow rate must be specified with due regard for product recovery, and a compromise must be reached between product quality, product recovery and throughput of a separator.

The selection of the slurry flow velocity in the application of matrix separation to concentration of minerals is dictated by whether the main emphasis is to be placed on the recovery or on the quality of the magnetic product. An inverse relationship between product grade and recovery implies that in order to achieve a high recovery, low flow rates should be used, while a high grade can be obtained at high flow rates. Magnetic separators can, therefore, be run in a regime of maximum recovery or in a regime of maximum grade of the concentrate, or any regime in between. Typical dependences of the recovery and grade upon the flow velocity are shown in Figs. 5.76 and 5.77.

The rate of decrease in recovery and increase in grade with increasing flow velocity are determined by the size distribution and differences in the values of

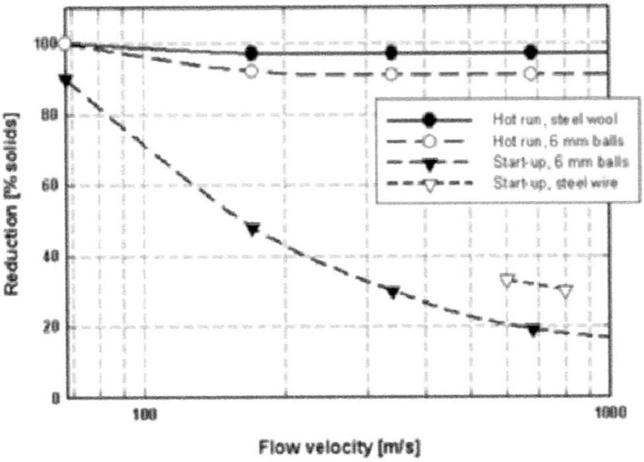

Figure 5.71: The filtration efficiency of PWR cooling water, as a function of flow velocity (adapted from [T7]).

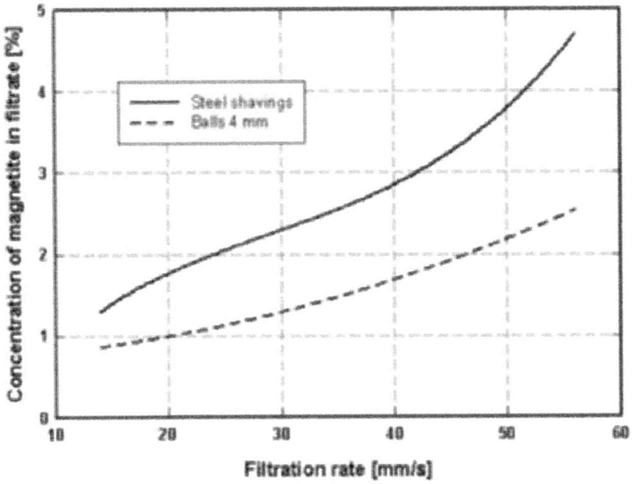

Figure 5.72: The concentration of magnetite (- 16 μm) in the filtrate from a chemical plant effluent, as a function of filtration rate (adapted from [S66]).

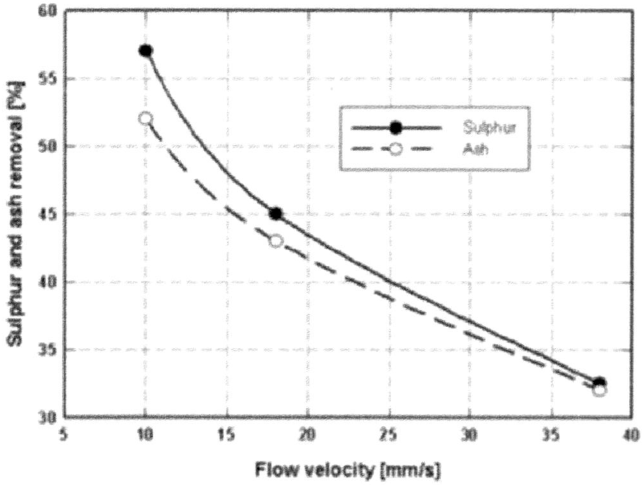

Figure 5.73: The effect of flow velocity of slurry on sulphur and ash removal from coal (adapted from [W24]).

magnetic susceptibilities of the minerals in the slurry, by the strength of the magnetic field and by the length of the matrix.

If the difference between the magnetic susceptibilities of a useful mineral and the gangue is small, an increase in flow velocity will result in a sharp drop in recovery, while the improvement in grade will be small. If the difference is great, the drop in recovery will only marginal, as only feebly magnetic gangue particles will be swept from the matrix, and the selectivity of separation will improve.

Similarly, if the applied magnetic field is sufficiently strong, an increase in flow velocity will not affect the captured magnetizable particles, while mechanically entrained gangue particles will be stripped from the matrix surface, thereby increasing the grade of the magnetic product.

When excessive amounts of slimes are present in the feed, recovery usually drops rather sharply with increasing flow velocity, while the grade of the concentrate does not improve appreciably.

Although flow velocity is an important parameter which considerably affects the performance of a magnetic separator, most industrial-scale continuous separators are not equipped with flow control and their operation is often below optimum.

5.3.6 Slurry density

Although the slurry density, or the percentage of solids, affects the throughput of a separator and the matrix loading, its effect on metallurgical results is usually

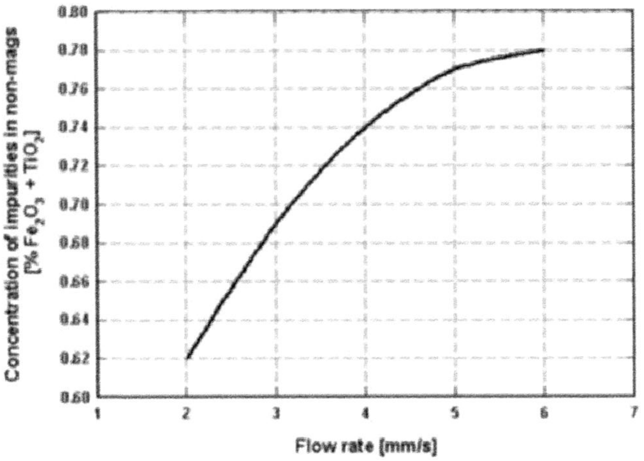

Figure 5.74: Concentration of impurities in the non-magnetic kaolin concentrate, as a function of flow rate. Matrix: 2 to 3 mm balls, $B_0 = 1$ T (adapted from [S62]).

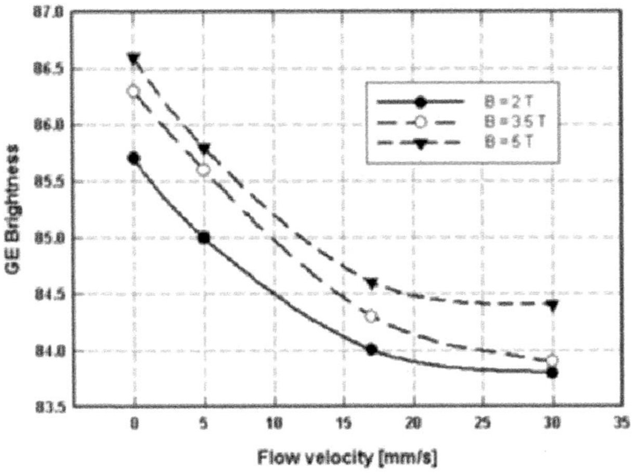

Figure 5.75: Brightness of the kaolin concentrate, as a function of flow velocity, at different values of the applied magnetic induction. Matrix: 75 μm steel wool, matrix length: 500 mm (adapted from [S60]).

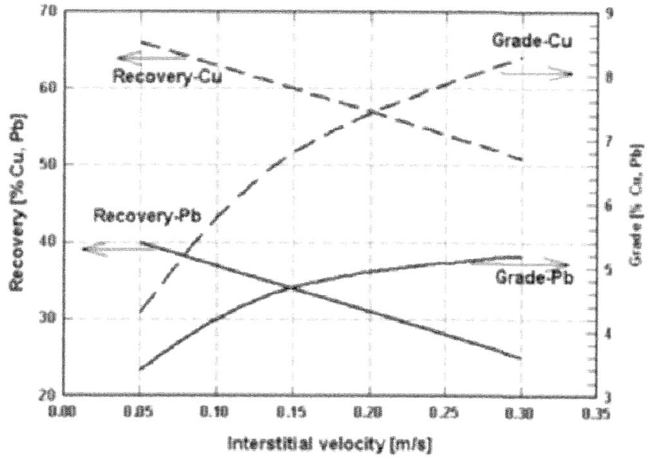

Figure 5.76: The effect of slurry velocity on the recovery of Cu and Pb, and the grade of the magnetic concentrate from Tsumeb flotation tailings. Matrix: mesh, $B = 1.2$ T (adapted from [S67]).

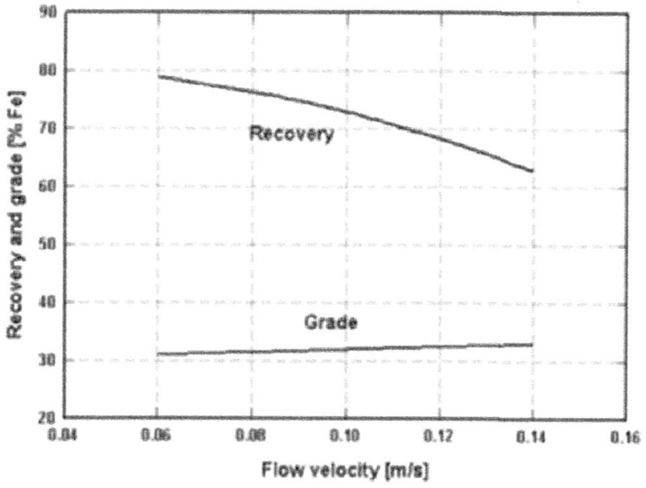

Figure 5.77: The effect of flow velocity on recovery and grade of magnetic concentrate of siderite ore (- 75 μm). Matrix: 3 mm rods, matrix length: 150 mm, $B = 0.5$ T (adapted from [H22]).

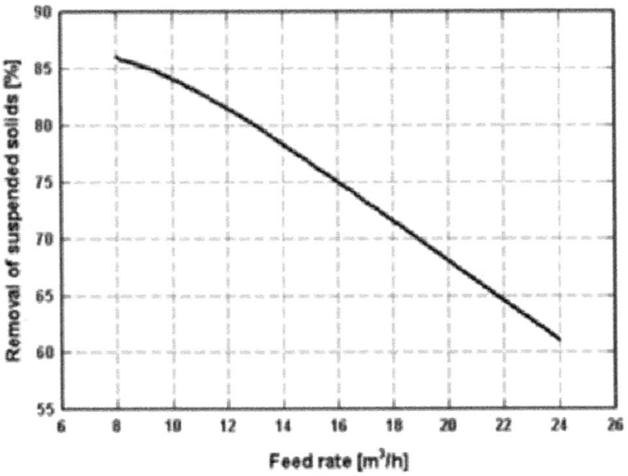

Figure 5.78: The efficiency of removal of solids from rolling mill effluent, as a function of feedrate (adapted from [F23]).

small. Recovery often does not depend on the slurry density over a wide range of densities, although in some cases recovery has been found to increase with increasing slurry density. The grade of the magnetic concentrate frequently, but not always, decreases with increasing slurry density.

As the effect of slurry density on the metallurgical performance of a magnetic separator is small, it is advantageous to use the highest possible slurry density, compatible with permissible matrix loading and feed rate. As a rule, in beneficiation of ores, it is possible to vary slurry density between 20% and 50% solids, while with fine clays the upper limit is probably about 30% solids.

5.3.7 Feed rate

The feed rate, which is closely associated with matrix loading, is an important variable in determining the cost-effectiveness of a magnetic separator. The higher the feed rate, the higher the tonnage treated, resulting in lower capital and operating costs per tonne treated. At low and moderate feedrates, where the matrix is not flooded and the continuity of the slurry flow has not yet been established, the interstitial velocity of the slurry and the erosion force acting on deposited particles are high. With increasing feed rate recovery into the magnetic fraction decreases and the grade of the magnetic product increases. This trend is shown in Fig. 5.78, which illustrates the effect of the volumetric flow on the filtration efficiency of rolling mill effluent water. Figure 5.79, on the other hand, shows the effect of feed rate on the reduction of impurities in fluorspar by a continuous WHIMS machine.

At high feed rates, when the matrix is flooded, interstitial velocity is reduced

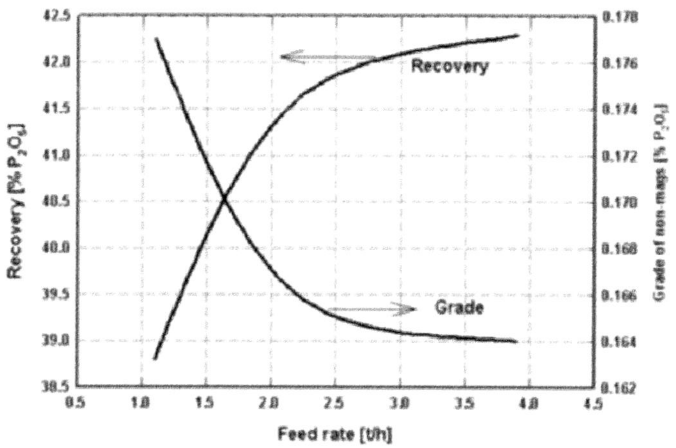

Figure 5.79: Recovery of P_2O_5 from fluorspar, and the grade of the non-magnetic product, as a function of feed rate. Matrix: 7 mm balls (adapted from [J6]).

due to the resistance of the matrix, and recovery can increase with increasing feed rate. However, the grade of the magnetic product decreases as a result of the reduced selectivity of the process and of the carry-over of gangue into the magnetic fraction.

5.3.8 The effect of particle size

In Chapter I, the performance of a magnetic separator was shown to be determined by the interplay of the forces acting on a particle, namely between the magnetic dipolar force on the one hand and the fluid drag, inertial, gravitational and centrifugal forces on the other. All these forces depend, in various ways, on particle size which is a crucial factor that limits the applicability of various types of magnetic separators. The approximate particle-size ranges of different classes of magnetic separators are depicted in Fig. 5.80. It can be seen that each magnetic separator operates in a limited particle size range and the upper and lower limits of this range are determined by the relative importance of the magnetic and competing forces.

The upper size limit of particles that can be treated is determined also by the construction characteristics of the separator and, in WHIMS/HGMS machines, by the type of matrix used. Of considerable relevance is the establishment of the lowest size limit of particles that can be recovered, and of the role of slimes since the efficiency of particle capture decreases with decreasing particle size. The lower size limit of particles that can be recovered affects not only the recovery and the selectivity of the magnetic separation process, but has a significant impact on the efficiency of the overall technological flowsheet, of

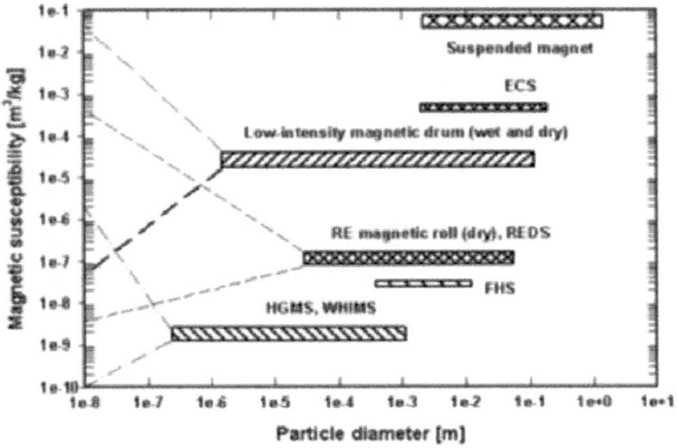

Figure 5.80: Particle size and magnetic susceptibility ranges of magnetic separators.

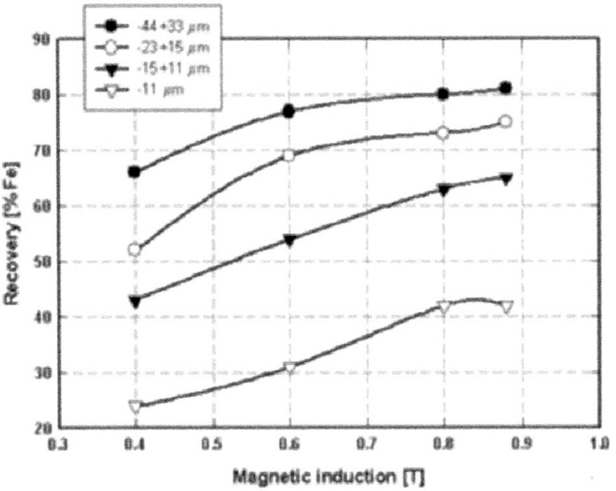

Figure 5.81: The effect of particle size on the recovery of iron from goethite ore, at different values of the magnetic induction, $\kappa = 2\times10^{-3}$ (SI), (adapted from [S68]).

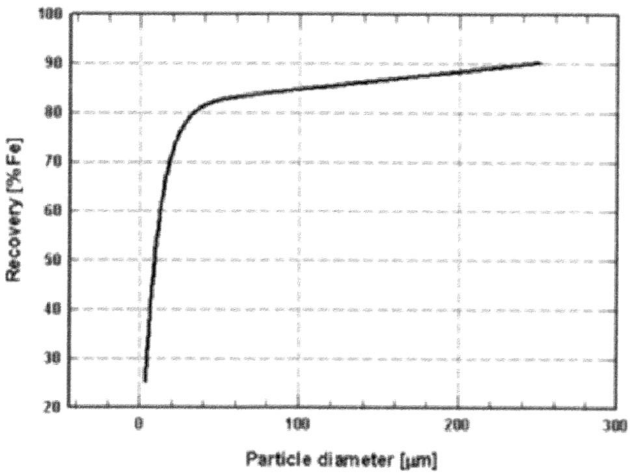

Figure 5.82: Recovery of iron by Readings WHIMS, from a hematite ore, as a function of particle size (adapted from [M29]).

which magnetic separation usually is a part. The reduced ability of a magnetic separator to recover fine particles is illustrated in Figs. 5.81, 5.82 and 5.83.

5.3.9 Pre-treatment of slurry

When an ore is to be beneficiated by magnetic separation, it often requires preliminary treatment in order to improve the efficiency of separation. The most useful operations used to prepare the ore are magnetic scalping, dispersion, preliminary magnetization, magnetizing roasting and desliming.

Magnetic scalping

Strongly magnetic impurities which are often present in ores, and other material mixtures, cause blockage of the matrix in WHIMS and HGMS machines and of the separation gap in induced magnetic roll separators. Adhesion of such impurities to the roll in belt roll separators causes misalignment and early wear of the belt. The presence of these impurities also impairs the grade of the magnetic concentrate as a result of entrainment of gangue and unliberated particles in the ferromagnetic clusters.

The detrimental effect of ferromagnetic impurities and tramp iron can be reduced by magnetic scalping, usually using low-intensity drum or roll magnetic separators. On a laboratory scale, magnetic scalping of very fine strongly magnetic impurities can be carried out using a high-gradient magnetic separator equipped with a paramagnetic matrix, e.g. aluminium.

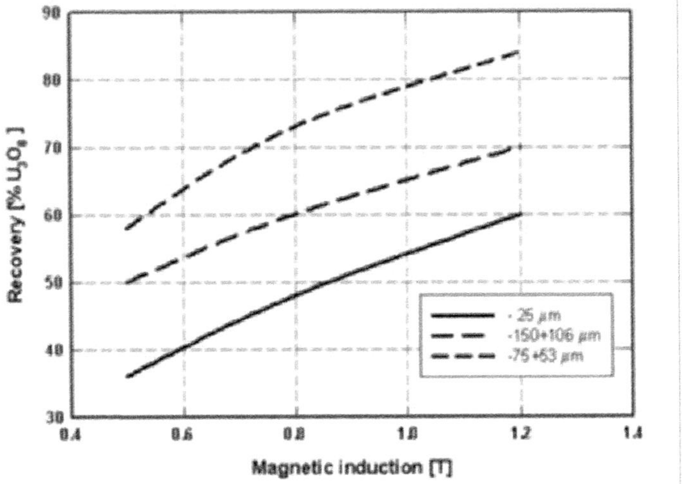

Figure 5.83: Recovery of U_3O_8 from leach residues, for different size fractions, as a function of magnetic induction (adapted from [S69]).

Dispersion

The degree of selective magnetic separation that can be achieved with finely dispersed particles depends on the liberation of individual particles from each other. Therefore, complete dispersion of all particulate components is an essential prerequisite for the selective separation that follows. In this way, non-selective flocs originally present in the slurry must be broken down and the colloidal stability of the dispersed state must be established.

As has been discussed in Section 3.6, selective interactions between particles of valuable material in a mixture with other material components are based on the following criteria:

- Particles of valuable material should carry an electrical charge of the same sign and magnitude as the gangue material, so that the probability of flocculation between them is minimized.
- The electrical charge on the particles of the valuable material should be such that the repulsive interaction energy between them will be smaller than the attractive energy.

Dispersion of natural ores is often difficult, owing to interference from dissolved ions and because it is not possible to draw reliable conclusions for selective aggregation or dispersion behaviour of a slurry from results obtained with pure minerals [S70]. Although dispersion is an inherent part of most flotation and flocculation flowsheets, its application to magnetic separation is limited mainly to kaolin purification by HGMS. Dispersion by sodium hydroxide and sodium silicate is a vital part of the pulp preparation.

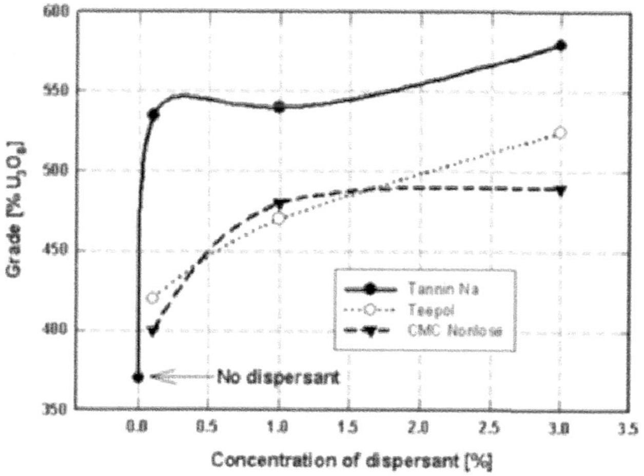

Figure 5.84: The grade of the magnetic concentrate from the -12 μm fraction of uranium-gold cyanidation residues, as a function of the concentration of dispersants (adapted from [S69]).

Dispersion has been found to be very beneficial in the upgrading of uranium-gold leaching residues by WHIMS [S69]. The use of a dispersant increased the grade of the magnetic concentrate by 35% for the full - 150 μm sample, and by 55% for the fraction smaller than 12 μm. Figure 5.84 illustrates the efficiency of various dispersants and the role in their concentration of the grade of the uraninite concentrate.

Preliminary magnetization

Several experimental studies have indicated that preliminary magnetization of magnetizable materials can result in improved recovery and selectivity of magnetic separation. This phenomenon is related to the increase or decrease of magnetization with time, in a constant magnetic field, the so-called magnetic viscosity. The change of magnetization M is generally proportional to the logarithm of time t that has elapsed since the last change in the field [D15].

The rate of increase or decrease of a magnetization is determined by the viscosity coefficient

$$S_v \simeq \frac{dM}{d\ln t} \tag{5.17}$$

which is strongly dependent on the particle size. It has been shown [D15] that for iron particles

$$S_v^{SD} \cong 300 S_v^{MD} \tag{5.18}$$

where S_v^{SD} and S_v^{MD} are single domain and multidomain viscosity coefficients,

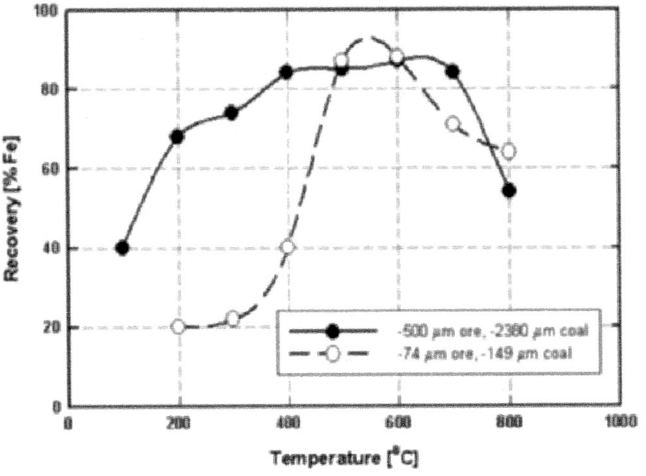

Figure 5.85: The effect of roasting temperature on the magnetizing reduction of two types of iron ore (adapted from [U5]).

respectively. A strong increase in the viscosity coefficient can thus be expected with a decrease in the grain size.

The viscosity coefficient is also dependent on the magnetic field strength. At low fields the coefficient is usually proportional to the field, while at moderate fields the coefficient becomes independent of the field.

Fraas [F24] exposed a wide spectrum of coarse (- 0.4 + 0.2 mm) magnetizable minerals to a homogeneous magnetic field before these minerals were separated in a belt separator. Fraas found that the pre-magnetization of samples resulted in increased recovery and improved selectivity of separation.

Pre-magnetization of magnetite ore by a magnetic induction of 0.03 T was implemented at the Magnitogorsk Metallurgical Plant (Russia) to improve the metallurgical performance of dry magnetic separators [R22]. By exposing a weakly magnetic iron ore from Krivoy Rog (the Ukraine), to a pre-magnetizing pulse magnetic field of 1.9 T and pulse duration of 1.2×10^{-3} s , the recovery increase and selectivity improvement were reported by Špaček and Dočkal [S74]. The effect was particularly noticeable in the - 10 μm size fraction.

Preliminary magnetization of fine fractions of hematite ore prior to HGMS resulted in an improvement in the recovery and grade of the magnetic concentrate [S75]. For a pre-magnetization time as short as 10 s, the iron recovery from the - 12 μm fraction increased from 44.5% to 49.2% Fe at 0.4 T, and the grade of the concentrate increased from 16% Fe to 21% Fe. Although a higher pre-magnetization magnetic field improved recovery, selectivity of magnetic separation was better at low fields.

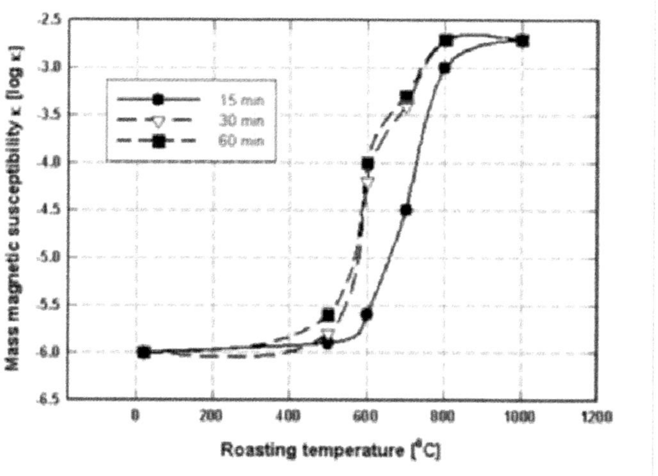

Figure 5.86: Magnetic susceptibility of roasted hematite as a function of roasting temperature. Mass magnetic susceptibility of hematite: 0.95×10^{-6} m^3/kg (adapted from [C21]).

Magnetizing roasting

Many weakly magnetic minerals can be converted to more strongly magnetic compounds by subjecting the ore to a reducing or oxidizing atmosphere at elevated temperatures. Magnetizing roasting is, for example, a very effective process in treatment of iron ores, which are poorly responsive to conventional beneficiation techniques such as flotation, gravity separation and magnetic separation. It involves conversion of feebly magnetic iron minerals in the ore to a strongly magnetic form. In most magnetizing reduction systems, the ore is reduced at a particular temperature and then quenched in water. The resulting maghemite (γ-hematite) is strongly magnetic and can then be recovered by low-intensity magnetic separation.

The effect of the roasting temperature on the recovery of iron from roasted iron ore (Agbaja) into the magnetic concentrate is shown in Fig. 5.85 [U5]. It can be seen that the optimum recovery was reached at a temperature of about 500 ^{0}C. A similar observation was made by Nasr and Youssef with Bahariya Oasis (Egypt) iron ore [N10].

The variation in magnetic susceptibility of hematite (Hoyt Lake, Michigan) as a function of reduction roasting time and temperature is illustrated in Fig. 5.86. It can be seen that at temperatures above 500 ^{0}C, the magnetic susceptibility increases significantly and reaches that of magnetite at 800 ^{0}C to 1000 ^{0}C, depending on the charcoal dosage [C21]. However, oxidizing roasting of hematite did not change its magnetic susceptibility.

The practice of oxidized roasting of ilmenite is well established and it is

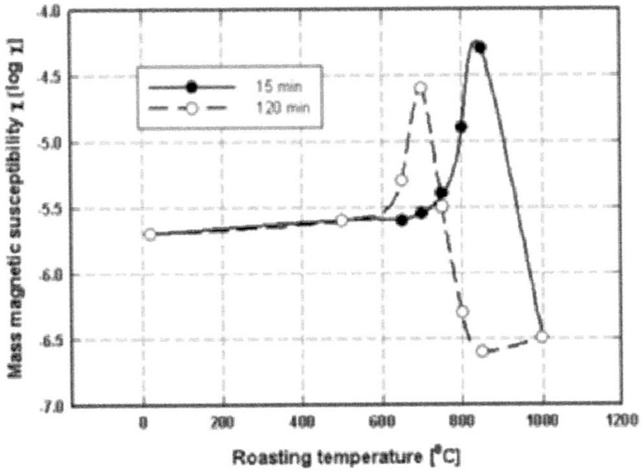

Figure 5.87: Oxidizing roasting of ilmenite: magnetic susceptibility of ilmenite as a function of roasting time and temperature (adapted from [C21]).

used routinely in the magnetic separation of mineral sands. For instance, ilmenite concentrates from heavy mineral deposits in northern Kwazulu-Natal (South Africa) and Mozambique often contain as much as 0.3% Cr_2O_3, which would contaminate the final high-titanium slag. While the magnetic susceptibility of iron-poor chrome-rich spinel is similar to that ilmenite, it is not significantly affected by oxidizing roasting [N11]. However, six-fold increase in the magnetic susceptibility of ilmenite, by oxidizing roasting, enabled the removal of chromium-bearing spinel from ilmenite concentrates by magnetic separation [N11]. This process is practiced at Richards Bay Minerals (South Africa) where the ilmenite concentrate from WHIMS is roasted in a fluidized bed at a temperature of 800 ^{0}C for 40 minutes. The mass magnetic susceptibility increases from 2.4×10^{-6} m^3/kg to 15×10^{-6} m^3/kg.

A detailed laboratory investigation of the roasting behaviour of ilmenite (St. Urbain, Quebec, Canada) was conducted by Ciu et al. [C21]. The test results indicated that oxidizing roasting of ilmenite ($\chi = 2.18\times10^{-6}$ m^3/kg) could increase its magnetic susceptibility within limited temperature and time ranges. As is evident in Fig. 5.87, at 500 ^{0}C, even extended heating (up to 2 hours) did not change the magnetic susceptibility of ilmenite. At higher temperatures between 650 ^{0}C and 700 ^{0}C, the susceptibility started to increase after about 30 minutes of roasting. After reaching its maximum at $\chi = 5.5\times10^{-5}$ m^3/kg, the susceptibility dropped sharply to 4×10^{-7} m^3/kg, lower than that of the original ilmenite.

However, reduction roasting of ilmenite exhibited a considerably more sig-

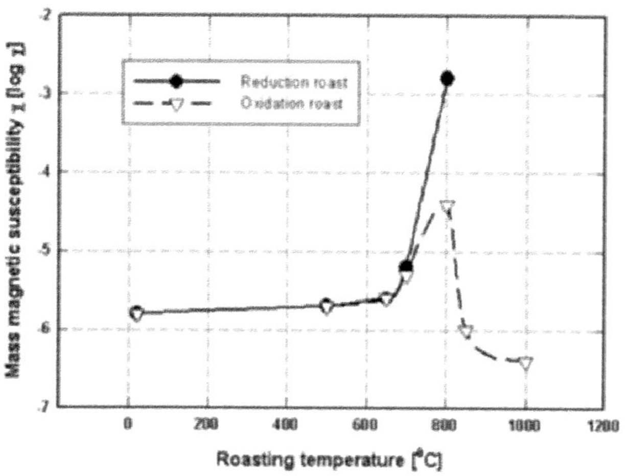

Figure 5.88: Magnetic susceptibility of ilmenite as a function of temperature, for oxidizing and reduction roasting. Roast time: 30 min. (adapted from [C21]).

nificant increase in magnetic susceptibility, approaching that of magnetite ($\chi = 2\times10^{-3}$ m^3/kg), as is illustrated in Fig. 5.88.

Allen [A31, A32] conducted a thorough experimental study of the effect of oxidizing roasting on the magnetic properties of ilmenite. He observed that the ilmenite particles behaved as though they contained discrete ferromagnetic components within a paramagnetic matrix. Magnetic changes at the roast temperature up to 450 ^{0}C were ascribed to small changes in size and shape of small magnetic cells. Although magnetic susceptibility was found to increase in the range around 500 ^{0}C, the differences were not sufficiently convincing. At temperatures above 550 ^{0}C, the growth of the magnetic cells and magnetic susceptibility appeared to be time and temperature-dependent.

The de-coppering of galena and molybdenite concentrate by the removal of chalcopyrite can be significantly improved by roasting the chalcopyrite. It was reported by Makarow et al. [M30] that by heating chalcopyrite in an oxidizing atmosphere to a temperature between 250 ^{0}C and 500 ^{0}C for 10 minutes, the magnetic susceptibility of the product increased by three orders of magnitude.

5.3.10 The selectivity of separation

It was shown in Section 1.2 that a separation process will be selective only when the magnitudes of the magnetic and competing forces are of comparable magnitude, compatible with conditions expressed by eqs. (1.1). The relative values of the magnetic and competing forces are thus critically important in determining the selectivity of magnetic separation. Because some of these forces have different dependence on particle size, their relative importance will vary with particle size, as has been outlined in Section 1.2.2. It is also important to recall, as shown in Section 1.2.2, that particle size is a more important discriminating factor in magnetic separation than magnetic susceptibility and that a magnetic separator is also a particle size classifier.

Some investigators have tried to determine the threshold ratio of magnetic susceptibilities of the components of a material mixture, for which their separation will be efficient. Such a threshold ratio is generally of limited value because of the significant effect of particle size distribution on the balance of forces. Classification or desliming can alleviate the negative effects of wide size distribution. In addition, magnetic susceptibilities of numerous materials depend on the magnetic field strength, to which they are exposed, as discussed in Section 1.4.7.

Of significant importance but particular difficulty is the selective separation of fine materials. Large amounts of fines and ultrafines are generated during the mining and milling of large tonnages of low-quality ores. A direct consequence of this situation is an increasing need for efficient technologies to beneficiate the fines. It has been shown in Section 3.6 and in [S1] and [S31] that the complexity of selective beneficiation of finely dispersed materials rests in the fact that volumetric forces present in a separator, viz. magnetic, gravitational and erosion forces, have to compete with surface forces. The question of whether a quasi-colloidal particle will be captured by a matrix element covered by particles previously deposited is thus determined, inter alia, by the surface characteristics of the particles.

In practice, therefore, fine particles to be separated should be colloidally unstable, i.e. they should not repel each other and thus prevent contact. On the other hand, particles of the gangue minerals should not coagulate with those of the valuable mineral lest they become entrained in the magnetic concentrate.

The effect of pH of the slurry

An experimental study of the effect of pH on magnetic separation of weakly magnetic iron ores [S38, S72] has shown that in order to achieve high recoveries of iron and a high grade of the magnetic concentrate, the pH of the slurry must be adjusted in such a way as to correspond to the point of zero charge (PZC) of the oxide mineral. This behaviour was ascribed to a combination of ionic and magnetic flocculation in the primary and secondary minima, and to a higher retention probability of particles on the matrix [S72, S1].

Similar observations were made in the magnetic separation of uranium and gold from cyanidation residues [S73]. As shown in Fig. 5.89, the grade of the magnetic concentrate, as a function of pH, reaches its maximum at a pH of 1.2, and decreases continuously to the alkaline region. It can be seen that pH adjustment from the usual operating value of about pH 9 can result in a gold content increase by 25%.

Control of surface interactions can also result in selective coating of mineral

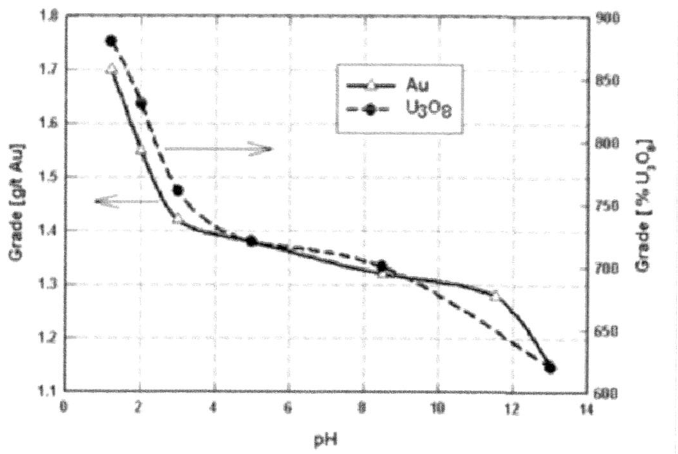

Figure 5.89: The grade of the magnetic concentrate from gold and uranium cyanidation residues, as a function of pH (adapted from Svoboda et al. [S73]).

and metallic particles with magnetite, followed by magnetic separation. It can be used as a method of particle separation, as demonstrated, for example, by Parsonage [P13].

Energetic and kinetic criteria for selective separation

In order to correlate quantitatively the rules for selective separation with the results of experimental investigations, energetic and kinetic criteria, based on DLVO theory, for selective magnetic separation of a useful mineral from a system of containing several mineral species, were derived by Svoboda et al. [S76]. It was shown that the onset of destabilization of a suspension and, consequently, the critical condition for selective magnetic separation, is determined from the energy condition $V_T = 0$, where V_T is the total energy of interaction of two particles and is given by eq. (3.123). At the same time, the residence time t of particles in the magnetic field must be such that $t > t_{1/2}$, where $t_{1/2}$ is the half time of coagulation, given by eq. (3.126). This condition imposes a restriction on stability ratio W, given by eq. (3.124), for a given concentration of particles in a slurry. The stability ratio for a suspension containing i and j mineral species can be re-written as:

$$W_{ij} = \int_0^\infty \exp(\frac{V_T^{ij}}{kT}) \frac{b_i + b_j}{(b_i + b_j + h)^2} dh \tag{5.19}$$

where b_i and b_j are the radii of particles of species i and j, respectively, and h is the distance between the surface of the interacting particles.

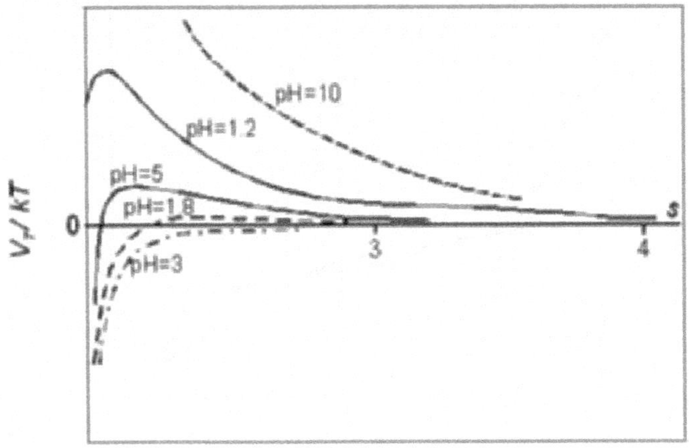

Figure 5.90: Total potential energy of interacting uraninite and quartz particles ($2b = 5\mu$m), as a function of pH, at $B_0 = 1$ T (adapted from [S76]).

If a particle of a useful mineral is denoted by 1, and the gangue particle as 2, high recoveries will be achieved if $V_T^{11} < 0$ and $W_{11} < 1$. Simultaneously, for the grade of the magnetic concentrate to be high, one must ensure that valuable and gangue particles do not coagulate mutually, and thus $V_T^{12} > 0$ and $W_{12} >> 1$. Furthermore, to avoid the mechanical entrainment of large clusters of gangue particles, it would be preferable to work under the condition $V_T^{22} > 0$ and $W_{22} > 1$. These conditions for selective magnetic separation of fine particles can be summarized as [S76]:

$$\text{High recovery:} \qquad V_T^{11} < 0;\ W_{11} < 1 \tag{5.20}$$

$$\text{High selectivity:} \qquad V_T^{12} > 0;\ W_{12} >> 1 \text{ and } V_T^{22} > 0;\ W_{22} > 1 \tag{5.21}$$

The validity of these conditions was confirmed experimentally by magnetic separation of uranium-gold cyanidation residues [S1, S73, S76]. The curves of total potential energy of interaction of uraninite and quartz particles are shown in Fig. 5.90. It can be seen that selective separation can be expected when $V_T^{12} > 0$, which occurs at pH 1.2. With further increase of pH, V_T^{12} decreases and the selectivity is reduced. This trend is confirmed by the experimental results shown in Fig. 5.89. Poor selectivity of separation in the alkaline region is a consequence of heterocoagulation into a deep secondary minimum [S76].

Carrier magnetic separation

Efficient recovery of fine particles can be aided by promoting mutual interaction between fine and coarse particles. In carrier (or piggy-back) magnetic separation, coarse particles serve as carriers for fine particles. Carrier magnetic

Table 5.29: Magnetic separation of mixtures of different size fractions of uranium-gold leach residues [S69].

Mixture	Head	Theoretical		Experimental	
5:1 [μm]	U_3O_8 [ppm]	Grade [ppm]	Recovery [%] U_3O_8	Grade [ppm]	Recovery [%] U_3O_8
-17+(-53+25)	185	820	30	1033	52
-17+(-75+53)	173	700	29	850	40
-17+(-150+106)	159	616	27	600	22

separation makes use of the fact that the frequency of particle collisions is much greater when there are particles of different sizes, which is the result of different trajectories in the accelerating fluid. The rate of adhesion of fine particles ($b \approx 2$ μm) to coarse ones ($b \approx 30$ μm) is 10^3 to 10^4 times higher than the rate of adhesion between fine particles [S77].

The results of the investigation of the possible effect of carrier magnetic separation on the efficiency of separation are summarized in Table 5.29 [S69]. It can be seen that the presence of fractions - 53 + 25 μm and - 75 + 53 μm considerably enhanced the recovery of uranium from the fine (- 17 μm) fraction, and the grade of the magnetic fraction.

The presence of the coarse fraction (- 150 + 106 μm) neither improved the selectivity of the process nor increased the recovery of uranium from the fine fraction. Moreover, no dependence of recovery or grade on pH was observed, in contrast to the mixtures of finer fractions, where pH played an important role. Since the value of pH did not affect the efficiency of separation of individual size fractions, it is obvious that pH influenced mutual interaction of fine and coarser particles.

A detailed investigation of carrier magnetic separation was carried out by Yanmin Wang et al. [W25] and the results confirmed the trends observed by Svoboda [S69, S1]. A purified chromite concentrate and chromite ore slime were used in that investigation. A very fine fraction (- 10 μm) was mixed with coarse fractions, - 53 + 38 μm, - 75 + 53 μm and - 106 + 75 μm. It was found that the fine particles coagulated onto the coarse ones and the cluster thus formed was recovered in the matrix of a magnetic separator. The fraction - 53 + 38 μm was found to be the most effective in improving the process efficiency, as shown in Fig. 5.91. The process of carrier separation was also found to be dependent on pH and the highest efficiency was achieved at a pH near the point-of-zero-charge of the valuable mineral.

The matrix vibration and slurry pulsation

The vibration of a matrix, either by mechanical or electromagnetic means, was proposed by several investigators, to assist in the flushing of the matrix, to propel the ore through the matrix in dry high-gradient magnetic separation and

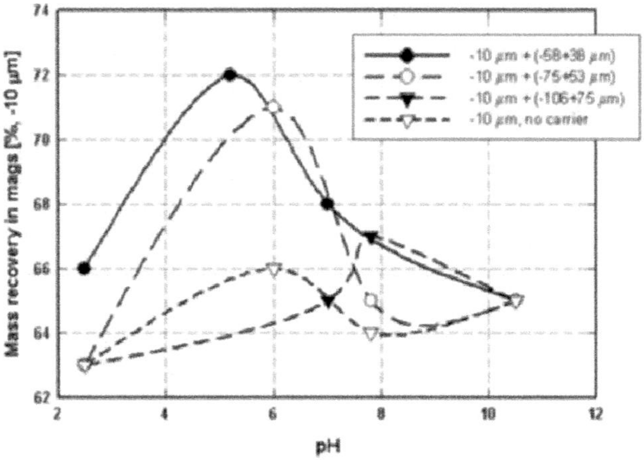

Figure 5.91: The effect of pH and carrier size on the mass recovery of chromite (-10 μm) by magnetic separation (adapted from Wang et al. [W25]).

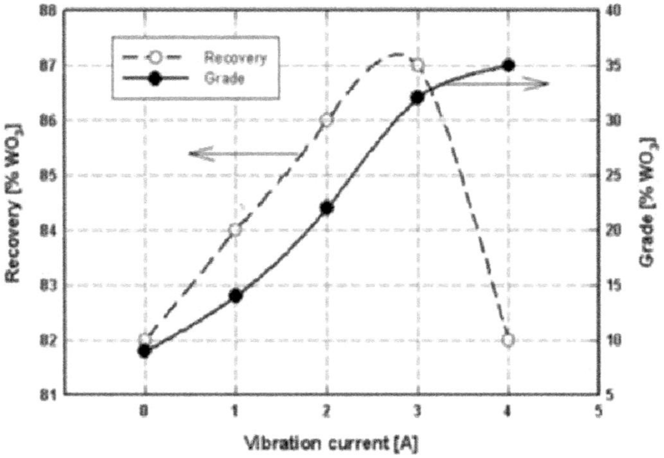

Figure 5.92: Vibration-assisted magnetic separation of wolframite-quartz mixture as a function of the vibration current (adapted from [S78]).

to improve the efficiency of separation [K5, L14]. Mechanical vibration can be provided, for instance, by an eccentric rotating shaft driven by a variable speed motor. Longitudinal electromagnetic vibration is provided by a coil mounted on the chamber and energized with alternating current. A vibratory force is exerted on the coil by the interaction of the d.c. field generated by the magnet of the separator, and of the a.c. field in the coil.

The vibration of the matrix reduces the mechanical entrainment of gangue particles in the magnetic product and facilitates their desorption from the matrix by fluid flow. For instance, matrix vibration was found to improve the selectivity of wet high-gradient magnetic separation of wolframite-cassiterite slimes and of wolframite-quartz mixtures, as is shown in Fig. 5.92 [S78], and of Cu-Pb and Cu-Mo flotation concentrates [Y3]. The application of matrix vibration also resulted in an increased recovery of kaolin in wet [L15] and dry [Y4] high-gradient magnetic separation.

Pulsation of a slurry proved to be a useful approach to improving selectivity of separation and reducing matrix blockage in HGMS. When applied to a low-intensity drum magnetic separator, the slurry pulsation resulted in an improved quality of the magnetic concentrate [K22]. Slurry pulsation, developed by Liu Shuyi et al. [Y2] is being successfully used, on production scale, in SLON high-gradient magnetic separators [X1].

5.3.11 Comparative tests with wet high-intensity magnetic separators

As discussed in Chapter 2, the development and commercialization of high-gradient magnetic separation as a viable technique for mineral beneficiation resulted in the design and manufacture of a wide range of wet high-intensity magnetic separators.

Most wet high-intensity magnetic separators differ in those technical and operational characteristics that can significantly affect their performance and cost-effectiveness. Any comparison of these separators should, therefore, take into consideration, in addition to the feed properties and the product requirements, the size of the operation and the capability of the machine to be scaled-up, if necessary.

Magnetic separation equipment, and particularly WHIMS and HGMS machines, cannot be readily miniaturized for bench-scale testing. The bench-scale testing is, therefore, often focused on characterization of the feed and of the separation products, as a function of those variables that can be controlled in pilot-scale testing. Without proper understanding of these characteristics it may be impossible, except by luck, to end up with a successful process.

On the other hand, pilot-scale tests are usually carried out using commercial equipment, at full feed rates and over an extended time scale. Reliable technical and economic assessment of the operation can then be made.

Comparative pilot-scale tests of several wet matrix high-intensity magnetic separators were described by Forssberg and Kostkevicius [F22]. Five models,

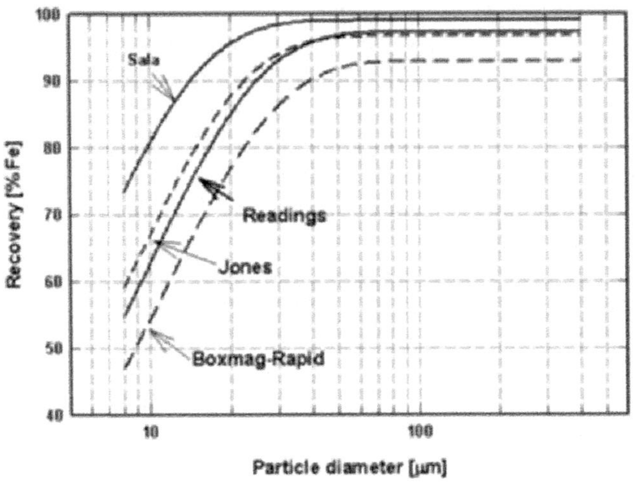

Figure 5.93: Recovery as a function of particle size for assorted wet matrix high-intensity magnetic separators (adapted from [F22]).

representing three distinct principles of magnetic circuit design, with three different types of matrices were tested. A mixture of hematite and quartzite, with 97% of the particles being smaller than 300 μm and 21% smaller than 15 μm, was used as a test material. The mean magnetic susceptibility of the mixture was approximately 2.5×10^{-6} m^3/kg.

The technical data of the separators used in the tests are summarized in Table 5.30. The effect of particle size on iron recovery is given in Fig. 5.93. For fine particles smaller than 50 μm, the Sala HGMS was most efficient, followed by the Jones and Reading machines. The efficiency on coarser material was comparable among these separators, although the Jones separator showed a fall-off in the recovery in coarse fractions associated with the middlings flushing. The energy consumption was lowest in the Jones separator (0.5 kWh/t), followed by the Reading and Boxmag-Rapid machines, although the value of this information is limited as the rating is likely to change for large-scale machines. The optimum metallurgical results obtained with these separators are shown in Table 5.31.

Although these results indicate the potential these various machines have, their direct translation into large-scale applications must be done with caution. Different separators have different potential for scale-up and the metallurgical performance, consumption of energy and water, and resistance to matrix clogging, will vary with the size of the machine. The assessment of the technological and economic potential of each machine must be based on a detailed analysis of their performance with a wide range of ores of different magnetic susceptibilities and size distributions, taking into account the anticipated magnitude of the operation.

Table 5.30: Specifications of magnetic separators used in tests [F22].

Parameter	**Jones P71**	**Boxmag SHW1**	**Reading 16 pole**	**Sala HGMS MkII**
Matrix	Grooved plates	Triangular bars	Salient plates	Expanded metal
Rotor dia. [mm]	710	800	1800	1720
Capacity [t/h]	1 - 2.5	1.5 - 6	15 - 30	1 - 3
No. of separation zones	2	1	8	1
Matrix width [mm]	75	25	66	120
Matrix length [mm]	220	150	205	140
Mass [t]	7.1	3.8	7.1	6

Table 5.31: Representative metallurgical results obtained with different wet matrix high-intensity magnetic separators [F22].

Separator	**Grade [% Fe]**	**Recovery [% Fe]**
Sala HGMS Mk II	66.0	94.0
Jones P71	66.0	92.0
Reading WHIMS-16	65.0	91.5
Boxmag-Rapid SHWI	64.5	85.5

5.4 Magnetic flocculation

5.4.1 Magnetic flocculation of ferromagnetic materials

As discussed in Section 3.6.1, magnetic flocculation of strongly magnetic materials is a process commonly encountered in wet low-intensity magnetic separation of finely ground ferromagnetic (*s.l.*) materials. It arose from a detailed study of mechanisms of particle capture in wet drum magnetic separators [R20, R23] that the dominant mechanism by which the magnetic particles are captured in wet drum magnetic separators is the formation of flocs by magnetic flocculation. This process, characterized by the rate of flocculation and by the flocculation time, is thus affected by the pulp density and the volumetric flowrate through the separation zone of the separator.

Magnetic flocculation is used to enhance the efficiency of numerous other processes. Magnetic flocculators are often used, for instance, to increase the settling rate of solids in dewatering equipment, to improve the performance of centrifuges that remove solids from steel effluent plants, and to reduce the addition of flocculating reagents.

Magnetic flocculation of strongly magnetic materials is largely an empirical technique. Owing to their simplicity, magnetic flocculators are generally used without any extensive research, although selection of variables such as magnetic field strength, solids concentration and particle size distribution is governed by a few basic rules.

Magnetic field strength

The onset of agglomeration of natural magnetite, as illustrated in Fig. 3.37, appears to take place at a magnetic induction as low as 10 mT (100 G), while 40 mT (400 G) seems to be necessary to generate stable flocs [B30]. In magnetite ores of various origin, the threshold magnetic induction was found to range from 8 mT (80 G) to 80 mT (800 G) [K14], and an empirical formula to describe a relationship between the threshold flocculation magnetic field strength H_F and the coercive force H_c was proposed:

$$H_F = 123\sqrt{H_c} \tag{5.22}$$

The strength of the magnetic field required to induce flocculation is also determined by the residence time of a suspension in the field and by the concentration of solids. The higher the flow velocity and the lower the concentration of suspension, the higher is the magnetic field strength required to form magnetic flocs. Benson et al. [B31] state that for effective flocculation, slurries with more than 1% of natural magnetite should be exposed to a 60 mT (600 G) magnetic field, for a period of not less than 0.07 s. The best results were obtained after a residence time of 0.2 s [B30]. Rayner and Napier-Munn [R23] observed that in a low-intensity drum separator the magnetic flocs were formed in the field of 0.2 T in less one second, a substantially shorter time than the typical residence time in the separator.

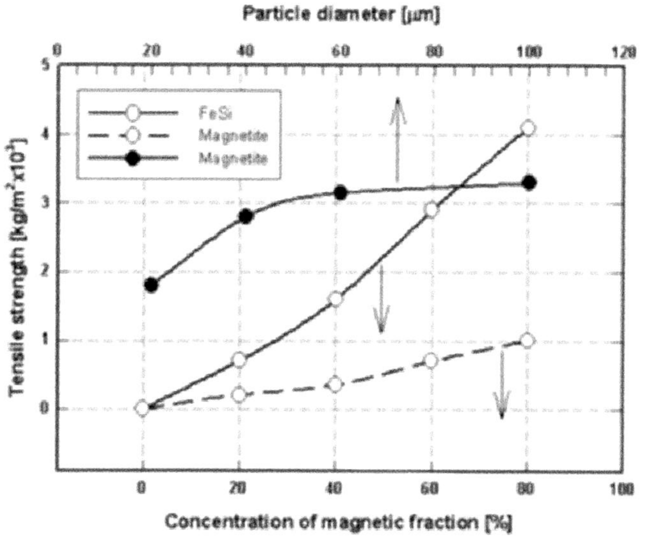

Figure 5.94: Tensile strength of FeSi and magnetite flocs as a function of particle size and concentration of the magnetic fraction in a suspension (after Karmazin et al. [K14]).

Concentration of solids

As the concentration of solids in a suspension increases, the characteristic time of flocculation $t_{1/2}$, given by eq. (3.126), decreases and the probability of particle collision and adhesion increases. If the retention time of particles is sufficiently longer than $t_{1/2}$, which is, in turn, determined by the concentration of solids in a suspension, then a high degree of agglomeration can be expected. It was observed [R23] that at a high solids content in the feed, the magnetic particles, independently of their size, began to flocculate rapidly on the entry into the magnetic field of a drum separator. At low slurry densities, for example in secondary drum magnetic separators used to recover dense media, flocculation is slow and the fines are increasingly getting lost.

The influence of the concentration of solids and of particle size on tensile strength is shown in Fig. 5.94. It is evident that the tensile strength of magnetic flocs increases with increasing pulp density and particle size, the latter dependence having the same pattern as the dependence of magnetic susceptibility on particle size. This is the consequence of the fact that the tensile strength of a floc is determined by the magnetic moment of the interacting components of the floc, rather than by coercive force, as claimed by Rainer et al. [R23].

5.4.2 Magnetic flocculation of weakly magnetic minerals

In contrast to the magnetic flocculation of strongly magnetic minerals, the application of this process to feebly magnetic materials has been limited to laboratory-scale investigations. The first experimental observation of the magnetic flocculation of weakly magnetic minerals was reported by Hencl and Svoboda in 1979 [H16]. It was observed that magnetic flocculation of fine siderite particles (51% - 20 μm), with a magnetic susceptibility of 0.75×10^{-6} m^3/kg, took place at a magnetic field as low as 0.4 T, and the rate of agglomeration increased with increasing field strength.

Experimental investigations into the magnetic flocculation of paramagnetic and diamagnetic particles were subsequently carried out by Watson [W26] and Parker et al. [P8, V5], while a more detailed systematic study of agglomeration of hematite was conducted by Ozaki and Matijevic et al. [O6, O7].

An extensive and thorough theoretical and experimental investigation of flocculation of weakly magnetic minerals in a magnetic field was conducted by Wang, Forssberg and Pugh [P10, W17, W18, W19]. The results illustrated in Figs. 3.43 and 3.44 confirm that flocculation of feebly magnetic materials takes place at moderate magnetic fields.

The possibility of industrial applications of magnetic flocculation

Selective chemical flocculation of mineral slimes provides an efficient means of recovering very fine particles. However, this method is rather costly and environmentally problematic. However, no chemical agents are required in magnetic flocculation and the flocculation time can be reduced by increasing the magnetic field. Magnetic flocculation would thus allow creation of clusters of very fine particles that could be readily recovered in a high-gradient magnetic separator. Detailed analysis of the potential of incorporating magnetic flocculation into magnetic separation circuits was presented by Svoboda [S79, S1]. Figure 5.95 illustrates two possible proposals for the inclusion of magnetic flocculation in a magnetic circuit for the recovery of very fine, weakly magnetic minerals. However, commercial large-scale exploitation of this technique will require involvement of industry with academia and their commitment to develop further innovative ideas and transfer them into production.

5.5 Demagnetization

Exposure of ferromagnetic *(s.l.)* materials to external magnetic fields often results in the formation of magnetic flocs that interfere with their further manipulation. In order to break the flocs and to improve dispersion of magnetic particles in suspension, the material must be demagnetized. The demagnetization process is often applied, for instance, to preparation of feed for sizing, filtration, magnetic separation and for heavy media suspensions. It is usually not possible to demagnetize fully a ferromagnetic *(s.l.)* material, such as mag-

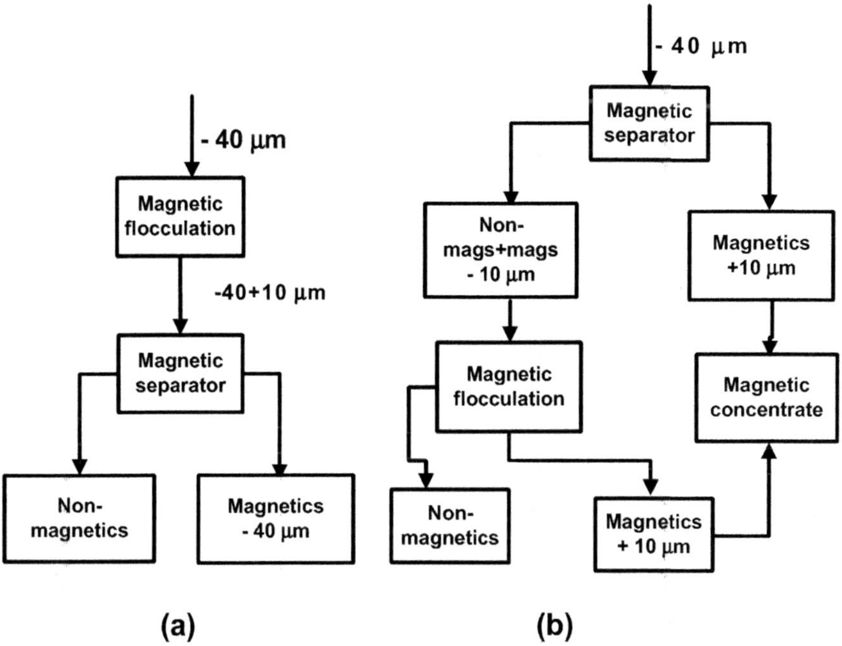

Figure 5.95: Possible applications of magnetic flocculation to the beneficiation of weakly magnetic ores (adapted from [S1]).

netite or ferrosilicon, at temperatures below the Curie temperature (570 ^{0}C for natural magnetite).

There are three main methods of demagnetization. Thermal demagnetization, commonly used with permanent magnet materials, consists of heating a material to a temperature above its Curie temperature and then allowing it to cool in zero magnetic field. Caution must be exercised to avoid chemical changes or sintering induced by temperature [L16, H23]. This rather impractical method is being used as a benchmark to compare the effectiveness of simpler demagnetization methods.

In d.c. field demagnetization, a particle must be exposed to a magnetic field greater than its intrinsic coercivity H_{ci} (Fig. 1.7), which, when the magnetic field is returned to zero, will result in zero magnetization. Practically, however, because the field inside a single particle depends in a rather complicated manner on the form of individual particles, the shape of the clusters, and the density of the slurry, it is not possible to demagnetize powdered material by use of a constant demagnetizing field [L16].

The practical method of demagnetization is by use of a time-varying magnetizing field of gradually decreasing amplitude, produced by a coil energized by an a.c. power supply. As a result, the particle magnetization traces a series of minor hysteresis loops and ultimately converges towards zero. The magnetizing field must start at a value appreciably greater than the coercive force and be gradually reduced with each alteration until it reaches a very low value. A well established rule of thumb is that the field amplitude should be five times the coercive force and the rate of decrease should be about 10% per cycle [H23]. For rough demagnetization, at least four alterations should be applied, although when demagnetizing difficult materials, as many as a hundred alterations may be needed [M19].

Demagnetization of magnetized particles by an a.c. field is based on the assumption that the body has a fixed position in the field. In practice, however, the particles and clusters are free to move in the slurry. When exposed to a rotating magnetic field, under suitable conditions, particles can rotate rather than become demagnetized. This problem becomes particularly pronounced for materials with a high value of intrinsic coercive force unless the strength of the demagnetizing field is greater than the coercive force and the particles are non-spherical. An approximate criterion for particle rotation is that if the external rotating field H is smaller than the coercive force H_c, the particles will rotate. If H is greater than H_c and H_{ci} , the magnetic moment of the particle will rotate but the particle will probably not. Whether a particle will rotate or not depends also on the viscosity of the slurry: high viscous drag can prevent particles from rotating fast enough to achieve a synchronous motion. This phenomenon will be even more pronounced at high frequencies of the applied demagnetizing field.

Several observations, essential for efficient demagnetization, can thus be summarized:

1. The flow rate may be high but should preferably be laminar. This follows from the observation that the probability of rotation of particles is smaller for laminar flow than for turbulent flow, as a result of rod-like clusters of particles being aligned along the streamlines in the laminar flow.

2. The minimum residence time of the slurry in the magnetic field should be of the order of 0.24 s to 0.28 s, which corresponds to 12 to 14 cycles of the a.c. magnetic field.

3. For feebly coercive materials, such as natural magnetite or ferrosilicon, a magnetic induction ranging from 0.03 T to 0.06 T is usually sufficient.

4. For highly coercive materials (e.g. maghemite, titano-magnetite ores) a field of high frequency (400 Hz to 600 Hz) and high magnitude (0.07 T to 0.1 T) should be used. Some magnetite ores could not be demagnetized even at a magnetic induction of 0.18 T [K14].

5. Demagnetization of very fine particles, close to single-domain size, is increasingly difficult. This observation is related to a trend of increasing coercive force with decreasing particle size, as shown in Fig. 1.13 and expressed by eq. (1.45).

Demagnetizing coils

The demagnetizing unit is usually an air-core a.c. solenoid wound on a pipe, preferably arranged vertically, through which the slurry flows. The pipe section through the coil is made from non-metallic material to eliminate the shielding of the slurry from the magnetic field and the eddy current generation. The sample to be demagnetized may be gradually withdrawn from the solenoid, or the sample can be left in a position with a damped sinusoidal field applied. In mineral processing, a method of continuously demagnetizing particles in a flowing slurry is preferred. The particles in the slurry experience a decreasing a.c. field as they flow out of the solenoid.

Semi-empirical equations were developed [B20] to calculate the design parameters of a demagnetizing coil:

$$H = \frac{1.03NI}{[(a+c)^2 + b^2]^{1/2}} \tag{5.23}$$

$$L_e = \frac{7.88N^2(a+c)^2}{(3a+9b+13c)} \times 10^{-6} \tag{5.24}$$

where H is the magnetic field strength (in [A/m]), L_e is the inductance (in [H]), a is the inside diameter of the winding (in [m]), b is the length of the coil (in [m]), c is the radial depth of the winding (in [m]), N is the number of turns and I is the current in the coil (in [A]).

5.6 Separation in magnetic fluids

Separation in magnetic fluids is one of those techniques that offer a unique approach to material manipulation over a wide spectrum of applications. This sink-and-float separation technique exploits differences in densities of the materials to be separated. As has been explained in Sections 2.8 and 3.8, a magnetic fluid placed in a non-homogeneous magnetic field exhibits an apparent density differing from its natural density. This apparent density can be controlled in a wide range of values, exceeding the densities of all known elements and materials. An important feature of this method is the very high selectivity of separation, whereby mixtures comprising particles with very small density differential can be separated from each other.

Density is a unique physical property of materials and, in contrast to other physical properties, it is, in most cases, well-defined for each mineral or material. Because of this, and in spite of the rather limited sophistication of gravity separation equipment, separation of materials based on density is a universal and fundamental technique, and is one of the oldest forms of mineral processing. In principle, density separation is a very accurate way of material processing, particularly if one can operate in a wide range of densities and if the technique is capable of distinguishing a small density difference.

The efficiency of separation of particles having several properties of similar magnitude (for instance density, magnetic susceptibility, electrical conductivity

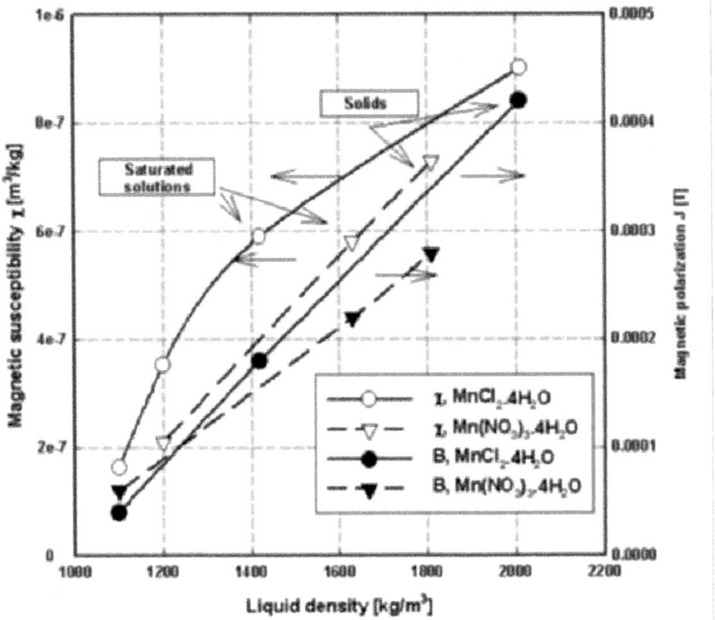

Figure 5.96: Mass magnetic susceptiblity and magnetic polarization of paramagnetic liquids, as a function of their densities. The polarization was determined at the external magnetic induction of 0.25 T [S8].

and others) can be increased by simultaneous exploitation of two or more of these properties. The precision and the separation rate can be increased in comparison with conventional processes, in which a single property is exploited. Separation in magnetic fluids is based on such a combination of two widely different physical properties of materials and offers a unique technique for material manipulation.

5.6.1 Magnetic fluids

Paramagnetic liquids

Magnetic fluids can be divided into two broad classes, namely solutions of paramagnetic salts and ferrofluids. Paramagnetic liquids, as the name indicates, are paramagnetic in their behaviour: their magnetization increases linearly with increasing magnetic field and their magnetic susceptibility is quite low, of the order of 5×10^{-7} m^3/kg (40×10^{-6} cm^3/g), and field-independent. The values of magnetic susceptibility and magnetic polarization, for two types of paramagnetic liquids, as a function of their density, are shown in Fig. 5.96. The values of magnetic polarization were measured at the applied magnetic induction of 0.25 T, which is a typical magnetic field strength used in laboratory-scale ferrohydrostatic separators to achieve the apparent density of approximately 4000

Table 5.32: Magnetic susceptibilities and viscosities of manganese chloride and holmium nitrate solutions (adapted from [A33]).

	Density 1400 kg/m^3	**Density 1850 kg/m^3**
$MnCl \times 4H_2O$		
χ [m^3/kg] η [Pa$\times$s]	6×10^{-7} 1.5×10^{-3}	
$Ho(NO_3)_3 \times 5H_2O$		
χ [m^3/kg] η [Pa$\times$s]	7.2×10^{-7} 2×10^{-3}	14.2×10^{-7} 12×10^{-3}

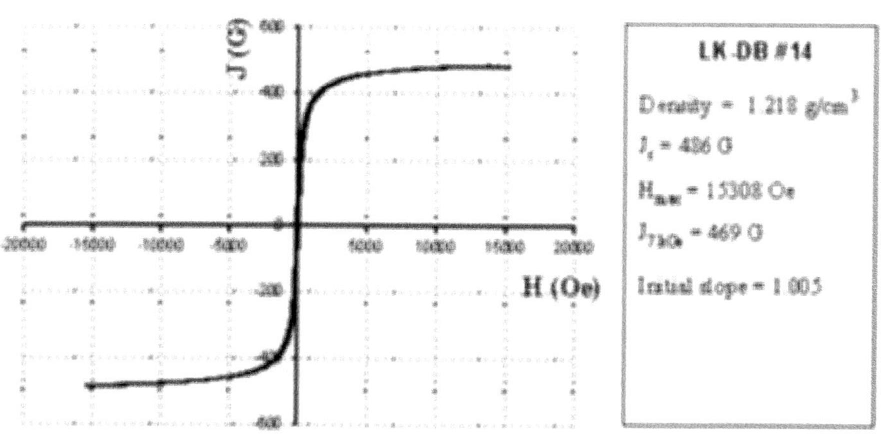

Figure 5.97: Magnetization curve of a kerosene-based ferrofluid, with average particle size of magnetite of 9 nm [K23].

kg/m^3. It can be seen that polarization is very low (2×10^{-4} T or 2 G).

Andres [A33] compared magnetic properties of aqueous solutions of rare-earth salts with those of manganese chloride. He observed that magnetic susceptibility of holmium nitrate, the most magnetic of the solutions of the rare-earth salts, at its limit of solubility, is approximately three times higher (1.6×10^{-6} m^3/kg) than that of manganese chloride (0.5×10^{-6} m^3/kg).

Although the solutions of rare-earth salts do extend the range of achievable apparent densities somewhat, their high cost compared to manganese chloride (by a factor of 50 for pure compounds) hardly justifies their industrial-scale application. A comparison of magnetic susceptibilities and viscosities of manganese chloride and holmium nitrate is given in Table 5.32.

In addition to low magnetic susceptibility, some paramagnetic liquids tend to degrade in the presence of light; others tend to crystallize at lower temper-

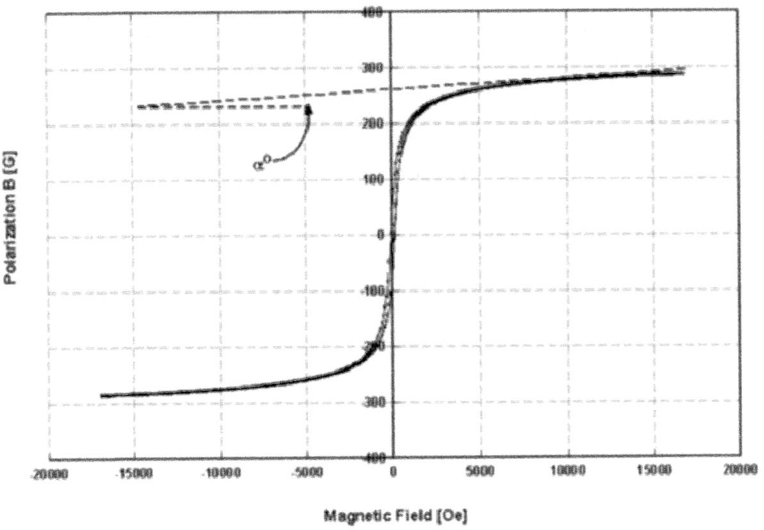

Figure 5.98: Magnetization curve of a ferrofluid with an average particle diameter 6 nm [K23].

atures, while others decompose at elevated temperatures. All such liquids are rather expensive and their recycling is essential, although not easy. The paramagnetic liquids have a high surface tension and do not wet the particles surface adequately.

Ferrofluids

A magnetic ferrofluid is a stable suspension of sub-domain magnetic particles, often magnetite, in a carrier fluid. Water, hydrocarbons and silicones are used as carrier liquids. The particles, which have an average size of 10 nm, are coated with a stabilizing dispersing agent, which prevents particle agglomeration, even when a strong magnetic field gradient is applied to the ferrofluid.

In the absence of an external magnetic field, the magnetic moments of individual particles are randomly distributed and the fluid has no net magnetization. When exposed to an external magnetic field, the ferrofluid becomes magnetized and reaches magnetic saturation at moderate magnetic fields. When the applied field is removed, the particles demagnetize rapidly and exhibit typical superparamagnetic behaviour characterized by the absence of coercivity and remanence. A typical magnetization curve of a ferrofluid is shown in Fig. 5.97, while Fig. 5.98 illustrates the magnetization curve of a ferrofluid with a smaller average particle size of 6 nm. It can be seen that a high magnetic field is required to magnetically saturate a ferrofluid containing very small magnetite particles. This is a consequence of the difficulty in aligning the magnetic moments of very small

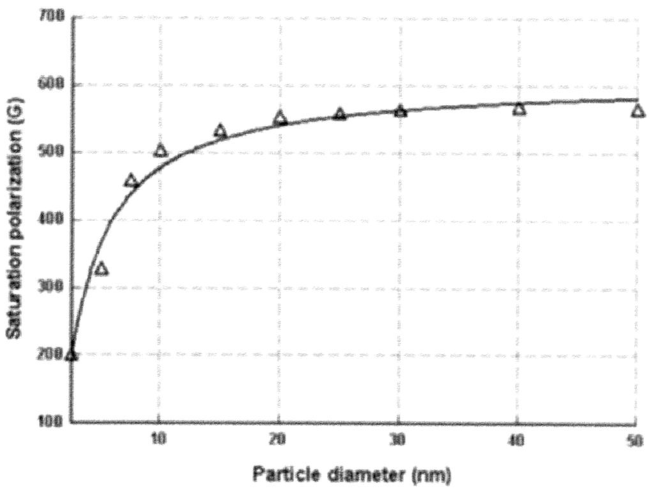

Figure 5.99: Saturation polarization of ferrofluids as a function of average particle size [K23].

particles [C23]. Although very small dimensions of magnetite particles improve the stability of a ferrofluid, the value of saturation magnetization decreases with decreasing particle size, as shown in Fig. 5.99.

In weak magnetic fields the main contribution to magnetization is made by the larger particles which are more easily oriented by a magnetic field, whereas the approach to saturation is determined by fine particles, the orientation of which requires large fields [R16, C22].

Viscosity of ferrofluids The viscosity of a ferrofluid is one of the most important parameters that affects selectivity of separation in ferrofluid. In the absence of a magnetic field, the viscosity of a ferrofluid is greater than that of the carrier liquid as a result of the perturbation of the streamlines by suspended particles. Einstein [E5] showed that the dependence of the viscosity of a suspension of rigid spheres on the volume fraction f is represented by

$$\eta = \eta_0(1 + 2.5f) \tag{5.25}$$

where η_0 is the viscosity of the carrier liquid. At volume fractions greater than 0.01, higher order terms in f have to be introduced into eq. (5.25) and the viscosity can be described by the Vand relation [V6]:

$$\eta = \eta_0[1 + 2.5f + 7.349f^2 + ...] \tag{5.26}$$

On application of a magnetic field, a torque $\mu_M \times B$ is introduced, which hinders the rotation of the particles about axes perpendicular to the magnetic

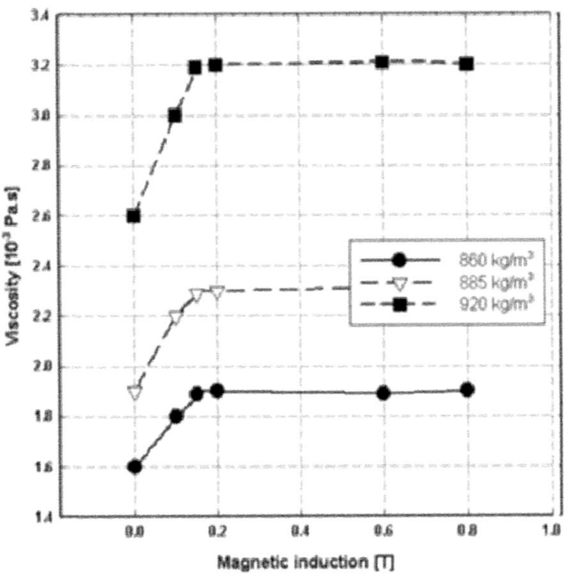

Figure 5.100: Viscosity of a kerosene-based ferrofluid, as a function of the applied magnetic field, for three densities of the ferrofluid. Viscosity of kerosene $\eta_0 = 1.75 \times 10^{-3}$ Pa×s (adapted from Gubarevich [G17]).

field. As a result, an additional frictional coupling between fluid layers is introduced thereby increasing the viscosity [C22]. The fractional increase in viscosity is larger when the field is applied parallel to the stream, e.g. in a capillary, than when it is applied perpendicular to the stream [C23].

By including the rotational component in the Einstein formula, several authors have shown that the saturation values of the viscosity parallel ($\eta_{\parallel}$) and perpendicular ($\eta_{\perp}$) to the field direction are given by:

$$\eta_{\parallel} = \eta(1 + \frac{3}{2}f), \qquad \eta_{\perp} = \eta(1 + \frac{3}{4}f) \tag{5.27}$$

where η is given by eq. (5.25).

The dependence of the viscosity of a kerosene-based ferrofluid on the applied magnetic field, for three values of the ferrofluid density, is shown in Fig. 5.100. The measurements were made at the constant viscosity of kerosene and shear stress. It can be seen that at a magnetic induction greater than 0.4 T, the viscosity of ferrofluid is constant.

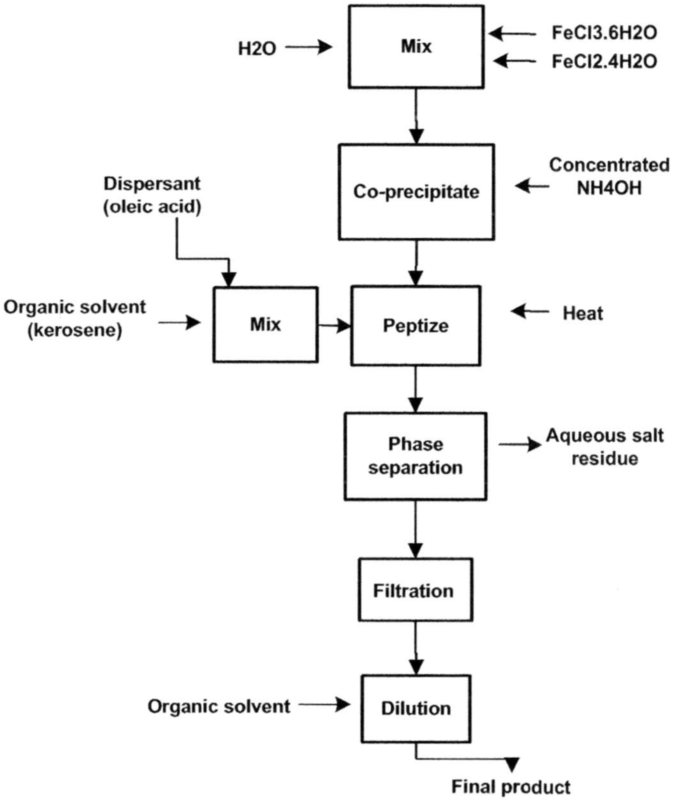

Figure 5.101: Preparation of a ferrofluid by chemical precipitation (adapted from [R16]).

Preparation of ferrofluids

There are two basic methods of preparing a ferrofluid - size reduction and precipitation. Size reduction by wet grinding of ferrite in a ball mill, for a period of 1000 hours or longer, in the presence of a surfactant, was originally investigated by Papell [P5], while Rosensweig et al. [R16] developed the technique further and produced kerosene-based ferrofluids. The technique has been extended even further by Kaiser and Rosensweig who succeeded in dispersing magnetic particles in other groups of solvents, including hydrocarbons, water, aromatics and esters [R16, C22, R24].

Magnetite particles can also be prepared by precipitating the magnetite from a solution of ferric and ferrous ions using an excess of an alkali hydroxide solution. The process is schematically shown in Fig. 5.101. In the peptization step, the particles are transferred from the aqueous phase to an organic phase containing a dispersion agent, such as, for instance, oleic acid. Subsequently, the particle dispersion in an organic phase is separated from the aqueous salt residue, filtered, and solvent adjusted to give the desired product concentration [R16, B32].

Stability of ferrofluids

The stability of ferrofluids in gravitational and magnetic fields is one of the fundamental parameters that ensures high selectivity of separation of materials in ferrofluids. Basic criteria for such stability can be obtained by investigating the balance of forces acting on the colloidal magnetic particles suspended in the carrying fluid. The relationship between the randomizing action of the thermal energy and the destabilizing effect of the gravitational field, surface forces, magnetic interaction between the particles and the non-homogeneous magnetic field, and between the particles themselves, determines the conditions under which a ferrofluid is stable.

It is fairly easy to show [R16] that while the force of gravity is of limited threat to the segregation of ferrofluids, the effect of magnetic agglomeration and of the field gradient can be eliminated only when the particle diameter is smaller than approximately 8 nm. However, molecular van der Waals forces that arise as a result of fluctuating electric dipole-dipole forces are always present in a colloidal system. It transpires from eq. (3.121) that infinite energy is required to separate a particle pair. Therefore, contact between individual particles must be prevented in order to obtain a stable colloidal suspension. Such a steric stabilization can be achieved by adding a chemical, for instance oleic acid, that can both adsorb on the surface of the particle and be solvated by the carrier liquid. Such long-chain molecules are chosen in such a way that their tails have dielectric properties similar to these of the surrounding carrier fluid. This results in the formation of a bound liquid sheath around each particle [K24]. The most commonly used surfactants give a surface layer thickness of 2 nm to 3 nm.

The stability of kerosene-based ferrofluids used in ferrohydrostatic separators extends over periods as long as one year or longer [G17, S8]. Rosensweig [R16] reported that the properties of a ferrofluid prepared in kerosene did not deteriorate after 18 years. However, water-based ferrofluids usually contain particles somewhat larger than the particles in organic carriers. The destabilizing effect of magnetic agglomeration and the influence of the gravitational field might reduce the long-term stability of such ferrofluids. Moreover, the highly polar nature of the carrier means that dispersants are not as effective as in organic carrier fluids.

A simple instrument that can be used to observe the long-term stability of ferrofluids was developed by Bissell et al. [B33]. In a scanning column magnetometer ferrofluid samples are introduced into glass tubes that are placed in a gradient magnetic field. The behaviour of particles as a function of time can be observed and the stability of the fluid determined.

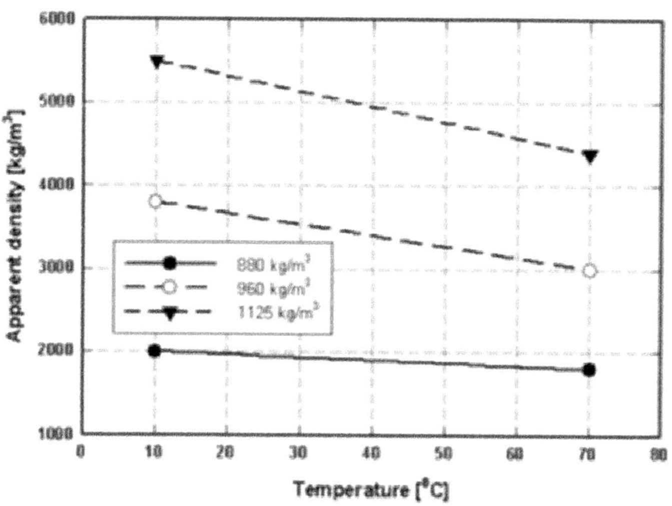

Figure 5.102: The dependence of the apparent density of ferrofluid on temperature, for various physical densities of the ferrofluid (adapted from [G17]).

The effect of temperature on the physical properties of a ferrofluid

A ferrofluid placed in a separation chamber of a ferrohydrostatic separator is, over a period of time, exposed to an increasing temperature as a result of the temperature increase in the windings of the electromagnet. The temperature increase results in an increase in the volume, and a decrease in the physical and apparent densities of the ferrofluid, at a constant concentration of magnetite. The change of the apparent density of the kerosene-based ferrofluid with temperature variation is illustrated in Fig. 5.102. It can be seen that when the temperature increased from 20 ^{0}C to 50 ^{0}C, the apparent density of 3800 kg/m^3 of a ferrofluid with a physical density of 960 kg/m^3 decreased by 14%.

In addition to density, magnetic susceptibility also undergoes changes when the temperature of the fluid changes. It was observed [G17] that a temperature increase by 15 ^{0}C results in the reduction of the magnetic susceptibility of the kerosene-based ferrofluid by 7% to 15%, depending upon the magnetic field strength to which the ferrofluid is exposed.

It is clear that temperature control of the ferrofluid and of the working environment in which a ferrohydrostatic separator operates, are essential for accurate performance of the equipment. Continuous detection of density variations and automated adjustment of the operating parameters of the separator are essential components of modern ferrohydrostatic separators.

Table 5.33: Typical physical properties of ferrofluids, at 300 K.

Property	Hydrocarbon	Water	Ester
Saturation polarization [T]	0.01-0.1	0.01-0.04	0.01-0.08
Density [kg/m^3]	850-1250	1100-1400	1150-1400
Viscosity [Pa$\times$s]	0.003-0.01	0.007-0.01	0.014-0.04
Boiling point [K]	350	299	422
Surface tension [10^{-3}N/m]	25-28	26	26
Thermal conductivity [W/m/K]	0.15	0.59	
Electrical resistivity [Ωm]	1.5×10^7	50-1000	

Review of physical properties of ferrofluids

The chemical, mechanical and physical properties of ferrofluids correspond closely to those of the carrier liquid. All commercially available ferrofluids utilizing organic carrier liquids are essentially electrically non-conductive. Their magnetic saturation can be proportionately varied by dilution with the carrier fluid. Most ferrofluids can be freeze-thawed without damage. Table 5.33 lists various physical properties of commercially available ferrofluids.

Selection of a magnetic fluid for material separation

When exposed to a non-homogeneous external magnetic field, a paramagnetic liquid will exhibit an apparent density given by eq. (3.144): $\rho_a = \rho_f + \frac{\kappa}{\mu_0 g} B\nabla B$. In order to achieve the apparent density equal, for instance, to the density of diamond, that is 3500 kg/m^3, in a saturated solution of manganese chloride (ρ_f = 1450 kg/m^3, $\chi = 5.65\times10^{-7}$ m^3/kg and $\kappa = 8.2\times10^{-4}$ (SI)), the product of the magnetic induction and its gradient $B\nabla B$ must be equal to 30.8 T^2/m. In a magnet that can generate 2 T, the field gradient must be 15.4 T/m or 1.54×10^3 G/cm. This is a very high value indeed. In a superconducting magnet, at the field induction of 20 T, the required field gradient is 1.54 T/m, or 154 G/cm.

The application of paramagnetic liquids is more feasible on a small laboratory scale [P14] using, for example, a Frantz laboratory isodynamic separator. In this separator, the product $B\nabla B$ can reach maximum value of 100 T^2/m and an apparent density of 10 000 kg/m^3 was achieved with maximum throughput of the separator at 40 g/h and a maximum particle size of 1 mm [P14].

It is clear, therefore, that for separation in paramagnetic liquids very high magnetic fields are required. This requirement results in high capital and running costs, and in the need to remove even very feebly magnetic material from the feed. In addition, the working volume of such a separator would necessarily be very small, which would limit the upper particle size, and its throughput would be very low. Even in a massive and expensive installation, the throughput would not exceed a few kilograms per hour.

These practical difficulties make separation in paramagnetic liquids (usually termed magnetohydrostatic separation - MHS) an intriguing curiosity rather

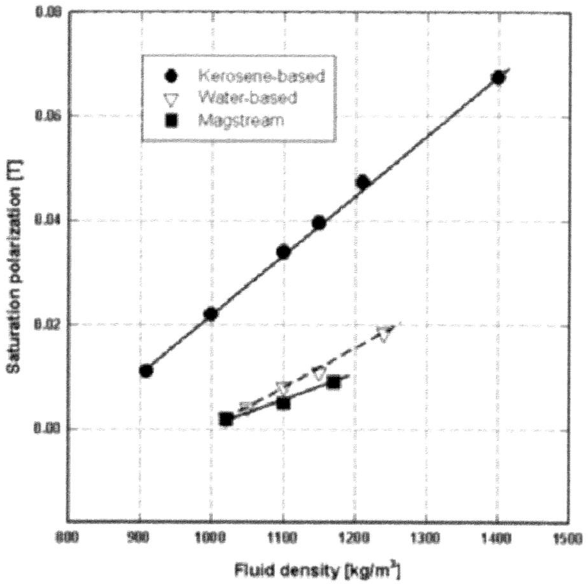

Figure 5.103: Saturation polarization of several types of ferrofluids, as a function of their physical density [S80].

than a viable technique. It was realized a long time ago [L17] and not without disappointment, that separation in paramagnetic liquids would never become a widely used processing technique. Only under special circumstances would production of a high-value concentrate from a small volume of coarse feed be economically justified.

In contrast to paramagnetic liquids, ferrofluids are moderately strongly magnetic and modest magnetic fields and gradients are required to generate apparent densities over a wide range of values. Figure 5.103 reviews the saturation magnetization of various ferrofluids used for material separation, as a function of their physical density.

When exposed to a non-homogeneous magnetic field, ferrofluids exhibit an apparent density that is given by eq. (3.145). For example, the apparent density of 3500 kg/m^3 can be obtained with a kerosene-based ferrofluid of density 1000 kg/m^3, having magnetic polarization of 0.02 T (200 G), at a field gradient of 1.54 T/m (154 G/cm). Such a gradient can be fairly easily generated by laboratory or pilot-scale FHS units at a magnetic field as low as 0.2 T. Table 5.34 compares conditions required to achieve an apparent density of 3500 kg/m^3 in paramagnetic liquids and ferrofluids

Table 5.34 illustrates the considerable advantage of ferrofluids over paramagnetic liquids, while Fig. 5.104 compares the operating density ranges of various sink-and-float techniques. The potential of ferrohydrostatic separation

Table 5.34: Typical parameters required to achieve an apparent density of 3500 kg/m3 in laboratory or pilot-scale separators with a paramagnetic liquid and a ferrofluid. The costs of liquids are for the same value of magnetization, i.e. 0.0016 T (16 G).

Parameter	**$MnCl_2 \cdot 4H_2O$**	**Kerosene FF**
Polarization @ 2 T	16×10^{-4} T	
Saturation polarization		200×10^{-4} T
Density [kg/m^3]	1400	1000
Apparent density [kg/m^3]	3500	3500
Viscosity @ B = 0	1.5×10^{-3} Pa×s	2.5×10^{-3} Pa×s
Magnetic induction	2 T or 20 T	0.2 T
Field gradient	15.4 T/m @ 2 T 1.54 T/m @ 20 T	1.54 T/m
Cost (US$/$\ell$)	10	0.7 (commercial) 0.07 (in-house)

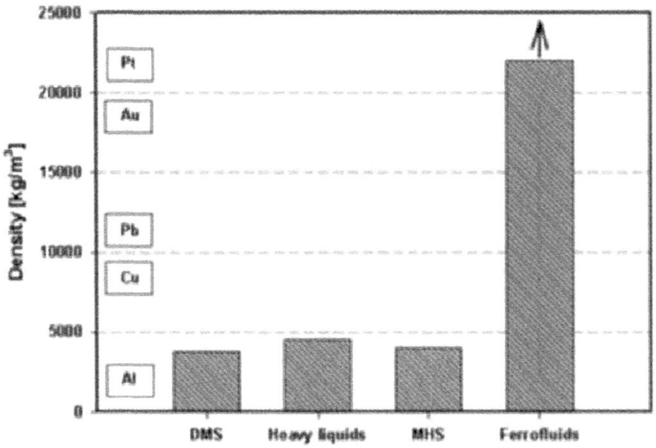

Figure 5.104: The operating density ranges of various sink-and-float separation methods.

Table 5.35: Comparison of MHS and FHS separation techniques.

MHS	FHS
High magnetic field (2 T or higher)	Modest magnetic field (0.2 - 0.5 T)
High field gradient (10 - 20 T/m)	Modest field gradient (1 - 2 T/m)
Maximum density: 10 000 kg/m^3	Maximum density > 22 000 kg/m^3
High-intensity magnetic scalping	Low-intensity magnetic scalping
Low throughput (max. 1 kg/h)	Easy scale-up to 1 t/h
Narrow size fractions	Limited sensitivity to particle size
Max. particle size approx. 1 mm	Max. particle size > 20 mm
High cost of liquid	Moderate cost of ferrofluid
High capital and running costs	Moderate capital and running costs
Large mass and floor space	Modest mass and floor space
Difficult handling of liquids	High stability of ferrofluids
Difficult re-concentration of liquids	Recovery and recycling of FF simple

in ferrofluids is obvious. The main advantages of ferrohydrostatic separation compared to magnetohydrostatic separation in paramagnetic liquids are summarized in Table 5.35.

Water-based or kerosene-based ferrofluid? For separation purposes, water-based and kerosene-based ferrofluids are commercially available. Generally speaking, ferrofluids with hydrocarbon carrier fluids tend to have a high degree of monodispersity and a minimum level of aggregation. A wide range of dispersants is available for such ferrofluids, the cost of which can be very modest as a result of a range of iron-containing low-cost waste materials available from the petrochemical industry. Given this wide range of available materials, it is possible to make relatively low-cost kerosene-based ferrofluids having a wide range of saturation magnetization and viscosity. These ferrofluids can be fairly easily removed from the products of separation, recycled and re-concentrated, and returned to the separation circuit.

As can be seen in Fig. 5.103, kerosene-based ferrofluids with saturation polarization as high as 0.05 T (500 G) can be produced. In a typical range of apparent densities from 3000 kg/m^3 to 9000 kg/m^3, a fluid with a saturation polarization of the order of 0.0150 T to 0.02 T (150 G to 200 G) is required. Such a fluid can be obtained by diluting the concentrated ferrofluid, which further reduces the viscosity and the cost of the process. Such a dilution must, however, be carried out with a certain level of caution. Excessive dilution could result in desorption of the protective surfactant from the surface of the particles in order to maintain the equilibrium between adsorbed and free surfactant. This could considerably reduce the colloidal stability of the fluid and change its magnetic and physical properties [C24].

Waste water resulting from the production of kerosene-based ferrofluid consists mainly of organic waste such as kerosene and oleic acid. Disposal of such

Table 5.36: Selected physical and chemical properties of kerosene.

Property	Value
Boiling point	200 ^{0}C to 260 ^{0}C
Freezing point	- 45.6 ^{0}C
Density	800 kg/m^3 (at 15 ^{0}C)
Reactivity with water	no reaction
Reactivity with common materials	no reaction
Flammable limits in air	0.7% to 5%
Ignition temperature	229 ^{0}C
Fire	combustible
Exposure	irritating to skin and eyes
Protective equipment	gloves, goggles

a waste either by incineration or by applying an aerobic reactor system, using microorganisms obtained from activated sludge, can be easily achieved [V7]. Some physical and chemical properties of kerosene are shown in Table 5.36.

The alternative separation medium is a water-based ferrofluid. In water-based colloids, the highly polar nature of the carrier renders dispersants less effective than in hydrocarbon fluids. Long-term stability and stability in high-gradient fields is, therefore, limited Furthermore, the inclusion of a higher concentration of dispersant, possibly of a more complex nature, tends to increase the viscosity of the water-based ferrofluid compared with a hydrocarbon-based material of the same concentration.

As illustrated in Fig. 5.103, because of the complexity of water-based fluids, it is somewhat difficult to obtain a wide range of magnetization, and the saturation polarization of 0.02 T (200 G) is typical for water-based ferrofluid used in material separation. Such a concentrated fluid must be used undiluted, in contrast to kerosene-based fluid. This results in an increased viscosity and cost of water-based ferrofluid. In addition, because of greater volatility of water, compared to hydrocarbons, the values of magnetization and other parameters are susceptible to greater variations with time and temperature. Recycling and re-concentration of water-based ferrofluids, for instance by evaporation, although seemingly simple, are rather costly, and time and space consuming.

The image of kerosene-based ferrofluids as hazardous and environmentally unfriendly, resulted, in some instances, in exclusion of the FHS technique from production-scale application. Nevertheless, certain obstacles that such fluids present can be fairly easily contained within a beneficiation circuit. Moreover, water-based ferrofluid, as a result both of the increased level of dispersants and their complexity, may not be as non-polluting as imagined from a simplistic point of view. However, the entire material separation circuit can be made simpler when water-based ferrofluids are used [F25].

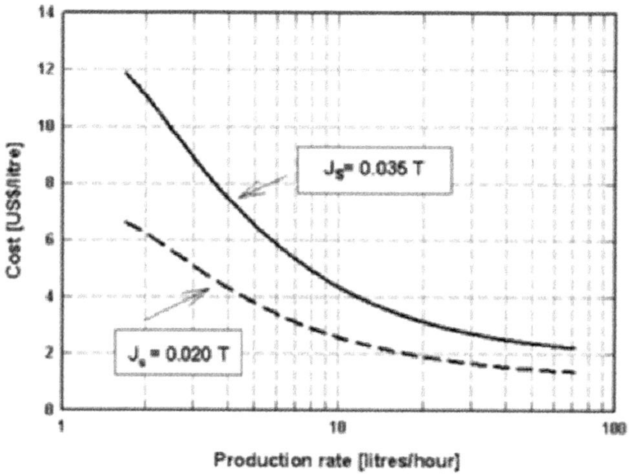

Figure 5.105: The 1983 cost of kerosene-based ferrofluid as a function of the production rate. Spent hydrochloric acid pickle liquor was considered as the source of iron ions (adapted from Farkas [F26]).

The cost of ferrofluid

If the ferrohydrostatic separation technology is to be successful on a production scale, ferrofluids must be available in sufficiently large quantities and at reasonable cost.

The current situation in the market is not conducive to the introduction of FHS on large scale. There are several small suppliers who provide speciality ferrofluids in small quantities and at high cost. Technical-grade kerosene-based ferrofluids for ferrohydrostatic separation are manufactured in larger quantities by one or two suppliers in Russia. The 2000 price for a 0.04 T (400 G) ferrofluid was US$40 or more per litre. Ferrotec (Japan), formerly Ferrofluidics Corp., supplies, on a larger scale, water-based ferrofluids for separation applications. The 2002 price for 0.017 T (170 G), 1230 kg/m^3 ferrofluid was US$45 per litre.

Although these prices are acceptable for laboratory and possibly for pilot-scale applications, a large industrial application of FHS requires reliable supply of low-cost ferrofluid of consistent quality. Union Carbide Corp., in the early 1970s, developed a pilot plant for kerosene-based ferrofluid for their separation application [F26]. The capacity of the plant was 4.3 ℓ/h of 0.035 T ferrofluid.

Based on manufacturing experience, sufficient information was obtained which permitted calculation of the cost of ferrofluid in an automated plant of expanded capacity. The 1983 cost data shown in Fig. 5.105 were based on the cost of materials, labour and supervision, and utilities [F26]. The diagram illustrates that, as the production rate increases, the unit labour cost diminishes and the production cost per litre, for large volumes, approaches the costs of the chemicals. At the production rate of 82 ℓ/h of 0.035 T fluid, which can be diluted to 143 ℓ/h of 0.02 T fluid, the 1983 cost of the 0.02 T ferrofluid would be less than US$1.00/$\ell$.

A batch pilot-plant with capacity of 6 ℓ/h was constructed in 1998 at De Beers Consolidated Mines (Pty.) Ltd. (South Africa) to evaluate the process scale-up for the manufacture of kerosene-based ferrofluid. The raw material cost for 1 litre of 0.04 T ferrofluid was, in 2002, US$1.25 [K23]. The manufacturing cost of this fluid, at the production rate of 6 ℓ/h, was approximately US$2.50 for the 0.04 T fluid or US$1.35 per litre for the 0.02 T fluid. Further cost reduction was obtained in the automated continuous plant with a capacity of 10 ℓ/h, as a result of cheaper chemicals and lower labour costs [V7].

5.6.2 Ferrohydrostatic separation

As has been discussed in Sections 3.8 and 4.8, the efficiency and selectivity of separation in a ferrofluid depends upon the pattern of the magnetic field in the working gap of a magnet. This pattern is determined by the shape of the pole tips. If the gradient of the magnetic field is not constant along the vertical axis of the system (see Figs. 4.51 and 4.54), the apparent density of the fluid along the vertical will also vary and several density fractions can be obtained, as is shown in Fig. 4.53. Such an approach to material separation is acceptable only if the density differential between the components of a mixture is large and the required accuracy of separation is limited.

For accurate separation into two well-defined density fractions the field gradient must be constant along the vertical. Such a pattern of the field gradient can be achieved using hyperbolic shapes for the pole tips, as shown in Section 4.8.2.

The fundamental relationship that defines the required field gradient is given by eq. (2.9). For the required range of apparent densities, and for the range of saturation polarizations of available ferrofluids, the field gradient can be determined from eq. (2.9). Once the gradient of the magnetic field has been determined, the shape of the pole tips can be designed using the procedure outlined in Section 4.8.3.

A typical relationship between the apparent density, magnetic field strength (or the electrical current in the coil) and the physical density of the ferrofluid is shown in Fig. 5.106. It can be seen that, in this specific case, the apparent density ranges from 2500 kg/m^3 to 5500 kg/m^3, depending on the density (and magnetization) of the ferrofluid. The lower limit of the apparent density that can be obtained in FHS, for a given shape of the pole tips, is determined by the magnetization of the ferrofluid that can still be held in the separation gap of the magnet. The upper limit of the apparent density is given either by the maximum magnetization of the ferrofluid and/or by the acceptable viscosity of the ferrofluid The effects of the ferrofluid polarization and of the field gradient on the apparent density, are shown in Figs. 5.107, and 5.108, respectively.

It can be seen, therefore, that for a given shape of the pole tips only a relatively narrow range of apparent densities can be generated. In order to

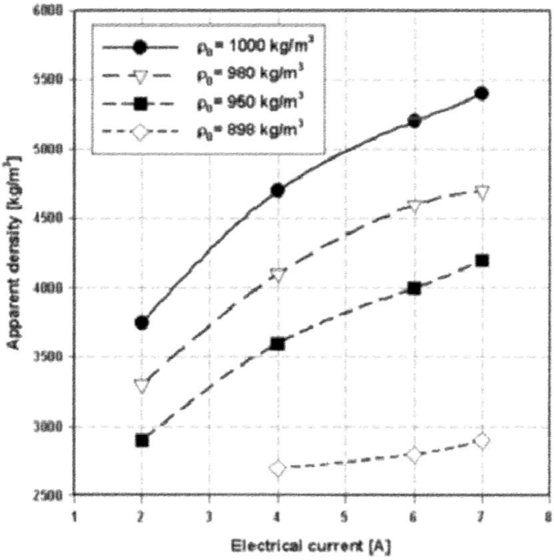

Figure 5.106: The apparent density of the ferrofluid as a function of the electrical current, for various physical densities of the ferrofluid and for a particular shape of the pole tips.

achieve higher (or lower) apparent densities, the shape of the pole tips has to be re-designed, as described in Section 4.8.3. To obtain low apparent densities, at which the ferrofluid is too weakly magnetic to remain in the working gap of the separator, special measures must be introduced to restrain the fluid in the separation space. In principle, it is essentially impossible to design a separator operating efficiently over the entire range of apparent densities [G4].

Optimization of the pole-tip profile

A simplified approach to the pole-tip design required that the vertical component of the field gradient be constant (eq. (4.60)). As has been discussed earlier, most ferrofluids are not likely to be saturated at the magnetic field usually used in ferrohydrostatic separators, (as shown in Fig. 5.98). In order to obtain the shape of a pole tip that would ensure truly constant apparent density of the ferrofluid along the vertical, it follows from eq. (2.9) that the product $J_f \nabla B$ must be constant, at the working magnetic induction of the ferrohydrostatic separator:

$$J_f \nabla B = const. \qquad (5.28)$$

It is obvious that the shape of the pole tips thus obtained will ensure a constant apparent density only for a ferrofluid of a specific density with known details of its magnetization curve.

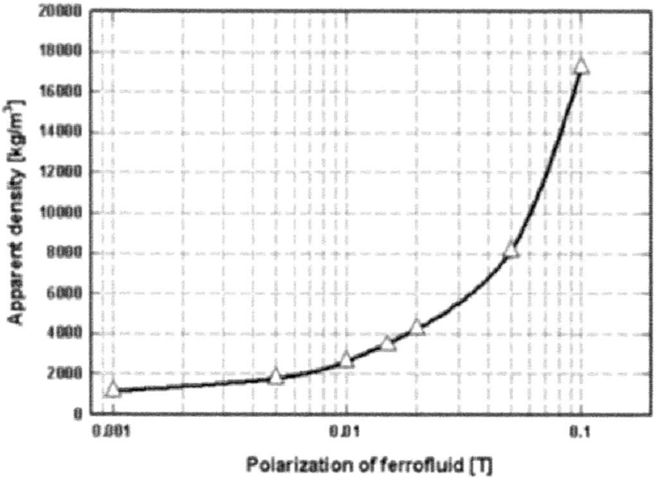

Figure 5.107: The effect of polarization of the ferrofluid ($\rho_0 = 1000$ kg/m^3) on its apparent density when exposed to the field gradient of 2 T/m.

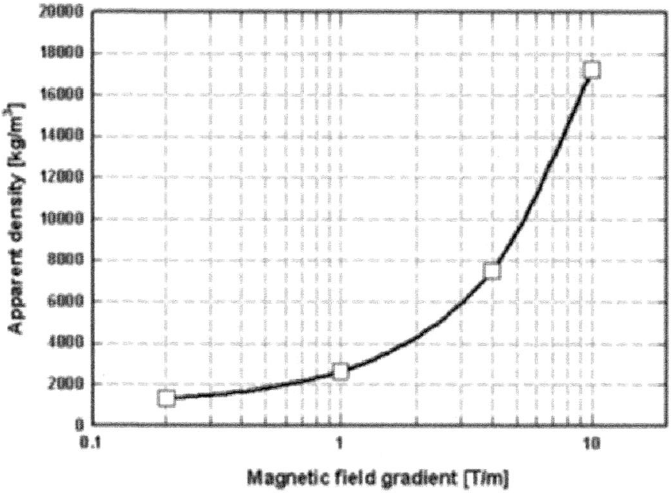

Figure 5.108: The effect of magnetic field gradient on the apparent density of ferrofluid ($J = 0.02$ T, $\rho_0 = 1000$ kg/m^3).

A well optimized shape of pole tips will generate an apparent density that will be constant along a limited region along the vertical, at the centre of the separation gap, as is shown in Fig. 4.57.

Feed preparation

Magnetic scalping In order to prevent blockage of the separation chamber, strongly magnetic components must be removed from the feed by magnetic scalping using, for instance, a permanent magnet roll. Selection of the type of the magnet in the roll and of the operating conditions, is determined mainly by the magnetic field strength generated by the ferrohydrostatic separator. FHS units operating at high densities, or high-throughput separators, in which the gap between the pole pieces is large, usually generate a magnetic field of the order of 0.5 T or higher and an RE magnet roll should be used. Smaller units, with throughputs up to 100 kg/h, usually use a magnetic field as low as 0.2 T to 0.3 T, and scalping using a strontium ferrite roll is sufficient. Material with a magnetic susceptibility as high as 6×10^{-6} m^3/kg (500×10^{-6} cm^3/g) can be fed into such a FHS machine without any problems.

As has been discussed in Section 3.8.4, materials with non-zero magnetic susceptibilities, immersed in a ferrofluid, exhibit an effective density that is greater than their physical density. This density difference, given by eq. (3.159), increases with increasing magnetic susceptibility and external magnetic field, as shown in Fig. 3.59. In those applications where accurate density separation of materials, which differ in their magnetic properties, is required, accurate magnetic scalping should be carried out. On the other hand, density differential as a result of differences in magnetic susceptibilities can be advantageously used to separate materials with similar physical densities. Using magnetic tracers, the cut-point magnetic susceptibilities can be set quite accurately. Cut-point magnetic susceptibilities down to 10×10^{-7} m^3/kg can be set with strontium ferrite rolls, while an NdFeB roll can remove materials with susceptibilities as low as 2.5×10^{-7} m^3/kg.

The effect of moisture The presence of moisture has an effect on the efficiency of separation and the degree of this influence depends on the type of ferrofluid used, and on the size, wettability and porosity of particles to be separated. When a kerosene-based ferrofluid is used, the presence of moisture reduces the effective density of the material, which is reflected in a decreasing amount of material reporting into the sink fraction, as can be seen in Fig. 5.109. For highly selective separation of relative fine materials with close densities, it is important, therefore, to feed the material into the FHS separator in a dry form. However, a mixture of coarse materials, particularly with a large density differential, can be treated wet.

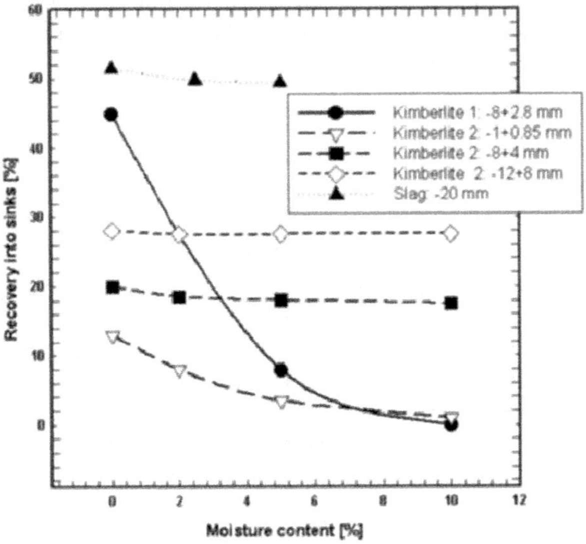

Figure 5.109: The effect of moisture on the FHS recovery of materials into the sink fraction. Apparent densities for kimberlite, $\rho_a = 3500\ \mathrm{kg/m^3}$, and for ferrochrome slag, $\rho_a = 5700\ \mathrm{kg/m^3}$ [S8, D16].

Separation chambers

Numerous designs of separation chambers have been used over the years and their description can be found in a review article by Fujita [F8]. More recent ferrohydrostatic separators usually use an open-ended stainless steel trough positioned between the hyperbolic pole tips of the magnet, as is shown in Figs. 2.103 and 4.56. Such a separation chamber shown in Fig. 5.110, is usually inclined at a small angle to facilitate movement of particles through the fluid. The chamber can also be vibrated to assist in propelling particles into usually two discharge outlets, where the float and sink fractions are collected. A splitter is often installed to separate the sink and float fractions, and its judicious position is important for accuracy of separation.

The most frequently employed separation chambers use a solid bottom, as shown in Fig. 5.111, while the front end of the chamber is opened to allow continuous removal of the separation products. Although the bottomless chamber, shown also in Fig. 5.111, allows simple removal of the sink fraction, the disadvantage is often a considerable sagging of the ferrofluid outside the working space of the separator. Gubarevich [G4, G17] showed that, in equilibrium, the balance of forces acting on the ferrofluid placed in a non-homogeneous magnetic field is expressed by

$$\rho_f g\ell = \frac{1}{\mu_0}\int_{B(y_1)}^{B(y_2)} J dB \tag{5.29}$$

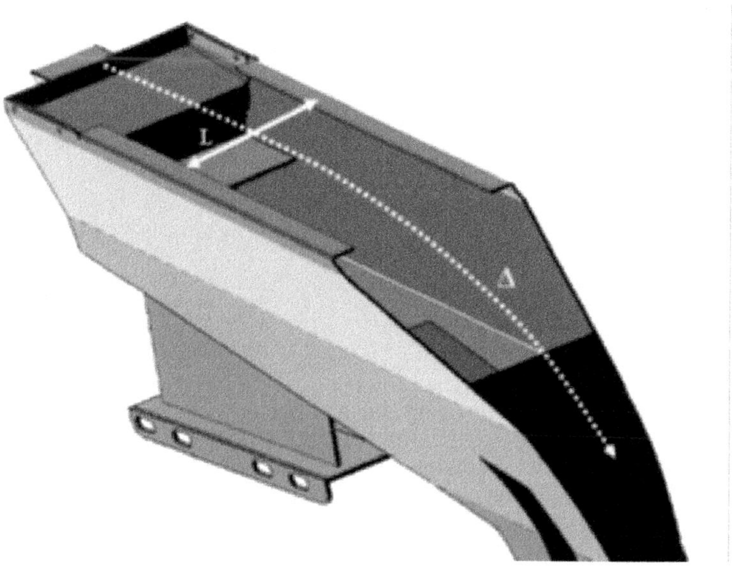

Figure 5.110: Separation chamber of a ferrohydrostatic separator.

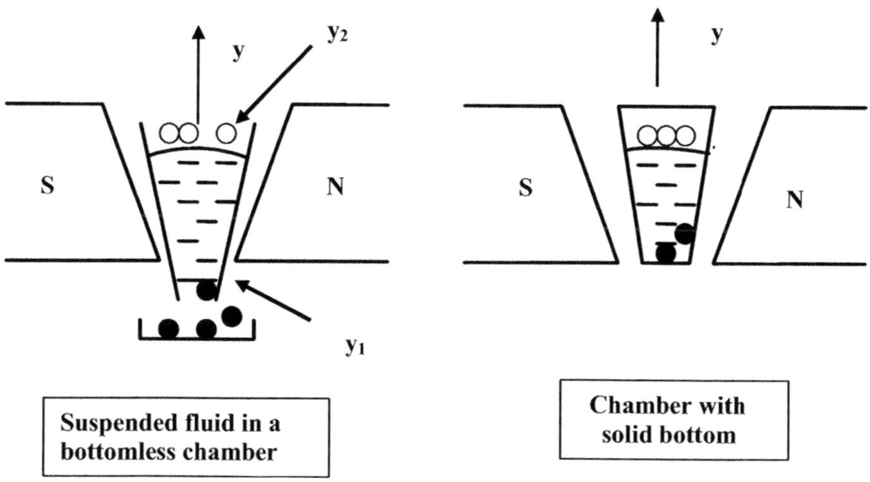

Figure 5.111: Designs of separation chambers in FHS.

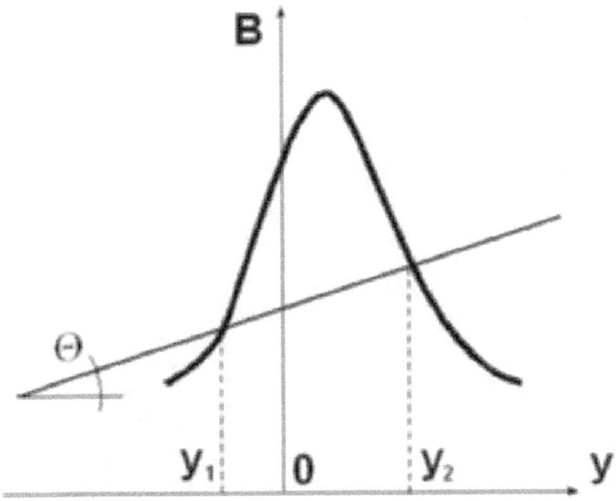

Figure 5.112: Graphical determination of the equilibrium length of the ferrofluid column. The slope of the straight line is given by: $tg\Theta = \mu_0 \rho_f g / J_s$(after [G4]).

where ℓ is the length of the column of the ferrofluid, ρ_f and J its density and polarization, respectively. $B(y_i)$ is the magnetic induction at position y_i ($i = 1$, 2) as shown in Fig. 5.111. While an analytical solution of the above equation is difficult, it is fairly easy to develop a graphical technique to solve the equation and to determine the length of the ferrofluid column that can be held within the magnetic field of the separator.

Figure 5.112 depicts a pattern of the magnetic induction along the vertical axis y with the origin at the lower edge of the pole pieces. If we assume that the polarization is independent of the magnetic induction and thus $J = J_s$, then eq. (5.29) can be re-written

$$\frac{\Delta B}{\ell} = \frac{\mu_0 \rho_f g}{J_s} \tag{5.30}$$

where $\Delta B = B(y_2) - B(y_1)$.

The tangent of the straight line in Fig. 5.112 is thus given by the right-hand side of eq. (5.30) and intersections of the straight line $B = y\mu_0 \rho_f g / J_s$ with curve $B = B(y)$ determine the coordinates y_1 and y_2 of the end points of the ferrofluid column. By shifting this line along the y axis it is then possible to determine the sagging of the ferrofluid, for a given length of the ferrofluid column.

It is clear that only a limited length of the ferrofluid column can be retained in a bottomless chamber before the fluid starts running out of the chamber. Several methods have been proposed to address this problem. A layer of a

Figure 5.113: Retention of a ferrofluid in a separation chamber by a more strongly magnetic ferrofluid.

strongly magnetic ferrofluid based on a carrier liquid (e.g. kerosene) that is immiscible with the working fluid (water) can be introduced into the separation chamber [G4]. As can be seen in Fig. 5.113, such a strongly magnetic fluid will form a "liquid bottom" that will support a sufficiently high column of the working fluid. Discharge of the sink fraction through the supporting layer is facilitated by the fact that it is positioned in a zone with an inverted magnetic field gradient, below the point at which $\nabla B = 0$, as can be seen in Fig. 4.57.

An alternative technique to prevent sagging of the ferrofluid is the application of a water column as shown in Fig. 5.114. This technique is particularly suitable for the separation of materials of low density where a relatively feebly magnetic ferrofluid is placed in a low-gradient magnetic field. The height ℓ_w of the hydraulic support column, for a given ferrohydrostatic separator, can be determined from a relationship

$$\ell_w = (\frac{1}{\mu_0} \int_{B_1}^{B_2} JdB + \rho_f g \ell_f)/\rho_w g \tag{5.31}$$

where ρ_w is the density of the supporting liquid.

The mode of feeding

It is well established that the way in which the particles are fed into the ferrofluid affects significantly the results of separation. Particles can be dropped into

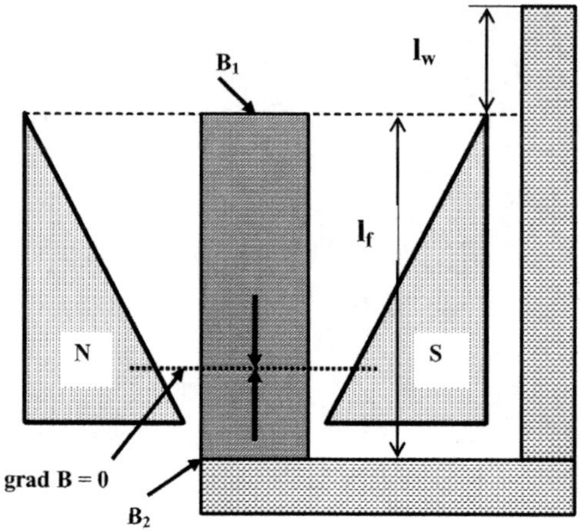

Figure 5.114: Schematic diagram of a hydraulic support of a ferrofluid column in FHS (adapted from [G17]).

the ferrofluid directly from a certain height or introduced indirectly into the ferrofluid at essentially zero initial velocity from a ramp that determines the level of the ferrofluid in the chamber. In the direct method, shown in Fig. 5.115, the selectivity of separation is strongly dependent on the drop height, particle size and on the depth of the ferrofluid pool, particularly for materials with small density differentials. Introduction of particles into the ferrofluid at non-zero velocity, on the other hand, improves considerably the wetting of the particles with ferrofluid, and partially eliminates the effect of surface tension, which is of importance in the separation of fine particles or particles of large aspect ratio.

In the indirect mode, the effect of particle size is less pronounced, particularly when the depth of the ferrofluid pool is sufficient. However, the adverse effect of surface tension of the ferrofluid does not allow selective separation of fine particles, particularly with a small density differential. The efficiency of separation, when the direct and indirect modes of feeding material onto the upper level of the ferrofluid are employed, is impaired by considerable interparticle interactions within the pool of the ferrofluid, particularly when high feedrates are used.

A comparison of trajectories of particles that were introduced directly into the ferrofluid from a height of 10 mm and from essentially zero height (2 mm) is given in Fig. 5.116. It can be seen that for a small drop height (2 mm), a particle with density of 3500 kg/m^3 will hover within the fluid of apparent density of 3500 kg/m^3, while a particle with density of 3400 kg/m^3 dropped into the fluid from 10 mm will sink.

Disadvantages of feeding material onto the top of the ferrofluid can be reduced by introducing the feed horizontally and directly into the centre of the pool of the fluid. Such an approach was employed in the early AVCO, NASA

Figure 5.115: Direct mode of feeding material into ferrofluid. Drop height: 40 mm [K23].

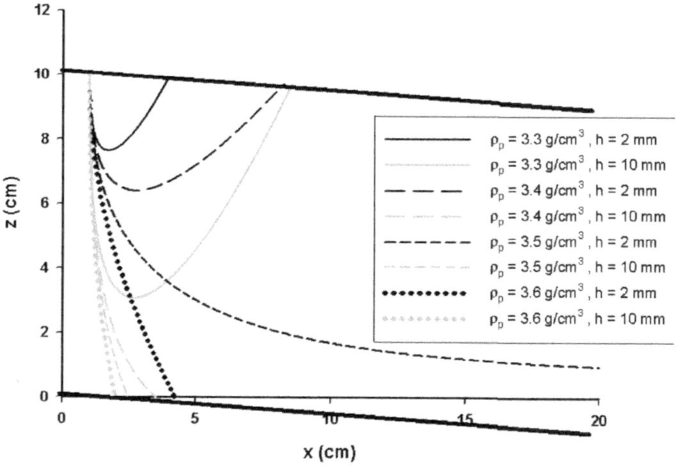

Figure 5.116: Comparison of trajectories of particles introduced in FHS from different heights. Apparent density: 3500 kg/m^3.

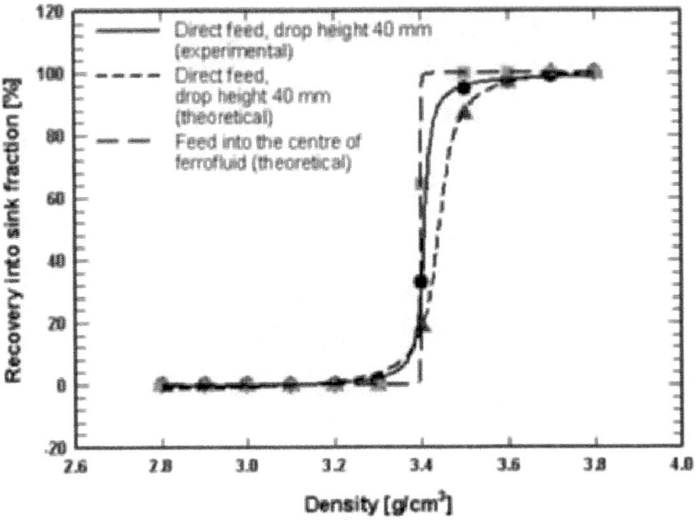

Figure 5.117: The experimental and theoretical partition curves for two modes of feeding: directly onto the top of the fluid and directly into the centre of the fluid [M14].

and US Bureau of Mines FHS machines [F8]. Particle flow modelling of such a feeding mode has shown that in addition to the control of the initial velocity of particles entering the fluid, interactions among particles can be reduced to a minimum [M14]. In this mode, float and sink particles move in opposite directions, their interactions, leading to misplacement, are reduced and selectivity of separation is improved, as illustrated in Fig. 5.117.

The selectivity of FHS

As has been discussed in Section 4.8.2, conditions of constant field gradient along the vertical axis and zero gradient in transverse directions, as expressed by eqs. (4.61), can be met only in the plane of symmetry passing through the centre of the interpolar gap. In a real situation the field gradient along the horizontal axis (x-axis, as denoted in Fig. 4.55) is always different from zero. The horizontal component of the magnetic field increases from the centre of the gap towards the pole tips and as a result of the field gradient thus created the surface of the ferrofluid forms a meniscus. The particles are pushed towards the centre of the gap, where the gradient and thus the apparent density, are lower. While, in principle, this phenomenon improves the accuracy of separation, it increases the crowding of the particles and the rate of interparticle interactions, it reduces the effective volume of the separation chamber in which separation takes place.

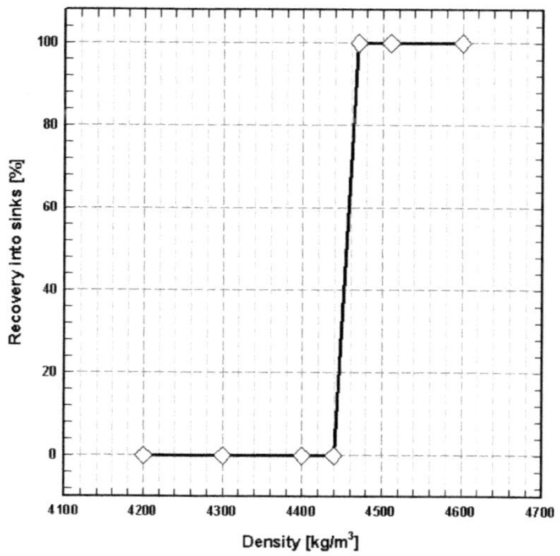

Figure 5.118: The experimental partition curve of Rhomag® ferrohydrostatic separator (De Beers) with hyperbolic pole tips, determined with 2 mm density tracers. Cut point density: 4450 kg/m^3 [S8].

It is evident, therefore, from the foregoing discussion that the upper limit of selectivity of separation can be achieved only when particles are fed into the centre of the interpolar gap at a very low feed rate. Figure 5.118 illustrates the partition curve of an FHS separator equipped with hyperbolic pole tips. The curve was obtained using 2 mm density tracers and kerosene-based ferrofluid. It can be seen that the separator could "see" a density difference as small as 30 kg/m^3 (0.03 g/cm^3), which corresponds to the probable error of separation Ep, defined by eq. (5.32), as low as 0.007.

$$Ep = \frac{\rho_{75} - \rho_{25}}{2} \tag{5.32}$$

where ρ_{75} and ρ_{25} are the densities of particles, 75% and 25% of which report into the sink fraction.

The effect of fluid level

From the analysis of particle trajectories in a ferrofluid outlined in Section 3.8, the accuracy of separation depends on the depth of the fluid. The degree of influence of the fluid depth depends on the mode of feeding, drop height, design and length of the separation chamber, position of the splitter, particle size, and density differential between the materials to be separated. Typical dependence of the efficiency of separation on the depth of the ferrofluid pool is shown in Fig. 5.119. It can be seen that the accuracy of separation of tracers of density

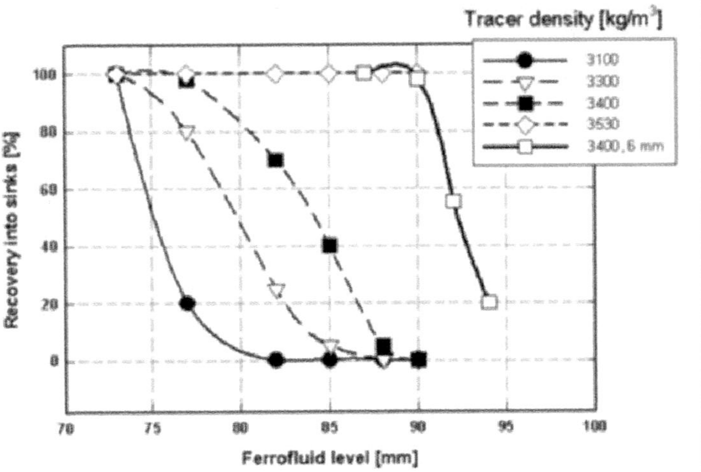

Figure 5.119: The efficiency of separation of 2 mm density tracers in a ferrofluid with cut-point density of 3450 kg/m^3, as a function of the fluid level. Comparison with 6 mm tracers of density 3400 kg/m^3 is also shown [D16, S8].

higher than the cut-point density is not affected by the fluid level. However, misplacement of the floating tracers increases with decreasing fluid level and this misplacement becomes more pronounced for near-density tracers.

Such a behaviour of particles in the ferrofluid can be explained by realizing that the gradient of the magnetic field, and thus the apparent density of the ferrofluid are constant only in a limited region along the vertical as shown in Fig. 4.57. In a region closer to the bottom section of the pole tips the field gradient rapidly decreases and so does the apparent density. Float particles that happen to reach this low-density region, as a consequence of their parabolic trajectories, will report into the sink fraction. This misplacement becomes more pronounced with increasing size of the particles, as shown in Fig. 5.119.

The effect of particle size

It is a well-confirmed fact that particle size plays an important role in any type separation, and ferrohydrostatic separation is no exception. As has been discussed in Section 3.8.5, the motion of particles in a ferrofluid is determined by the interplay of assorted forces acting on the particles, and the balance of these forces determines the trajectory of a particle in the ferrofluid and ultimately the fraction into which the particle will report. The force of gravity, and the magnetically and gravitationally-derived buoyancy forces, depend on the cube of the particle size, while the hydrodynamic drag depends only on the first power of the diameter. It is clear, therefore, that the separation efficiency will depend, to some extent, on particle size.

It has been shown in Section 3.8.3, that for particles smaller than 1 mm, the effective cut-point density is dependent on particle size, and the selectivity of separation is, therefore, impaired. For particles smaller than 0.1 mm the effect of hydrodynamic drag becomes so pronounced that accurate separation of materials with a narrow density difference is virtually impossible.

For particles greater than 1 mm, the influence of hydrodynamic drag is negligible and separation should be fairly accurate. The performance of a separator, however, becomes affected by design parameters, such as depth of the fluid, length of the separation chamber, position of the splitter and mode of feeding. As in any separation technique, classification into reasonably narrow size fractions improves the efficiency of separation.

The upper limit of particle size that can be treated by FHS is determined by the width of the bottom part of the tapered separation chamber which, in turn, is determined by the width of the air gap of the magnet. Ferrohydrostatic separators designed to treat particles up to 100 mm in diameter have been built [G4].

The effect of feed rate

The processing capability of a ferrohydrostatic separator is of considerable economic importance. The feed rate affects the separation efficiency and the optimum value of the feed rate is in turn a function of particle size and shape, density difference between the materials to be separated, mass split between the sink and float fractions, design of the separation chamber and of the feeding system, and on metallurgical requirements.

In general, as the feed rate through the separator increases, the separation efficiency decreases. Interparticle collisions, particularly among the sinking and floating particles, result in the change of their trajectories and thus in their misplacement. In addition, an increase in the feed rate shortens the residence time of particles in the fluid pool and leads to further misplacement of the particles. These effects are more pronounced for feeds with a narrow size distribution, for particles with a small density, difference and for particles with densities close to the apparent density of the ferrofluid.

The maximum feed rate acceptable from a metallurgical point of view is thus determined by so many variables that no general rule can formulated and only experimental tests can give information on the optimum operation of a ferrohydrostatic separator. Experience with separation of a wide range of materials in a separation chamber, shown in Fig. 5.110, with direct feed onto the top of the fluid, indicates that the approximate maximum feed rate is frequently of the order of 400 kg/h per 100 mm of width of the feeder tray [S8].

A similar feed rate was used in the Hitachi FHS [N4] for an easy separation of aluminium ($\rho = 2700$ kg/m^3) from zinc ($\rho = 7140$ kg/m^3) and copper ($\rho = 8960$ kg/m^3) from metal scrap, in the particle size range - 50 + 6 mm.

The Tohoku University and Nittetsu Mining Co. (Japan) ferrohydrostatic separator for metal scrap recovery was operated at a feed rate of 700 kg/h/100 mm [S50, F8], while a feed rate as high as 1000 kg/h per 100 mm width of the feeder tray was employed in the US Bureau of Mines FHS [K17], in a similar application. The width of the feeder trays and of the separation chambers ranged from 100 to 200 mm in these applications, corresponding to throughputs between 500 kg/h and 2t/h.

Density stability and its control

For the accurate and selective separation of materials with small density differences it is necessary to maintain accurate control over the apparent density. Over a period of operation of FHS, the apparent density and the level of ferrofluid changes, as a result of temperature variations, ferrofluid evaporation and changes in the operational characteristics of an electromagnet. One approach to achieving high stability of the apparent density is to use a PC-based control system that allows an operator to select the desired cut-point density from a computer screen and the system then automatically controls the separator at the set point. The apparent density can be continuously monitored, for example, with the aid of a strain gauge. Any change in the apparent density, sensed by the strain gauge, is used as an input into the closed loop control system that manages the power supply operation so that a constant apparent density of the ferrofluid is maintained.

Such an automated control of the system requires the separator to be accurately calibrated. The calibration procedure is carried out using density tracers. Once the system is calibrated and the apparent density has been determined, the closed loop system automatically adjusts the apparent density to suit the required cut point density.

5.6.3 Ferrofluid recovery and recycling

In order to keep the running costs of ferrohydrostatic separators low, it is essential, for most applications, that the ferrofluid be recovered and recycled. At the same time, from the environmental point of view, it is imperative to remove tracers of ferrofluid from the products of separation. Several methods, usually efficient and cost-effective, have been developed for the recovery and recycling of water-based and kerosene-based ferrofluids.

Recovery of a water-based ferrofluid

A water-based ferrofluid adhering to the products of separation can be easily removed by washing the material with water. Fujita et al. [S50, F8] developed a process for ferrofluid recovery outlined in Fig. 5.120. After the products of separation are washed with water, the diluted suspension is acidified to change the surfactant (sodiumdodecyl-benzene sulphonate), providing the secondary adsorption layer, to free acid. A thickener and a centrifugal separator can then be used to separate flocs from the excess water due to the washing process. The

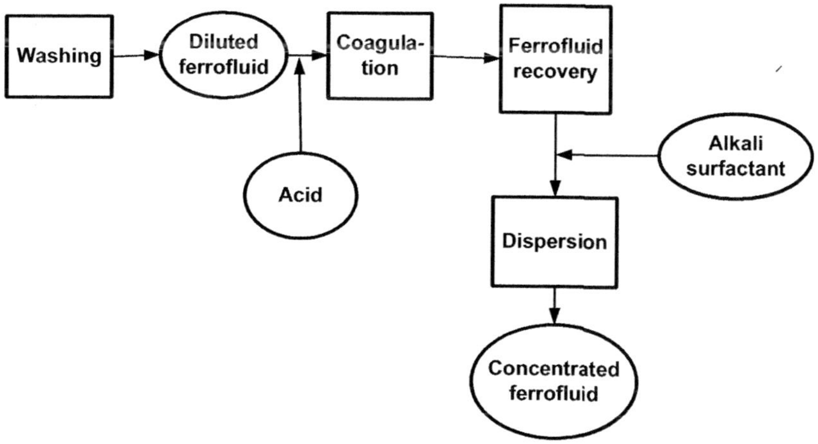

Figure 5.120: Recovery of water-based ferrofluid (adapted from Fujita [F8]).

concentrated flocs thus obtained are re-dispersed as a concentrated ferrofluid by adding alkali and the lost surfactant. The loss of ferrofluid in this process is claimed to be about 0.1% of the feed by weight. Diluted water-based ferrofluid can be re-concentrated also by ultrafiltration, a slow and potentially onerous process.

Recovery of a kerosene-based ferrofluid

A well-established procedure for recovering kerosene-based ferrofluid from the products of separation is to wash particles with water. As this ferrofluid is immiscible with water, the resulting emulsion can be separated either by settling [R9, F27] or in a magnetic field [G17, D16, S81]. The latter approach is very efficient and quick, and numerous designs of the magnetic-field-based ferrofluid recovery systems have been proposed [G19, G20, G21, S81]. A schematic diagram of such a recovery system is shown in Fig. 5.121.

After the water wash, the emulsion is passed over a permanent magnet circuit where the concentrated ferrofluid is separated from the wash water. A chamber situated between the magnet pole pieces allows water to be displaced to the top of the chamber, while the ferrofluid is attracted to the bottom part where the magnetic field is strongest. The rate of drainage of the ferrofluid from the chamber is determined by the ratio between the weight of the ferrofluid and the holding strength of the magnetic system. The efficiency of recovery of the ferrofluid by this method is, for particles bigger than 0.5 mm, usually greater than 99%, even for rather porous materials,

In those applications where the final product(s) of separation have to be very

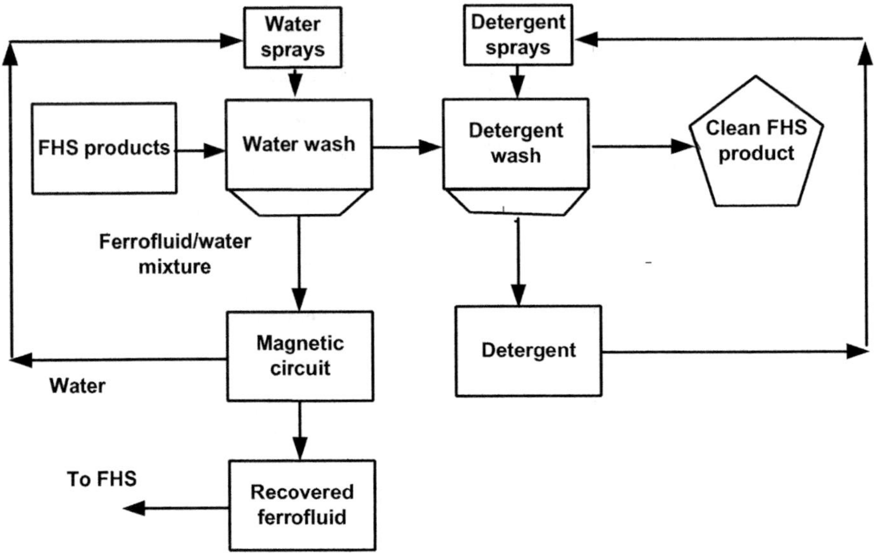

Figure 5.121: Schematic diagram for recovery of kerosene-based ferrofluid.

clean, for instance for visual inspection, or where the value of the final product is to be enhanced, the traces of hydrocarbon and magnetite must be removed from the surface of the grains. This can be accomplished, for instance, by using a biodegradable detergent [K23]. The inclusion of the detergent washing stage in the ferrofluid recycle circuit is shown in Fig. 5.121. This additional step is also essential for the cleaning of small particles, as the water wash alone is not sufficiently efficient for particles smaller that 0.5 mm [F27].

Although the kerosene-wash is more efficient than the water wash, particularly for small particles [F27], its efficiency is still unsatisfactory, mainly for porous materials. It has been observed that, for instance, petrol can recover more ferrofluid, by at least a factor of two, compared to kerosene [S8]. Straight-chain aliphatics including petrol, are much less sterically hindered and have enhanced solvating power. This results in better dissolution of the ferrofluids from the pores and crevices for petrol compared with kerosene.

It is well-established experimentally, that the temperature of water does not affect the efficiency of ferrofluid recovery in the temperature range from 20 ^{0}C to 60 ^{0}C [S8, K25, F27]. Similarly, a high water flow rate was not found to be advantageous in removing ferrofluid from particles [F27]. However, pre-treatment of grains, for example by pre-wetting with water or other fluids, resulted in more efficient removal of ferrofluid from the material and thus in reduced losses of the fluid [G17].

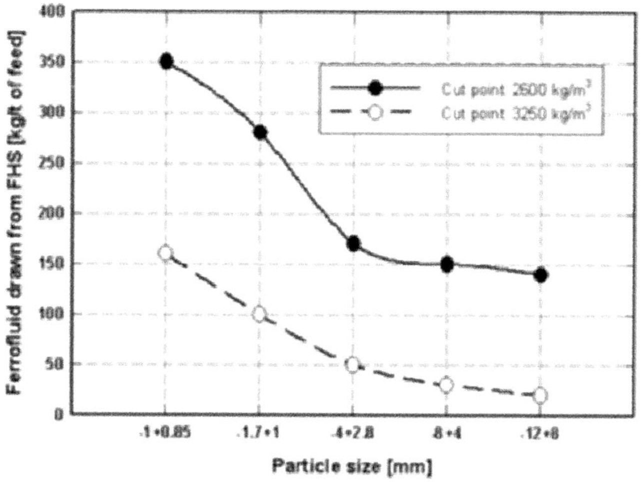

Figure 5.122: Kerosene-based ferrofluid drawn from FHS at different apparent densities (adapted from [D16]).

Losses of ferrofluid

Although the methods of ferrofluid recovery, which have been discussed, are very efficient, some ferrofluid is always lost during the separation and recovery processes. This loss, and the efficiency of the ferrofluid recovery, depend mainly on particle size, moisture content, surface preparation and porosity of particles. Figure 5.122 shows that the amount of ferrofluid drawn by the products of separation from the FHS separator depends mainly on the particles size and on the operating apparent density of the ferrofluid. It can be seen that the amount of ferrofluid drawn decreases with increasing apparent density, for a ferrofluid of constant magnetization. At a higher apparent density, and thus higher magnetic field, more fluid is held back by the magnetic field as the products of separation emerge from the separator.

Typical losses of ferrofluid for various sizes of kimberlite are summarized in Table 5.37. These values agree well with those reported by Gubarevich [G17] for various non-ferrous ores and coal, at a size range, as a rule, of - 25 + 1 mm. For materials with low porosity, such as metals, the losses are much lower.

Solodenko and Karmazin [S71] investigated the effect of pre-treatment of the feed material on losses of ferrofluid. They observed that by wetting the material with water, losses of kerosene-based ferrofluid can be reduced by a factor 4 to 7. In addition, a novel method of material pre-wetting was proposed, and it was found that ferrofluid losses as low as 0.005 to 0.01 kg per tonne of the feed can be achieved.

Table 5.37: Loss of kerosene-based ferrofluid after water-wash of kimberlite [D16, F20, S8, S82]).

Size fraction [mm]	Loss of ferrofluid [kg/t of feed]
-1 + 0.85	0.7
- 1.7 + 1	0.58
- 4 + 2.8	0.28
- 8 + 4	0.27
- 12 + 8	0.12

5.6.4 Separation in a rotating ferrofluid

Magstream

It has been discussed in Sections 2.8.3 and 3.8.6 that a serious limitation of ferrohydrostatic separation is the inefficient separation of fine particles. A proposal to replace the force of gravity by a centrifugal force as the competing force to the magnetically derived buoyancy force resulted in the development of the rotation-based Magstream process [W7].

Numerous experimental investigations into the operational characteristics of Magstream showed that the selectivity of this separator is 100 kg/m^3 or worse [K26]. In the application of Magstream to the separation of beach sands [K26] the probable error *Ep*, defined by eq. (5.33) ranged from 0.14 to 0.43.

$$Ep = \frac{\rho_{90} - \rho_{10}}{2} \tag{5.33}$$

Based on experimental results with materials in the size range from 1 mm to 45 μm, Kojovic [K26] proposed that the efficiency of separation *Ep* of Magstream could be approximated by an expression

$$Ep = 0.04 + 0.05 \frac{4\pi(\rho_{cp} - \rho_f)^2}{J_s} 10^{-10} \tag{5.34}$$

where ρ_{cp} is the cut-point density and ρ_f is the ferrofluid density, both expressed in kg/m^3, while the saturation polarization of the ferrofluid J_s in given in T.

It is clear from eq. (5.34) that a greater precision of separation is achieved for lower cut-point density and with a fluid of higher magnetic polarization. It can also be seen that selectivity of Magstream is at least an order of magnitude lower than that of ferrohydrostatic separation.

An important disadvantage of Magstream is that it is not possible to measure the apparent density of the ferrofluid and thus to calibrate the instrument. The apparent density and thus the cut-point density can be only calculated from eq. (3.173) and their determination is thus as accurate as the theory from which such an equation was derived.

A significant problem that affected the accuracy of Magstream was the low magnetization and high viscosity of the biodegradable water-based Magstream ferrofluid. A rather steep dependence of the fluid magnetization on its density resulted in a high sensitivity of the fluid density and thus of the cut-point density to the ambient temperature and to the temperature of the ferrofluid within the separation space.

In view of the fact that the Magstream concept is potentially advantageous compared to static techniques such as ferrohydrostatic separation, exploitation and further development of this concept would be worthwhile and it could result in a really accurate separation technique.

Separation in a hydrocyclone using magnetic fluids

Another innovative concept how to eliminate the poor separation efficiency of fine particles in ferrohydrostatic separators was proposed by Lin and Fujita [L18, L19]. A cyclone was placed between the pole tips of a magnet that generated a non-homogeneous magnetic field. The apparent density of a ferrofluid introduced into the cyclone could be controlled by pressure differential and flow rate. This concept combined the principles of centrifugal and ferrohydrostatic levitation. Selective separation of material as fine as 38 μm into the overflow and underflow fractions was achieved in a continuously variable density range up to 5000 kg/m^3.

5.7 The application of magnetism in other areas of material handling

The efficiency of separation of a mixture of materials into individual components which have properties of similar magnitude (for instance density, magnetic susceptibility, electrical and thermal conductivity, and others) can be increased by simultaneous exploitation of two or three of these properties. By providing such additional external forces, which can vary over a wide range of values, it is possible, in principle, to manipulate a material more efficiently under a wide spectrum of experimental conditions.

Magnetic forces offer a unique approach to material manipulation. One of the main advantages of material treatment in a magnetic field is that the magnetic force can be superimposed on other physical forces and several physical properties of materials can, therefore, be exploited simultaneously. There are numerous examples where innovative combinations of forces, including magnetic force, were attempted, albeit with varying degree of success. The most natural approach to synergetic application of multiple forces in material handling is a combination of a magnetic force with the force of gravity. Magnetic force has been applied to spirals, tables, jigs and hydrocyclones. Attempts have also been made to apply magnetism to vacuum filtration and column flotation. A brief review of these attempts is given below.

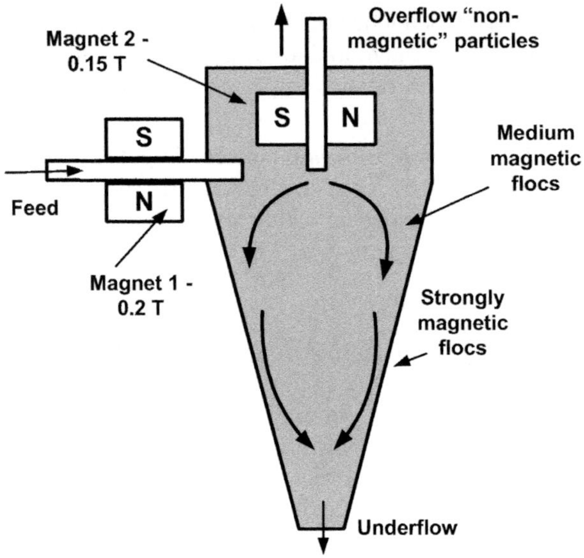

Figure 5.123: Magnetic cyclone for beneficiation of iron ores (adapted from Yurov [Y5, Y6]).

5.7.1 Magnetic cyclone

The concept of a magnetic cyclone was developed in the late sixties of the last century as a natural extension of a conventional hydrocyclone with the objective of providing an additional external force to supplement the gravitational and centrifugal forces that cause classification and separation. Iron ore beneficiation and heavy media manipulation are two areas in which the magnetic cyclone was applied on production scale. A review of some of the concepts and applications of a magnetic cyclone can be found in [F29].

Iron ore beneficiation

Probably the first magnetic cyclone was developed and applied in mineral processing on a production scale in the former Soviet Union [Y5, Y6]. A schematic diagram of such a magnetic cyclone is shown in Fig. 5.123. A slurry of roasted magnetite passing through the first ferrite permanent magnet system became magnetized and the flocs thus formed reported into the underflow. In the vicinity of the overflow, moderately magnetic unflocculated particles were exposed to another magnet. Additional flocs thus created also reported into the underflow. The diameter of the cyclone was 2500 mm and a throughput of up to 190 t/h of solids was achieved. At least 24 such cyclones were in operation in 1986 at the CGOK plant.

Freeman et al. [F30] emulated the above approach simply by replacing the

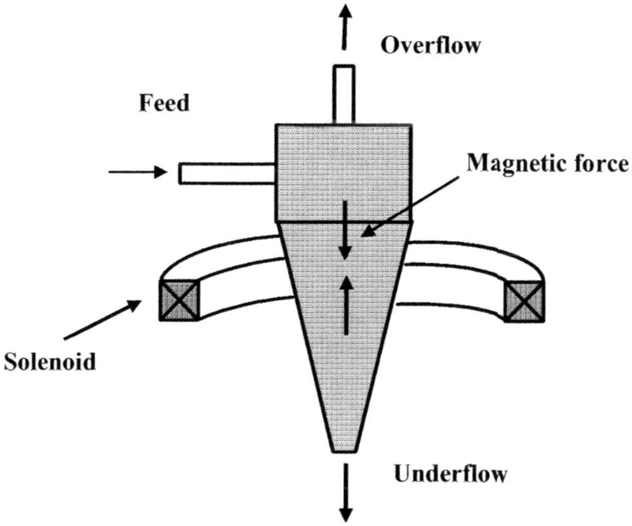

Figure 5.124: Schematic diagram of a heavy media magnetic cyclone for the control of density differential [S84].

ferrite magnets with rare-earth NdFeB permanent magnets. Improved recovery of magnetite, compared to a standard hydrocyclone, was reported.

A straightforward improvement of the Russian concept was proposed by Watson et al. [W27]. The magnetic circuit used in this cyclone consisted of a two-coil 0.2 T electromagnet which attracted the magnetizable particles to the walls of the cyclone from where they were directed to the underflow. The system was found to separate magnetic particles efficiently into the underflow and reject "non-magnetic" particles to the overflow only for small-diameter (< 100 mm) cyclones. For larger-diameter cyclones, the results were disappointing.

Fricker [F28] proposed a different system for the magnetic circuit with the objective of attracting magnetizable particles to the centre of the cyclone and then through the overflow. An electromagnet of horseshoe cross-section, with the magnetic field gradient increasing radially inwards, was used and the hydrocyclone section was fitted between the two poles of the magnet. The device was applied, with modest success, to the beneficiation of iron sands.

Boxmag-Rapid Ltd. have improved the above described designs by employing a quadrupole electromagnet [A34, A35]. The quadrupole arrangement generates a higher magnetic field gradient and thus a stronger magnetic force compared to the dipole configuration. However, theoretical analysis of the multipole magnetic cyclones [G22] revealed that the quadrupole arrangement is not sufficiently powerful for successful separation and the eight-pole configuration is optimum.

The Boxmag-Rapid device was designed as a thickener for de-watering the dilute medium from the rinse screens in heavy media recovery plants, where the higher recovery of the magnetic cyclone can lead to a smaller size of the wet drum magnetic separators. In iron-ore treatment the magnetic cyclone was intended to replace conventional large-diameter thickeners and cyclones, to give higher unit throughputs and magnetic recoveries.

In order to increase the magnetic force in a cyclone, Savitzky et al. [S83] proposed to incorporate ferromagnetic rings on the outside of the conical part of the cyclone, along its vertical axis. These rings become magnetized by the externally positioned coils and generated, therefore, a higher magnetic field and field gradient. The presence of several coils and the ferromagnetic rings, allowed optimization of the patterns of the magnetic force in different zones along the axis of the cyclone. A similar design, based on permanent magnets, was used, on laboratory scale, for beneficiation of beach sands [P24].

Despite these efforts, the concept of the magnetic cyclone as applied to iron ore beneficiation and thickening has not been accepted by the mining industry. Insufficient understanding of the theoretical principles of magnetic cycloning and rather unsophisticated design of the magnetic circuits are the main reasons for the disappointing performance of such devices. Typically, the performance of magnetic cyclones has been characterized by inefficient mineral recovery, undesirable flocculation of magnetic particles, poor concentrate grades, and product accumulation in the cyclone.

Heavy media manipulation

In order to affect the distribution of the particles of the magnetic medium and thus the density differential (difference between the underflow and overflow medium densities) within the cyclone, Svoboda [S84] proposed to apply a vertically oriented magnetic field generated by a solenoid wound around the cyclone. The arrangement is shown in Fig. 5.124. By adjusting the magnetic field strength and by suitable positioning of the magnet, it is possible to control the density distribution of the heavy media and to set an optimum density differential, cut-point density and *Ep* for the Tromp curve of the cyclone. Figure 5.125 illustrates the effect of the magnetic field on the density differential in the cyclone. It can be seen that, by applying the magnetic field, the density differential between the overflow and underflow decreases until it reaches a minimum. With a further increase in the field, the density differential begins to rise. This is the result of the onset of magnetic flocculation of the ferrosilicon particles, increased settling of the magnetic flocs and distortion of the flow pattern within the cyclone. The greatest reduction in the density differential was achieved with the magnet close to the overflow because the closer the magnet is to the overflow, the more uniform the distribution of ferrosilicon within the cyclone. It is even possible to induce inversion of the density, resulting in a negative density differential.

The ability to vary the density differential enabled the density differential to be related to the mean probable error. By determining the minimum *Ep* it was possible to identify the density differential at which the cyclone should

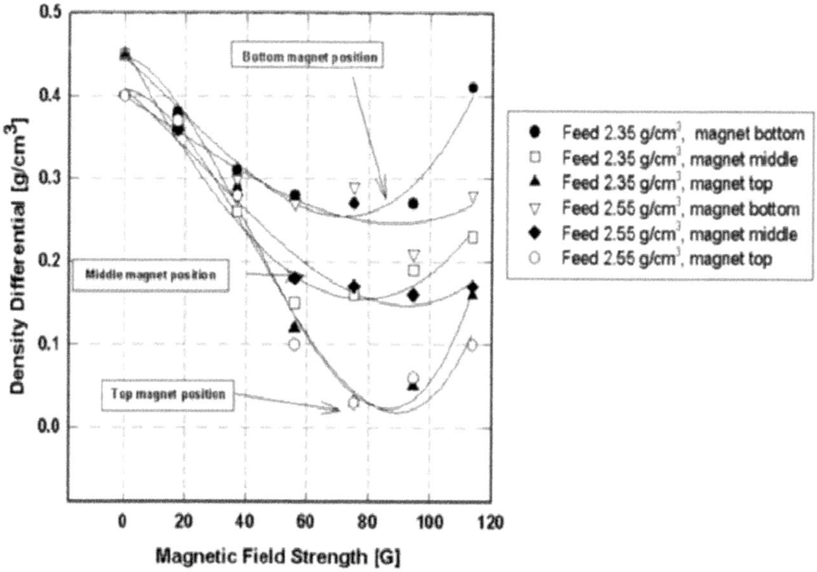

Figure 5.125: Density differential as a function of the magnetic field strength, for various positions of the magnet and for two feed densities. Ferrosilicon: 270D (adapted from [S85]).

operate to provide the highest selectivity of separation. It can be seen in Fig. 5.126 that the minimum value of *Ep*, for 270D ferrosilicon, occurred at a density differential of 250 kg/m^3, in agreement with the plant experience [F31].

Subsequent test work was carried out on production scale by Myburgh [M31]. She observed that the magnetic field had a similar effect on the operation of a small (100 mm) and large (510 mm) cyclone, shown in Fig. 5.127. It was also found that the effect of the field was the same whether medium alone or medium plus diamondiferous gravel feed was used. It was observed that by applying a magnetic field, the medium was stabilized, which resulted in a reduction in the underflow density with increasing magnetic field strength. It was also found that by direct manipulation of the underflow density, the cut point can be directly controlled. As a consequence, the yield to the concentrate was reduced with concomitant reduction of the volume of material to be processed downstream.

Additional pilot-plant scale test work was conducted at De Beers [V8], in a well-defined fully instrumented dense medium system, shown in Fig. 5.128, to verify observations acquired in a long-term industrial plant environment, The feed, underflow and overflow densities were recorded on-line and the effect of the magnetic field on the latter two densities is clearly noticeable in Fig. 5.129.

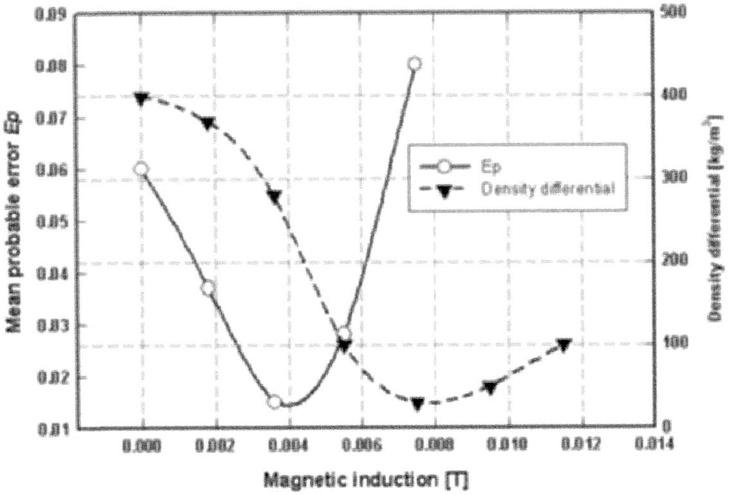

Figure 5.126: Mean probable error Ep and density differential, as a function of magnetic induction, for 270D ferrosilicon (adapted from [S85]).

Figure 5.127: A cyclone with solenoid in a bottom position.

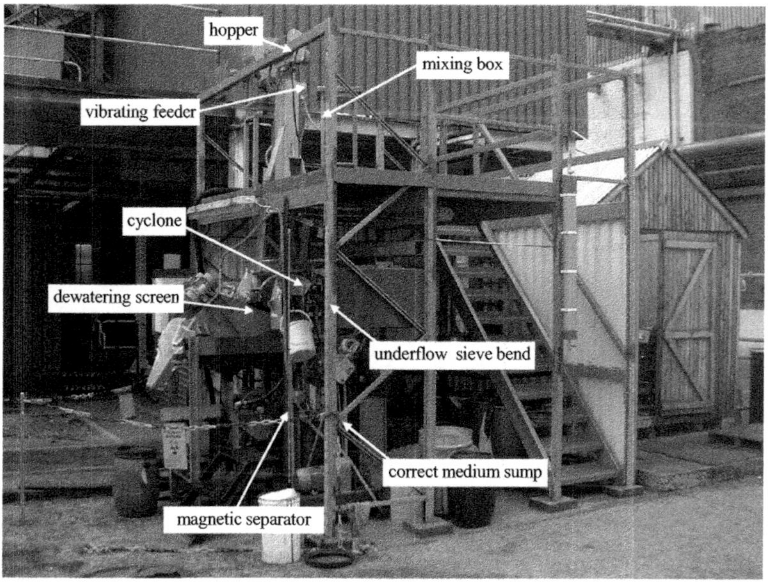

Figure 5.128: The fully instrumented DebTech (De Beers Consolidated Mines, Ltd.) pilot-scale DMS plant, equipped with 100 mm cyclone [V8].

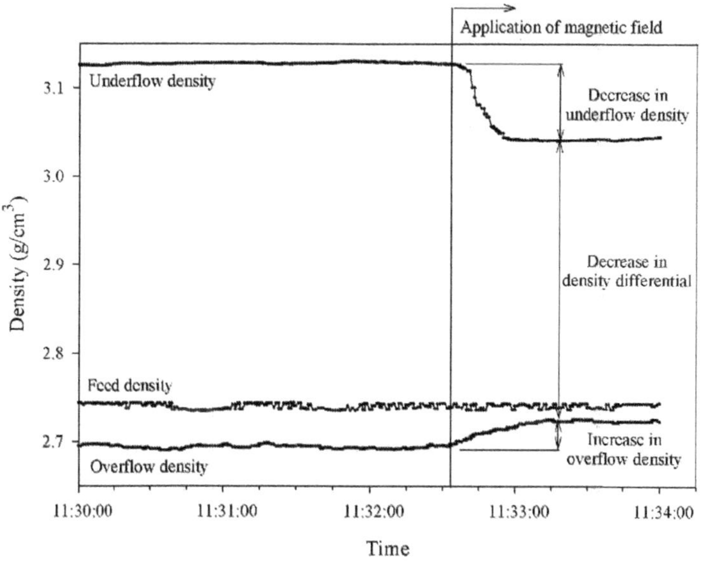

Figure 5.129: The on-line observation of the effect of the application of the magnetic field on the underflow and overflow densities.

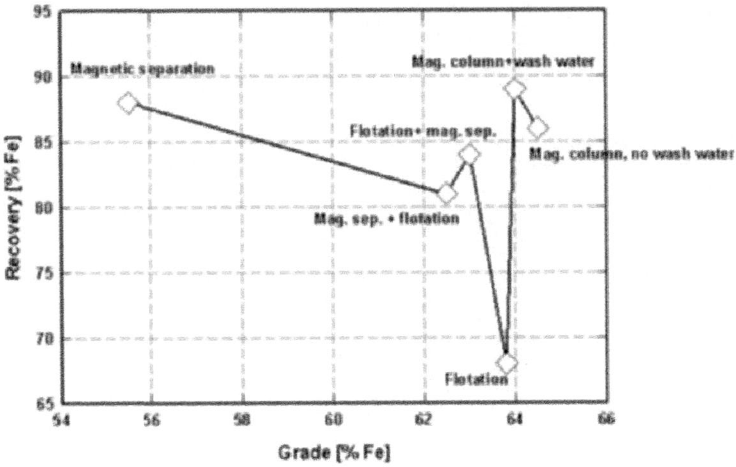

Figure 5.130: Comparison of performance of various processes for the beneficiation of magnetite ore (30.4 % Fe, 94% - 44 μm), (adapted from Yalcin [Y7]).

The results confirmed the previous pilot-plant and production-scale observations of the effect of the magnetic field on the density differential, cut-point density and *Ep*. It was also observed that for a specific selection of the magnetic field strength, to give a suitable density differential, the yield to the concentrate was reduced [V8].

5.7.2 Magnetic gravity separation, flotation and vacuum filtration

Although the initial efforts to combine the force of gravity with a magnetic force in spirals and shaking tables date back to the middle of the last century, success has been limited. The modification of a gravity separator usually consists of placing permanent magnets under the separating surface. The magnets can be arranged so that an intermittent or a moving magnetic field permits removal of the magnetic material to the concentrate discharge slot [M32]. In spite of obvious advantages and often promising results, the technique has not found a wide-spread application in material processing. A more sophisticated approach to the understanding of the physical principles of the process and to magnetic circuit design is required for this method to fulfil its potential.

The application of a magnetic force to column flotation was investigated by Yalcin [Y7]. The device consisted of a conventional flotation cell with a column attached to one side. The column was fitted with an array of rare-earth permanent magnets. The top of the flotation cell was sealed off and the froth was forced to flow over the magnets. Magnetic particles that may be carried with the froth were then captured and returned to the cell. This technique, designed to prevent the recovery of magnetic materials to froth products while non-magnetic minerals were floated, produced a high-grade magnetite concentrate at high recovery, compared to alternative processes, as illustrated in Fig. 5.130.

Watson et al. [W28, W29] demonstrated that a magnetic force can enhance solid/liquid separation of magnetic slurries in drum or disk vacuum filters. A laboratory vacuum drum filter, equipped with rare-earth permanent magnets (with a magnetic induction of 0.16 T at the cloth), was applied to the filtration of magnetite pulp and steel plant waste sludge. It was observed that the capacity of the filter could be increased by up to 300% without significantly increasing the moisture content of the cake.

The filtration results using an industrial disk filter equipped with ferrite permanent magnets (magnetic induction at the filter cloth of 0.035 T), demonstrated that the filter capacity could be increased by a factor of two, with a slight increase of 1% to 4% in the cake moisture content.

5.7.3 Magnetic jigging

The basic idea of combining gravitational and magnetic fields in the jigging process for separation of fine particles was proposed and the physical principles of such a process outlined by Lin et al. [L20]. A mathematical model of a batch magnetic jig was developed, and the bed dynamics were analyzed and compared with the theoretical predictions [L20]. Lin et al. [L21, L22] further expended the theory of magnetic jigging and described particle motion and separation in a continuous magnetic jig

This analysis indicated that the superposition of gravity and relatively weak magnetic forces principally changes the character of jigging, and enables the separation of mixtures that are difficult to beneficiate. It was shown that the introduction of additional magnetic force F_m in the range of 0.2 to 0.4 of the magnitude of the gravitational force F_g (i.e. $F_m = 0.2$ to $0.4F_g$) allows separation of particles of small density differential for particle diameter smaller than 0.04 mm. For larger particles, with diameters in the range - 0.2 + 0.04 mm, the additional magnetic force providing the separation increases to 0.5 to $0.8F_g$.

Chapter 6

Industrial Applications of Magnetic Methods of Material Treatment

During the last forty years of the 20th century, magnetic separation evolved from a simple technology for manipulation of strongly magnetic coarse materials into a powerful technique for the treatment of weakly magnetic, finely dispersed particles. Development of magnetic separators based on powerful new permanent magnets, construction of reliable superconducting magnets, design of efficient eddy-current separators and production of magnetic carriers and low-cost magnetic fluids enabled the extension of this technology from mineral processing to diverse areas of science and engineering. The spectrum of applications of the magnetic techniques to material manipulation is formidable. The concentration of various ferrous and non-ferrous minerals has become an important application, as has the removal of low concentrations of magnetizable impurities from industrial minerals. Removal of impurities from waste water, the recycling of metals from industrial wastes and the concentration or removal of biological objects in medicine and biosciences, have become important areas of application of magnetic technology. In view of the enormous wealth of information on various current applications of magnetic techniques to material handling, only a cursory review of these applications can be given in this monograph. A detailed description of some of the earlier industrial applications of magnetic separation to minerals beneficiation can be found in [S1].

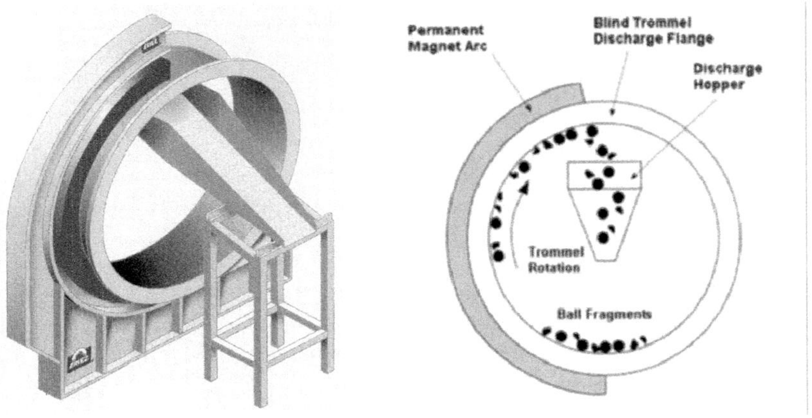

Figure 6.1: Schematic illustrations of the trommel magnet system [S86] (courtesy of Eriez Magnetics, Inc.).

6.1 Treatment of minerals

6.1.1 Recovery of grinding ball fragments from SAG and ball mills

Grinding ball fragments from SAG and ball mills cause serious wear to downstream processing equipment. These fragments, recirculating in a milling circuit, cause wear to sumps, pumps, hydrocyclones and interconnecting piping. Magnetic separation systems designed to remove grinding ball fragments from the mill discharge have been developed and successfully applied in milling circuits.

Trommel and trunnion magnet systems developed by Eriez Magnetics were applied at the Escondida Copper Concentrator in Chile and subsequently at other concentrators, including Los Pelambres [S86].

The trommel magnet shown in Fig. 6.1 consists of an arc of permanent magnets mounted at the discharge end of an existing trommel screen. The magnet removes the grinding ball fragments from the oversize material allowing the oversize ore to be returned to the circuit for processing. As can be seen in Fig. 6.1. the system consists of a blind trommel, magnet sector, support structure and a discharge hopper. The function of the blind trommel is to provide a flange to transport the trommel oversize material through the magnetic field.

During operation, the trommel screen oversize material moves across the blind trommel prior to discharging into a sump. Any strongly magnetic material will be attracted to the magnetic arc and held to the rotating blind trommel. This material is then released as it rotates past the end of the magnet sector. Figure 6.2 shows an installed trommel magnetic system mounted on the discharge end of a trommel screen.

After the installation of the trommel magnet in the Escondida concentrator

Figure 6.2: Trommel magnet (foreground) mounted at the discharge end of a trommel screen (courtesy of Eriez Magnetics, Inc.).

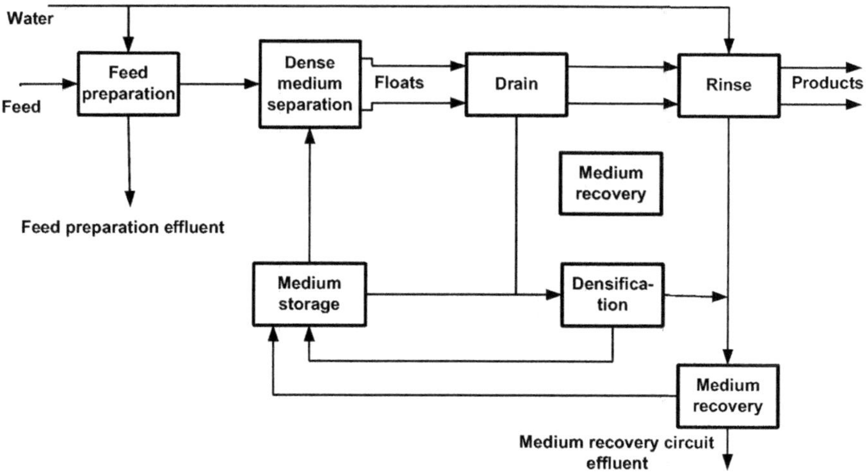

Figure 6.3: Block flowsheet of a dense medium plant.

126 tonnes of grinding ball chips were removed on the first day. This amount was progressively decreasing with each subsequent day and after the fifth day the recovered fragments stabilized at about 7 tonnes per day [S86]. When installed at the Los Pelambres Copper Concentrator (Chile) on a SAG mill, on average 16.6 tonnes of ball chips per day were collected.

The trunnion magnet was subsequently designed to collect magnetically the grinding ball fragments directly from the mill discharge stream without the use or requirement of a trommel screen. The magnet installed at a ball mill at Los Pelambres collected, on average, 34.8 tonnes of ball chips per day.

Installation of these magnetic systems for the removal of grinding ball fragments resulted in extension of the lifetime of pumps by 300%, the milling throughput increased by 5% and the energy consumption decreased by 10%.

6.1.2 Heavy medium recovery circuits

Dense medium separation has been practised for a long time in the coal, iron ore and diamond industries and in spite of its long history and relative maturity the process is being steadily improved. The DMS plant block diagram is shown in Fig. 6.3, while Fig. 6.4 illustrates the medium recovery circuit with water recycling. Since the major component of the total loss of heavy medium is the magnetic loss, significant emphasis has been placed on the development of the magnetic circuits and their operation. Improved drum and tank designs provide better recoveries and correct adjustment and optimization of variables affecting operation of magnetic separators also provide more efficient and economical operation of the circuit. Implementation of rare-earth drum magnetic separation, with important consequences for better recoveries of heavy media and for simplification of the entire circuit, will be the next step for improvement of this mature process.

6.1.3 Beneficiation of iron ores

EVTAC Mining Co., Minnesota, USA

EVTAC Mining, the magnetite concentrate and pellet producer, has recently faced a problem of the reduced feed grade and the need to move into mining areas of lower-grade ore with a higher silica content and lower magnetic susceptibility iron minerals. In order to reverse the losses of the magnetic minerals, low-intensity magnetic separators developed during the 1960s were replaced by more efficient Metso Minerals machines.

The original 98 concurrent Stearns low-intensity magnetic drum separators (900×2400 mm) in the rougher circuit were replaced in 2001 by 20 Metso Minerals counter rotation LIMS, shown in Figs. 6.5 and 6.6. In addition to considerable reduction in floor space, the replacement separators offered advantages such as a longer pick-up zone, increased retention time and self-adjusting level control. The metallurgical performance of the new separators exceeded not only

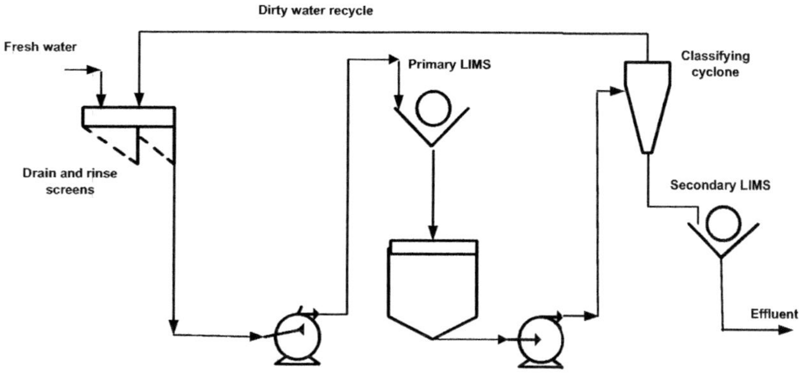

Figure 6.4: Dense medium recovery circuit with water recycling (adapted from Dardis [D13]).

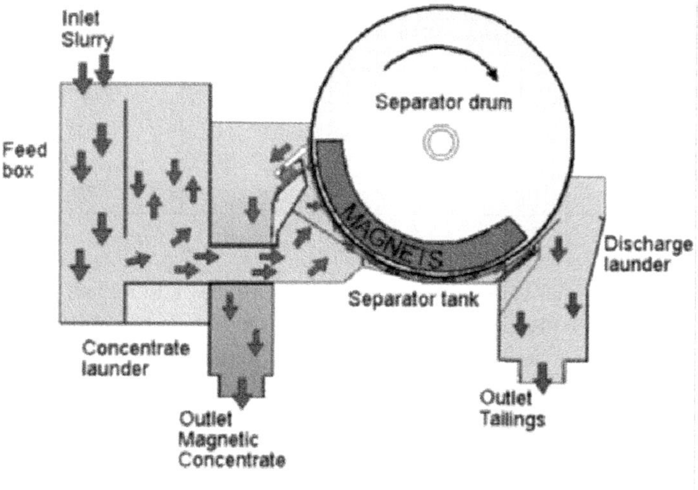

Figure 6.5: Metso Minerals low-intensity counter-rotation permanent magnet drum separator (courtesy of Metso Minerals).

Figure 6.6: Metso Minerals low-intensity permanent magnet drum separator (courtesy of Metso Minerals).

Table 6.1: Metallurgical results of rougher optimization at EVTAC Mining [E6].

Parameter	Metso LIMS	Stearns LIMS
Feed [% solids]	30 - 35	42.3
Davis Tube concentrate [%]	6.5	13.1
Recovery [% Fe]	99.6	96.9
Davis Tube total conc. [%Fe]	68.9	68.5
Davis Tube total tails [%Fe]	54.9	57.5
Silica in conc. [%Si]	19.3	18.0
Silica rejection [%]	26.8	27.3

that of the original separators but also the set target. The metallurgical results from the optimized rougher circuit are summarized in Table. 6.1.

Following the success of the rougher circuit optimization, the Metso LIMS machines were tested in the cobber position. The new cobbers improved the recovery and grades in both the concentrate and tails streams and their installation resulted in a reduction of 75% in the original equipment [E6]. The metallurgical improvements are recorded in Table 6.2, while Fig. 6.7 illustrates the pattern of the iron recovery and the feed grade as a function of time.

Iron ore beneficiation by the Ferrous Wheel magnetic separator

Ferrous Wheel magnetic separators, described in Section 2.4 have been applied to the beneficiation of weakly magnetic iron ores at several mining operations.

Table 6.2: Metallurgical results of cobber optimization at EVTAC Mininig [E6].

Parameter	Metso LIMS	Stearns LIMS
Feed [% solids]	35 - 40	43.7
Feed grade [% Fe]	21.6	20.8
Concentrate grade [% Fe]	33.1	32.4
Recovery [% Fe]	97.2	92.4
Mass recovery [%]	63.9	59.5
Davis Tube total conc. [% Fe]	68.8	69.1
Davis Tube total tails [% Fe]	53.8	61.9
Silica in conc. [% Si]	33.7	34.2
Silica rejection [%]	22.2	22.7

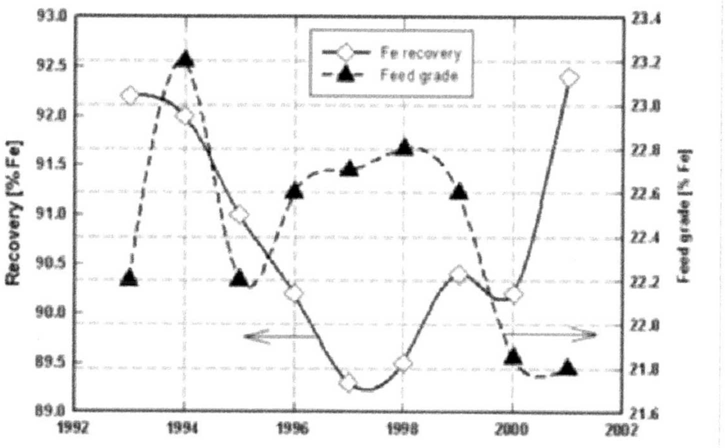

Figure 6.7: Magnetic iron recovery and the feed grade at EVTAC (adapted from [E6]).

Figure 6.8: Installation of sixteen Ferrous Wheel magnetic separators to recover hematite from tailings ponds (courtesy of Eriez Magnetics, Inc.).

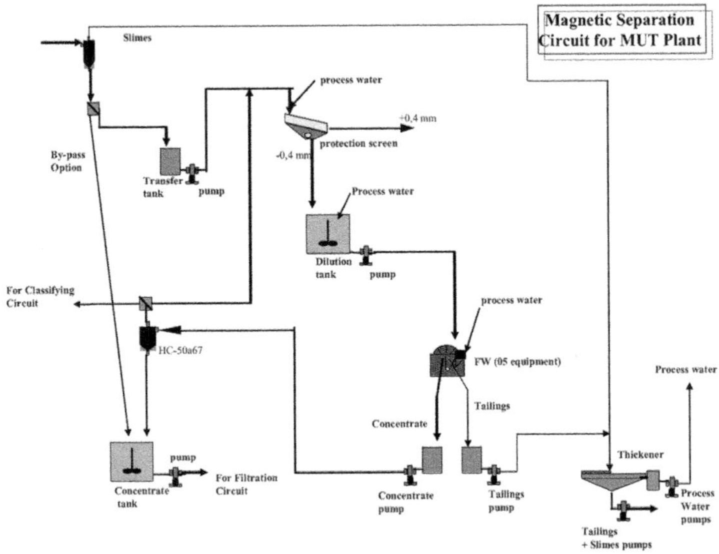

Figure 6.9: Magnetic separation circuit at the Mutuca Plant, Minas Gerais, Brazil (courtesy of S.C. Amarante).

Table 6.3: Production-scale results at the Hercules Mine [P15].

Product	Mass [%]	Grade [% Fe]	Recovery [% Fe]
LIMS	0 - 8	55 - 61	0 - 9
FW Rougher magnetics	18 - 35	57 - 60	22 - 41
FW Scavenger magnetics	13 - 34	55 - 57	15 - 38
Combined concentrate	49 - 58	56 - 58	58 - 68

Hercules Iron Mines, Coahuila, Mexico In 1995 the Grupo Acerero del Norte's Hercules Mine expanded its concentrator to recover iron values from mill tailings. The expansion incorporates low-intensity drum magnetic separators and matrix magnetic separators as a pre-concentration stage. Flotation is then used to produce the final iron concentrate for pelletization.

Approximately 6 million tonnes of flotation tailings, averaging 50% Fe, mostly hematite, are contained in tailings ponds. The tailings are relatively fine at 93% -75 μm and 72% - 25 μm. The non-magnetic fraction from the low-intensity drum magnetic separator is treated by 15-ring Ferrous Wheel (FW) separators, with 2.44 m diameter rings, equipped with the 14 mesh screen cloth matrix. The ferrite permanent magnets generate a magnetic induction of 0.1 T in the open air in the upper rougher stage, while the magnets in the lower scavenger zone produce a background magnetic induction of 0.22 T.

The results of the production-scale tests at Hercules are summarized in Table 6.3. The best combined results of LIMS and Ferrous Wheel separators were obtained at a feed rate of 26 t/h at 37% to 38% solids. As a result of successful pilot-scale and production-scale tests, additional three low-intensity drum magnetic separators (1220 mm diameter and 3170 mm drum width) and sixteen Ferrous Wheel separators, shown in Fig. 6.8, were installed [P15].

Mutuca Plant, Minas Gerais, Brazil Four 15-ring Ferrous Wheel separators were installed in 2001 at the Mutuca plant of MBR, Minas Gerais, the second largest Brazilian iron ore producer, to treat 240 t/h of cyclone underflow at the iron ore beneficiation plant. A schematic diagram of the magnetic separation circuit of the MUT plant is shown in Fig. 6.9. The plant expansion followed extensive laboratory and pilot-scale tests [A36]. The results of the tests confirmed that significant reduction in concentration of aluminium-bearing minerals and of silica could be achieved at high iron oxide recovery. It was also observed that loss on ignition and concentration of manganese and phosphorus also decreased significantly. Figure 6.10 illustrates the performance of the Ferrous Wheel separator at the Mutuca plant. The Ferrous Wheel separator installed at the Mutuca Iron Ore Plant is shown in Fig. 6.11, while Fig. 6.12 illustrates the positioning of the magnetics/non-magnetics splitter within the Ferrous Wheel machine.

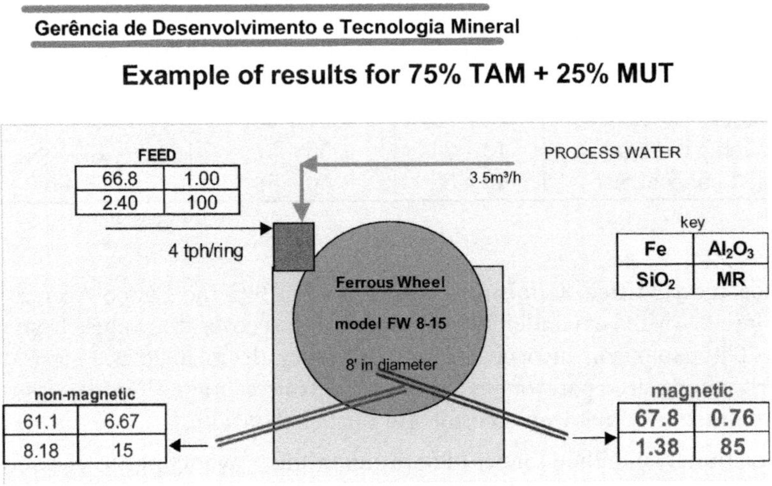

Figure 6.10: Performance of Ferrous Wheel magnetic separators at the Mutuca Iron Ore Plant, Brazil (courtesy of S.C. Amarante).

Figure 6.11: Ferrous Wheel magnetic separator installed at Mutuca Plant, Brazil (courtesy of S.C. Amarante).

Figure 6.12: The positioning of a magnetics/non-magnetic splitter in the Ferrous Wheel separator at Mutuca Plant (courtesy of S. C. Amarante).

Fábrica Mine, Forteco Mineração SA, Brazil

The concentrator plant of Fábrica Mine, which started beneficiating itabirite ore in 1977, was using nine Jones wet high-intensity magnetic separators DP317, with 1.5 mm gap and a magnetic induction of 1.2 T, to produce a concentrate for pelletization. These WHIMS units were operating in parallel at a feed rate of 110 t/h, with a feed grade of 47% Fe, of which maximum 0.8% was magnetite.

The increasing production rate and decreasing grade of the ore in the nineteen eighties resulted in the increase of the magnetite content, up to 3%, in the feed to magnetic separators. This, combined with the need to increase the magnetic field strength to treat finely disseminated hematite, resulted in partial clogging of the separation boxes.

After extensive pilot-scale testwork the WHIMS plant was modified into a two-stage process using the existing nine Jones separators. The width of the gap of the first stage, with four machines, was increased to 3.5 mm, which corresponded to a magnetic induction of 0.7 T. This first medium intensity stage was followed by a higher intensity WHIMS stage, with five machines operating at 1 T and 1.5 mm gap width. Feed rates of up to 320 t/h in the first and 240 t/h in the second stages could be achieved [H24].

The new circuit, the flowsheet of which is shown in Fig. 6.13, allowed an increase in the overall production rate from 832 t/h to 1200 t/h, and an increase in the recovery of iron from 65% to 70%.

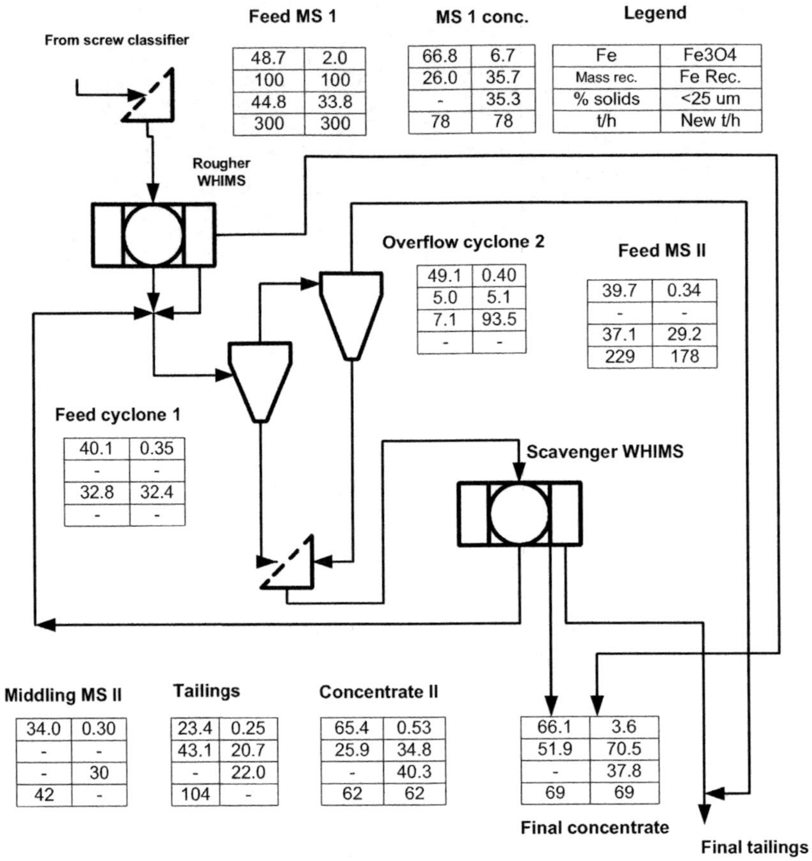

Figure 6.13: Flowsheet of a two-stage WHIMS circuit at Fabrica Mine, Brazil (adapted from [H24]).

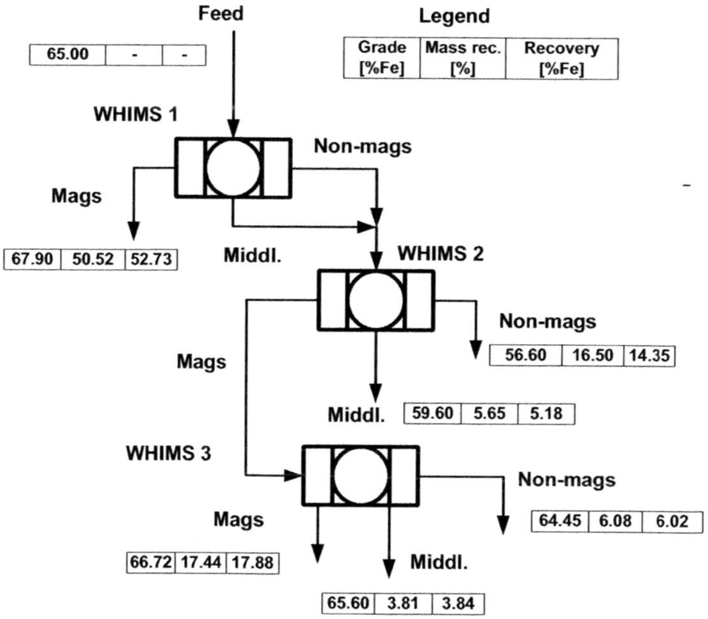

Figure 6.14: Flowesheet of the BHP (Australia) three-stage WHIMS plant for the beneficiation of hematite ore (adapted from [H25]).

BHP, Australia

A three-stage Jones WHIMS separation circuit, shown in Fig. 6.14, beneficiates hematite ore at the rate of 190 t/h at one of the BHP plants in Australia [H25]. The grade of the feed into the circuit is approximately 63% Fe and 8% SiO_2+Al_2O_3. The grade of the final product from the circuit, at a recovery of 71% Fe, is 67.7% Fe and the content of SiO_2+Al_2O_3 is less than 2%.

At another BHP Iron Ore (Goldsworthy) plant, three Reading WHIMS machines have been used to recover iron from - 100 μm hematite ore [M29]. The upgrading of the feed (50.5% Fe, 19.7% SiO_2 and 4.80% Al_2O_3) is achieved by rejection of silica and alumina into the tailings. At an overall recovery of 73.7% Fe the grade of the final concentrate amounts to 61.5% Fe, 6.2% SiO_2 and 2.09% Al_2O_3. The flowsheet of the process is shown in Fig. 6.15.

The application of SLON WHIMS to the beneficiation of iron ores in China

The first SLON wet high-intensity magnetic separator was developed in 1988 and since its introduction into the market it has been successfully installed at numerous mineral processing plants in China. At present SLON separators of various sizes are operating at more than 30 beneficiation plants.

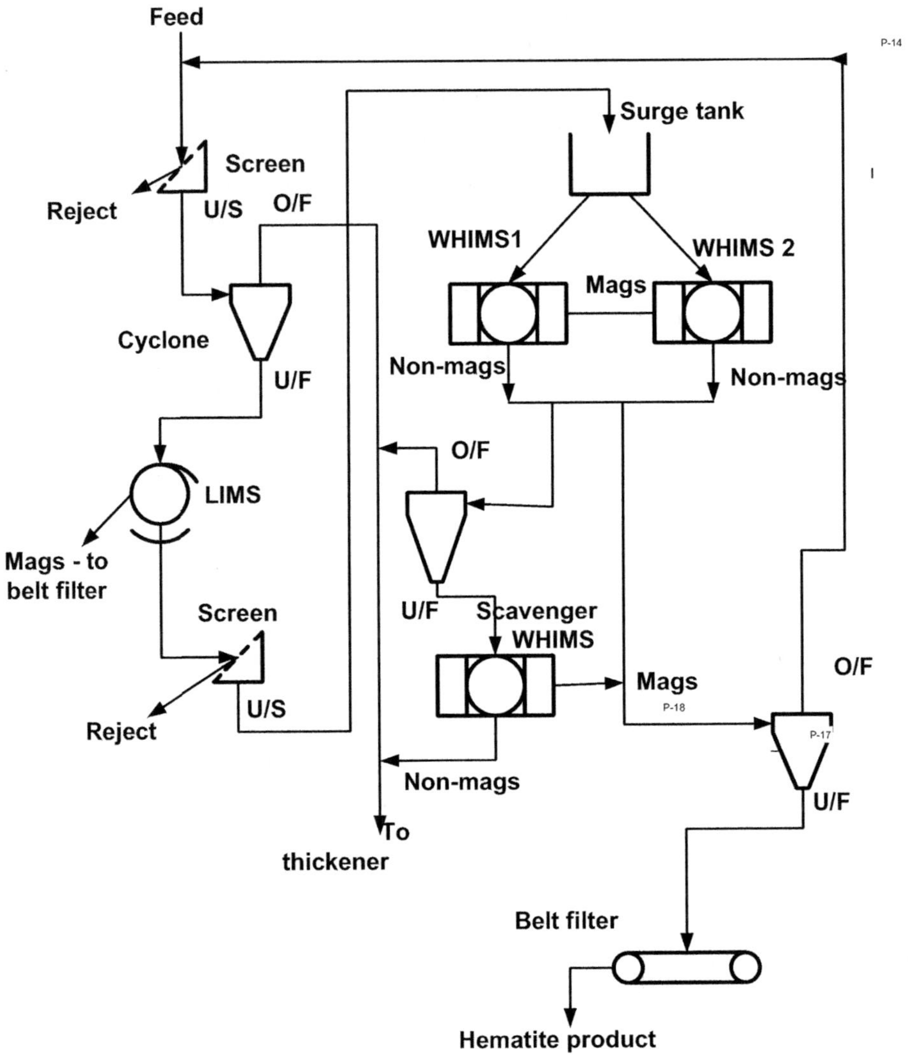

Figure 6.15: Flowsheet of the Reading WHIMS circuit at the BHP Iron Ore (Goldsworthy) Plant (adapted from [M29]).

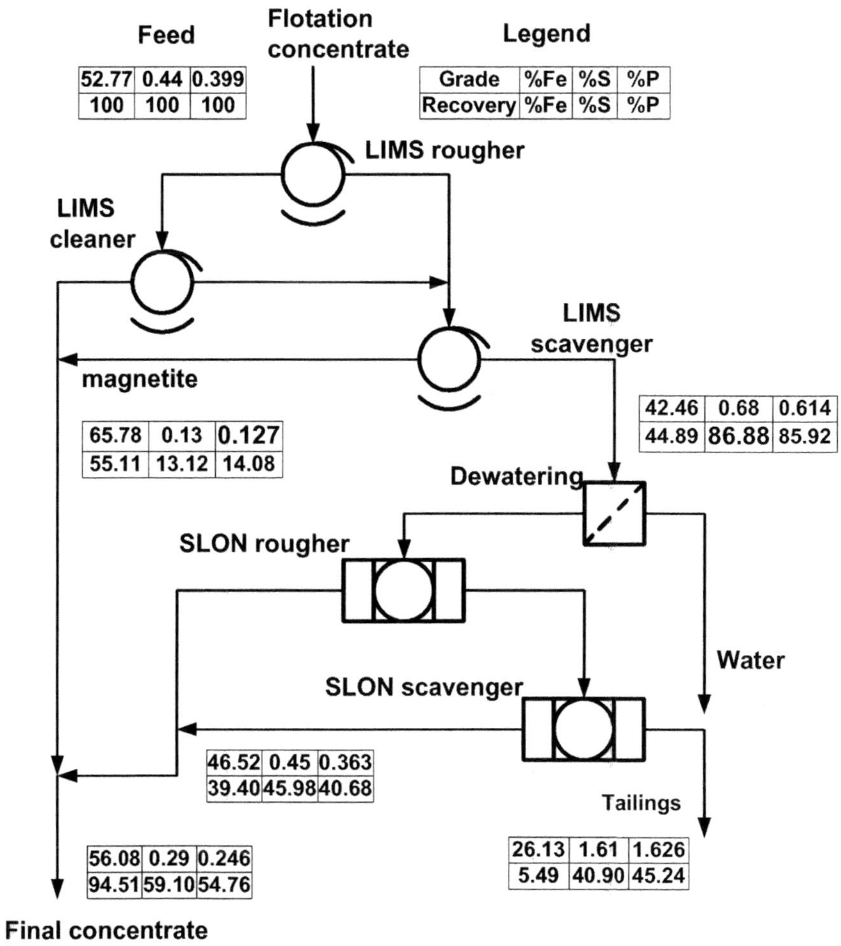

Figure 6.16: Flowsheet of the magnetic separation circuit at the Meishan Iron Ore Mine, China (adapted from [X4]).

Figure 6.17: SLON-1500 separators operating in Meishan Iron Ore Plant (courtesy of the Ganzhou Non-ferrous Metallurgy Research Institute, China).

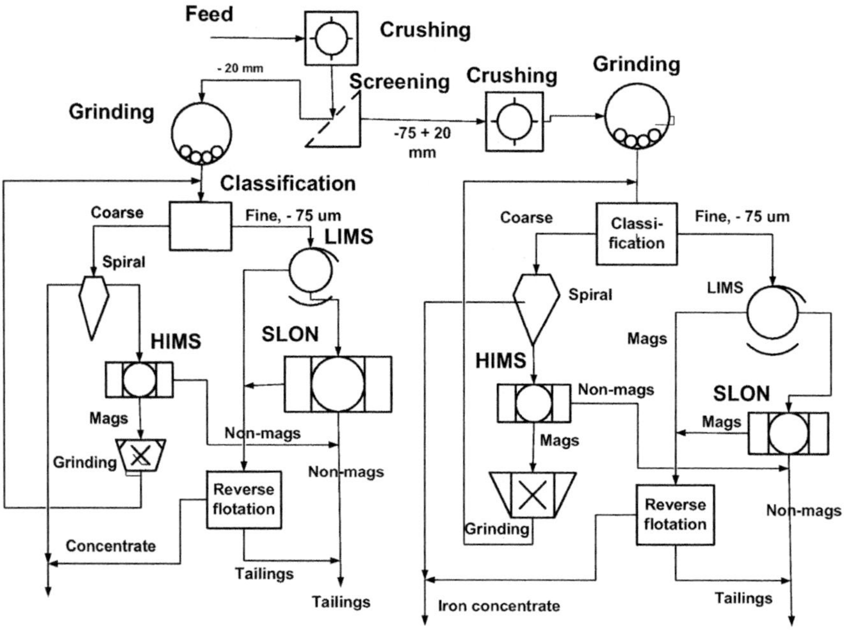

Figure 6.18: Flowsheet of the Qi Dashan Iron Ore Processing Plant, China (adapted from [X5]).

Figure 6.19: SLON 1750 wet high-intensity magnetic separators operating at the Qi Dashan Beneficiation Plant, China (courtesy of Xiong Da-he).

Meishan Iron Ore Mine, China Meishan Iron Ore Mine processes 4 million tonnes of iron ore and produces about 2.6 million tonnes of iron concentrate annually. Because of a relatively high content of phosphorus, the iron concentrate did not meet market requirements. In recent years 18 SLON Model 1500 WHIMS machines have been applied to reduce the concentration of phosphorus. The Meishan Mine circuit is the largest application of SLON separators in China.

In the flowsheet shown in Fig. 6.16, the iron concentrate from sulphide flotation (70% - 75 μm) is first exposed to low-intensity magnetic separators to remove magnetite. SLON 1500 machines, shown in Fig. 6.17, are then used as roughers and scavengers to recover hematite and siderite.

Qi Dashan Mineral Processing Plant, China Qi Dashan Mineral Processing Plant in Liao Ning Province, China, processes 8 million tonnes of low-grade iron ore per annum. Hematite, magnetite and limonite are the main iron minerals. The original beneficiation process for the coarse fraction (- 75 + 20 mm) was based on roasting, grinding and low-intensity magnetic separation. Since the final concentrate could not meet the metallurgical requirement of 67% Fe, the flowsheet was re-designed and six SLON separators Model 1750 were implemented, as is shown in Fig. 6.18. Figure 6.19 depicts SLON separators in operation.

In the new flowsheet the ore is ground to 60% - 75μm and classified in a hydrocyclone. While the underflow is treated by spirals, the overflow is upgraded by LIMS, SLON WHIMS and flotation. Compared to the original roast-LIMS

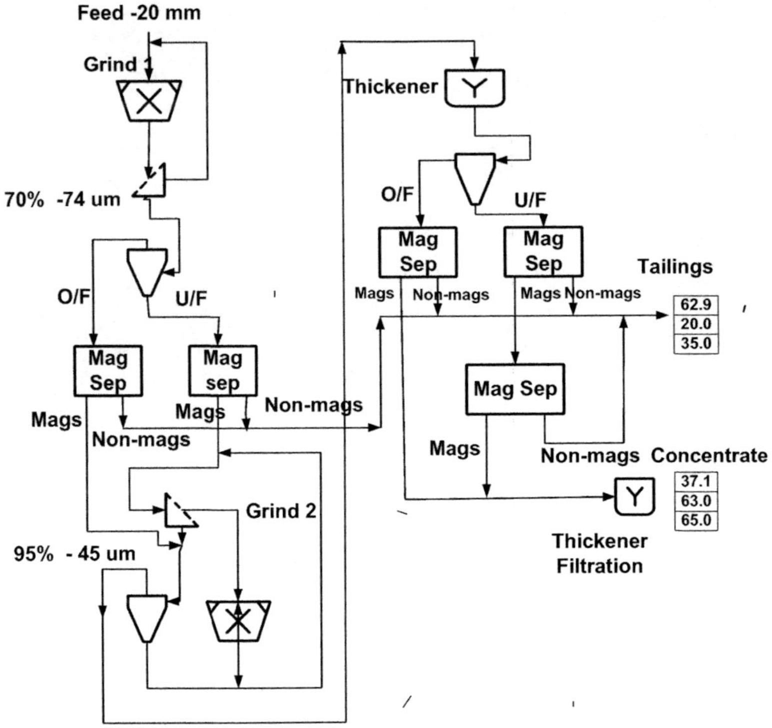

Figure 6.20: Flowsheet of the KGOKOR (the Ukraine) Iron Ore Beneficiation plant (adapted from [M33]).

flowsheet the grade of the concentrate is higher by 3.6% and the recovery of iron is higher by 5.6% [X5].

Beneficiation of iron ore in Krivoy Rog, the Ukraine

Oxidized ferruginous quartzites of deposits in the Krivoy Rog Basin consist mainly of magnetite (4.5%), hematite and martite (38.4%), iron hydroxides (9.5%) and quartz (44.9%). The mineral grains are liberated at - 45 μm, although hematite is liberated at - 15 μm. Various beneficiation methods were investigated when the Krivoy Rog Mining and Beneficiation Complex (KGOKOR) was designed [M33, M34,U3]. The ultimate flowsheet includes 11 sections, each having a capacity of 2.4 Mt of ore per annum. High-intensity three-rotor separators 6-ERM 35/315 with grooved plate matrix form the core of the beneficiation circuit [M33]. Throughput of the separator is 100 t/h, rotor diameter 3150 mm, magnetic induction 1.35 T and the gaps between the grooved plates 4 mm. The non-magnetic fractions from the top rotor are re-cleaned in the middle and lower rotors. The flowsheet of the KGOKOR beneficiation plant is shown in Fig. 6.20

Figure 6.21: Geographical positions of the Krivoy Rog, Nikopol and Marganets (the Ukraine) iron and manganese ore deposits.

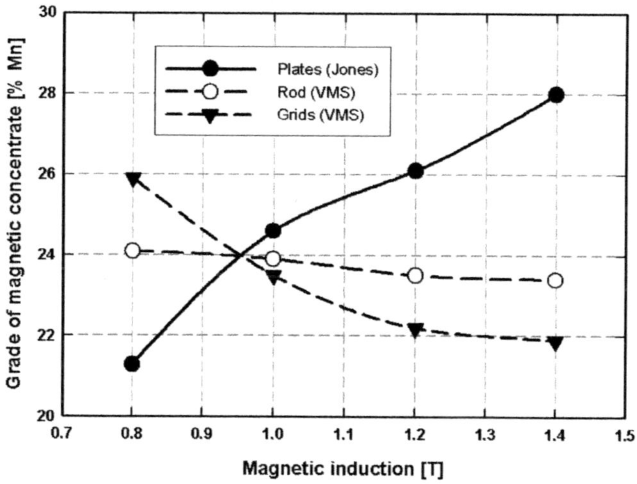

Figure 6.22: The grade of the magnetic concentrate from the Ordzhonikidze Basin manganese slimes, as a function of the magnetic field strength, as obtained with Jones and VMS separators (adapted from [V9]).

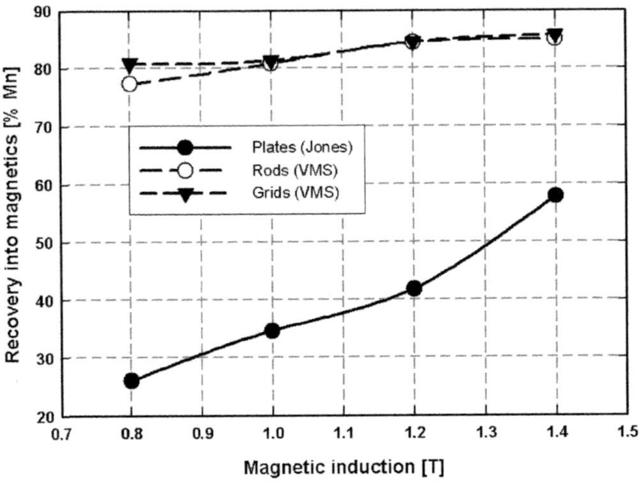

Figure 6.23: The recovery of manganese into the magnetic concentrate of the Ordzhonikidze Basin slimes, as a function of the magnetic field strength, obtained with Jones and VMS separators (adapted from [V9]).

and the geographical location of the plant in Fig. 6.21.

Treatment of tailings from the Krivoy Rog GOK The Krivoy Rog beneficiation plant generates more than 60 million cubic meters of tailings annually. By employing wet high-intensity magnetic separation iron-rich and iron-poor products are obtained. These products are then used in the building industry: the iron-rich fraction, containing up to 38% Fe is used as a correction component in the production of portland cement clinker, while the non-magnetic fraction with an iron concentration of 7% is used for the manufacture of bricks [K27].

6.1.4 Beneficiation of manganese ores

The exhaustion of high-grade oxide manganese ores and gradual introduction of the increasing volumes of low-grade oxide and carbonate manganese ores into the beneficiation process result in the reduction of the efficiency of their upgrading. In addition, during the beneficiation process, large quantities of tailings are generated, with manganese minerals usually smaller than 20 μm. The slimes are the main source of losses of manganese.

Extensive research at the Mekhanobrchermet Institute (the Ukraine) demonstrated wet high-intensity magnetic separation as a promising method of increasing the efficiency of beneficiation of refractory manganese ores and their slimes from the Nikopol basin in the Ukraine. Laboratory and pilot-scale tests were

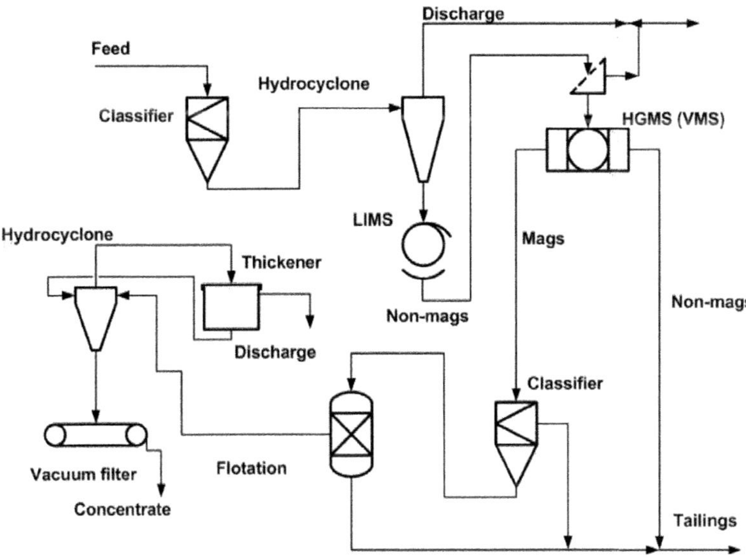

Figure 6.24: Flowsheet of the Chkalovsky manganese ore beneficiation plant, the Ukraine (adapted from [V9]).

conducted using Jones WHIMS with the plate matrix and VMS high-gradient separator with rod and grid matrices. The results of the tests are summarized in Figs. 6.22 and 6.23. The results confirm that the VMS separator equipped with rod and grid matrices produced the manganese concentrate with a considerably higher recovery [V9].

The production-scale circuit for the treatment of the manganese slimes, which included full-scale VMS 100/2 magnetic separators was installed in 1988 at the Tchkalov beneficiation plant of the Ordzhonikidze GOK in the Nikopol Basin (the Ukraine) [L23, L24]. The geographical position of the basin is shown in Fig. 6.21 and the flowsheet of the 400 kt per annum pilot plant at the Tchkalov beneficiation plant is illustrated in Fig. 6.24. The VMS separator installation at the Tchkalov plant is shown in Fig. 6.25. Two VMS magnetic separators are also in operating at the Marganets plant.

6.1.5 Beneficiation of ilmenite ores

Pang Zhi Hua Processing Plant, China

The Pang Zhi Hua ilmenite processing plant is the biggest producer of ilmenite in China. The deposit contains vanadium, titanium and iron minerals. The processing plant recovers magnetite and vanadium and the tailings from the iron recovery plant represent the feed into the ilmenite processing plant. In the original process the coarse fraction (+ 45 μm) of the ore was beneficiated using

Figure 6.25: VMS-100 separators at the Tchkalov manganese ore beneficiation plant (the Ukraine), (courtesy of Ore Research Institute, Czech Republic).

spirals and electrostatic separators. The fine fraction (- 45 μm) was discarded for the lack of suitable beneficiation technology. The total recovery of ilmenite amounted to only 17% to 20%.

To increase the ilmenite recovery, a magnetic separation-flotation flowsheet was tested and applied in production in 1997. A SLON-1500 magnetic separator was installed as a rougher, to treat the - 45 μm fraction, with a grade of 11% TiO_2, followed by flotation as a cleaner. The process flowsheet is shown in Fig. 6.26.

As can be seen, the ilmenite concentrate with a grade of more than 47% TiO_2 is being obtained at a recovery of 44%. In 1999 20 kt of the ilmenite concentrate were produced from the fine fraction, and another 20 kt/a of the concentrate were recovered from the tailings dump [X3]. After installation of four more SLON separators to complete five ilmenite processing lines, the production has increased to 80 kt/a.

Rooiwater ilmenite ore (South Africa)

Magnetic separation was found to be an efficient technique for concentration of ilmenite and for rejection of silica from Rooiwater (Limpopo Province, South Africa) ilmenite ore, as is demonstrated in a flowsheet shown in Fig. 6.27. The coarse fraction (+ 106 μm) was upgraded from the head grade of 13% TiO_2 to 44.3% TiO_2 at an overall recovery greater than 68%. The concentration of silica was reduced from 53.1% SiO_2 in the feed to about 7% in the final concentrate. The role played by magnetic separation at different stages of the process is well

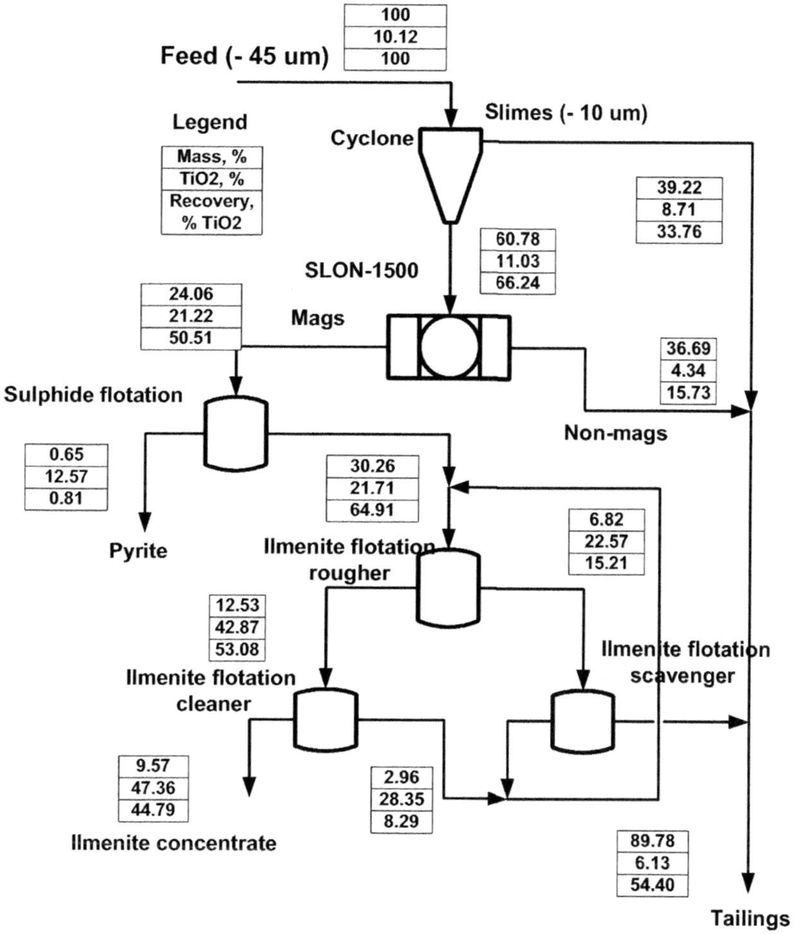

Figure 6.26: Flowsheet for the treatment of - 45 μm ilmenite fraction at Pang Zhi Hua, China (adapted from [X3]).

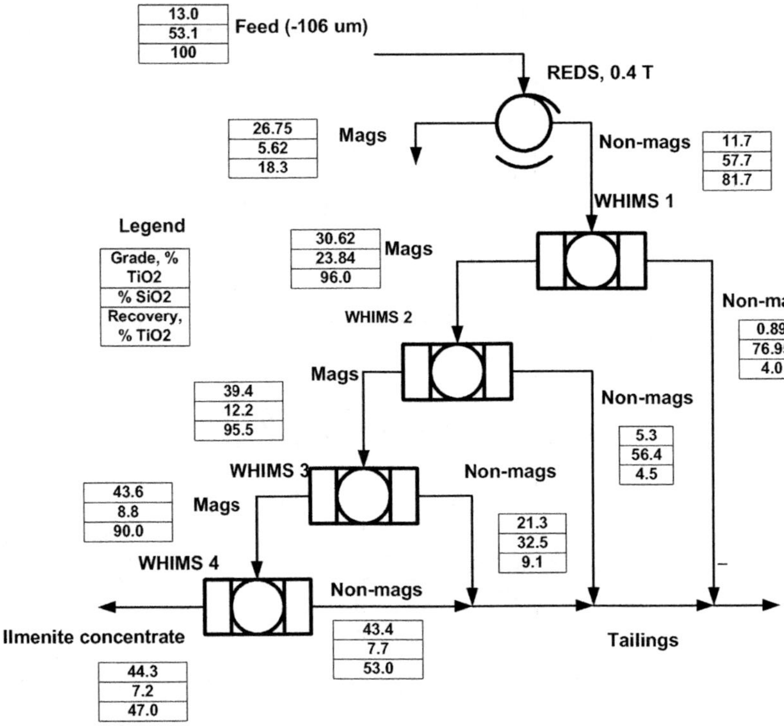

Figure 6.27: Flowsheet for the treatment of + 106 μm fraction of the Rooiwater ilmenite sands by magnetic separation.

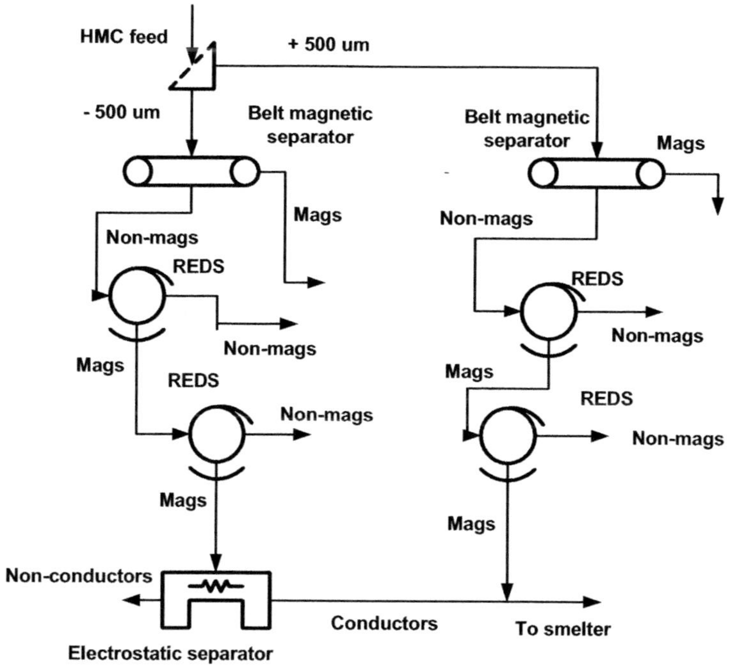

Figure 6.28: Dry magnetic separation flowsheet for the beneficiation of the Rooiwater ilmenite ore (adapted from [E7]).

demonstrated. Since wet high-intensity magnetic separation failed to produce smelter-quality ilmenite concentrate, mainly because of a high concentration of silica, dry magnetic beneficiation was investigated [E7]. Figure 6.28 shows the dry beneficiation flowsheet applied to the treatment of the heavy mineral concentrate, assaying 43% TiO_2 and 6% to 8% SiO_2. Rare-earth drum magnetic separators were used in two stages. The grade of the combined + 500 μm magnetic fraction and - 500 μm conductors from electrostatic separators was 48.5% to 50.0% TiO_2 and less than 1% SiO_2. The overall recovery of TiO_2 was greater than 90%.

The grade of the fine fraction - 106 μm was increased from 9.5% TiO_2 in the feed to 42.1% TiO_2 in the final concentrate, while silica was rejected from 47.9% SiO_2 to about 9.6% in the concentrate.

A hard ore from another Rooiwater deposit was upgraded using a combination of WHIMS and spirals, as can be seen in the flowsheet shown in Fig. 6.29. The grade of the final ilmenite concentrate was 44.69% TiO_2 and 0.40% SiO_2 at an overall TiO_2 recovery of 50%.

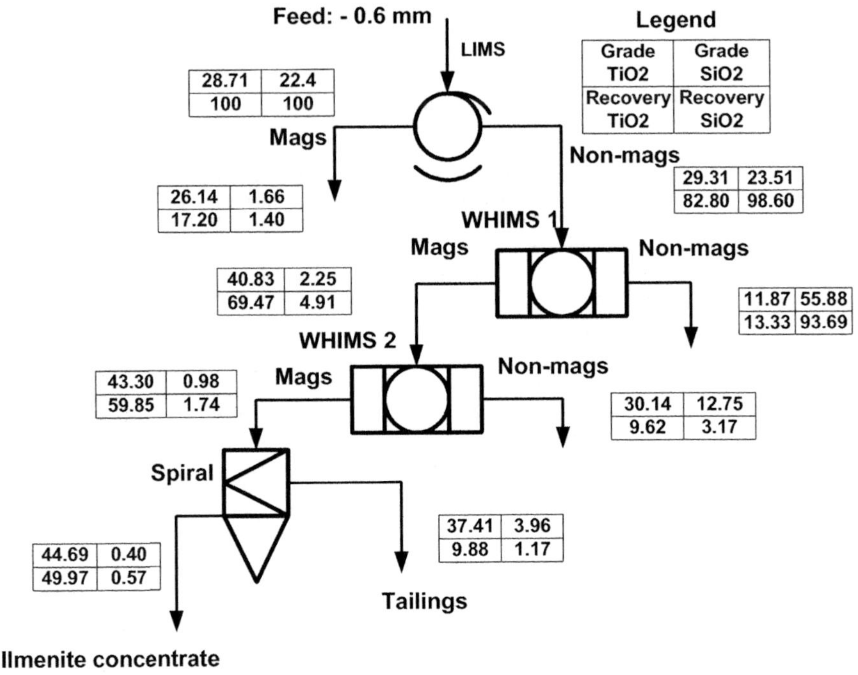

Figure 6.29: Flowsheet for the treatment of the - 0.6 mm fraction of the Rooiwater ilmenite ore by magnetic separation.

6.1.6 Beneficiation of ores of non-ferrous metals

Boliden Mineral AB, Sweden

Boliden Mineral AB is a producer of concentrates of sulphide minerals and base metals. The metals are recovered, as a rule, by flotation. However, the flotation concentrates usually contain small concentration of other metal sulphides. The lead concentrate, for instance, contains copper and zinc. To recover zinc lost to the lead concentrate, a Sala carousel HGMS is used as a cleaner, as is shown in Fig. 6.30. The non-magnetic product from the HGMS is the final lead concentrate, while the magnetic fraction, containing mainly sphalerite, chalcopyrite, galena and iron-rich silicates, is further treated by flotation [W30].

It can be seen that HGMS reduces the zinc concentration in the lead concentrate from 10% to 6%, with little loss of silver. The removal of sphalerite and the fact that the iron-containing silicates are also recovered, reduces the amount of material in the galena concentrate.

High-gradient magnetic separation was also successfully applied, on pilot-plant scale, to the upgrading of complex sulphide flotation concentrates from a Chinese beneficiation plant [Z8]. A Cu-Pb-Zn bulk flotation concentrate from a polymetallic sulphide ore was treated in a high-gradient magnetic separator with

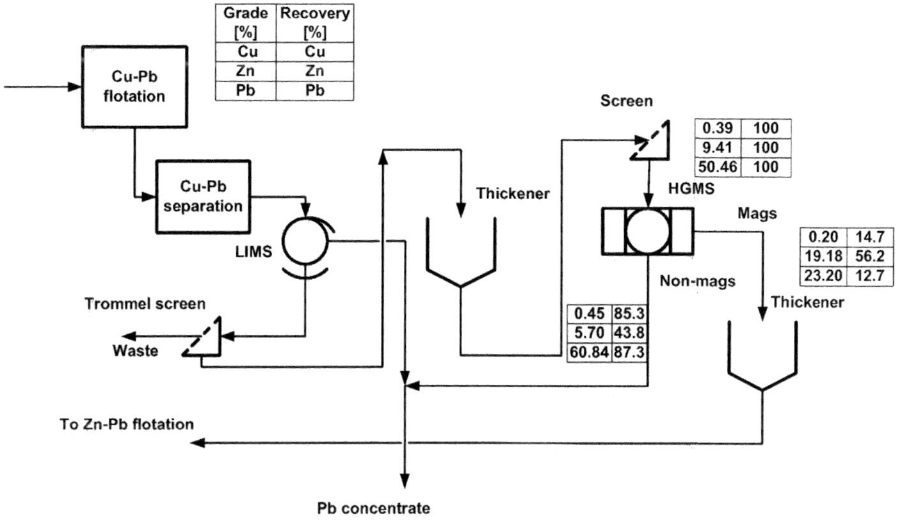

Figure 6.30: Flowsheet of the HGMS circuit at the Boliden Garpenberg concentrator, Sweden (adapted from [W30]).

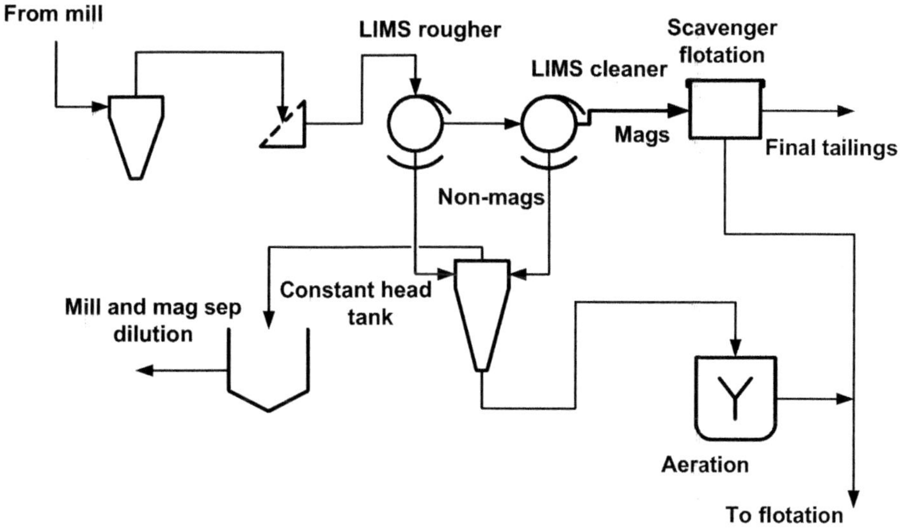

Figure 6.31: Magnetic separation flowsheet at the Black Mountain, South Africa, beneficiation plant (adapted from [H26]).

Table 6.4: HGMS of a sulphide flotation concentrate (China) [Z8].

Parameter	Feed	Magnetics	Non-magnetics
Grade [%]			
Cu	8.97	18.68	3.24
Pb	3.64	0.72	5.36
Zn	40.30	18.47	53.18
Recovery [%]			
Cu	100	77.28	22.72
Pb	100	7.34	92.66
Zn	100	17.01	82.99

Table 6.5: HGMS of the black ore Pb concentrate, Dowa Mining, Japan [K28].

Parameter	Feed	Magnetics	Non-magnetics
Grade [%]			
Cu	7.6	46.8	3.2
Pb	58.2	3.5	64.3
Recovery [%]			
Cu	100	62.1	37.9
Pb	100	0.6	99.4

a pulsation mechanism. Typical results of the investigation are shown in Table 6.4. It can be seen that, in one pass, the copper concentrate was upgraded from 8.79% Cu to 18.68% Cu at a recovery of 77.28% Cu. Lead was concentrated into the non-magnetic fraction with a recovery of 92.66%, while the recovery of zinc was 82.98%.

Pb flotation concentrate, Dowa Mining, Japan

In the flotation process of black ore, the lead concentrate contains a small amount of unfloatable copper minerals. Kim et al. [K28] found that the Pb flotation concentrate, containing 5% to 8% of Cu, can be upgraded by removing copper by HGMS. The removal of copper minerals is facilitated by the relatively high values of magnetic susceptibility of copper minerals compared to diamagnetic galena. Typical results of the removal of copper by HGMS are summarized in Table 6.5. It can be seen that the Cu grade in the non-magnetic Pb concentrate decreased from 7.6% to 3.2% while the losses of lead into the magnetic fraction were as low as 0.6%.

Black Mountain Mineral Development Co., Aggeneys, South Africa

The concentrator of Black Mountain Mineral Development Company produces three concentrates, namely copper, lead and zinc in a differential flotation process. A large portion of the Broken Hill ore body consists of magnetite

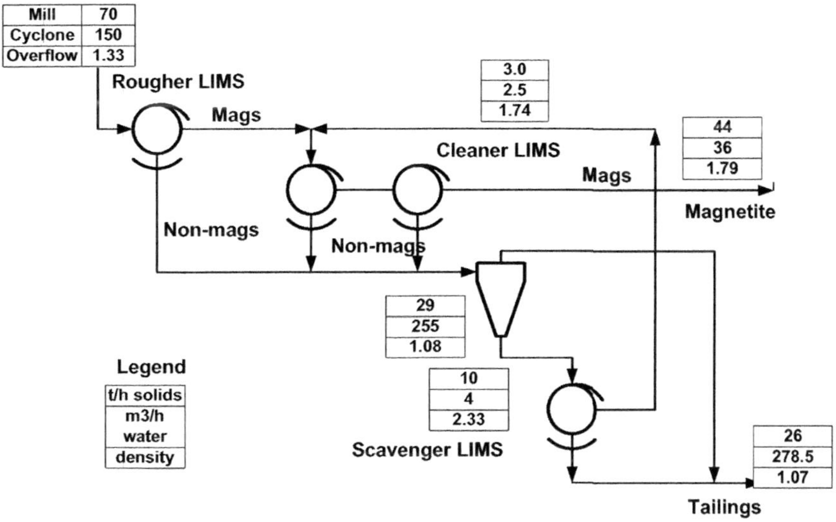

Figure 6.32: Magnetic separation plant for the recovery of vanadium from vanadiferous magnetite (Limpopo Province, South Africa).

(20% to 30%) and pyrrhotite (10% to 15%). Since a significant amount of the magnetic material can be removed by magnetic separation prior to flotation, a magnetic separation plant was installed during 1990.

Inclusion of the magnetic separation circuit, shown in Fig. 6.31, allowed an increase in the monthly milling rate of 24 kt, from 102 kt to 126 kt. As a result of the increased tonnage milled, the concentrate production increased and the operating costs per tonne milled decreased.

Vanadiferous magnetite ore, South Africa

A three-stage low-intensity magnetic separation circuit is used to upgrade vanadiferous magnetite at a beneficiation plant in Limpopo Province of South Africa. The flowsheet of the magnetic separation plant is shown in Fig. 6.32. The size distribution of the feed into the plant is 100% - 300 μm and 35% - 74 μm. The mass recovery into the magnetic concentrate amounts to 65% while the recovery of vanadium exceeds 97%.

Chromite gravity tailings, Üçköprü, Turkey

While most chromite ores are easily beneficiated using gravity concentration methods, magnetic separation is usually unsuitable as a result of close values of magnetic susceptibilities of chromite and the gangue minerals. Nevertheless, close control of the operating conditions of a magnetic separator can result in efficient upgrading of the chromite tailings by removing e.g. olivine and serpen-

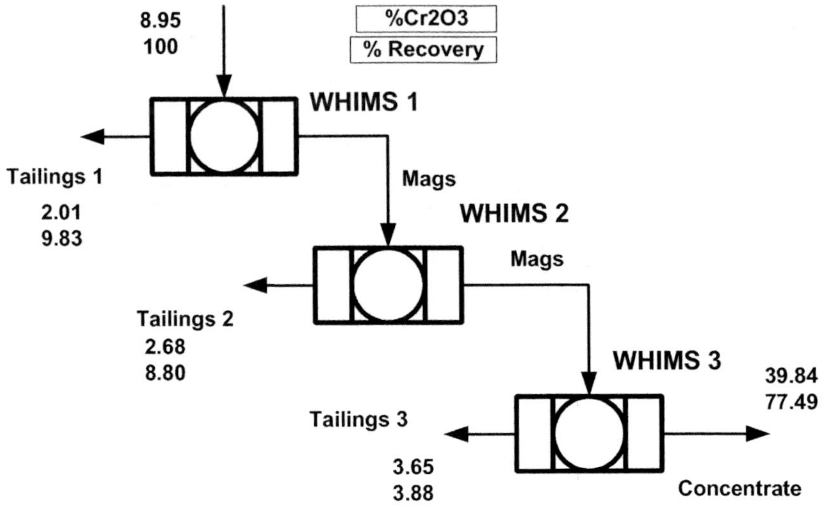

Figure 6.33: Flowsheet of the magnetic separation circuit for the beneficiation of the Üçköprü chromite gravity tailings (adapted from [A37]).

tine minerals [A37]. A three-stage wet high intensity magnetic circuit, shown in Fig. 6.33, was applied to gravity tailings assaying 8.94% Cr_2O_3. After a roughing stage and two cleaning stages the final concentrate containing 39.84% Cr_2O_3 was obtained with 77.49% recovery. A combination of magnetic separation with flotation yielded the final concentrate assaying 46.17% Cr_2O_3 with 64% recovery [A37].

6.1.7 Mineral sands

Mineral sands deposits are mixtures of commercial minerals such as ilmenite, rutile, leucoxene, zircon, monazite and xenotime coexisting with sub-economic minerals such as kyanite and staurolite and gangue such as quartz and clay. Magnetic properties of the major heavy minerals are given in Table 6.6. The heavy minerals content of the commercially viable ore bodies varies between 5% to 20%. The principal producers of these minerals are South Africa, Australia and the USA.

Richards Bay Minerals, South Africa

Richards Bay Minerals (RBM) in KwaZulu-Natal Province is a major world producer of zircon and rutile raw materials together with titanium slag produced from ilmenite by the on-site smelter. To feed the smelter, ilmenite is produced in a wet high-intensity magnetic separation circuit.

RBM operates two mining plants, both incorporating dredging and gravity concentration, as is shown in Fig. 6.34. The gravity concentrate, which contains

Table 6.6: Typical values of mass magnetic susceptibility of heavy minerals.

Mineral	Composition	Mag. susc. $[m^3/kg] \cdot 10^{-7}$
Ilmenite	$FeTiO_3$	11.3
Leucoxene	Altered ilmenite	6.2
Rutile	TiO_2	0.16
Zircon	$ZrSiO_4$	- 0.03
Monazite	$(Ce,La,Y,Th)PO_4$	1.9

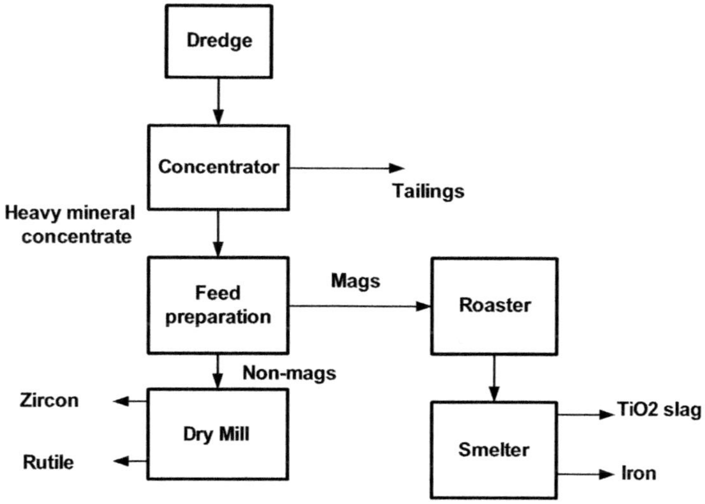

Figure 6.34: Mineral sand beneficiation process at Richards Bay Minerals [B34].

Figure 6.35: Installation of Reading wet high-intensity magnetic separators (courtesy of Roche Mining).

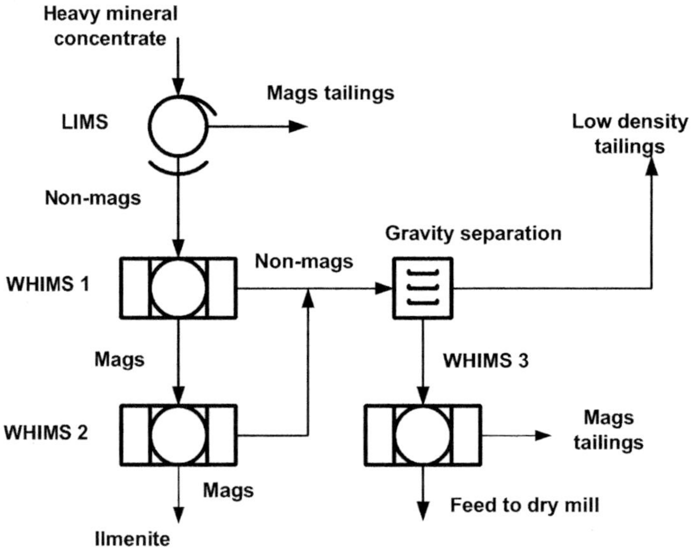

Figure 6.36: The dry mill feed preparation circuit at Richards Bay Minerals.

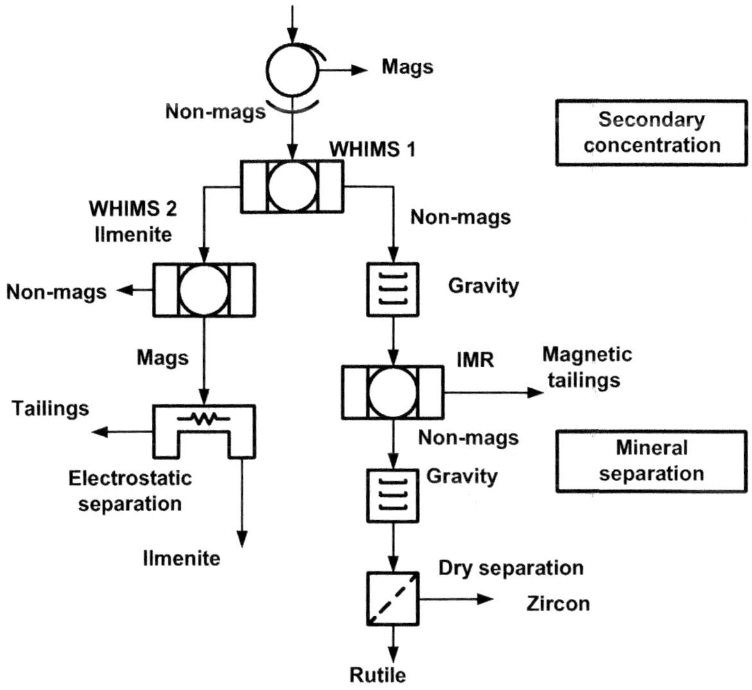

Figure 6.37: The concentration and separation circuits at Namakwa Sands (South Africa).

85% of heavy minerals, is first passed through a low-intensity wet drum magnetic separator to remove magnetite. The non-magnetic fraction from this stage is re-treated with Reading wet high-intensity magnetic separators in a complex rougher-cleaner circuit to produce a concentrate that consists largely of ilmenite containing 46% TiO_2. An installation of banks of Reading WHIMS machines is shown in Fig. 6.35.

The non-magnetic fraction from WHIMS is again gravity-concentrated and then further upgraded by the use of WHIMS to remove feebly magnetic material. The resulting concentrate is dried and fed into the "dry mill", where high-tension roll separators and induced magnetic separators are used to produce saleable rutile and zircon products.

A diagram of this feed preparation circuit for the dry mill is shown in Fig. 6.36. A total of 31 Reading 16-pole WHIMS are used at RBM. The feed rate into each WHIMS machine ranges from 23 to 31 t/h at pulp density of 40% solids. The magnetic fraction from the WHIMS machines contains, in addition to ilmenite, small quantities of weakly magnetic impurities. These are removed prior to feeding the material to the smelter by dry magnetic separation, after magnetizing roast in a fluidized bed.

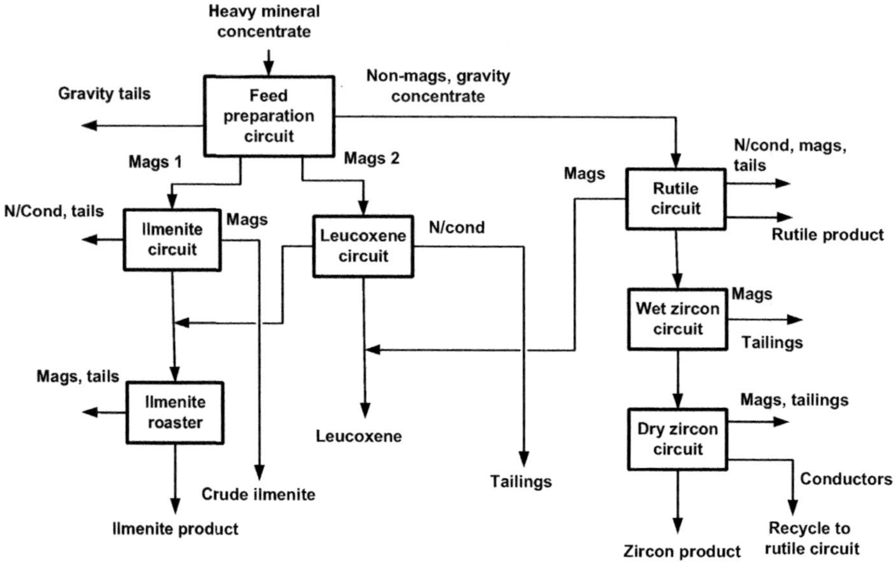

Figure 6.38: Schematic flow diagram of the Ginkgo (Broken Hill, Australia) mineral separation plant (adapted from [E8]).

Namakwa Sands, South Africa

Namakwa Sands (Anglo-American Corp.) produces heavy minerals from mineral sands from the Graauwduinen deposit in the Western Cape Province of South Africa. This fairly new operation, that was commissioned in 1994, uses a conventional concentration flowsheet based on gravity, magnetic and electrostatic separations. A simplified process flowsheet is shown in Fig. 6.37.

The primary concentration plant consists of a four-stage spiral circuit that treats - 1 mm + 45 μm feed and produces a heavy mineral concentrate assaying 90% of total heavy minerals. Low-intensity drum magnetic separators remove magnetite from this concentrate in the secondary concentration plant and the non-magnetic fraction is then processed in a four-stage wet high-intensity magnetic separator circuit. The Reading WHIMS machines produce a magnetic fraction containing ilmenite and garnet and a non-magnetic fraction containing zircon, rutile, leucoxene and other minerals. This fraction is further upgraded in a spiral circuit.

The ilmenite concentrate and non-magnetic fraction from the secondary concentrate plant are further processed in the mineral separation plant. The ilmenite concentrate is produced using electrostatic separators while separation of rutile and zircon takes place in a process consisting of magnetic, gravity and electrostatic separation and acid cleaning circuits. Permanent magnetic rolls and rare-earth drum magnetic separators are employed in this circuit.

Table 6.7: Composition of Ginkgo (Australia) heavy mineral assemblage [E8].

Mineral	Assay range [%]
Rutile (+ 95% TiO_2)	8.8 - 12.5
Zircon	6.7 - 12.6
Primary ilmenite (< 54% TiO_2)	0.7 - 21.2
Altered ilmenite (54% - 66% TiO_2)	32.4 - 49.0
Leucoxene (66% - 85% TiO_2)	4.5 - 12.6
Monazite	0.3 - 0.8
Chromite	0.7 - 1.1

Murray Basin, Australia

Extensive reserves of heavy minerals exist in the Murray Basin, an area covering parts of South Australia, New South Wales and Victoria. Several of the deposits are of potential economic size and of these the Ginkgo deposit has been identified for establishment of the first commercial operation [E8].

The mineral assemblage within the Ginkgo ore body consists primarily of ilmenite and its alteration products, rutile and zircon, with minor amounts of other minerals. The mineral composition is shown in Table 6.7.

Extensive bench-scale and pilot-plant testwork has been carried out on samples of ore from the deposit. The results of the tests and mathematical modelling of the beneficiation circuits resulted in a proposed design for the wet plant, mineral separation plant and the roaster.

As a result of wide mineralogical variations within the ore body, the wet plant is based on a combination of roughing, scavenging and cleaning spiral gravity separators and is expected to produce up to 130 t/h of heavy mineral concentrate. This concentrate will be treated at the proposed mineral separation plant to be located at Broken Hill. The block diagram of the flowsheet for this plant is shown in Fig. 6.38.

This flowsheet consists of a number of discreet circuits with minimal interaction between them. The feed preparation circuit is based on a two-stage WHIMS circuit. The magnetic product will be the feed into the ilmenite and leucoxene dry mill circuits while the non-magnetic fraction will be treated in the zircon and rutile circuits.

In the ilmenite reduction roasting circuit the concentration of the chromite contaminant in the ilmenite concentrate will be reduced from > 1.5% Cr_2O_3 in the unroasted ilmenite to < 0.3% Cr_2O_3. The non-magnetic rejects from the ilmenite roasting can be upgraded by magnetic and electrostatic separation to yield a chromium-rich product with a grade similar to commercial chromite ores [F32].

In the initial years of operation, 10.4 Mt of ore will be mined per annum and 58 kt of zircon and 41.5 kt of rutile per annum are planned to be produced. The scheduled date for the first shipment of product is January 2005 [E8].

Table 6.8: Kaolin product specification [E9].

Specification	**Brightness**		
	Premium	Standard	Filler
TAPPI brightness [%]	89.5 - 92.0	85.5 - 89.0	80.0 - 85.5
Particle size < 2 μm [%]	80 - 100	80 - 97.0	62.0 - 95.0
Particle size + 45 μm	0.01	0.01	0.30 - 0.40

6.1.8 Industrial minerals

Kaolin beneficiation

One of the most successful applications of HGMS technology is in the brightening of kaolin clay. Essentially all economically viable deposits of kaolin are contaminated by small amounts (usually from 0.5% to 3%) of ferruginous minerals, such as iron oxides and rutile. Since these contaminants are deleterious to clay brightness, considerable effort has been expended to remove them by physical and chemical methods. Under favourable conditions chemical bleaching usually removes less than half of the iron present in clays [I3]. Although discolouring contaminants are usually very weakly magnetic and very fine (e.g. 80% smaller than 2 μm), they can be efficiently removed by HGMS, if a sufficiently large magnetic field strength, fine matrix and low flow velocity of the slurry are used [I6, L25, B35].

The high-gradient magnetic separation of kaolin clay became a practical economic process that has been thoroughly proven on a large scale worldwide. Moreover, magnetic beneficiation of kaolin became one of the rare production-scale applications of superconductivity, with Eriez Magnetics and Outokumpu-Carpco superconducting HGMS machines now well established in the minerals industry. Iron-clad resistive solenoid separators or superconducting cyclic or reciprocating magnet separators are used in several operations, with the operating magnetic induction ranging from 2 T to 5 T. [S18, R4, W5, R25].

A brief summary of the kaolin product specifications is given in Table 6.8, while Tables 6.9 and 6.10 illustrate the metallurgical performance and technical data of Cryofilter HGMS as applied to two different types of kaolin.

Limestone beneficiation

In the Gåsgruvan beneficiation plant of Svenska Mineral AB, Sweden, a two-head carousel SALA HGMS is used to remove iron oxide impurities from the limestone flotation concentrate. The quality of the non-magnetic concentrate meets the market specifications for pigment and paper filler applications [W30]. Typical production results are listed in Table 6.11.

Table 6.9: Performance of Cryofilter HGMS as applied to the beneficiation of two types of kaolin [E9].

Kaolin grade	**Feed specs**		**Product (unbleached)**	
	Grade	% Solids	Grade (gain)	Throughput (t/h)
Coating grade	54 to 84 ISO	23 to 28	67 (13) to 88 (4) ISO	5 to 40
Ceramic	77 ISO	23	81 (4) ISO	5
			80 (3) ISO	12
			79 (2) ISO	20

Table 6.10: Cryofilter HGMS performance in kaolin purification [E9].

Parameter	**Performance**
Power input	0.1 kW/t
Throughput	up to 100 t/h
Brightness gain	7 points on feed with 70 to 75 brightness
	5 points on feed with 80 to 85 brightness
	2 points on feed with + 90 brightness

Table 6.11: Limestone beneficiation at Svenska Mineral AB [W30].

Product	**Yield [%]**	**CaO [%]**	**MgO [%]**	**Fe_2O_3 [%]**
Feed	100	50.5	3.96	0.68
Flotation concentrate	85.5	53.4	1.90	0.24
HGMS concentrate	77.5	54.8	1.01	0.12

Table 6.12: Comparison of the performance of WHIMS and HGMS separators in beneficiation of glass sand [S29].

Parameter	WHIMS	HGMS
Slurry velocity [cm/s]	> 25	4.0
Target Fe_2O_3 [%]	< 0.012	
Achieved Fe_2O_3 [%]	0.0172	0.0116

Table 6.13: Comparison of 1998 costs of glass sand beneficiation by WHIMS and Eriez HI Filter HGMS [D8].

Parameter	WHIMS	HGMS
Capital cost	£200k	£220k
Running costs (power)	70 kW	10 kW
Matrix cost per month	£500 - £1000	< £50
Labour cost	Manual operation	Fully automatic

Glass sand beneficiation

Sand deposits exploited to supply the glass and ceramics industries, have, until recently, been of a low iron content to enable production with little or no processing of the sand. Screening and attrition techniques have been used, often in combination with spiral gravity concentrators. With increasing demand from the glass industry for high quality material and with diminishing high-quality reserves, more sophisticated techniques are becoming more common to meet the specifications of the glass making industry. Dry permanent magnetic rolls, wet high-intensity magnetic separators and cyclic high-gradient magnetic separators have been applied, on production scale, to beneficiation of glass sand.

Comparison of the performance of WHIMS and Eriez Magnetics HI Filter HGMS machines as applied to the reduction of iron oxides, is shown in Table 6.12, while Table 6.13 compares the 1998 costs of these two types of magnetic separators. Superior performance of the HGMS machine is a result of controlled flow velocity of the pulp through the matrix. Since HGMS is a cyclic separator, its throughput is limited, typically 10 t/h or less, while a WHIMS can operate at throughputs of the order of 50 t/h or more. The problem of the HGMS being a batch separator with low duty cycle (e.g. 50%) can be solved by installing two units operating as a pair.

Dry magnetic separation using either induced magnetic roll or permanent magnetic roll separators has been confined to relatively specialist grades produced in small quantities. Typical throughput of an installation of dry magnetic separators for sand purification is 2 to 3 t/h per meter width of the roll [F33]. Table 6.14 illustrates the efficiency of beneficiation of glass sand in a two-stage dry high-intensity magnetic separation circuit [D17].

By incorporating spiral concentrators into the separation process, the non-magnetic titanium and aluminium-bearing particles were removed even further.

Table 6.14: Comparison of glass sand ore material and a product from two-pass dry high-intensity magnetic separation [D17].

Compound	Glass sand ore	Glass sand product specs	Mag sep product
Fe_2O_3 [%]	0.035	< 0.010	0.007
TiO_2 [%]	0.105	< 0.020	0.023
Al_2O_3 [%]	0.852	< 0.060	0.031

Table 6.15: Comparison of glass sand products from three process circuits with capacity of 50 t/h [D17].

Parameter	Mag sep product	Spiral+mag sep	Floatex+spiral
Fe_2O_3 [%]	0.007	0.007	0.007
TiO_2 [%]	0.023	0.014	0.014
Al_2O_3 [%]	0.031	0.023	0.020
Recovery [%]	94.3	93.1	93.8
Equipment cost [k$], 2001	600	720	250

This improvement was, however, accompanied by higher capital and operating costs, as the feed into the magnetic circuit had to be dried.

A process circuit comprised of a Floatex density separator and spiral concentrators provided equally efficient metallurgical performance. However, this all-wet gravity circuit produced a product equal to that obtained in the spiral-magnetic process, with similar recovery and at less cost, as is shown in Table 6.15.

6.1.9 Magnetic beneficiation of coal

The emission levels of sulphur dioxide and ash particles from coal-burning facilities must meet stringent environmental standards. In order to reduce these emissions, numerous techniques for coal cleaning, including magnetic methods, have been tested over the last twenty years

Coals generally contain 1% to 5% of sulphur, of which over a third is carbonaceous in form and is referred to as organic sulphur. The rest is mineral sulphur consisting of weakly paramagnetic pyrite or marcasite FeS_2. Pyrite found in coal often has a much higher magnetic susceptibility than the pure mineral; this is presumably due to impurities or to its alteration to pyrrhotite. On the other hand, organic sulphur is closely bonded to coal so that its removal by physical methods is not possible. Consequently, virtually all work on the desulphurization of coal has been directed towards the removal of inorganic sulphur.

Table 6.16: Physical properties of coal constituents.

Property	Coal	Mineral matter	Pyrite
Mohs hardness [MHo]	2.2 - 2.5	2.0 - 4.5	6.0 - 6.5
Electrical conductivity [S/m]	10^{-6} - 10^{-11}	10^{-4} - 10^{-6}	1 - 10^{-2}
Dielectric constant	2.0 - 2.5	4.5 - 7.8	5.2 - 8.6
Magnetic susceptibility [m^3/kg]·10^{-8}	- 0.5	20 - 200	0.34 - 1.5
Density [kg/m^3]	1150 - 1500	2400 - 3900	4800 - 5000

Various techniques have been investigated for the extraction of pyrite, ranging from froth flotation to electrostatic separation to microbiological desulphurization, with varying degree of success. The physical techniques are based on often large differences between the values of physical properties of coal and pyrite, as is shown in Table 6.16. Magnetic separation has been recognized as a potentially cost-effective and efficient method of desulphurization and considerable research and development work has been conducted in this area, particularly with the advent of HGMS, OGMS and rare-earth-based permanent magnetic separators.

Although it has been extensively demonstrated that both wet and dry magnetic separation methods are able remove up to 90% of pyritic sulphur, and a significant amount of ash (up to 60%), at combustible recovery greater than 85%, no magnetic desulphurization process has proven to be sufficiently cost-effective and efficient, on an industrial scale, in removing pyrite and ash. Further research into magnetic methods of desulphurization was considerably scaled down in the nineteen nineties. This scale-down coincided with identification of cleaning options for environmental control strategies by the US Department of Energy, magnetic methods not being among them [B36].

Wet HGMS of coal

Wet high-gradient magnetic separation used to be a popular technique for laboratory and pilot plant investigations into coal desulphurization. Numerous research papers and reports were published in the seventies and eighties of the last century. They are often associated with the names of Trindade [T8], Luborsky [L14], Wechsler [W24] and Male [M35, M36]. The results show that the efficiency of sulphur and ash removal is strongly dependent on the origin of the coal and on the recovery of clean coal. It can be seen in Fig. 6.39 that with increasing reduction of ash and sulphur concentrations, the losses of clean coal into the magnetic fraction become unacceptably high. More recent reports [K29, D18] on the subject confirmed the potential of HGMS to recover high percentage of sulphur and ash, but at the cost of high losses of combustible matter.

Even though most studies suggest that desulphurization by wet HGMS is

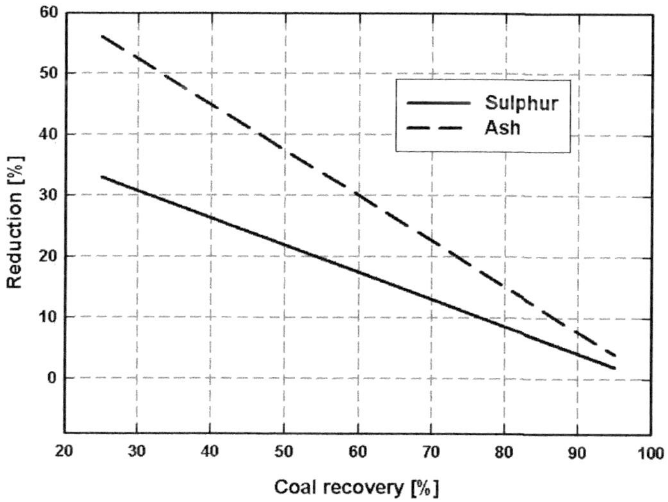

Figure 6.39: The reduction in sulphur and ash as a function of the recovery of clean coal (adapted from [L14]).

technically feasible, there are major problems that must be addressed before a commercially viable system can be realized. The first is the cost of preparation of the coal and subsequent dewatering and handling of the product. The second is the non-selectivity of separation, causing an excessive amount of coal to be lost in the magnetic fraction. Because of the high cost of the equipment and loss of coal, wet HGMS does not appear economically feasible at present. A detailed description of the research and development effort in coal beneficiation by magnetic means, up to 1985, was given by Male [M36].

Dry magnetic separation of coal

Studies of dry HGMS of unclassified coal have generally shown that the separation of sulphur and ash from clean coal is non-selective, mainly due to the presence of the fines. Classification and removal of fines, particularly of particles smaller than 10 μm, considerably improved the performance of magnetic separation [L14, H27].

More encouraging results have been obtained using rare-earth permanent magnet roll separators. Feasibility tests on Western Kentucky (USA) coals resulted in ash reduction up to 75% and total sulphur reduction up to 27%. The recovery of the combustible matter ranged from 85% to 95% [P16]. Similar reduction in the ash concentration was achieved in several UK coals [B37, S87], although the technique proved again to be rather inefficient in removing sulphur.

Dry OGMS of coal The application of a split-coil superconducting magnet to the beneficiation of coal was investigated by numerous authors [H28, C25, P17]. The efficiency of removal of ash and sulphur was generally lower compared to that obtained in wet HGMS, at a similar level of losses of the combustible matter. The selectivity of separation improved considerably when the fines smaller than 40 μm were removed, while with wet HGMS the limiting minimum particle size was found to be 20 μm [P17].

Enhancement of magnetic properties

The efficiency of the magnetic separation process could be improved by increasing the magnetic susceptibility of pyrite. Numerous means of enhancing the magnetic properties of pyrite have been proposed over the years, few of the them efficient enough, and probably none of them economically justifiable on an industrial scale.

In the *Magnex* process [K30], dry pulverized coal is heated to 170 ^{0}C and then subjected to iron pentacarbonyl vapour. The iron is selectively deposited on the mineral matter, which is removed by a magnetic separator. Reductions of up to 85% of pyritic sulphur and 26% of ash were reported. Distinct drawbacks of the process are toxicity and high consumption of the chemical agent.

Conversion of pyrite to ferrimagnetic pyrrhotite takes place during semi-coking and pyrolysis of coal, in the temperature range from 500 ^{0}C to 700 ^{0}C. Pyrolysis followed by dry magnetic separation is thus a possible mechanism for coal cleaning [Y8, K31]. Reduction of 61% of sulphur with 57% of combustible recovery was obtained after a 30 minute exposure of coal to a temperature of 450 ^{0}C [K31].

To avoid heating coal and wasting energy, it is possible to convert pyrite selectively to pyrrhotite by dielectric heating [B38, B39]. As a result of differences in dielectric properties of coal and pyrite, pyrite heats more rapidly than coal [B38]. Although significant increases in the magnetic susceptibility of pyrite can be expected, the results of the microwave treatment of coal are often disappointing. Although ash removal of the order of 20%, at the loss of calorific value of 17% has been obtained, sulphur tends to report into the non-magnetic product of clean coal. However, when coal was mixed with aqueous NaOH solution and treated with microwaves, sulphur reduction as high as 40% was achieved [B39].

6.2 Nuclear industry

6.2.1 Removal of solids from fuel dissolver liquor

The reprocessing of the spent irradiated fuel from a nuclear reactor consists of separating the fuel into three ingredients: unburnt uranium, highly radioactive fission products and plutonium. Concentrated nitric acid is usually used to dissolve the fission products; however, not all of these are soluble in nitric acid. In the dissolution of oxide fuels, a significant amount of insoluble matter arises

Table 6.17: Mass magnetic susceptibilities of insoluble fission products [M37].

Product	χ_∞**[m^3/kg]·10^{-8}**	χ **at 1.2 T [m^3/kg]·10^{-8}**
UO_2fuel	1.29	6.43
15% PuO_2 fuel		10
40% PuO_2 fuel	6.13	6.39

(0.1% to 0.3% by mass) which contains the fission products Ru, Rh, Mo, Te, Zr, Pd, with particle size ranging from 0.1 μm to 10 μm.

Although the centrifuging of the dissolution streams is a well-established and efficient technique [W31], as a result of some mechanical and thermal problems, HGMS was proposed [W32] as an alternative method for removing insoluble fission products from the reprocessing streams.

Almost all solids that arise at different stages in the reprocessing of nuclear fuels are transition elements or compounds thereof, and have moderately high magnetic susceptibilities, as shown in Table 6.17. The removal efficiency of fission products with mean particle size of 2 μm, by HGMS operating at 1.5 T was found to range from 85% to 98% [W33].

6.2.2 Removal of ferric hydroxide flocs

Waste-water streams from nuclear power plants often contain significant portions of iron. If the pH is adjusted to a value greater than 3, the iron precipitates out as ferric hydroxide flocs. It has been found that about 60% of total radioactivity and approximately 99% of the α-emitting radioactive contaminants are precipitated with ferric flocs and can therefore be removed with it [W32].

Precipitates of this type are difficult to separate with conventional equipment, such as filters or centrifuges, and the mechanical complexity of such equipment does not make it suitable for operating in a radioactive environment. It has been shown by Harding and Baxter [H29] that iron and radionuclides can be removed using HGMS with an efficiency of 93% to 94%.

6.2.3 Soil decontamination

DOE Defence Complex, Nevada, USA

It is estimated that more than two million cubic meters of soil are contaminated with plutonium, uranium and fission products in the Department of Energy (DOE) Defence Complex (USA). The US Department of Defence has additional contaminated sites. In order to remediate these sites, several physical techniques for extraction of radioactive materials were tested by Lockheed Environmental Systems and Technologies Company (LESAT), in cooperation with Los Alamos National Laboratory (LANL).

Since many fission products and all plutonium and uranium compounds are paramagnetic, while most soil components are diamagnetic, magnetic separation

Table 6.18: Volume magnetic susceptibility of selected compounds and elements [A38].

Compound or element	Susceptibility $\times 10^6$[SI]
UO_2	1204
Cr_2O	3844
Pd	805
Am	707
Pu	636
U	411
PuO_2	384
RuO_2	107
UO_3	41
ThO_2	-7
ZrO_2	-8
MgO	-11
SiO_2	-14
Graphite	-14
Al_2O_3	-18

of paramagnetic contaminants from soil is possible. Table 6.18 summarizes volume magnetic susceptibilities of selected compounds and elements.

In many situations, the highest concentration of contaminants is found in size fractions of soils smaller than 50 μm and often smaller than 10 μm. While most physical separation techniques are inefficient in this size range, HGMS can treat particles down to quasi-colloidal size. High-gradient magnetic separation was thus identified as a technique that can selectively extract numerous radioactive materials from soils while producing a minimum of secondary waste [A38].

Under a detailed experimental programme at LANL, tests were conducted with slurries of CuO with 4 μm mean diameter and volume magnetic susceptibility of 242×10^{-6} (SI), at a concentration of 400 ppm. The slurry was passed through an ultrafine steel wool matrix 300 mm long placed in the warm 150 mm diameter bore of a cyclic superconducting magnetic separator. The magnet generated a background magnetic induction of 2 T or higher. The separation efficiency of CuO particles was found to be generally greater than 98% [P18]. In addition, less than 6% of the soil smaller than 45 μm reported into the magnetic concentrate, so that the volume of soil that must be stored in a repository, at very high cost, is considerably reduced. The results are shown in Table 6.19.

In spite of promising results, the soil remediation project was terminated, as a result of budget and schedule constraints [P25].

Chernobyl Nuclear Power Plant, The Ukraine

After the Chernobyl Nuclear Power Plant accident in 1986, large areas in the 30-km zone were uninhabitable as a result of radioactive contamination. In order

Table 6.19: Mass yield into the HGMS magnetic concentrate of the - 45 um fraction of various soil samples. The applied magnetic induction: 2 T [A38].

Soil sample	**Soil fraction (<45 μm) [%]**	**Magnetic fraction (< 45 μm) [%]**
Fernald (0.5 T)	66.1	4.9
INEL (Idaho)	63.9	5.2
Los Alamos	26.2	3.4
Oak Ridge	53.2	5.5
Rocky Flats	47.8	5.8
Savannah River Site	21,8	5.0

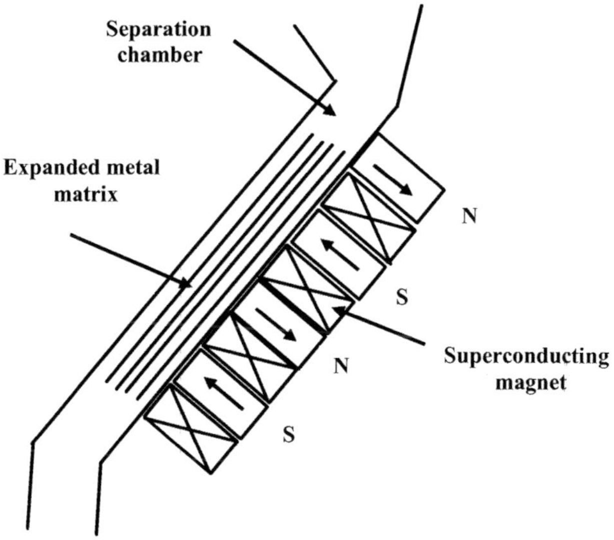

Figure 6.40: A dry magnetic separator for decontamination of radioactive soils at the Chernobyl Nuclear Power Plant (adapted from [P19]).

Table 6.20: Decontamination of Chernobyl soil samples by dry magnetic separation. K = initial activity/activity of non-magnetics, E = activity of magnetics/initial activity [P19].

Soil	Radionuclide	K	E [%]
Clay	^{137}Cs	2.1	52
	^{134}Cs	2.3	56
	^{144}Ce	3.5	71
	^{106}Ru	3.1	68
Sand	^{137}Cs	7.1	86
	^{134}Cs	7.7	87
	^{144}Ce	12.1	92
	^{106}Ru	10.3	90

to investigate the possibility of soil decontamination, a planar superconducting magnetic system shown in Fig. 6.40. was tested [P19]. A separation chamber filled with an expanded metal matrix was placed on the surface of the magnet, which generated a magnetic induction of 1.2 T inside the chamber. In order to propel dry material through the matrix and to eliminate entrainment of non-magnetic particles in the matrix, the inclined chamber was vibrated at a frequency of 100 Hz.

It was observed that the efficiency of the removal of contaminants depended on the types of soil and decontaminants and not on the degree of contamination. Typical results of removal of radionuclides ^{137}Cs, ^{134}Cs, ^{144}Ce and ^{106}Ru, which contribute approximately 98% of the total radioactivity, are listed in Table 6.20.

6.3 Waste water treatment

6.3.1 Steel industry

In modern steel plants, the volume of water used for metal production varies from 100 to 200 m^3 per tonne of raw steel produced. About one half of the total volume of dirty water produced originates in hot mills, the principal contaminants being mill scale and oils. Because many of the fine suspended particles in these waters are magnetic, high-gradient magnetic separation offers an alternative to conventional sedimentation and flocculation in clarifiers. The early studies by McNallan et al. [M38] and Harland et al. [H30] showed that HGMS provides comparable efficiency at substantially higher flow rates, lower space requirements and considerably lower running costs.

A detailed investigation of the performance and cost effectiveness of the HGMS as applied to the filtration of steel mill waste water was carried out by Yano and Eguchi [Y9]. They found that while the installation costs of HGMS and a sand filter are almost the same, HGMS surpasses the sand filter with regard to operating costs and space requirements. Comparison between HGMS

Table 6.21: Comparison between HGMS and sand filter for waste water treatment from the scubber for an oxygen converter [Y9].

Parameter	HGMS	Sand filter
Inlet concentration of solids	150 mg/ℓ	150 mg/ℓ
Target solids in effluent	< 15 mg/ℓ	< 15 mg/ℓ
Throughput	20 m^3/h	20 m^3/h
Flow rate	$6{\cdot}10^{-2}$ m/s	$8{\cdot}10^{-3}$ m/s
Feed time	20 min	8 hours
Flush time	1 min	15 min
Back-flush water volume	16 m^3/d	30 m^3/d
Space requirement	4.5 m^2	8 m^2
Relative operating cost	0.7	1
Relative installation cost	1.1	1
Magnet input power	5 kW	-

and sand filter as applied to the filtration of waste water from an oxygen converter scrubber is shown in Table 6.21.

The first full-scale HGMS device for steel mill water treatment was installed in 1977 at the Chiba Plant of the Kawasaki Steel Corporation (Japan) to treat scrubber waste water from the vacuum degassing process. The operating conditions and the results of magnetic filtration are summarized in Table 6.22.

In subsequent years several HGMS machines were installed in Japan and other countries, to remove suspended solids from waste water from hot rolling mills, continuous casting plants and others. Figure 6.41 illustrates the installation of a Metso Minerals high-gradient magnetic filter at a steel plant.

More recently a high-gradient magnetic filter has been installed at the Chiba Plant to remove fine iron particles from a coolant used in a cold rolling plant [T10]. The coolant that is used in cold rolling processing of steel sheets usually contains 1% to 10% of emulsified oil. With increasing concentration of fine solid particles in this recycled coolant, surface defects develop on the steel sheets. By removing iron particles from the coolant by HGMS, viscous oil is captured, together with iron particles, on the matrix. In order to reduce the loss of oil and to eliminate matrix blockage, the viscosity of oil is reduced with steam and the matrix is then flushed with hot water and compressed air. [T10]. With this procedure, the matrix can be kept clean for two to six months. The filtration efficiency is a function of the solids concentration, as is shown in Fig. 6.42.

In spite of considerable advantages of HGMS, most steel plants are still using conventional filtration techniques such as sand filters. One of the reasons is relatively high energy consumption of the magnet. In order to overcome this problem, Franzreb et al. [F23] developed an HGMS based on rare-earth permanent magnets. The magnetic circuit is designed in such a way that the magnetic field in the separation chamber can be switched on and off by rotating the permanent magnet system in and out of the iron yoke which magnetizes the matrix. Pilot-plant tests at several German steel mills are reported to show

Table 6.22: Magnetic filtration results and operating conditions of HGMS as applied to gas scrubber waste water treatment at Chiba Works [T9].

Parameter	**Specification**
Solid constituents	Fe and Mn oxides
Particle size	< 100 μm, 85% - 20 + 5 μm
Concentration of solids	100 mg/ℓ
HGMS effluent	< 20 mg/ℓ
Back flushed sludge	1000 to 3000 mg/ℓ
Filtration efficiency	> 80%
Magnetic induction	0.3 T
Flow velocity	max. 0.08 m/s
Matrix	Steel wool, 100 μm
Matrix height	150 mm
Canister diameter	2.1 m
Throughput	900 m^3/h
Input power	20 kW
Feed time	40 to 90 min
Flush time	1 min
Mass	40 t

Figure 6.41: Metso Minerals high-gradient magnetic filter installed at a steel plant (courtesy of Metso Minerals).

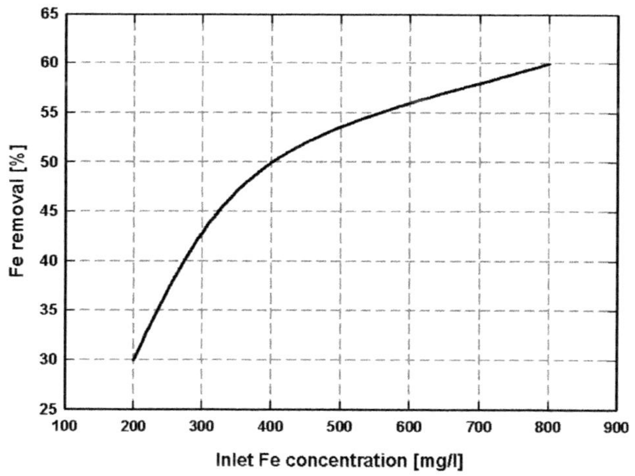

Figure 6.42: The iron removal from a cold rolling mill coolant, as a function of inlet iron concentration. B = 0.3 T, v = 400 m/h (adapted from [T10]).

performance comparable to that of electromagnet-based HGMS. The efficiency of solids removal from a rolling mill effluent, as a function of the feed rate, obtained using the Franzreb's separator, is shown in Fig. 5.78.

6.3.2 Thermal power plants

Corrosion products (crud), in the primary coolant of thermal power plants, are deposited on the surfaces of circuits, where they interfere with fluid flow and heat transfer. Corrosion products can also become radioactive in nuclear reactors, thereby leading to the irradiation of personnel and the restriction of access. HGMS was found to be effective for the removal of these fine, predominantly ferromagnetic particulates.

Heitman [H31] described the operational experience with magnetic filters in conventional thermal power plants. The efficiency of removal of iron oxide from clean turbine condensate ranged between 90% and 99%. Table 6.23 describes the filtration efficiency and selected technical data of these magnetic filters.

In nuclear power plants, crud composition varies between pressurized water reactors (PWR) and boiling water reactors (BWR). The principal dose-associated radionuclides are Co^{58}, Co^{60}, Fe^{59} and Mn^{54}. Under normal PWR operating conditions of 300 ^{0}C, the only stable oxides are magnetite and other substituted spinel ferrites of Co, Ni and Mn, which are ferromagnetic.

In BWRs, system conditions vary from approximately 300 ^{0}C at the core to 50 ^{0}C to 100 ^{0}C at the feed water purification system. Under these conditions, about 30% of the oxides are hematite, the balance being formed by Fe_3O_4 spinels

Table 6.23: Water filtration by HGMS in conventional power plants [H31, M39].

Parameter	**Specification**
Filtration efficiency	90% (total iron)
Influent Fe concentration	< 1 mg/kg
Effluent Fe concentration	< 0.02 mg/kg
Magnetic induction	0.2 T
Flow velocity	0.3 m/s
Feed time	3 to 7 days
Matrix	6 mm balls
Matrix height	1000 mm
Throughput	100 to 1000 m^3/h
Input power	35 kW at 700 m^3/h
Diameter of magnet	1.5 m at 700 m^3/h

as found in the PWR [E10].

In the USA, the application of magnetic filtration to the primary and secondary sides of PWRs was studied jointly by EPRI and Westinghouse [T11]. The results indicate that the treatment of 10% of the steam generator secondary flow can reduce sludge accumulation by 70% to 80%, and that treatment of 0.5% of the primary coolant has the potential to reduce radioactivity by a factor of 2 to 5.

Heitmann et al. [H32] studied the performance of various types of magnetic filters in the purification of feed water in PWR with Fe solids concentration of the order of 10 to 20 ppb. It was observed that, depending on the type of matrix, the efficiency of filtration ranged from 85% to 99%, while the maximum time of operation amounted to 650 days with a combination of steel ball and steel wool matrix.

Filtration of the cooling water from the multiple force circulation circuit (MFC) and of a condensate of the steam separator of the St. Petersburg (Russia) nuclear power plant was investigated by Alikhanov et al. [A39]. The filtration efficiency of a permanent magnet-based magnetic filter generating 0.2 T in the expanded metal matrix ranged from 67% to 73% with the solids concentration in the influent ranging from 9 to 29 $\mu g/\ell$. The dose rate based on the measurement of γ-radiation at various sections of the matrix, and shown in Fig. 6.43, illustrates that the large majority of radionuclides were retained by the upper part of the matrix. The measurement of the specific activity of the coolant also revealed that the highest retention, as high as 99%, was obtained with the radionuclides present at high concentrations, such as Zr^{54} and Nb^{95}.

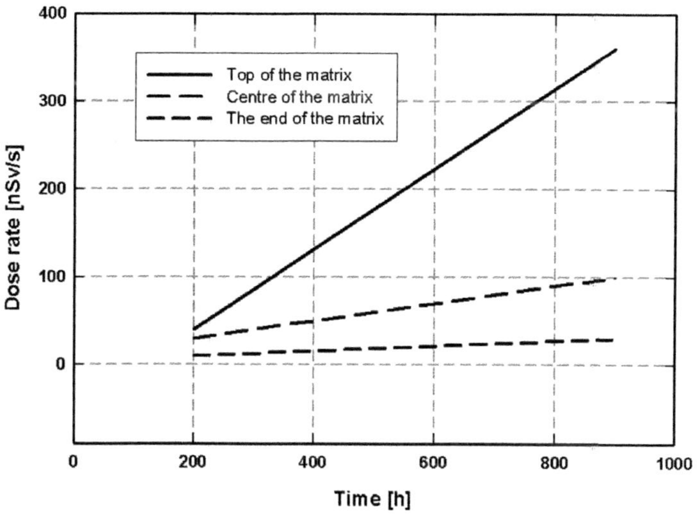

Figure 6.43: The γ-radiation dose rate from the matrix of a magnetic filter, as a function of the duration of filtration (adapted from [A39]).

6.4 Magnetic carrier techniques

6.4.1 General principles

Magnetic separation can be extended to materials that are not naturally magnetic by introducing a material (a carrier) which can selectively enhance the magnetic properties of a non-magnetic material (a target) that needs to be separated. Magnetic carriers are of particular interest for the separation of fine particles, colloids and organics, which are slow or difficult to separate by conventional methods. The magnetic carrier technique has been applied in fields as diverse as mineral separation, effluent treatment, biological cell separation and drug delivery.

The action of magnetic carrier particles is to enhance the magnetic properties of a non-magnetic target material so that its magnetic properties are sufficient to make the target material separable in conventional magnetic separators. In order to achieve that, only a small amount of a strongly magnetic material is necessary. The initial mass magnetic susceptibility of ferrosilicon or magnetite is approximately 3×10^{-4} m^3/kg, while weakly magnetic materials, with susceptibility of the order of 2×10^{-7} m^3/kg can be fairly easily separated in a variety of high-intensity magnetic separators. Magnetic polarization of such a weakly magnetic material, of density 5000 kg/m^3, placed in an external magnetic field of 0.2 T, is 0.2×10^{-3} T. Since the saturation polarization of the industrial-grade magnetite is 0.45 T (see Table 5.16), only 0.05% of magnetite is thus required to cause a magnetic polarization in the target material sufficient for its recovery

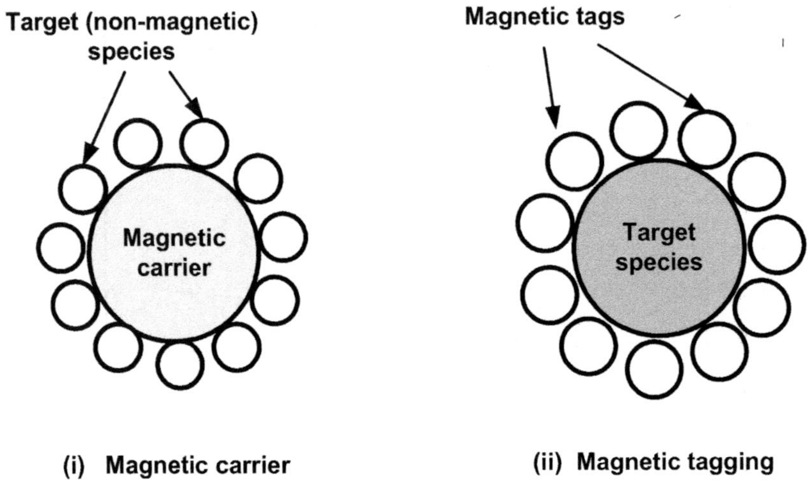

Figure 6.44: Schematic representation of magnetic carrier and magnetic tagging techniques.

by a magnetic separator operating at 0.2 T. It can thus be seen that a volume fraction of the magnetic carrier material in the range 0.05% to 1% is enough to make the target material sufficiently magnetic for easy separation.

There are two basic mechanisms for using a magnetic particle so that it renders a non-magnetic particle magnetic, namely as a magnetic carrier or as a magnetic tag. These mechanisms are schematically shown in Fig. 6.44.

Magnetic carriers are usually 10 to 1000 times larger than the target species [M40]. By varying the surface characteristics, selective recovery of non-magnetic colloidal and macromolecular species can be achieved by attaching the desired species onto the surface, or entrapped within, magnetic carrier particles, as is illustrated in Fig. 6.44. Fine magnetite and polymer-based beads coated with magnetite are frequently used as magnetic carriers, depending on application.

Magnetic tags, on the other hand, are usually smaller than the target non-magnetic particles. These particles, often very fine magnetite or ions (e.g. Y^{3+}), coat non-magnetic species in order to allow them to be manipulated by an external magnetic field. In view of their small size the mean magnetization of the cluster is low and large magnetic fields are often required for their recovery.

6.4.2 Effluent processing and metal ion removal

Magnetic seeding

The earliest examples of the use of magnetic carriers are in effluent treatment using ground magnetite to remove organic impurities from industrial waste waters. In this process a coagulant (e.g. alum) was added to the feed stream,

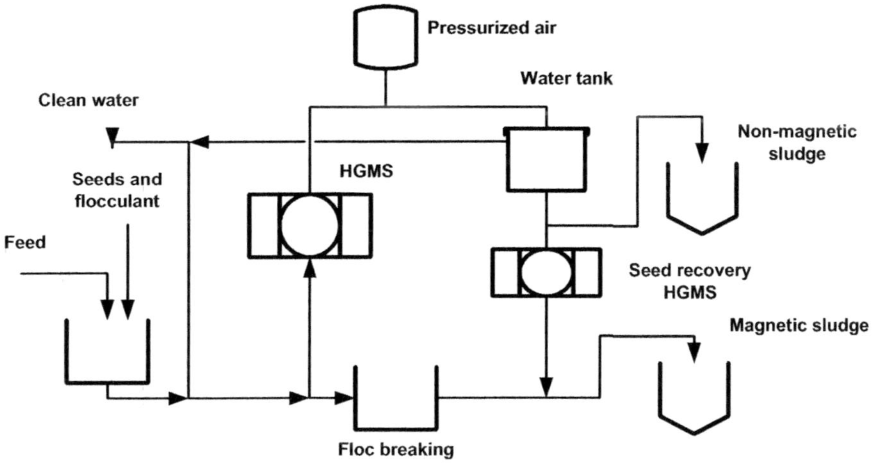

Figure 6.45: Magnetic filtration process with a magnetic seeding technique (adapted from [K32]).

at concentrations ranging from 10 to 200 mg/ℓ, followed by the addition of the magnetite seed in a concentration from 50 to 5000 mg/ℓ. The magnetic particles coagulated with the non-magnetic impurities, thereby allowing the flocs to be recovered by a magnetic separator. The flowsheet of the process is shown in Fig. 6.45.

De Latour and Kolm [D19, D20] and Bitton and Mitchell [M41, B42] conducted some pioneering work on magnetic filtration of river water using the magnetite seeding technique. The results demonstrated the efficiency of this technique for the removal of bacteria, viruses, algae and non-magnetic solids and for improving turbidity and colour of the water. The removal efficiency was around 95% and was strongly dependent on the pH.

The development of an industrial-scale process was carried out in the late nineteen seventies at Sala Magnetics. The applicability of magnetic seeding filtration to sewage and combined storm and sewage overflow was studied on pilot-plant scale and a comparison between HGMS and other sewage treatment processes is shown in Table 6.24. [A40, A41].

Selective removal of oil, phosphorus and heavy metals from waste water by a combination of magnetic seeding, chemical coagulation and high-gradient magnetic separation has been demonstrated in several reports [B43, O9,T12].

In order to overcome limitations of a cyclic operation of high-gradient magnetic filters and high energy consumption, Isogami et al. [I5] developed a continuous superconducting filter for the treatment of sewage water. By the application of the magnetite seeding technique, the level of biological oxygen demand was reduced by 90%, chemical oxygen demand by 79% and concentration of suspended solids by 97%. Figure 6.46 shows the pilot plant at Hitachi, Ltd.

Table 6.24: Comparison between HGMS and other sewage treatment processes [O8].

Process	**Removal** [%]		
	Solids	COD	Bacteria
HGMS	88 - 95	60 - 75	99 - 99.9
Chemical clarification	60	-	-
Activated sludge treatment	55 - 95	70 - 80	90 - 98
Physical chemical treatment	99	80	99

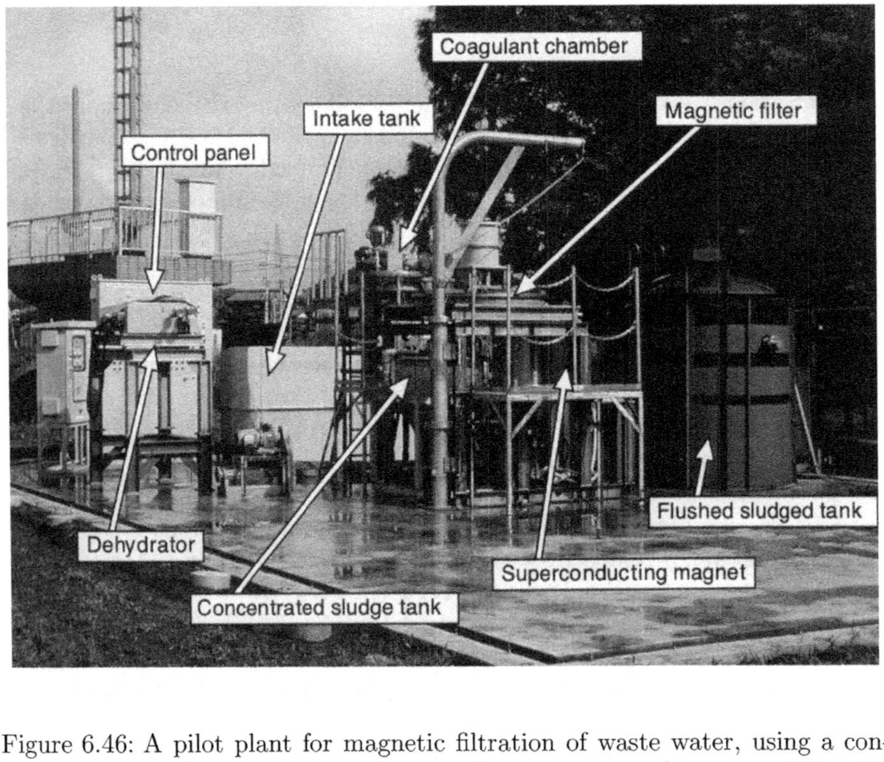

Figure 6.46: A pilot plant for magnetic filtration of waste water, using a continuously operating superconducting magnet (courtesy of Hitachi, Ltd.).

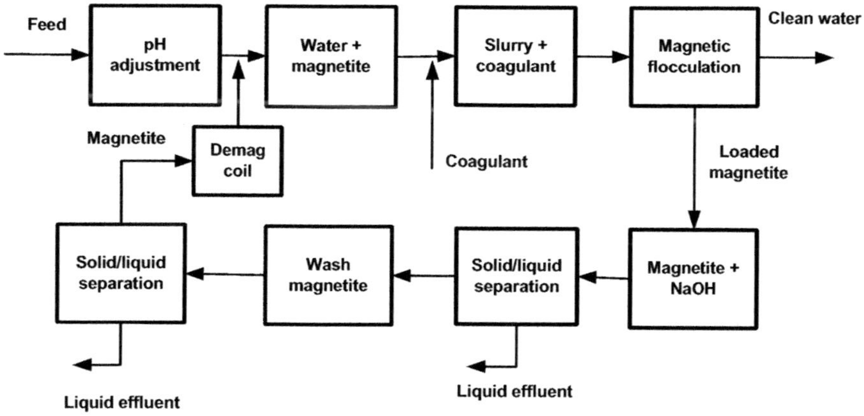

Figure 6.47: Flowsheet diagram for water treatment with reusable magnetite particles (adapted from [A42]).

(Japan).

Magnetic coating

This process relies on the selective adhesion of fine magnetic particles to the surfaces of target materials, as is shown in Fig. 6.43(ii). Fine particles, usually smaller than 2 μm, are required to form resilient coatings on the minerals [P20, P21]. The selective removal of calcite and dolomite, from mixtures of apatite, barite and scheelite, by selective magnetic coating, or tagging, has been demonstrated at laboratory and pilot plant scale [P20]. Other mixtures that were separated are quartz/fluorite and metallic lead/copper.

Coating of particles with magnetic tags has not yet been exploited to any significant extent on industrial scale. However, tags can be potentially used in recovering suspended particulates from a wide spectrum of materials and magnetic coating can provide a basis for selective separation of minerals, effluent particles and organic wastes [M40].

Sirofloc^{TM} process

The Sirofloc^{TM} process developed at CSIRO (Australia) [K33] employs the magnetite seeding technique. It is established that a surface of magnetite particles activated in acid environment is positively charged, which enables negatively charged particles of the impurities contained in the water to be captured on the surface of the magnetite particles. On the other hand, in an alkaline environment, the surfaces of the magnetite particles are charged negatively and thus repulsion of charged particles from the magnetite surface takes place. Magnetite thus acts as an impurity adsorbent in acid environment and as a desorbent in alkaline medium.

Table 6.25: River water treatment by chemically precipitated magnetite and HGMS [H33].

Parameter	Untreated water	Treated water
Colour [mg/ℓ Pt]	40	8
Turbidity [ZF]	11.5	1.5
COD [mg/ℓ]	11.5	1.5
Fe total [mg/ℓ]	0.23	0.01
No. of psychrophylic bacteria/mℓ	1000	130
No. of coliform bacteria/mℓ	13	0.02

Kolarik's results [K33] were tested on pilot-plant scale, the process flowsheet of which is shown in Fig. 6.47. It was demonstrated in several plants that have been in operation in Australia and UK that the process can produce water which complies with the EC Drinking Water Directive [B40].

The magnetite particles used in the Sirofloc process are rather coarse (5 to 10 μm), which results in their low specific surface area and in the need to use a high concentration of magnetite (e.g. 10 to 12 g/ℓ). Moreover, preparation of magnetite in this size range is rather costly and complex. In order to overcome this drawback, Hencl and Mucha [H33] proposed to employ chemically precipitated magnetite [O9] prepared from waste material, such as acid pickling steel liquor or waste ferrous sulphate arising from production of titanium white. Magnetite prepared in this manner, with 90% of particles smaller than 1 μm and 54.3% smaller than 0.4 μm, was used to purify the Vltava River (Czech Republic) water. The impurity-coated magnetite was concentrated by HGMS operating at 0.5 T at a flow velocity of 0.1 m/s. As can be seen from Table 6.25, colour was improved by 79%, while turbidity and COD were reduced by 85% and 76.4%, respectively. The concentration of coliform bacteria was reduced by 99.9%.

6.4.3 Biomagnetic extraction of heavy metals

The phenomenon of bioaccumulation of metal ions from solution by a wide variety of micro-organisms has been well established for some time. A number of micro-organisms can precipitate heavy metals onto their surfaces as a consequence of their metabolism and growth. This, therefore, gives them the ability to accumulate such metals in large quantities from external surroundings.

Microbial accumulation is achieved by growing a variety of micro-organisms in solutions containing the ions of the metal to be extracted and some added nutrients. As the microorganisms grow, they accumulate the metallic species onto their surfaces with a corresponding decrease in the concentration of the metal ion in solution. If the ions are paramagnetic, the resultant coating on a microorganism makes it magnetic and magnetic separation can be used to extract the biomass from the solution. A block diagram of the process is shown in Fig. 6.48.

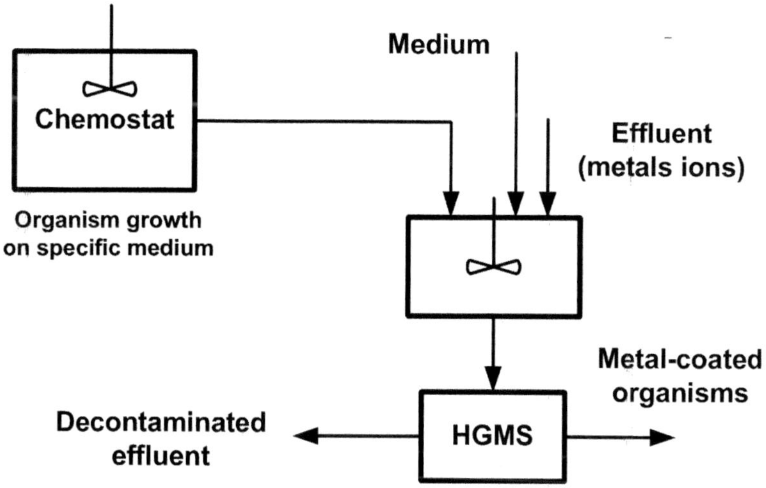

Figure 6.48: Block diagram of growth and separation of microorganisms loaded with heavy metals (adapted from [B44]).

Watson et al. [W34, B44] investigated a large number of magnetic ion species and of microorganisms and their ability to ingest or precipitate ion species onto their surfaces. The work was directed towards a study of effluents from nuclear and other industries. In most cases, starting concentrations of heavy metals of the order of 10 ppm were reduced to the 1 ppb level. Successful tests were carried out on solutions containing U, Cu, Pt, Pd, Rh, Rd, Fe, Ni and other metals.

6.5 Magnetic carriers and magnetic separation in biosciences

6.5.1 Principles of biomagnetic separation techniques

With the exception of a few naturally occurring paramagnetic or ferrimagnetic molecules or cells, such as ferritin, deoxygenated erythrocytes and magnetotactic bacteria, biological material is diamagnetic and has to be magnetically labelled or immobilized on magnetic particles. The magnetic labelling or immobilization can be performed using magnetic carriers, magnetic nanoparticles, magnetoliposomes and molecular magnetic labels (such as erbium ions) [S88].

Magnetic carriers, in the form of either magnetite or magnetic polymer particles, or beads, are being increasingly used to separate, analyze and diagnose cells or biomolecules such as antibodies, DNA molecules, proteins, hormones or antigens. Drug and reagent delivery are another area where magnetic carriers or tags are finding their application.

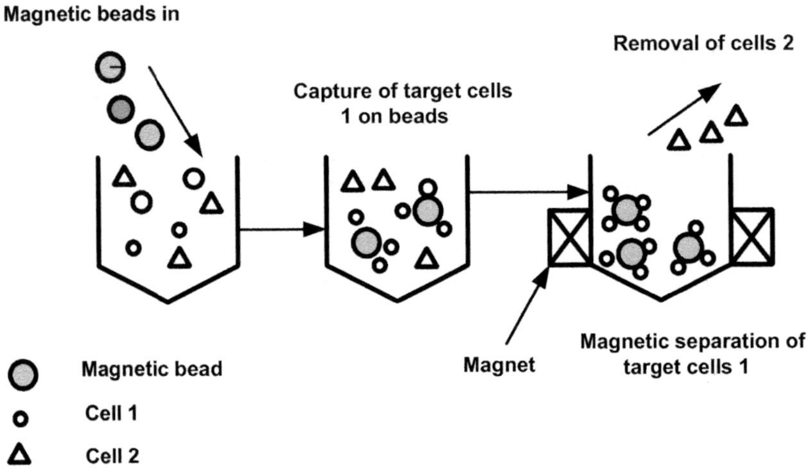

Figure 6.49: Schematic diagram of magnetic separation of cells using magnetic beads.

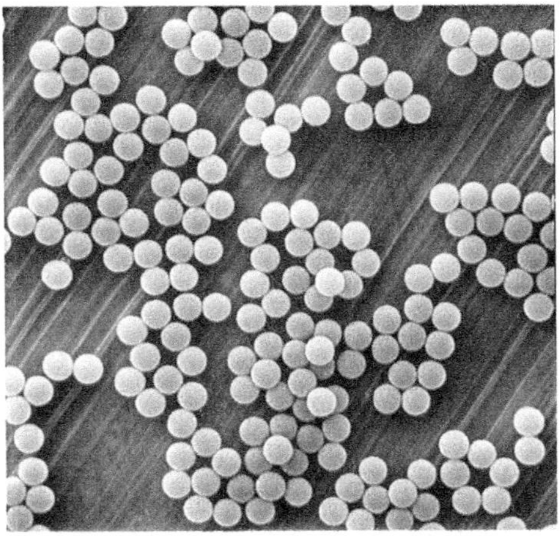

Figure 6.50: A scanning electron microscope image of 2.8 μm Dynabeads® superparamagnetic monosized polymer magnetic carriers (courtesy of Dynal Biotech ASA).

Table 6.26: Physical characteristics of magnetic carriers Dynabeads.

Head	**Dynabeads®**		
	M-450	**M-280**	**MyOne**
Diameter	4.5 μm	2.8 μm	1.05 μm
Density	1600 kg/m^3	1400 kg/m^3	1800 kg/m^3
Fe content	20%	12%	26%

The magnetic beads are usually spherically shaped particles of diameter ranging from 1 to 20 μm, and consist of a polymer matrix, often polystyrene, polyderivatives and polyvinylalcohol, in which a magnetic colloid, such as magnetite, is encapsulated. Specific affinity ligands are chemically coupled, often selectively, to the polymer matrix to separate biomolecules or cells. These ligands display complementary structures to specific sites of biomolecules or cell receptors so that these target substances specifically bind to the magnetic particles. These target structures can then be separated from the substance mixture in a magnetic field. Many magnetic carriers are biocompatible and some of them are biodegradable. The basic principle of cell separation is shown schematically in Fig. 6.49. A scanning electron microscope image of 2.8 μm magnetic carriers Dynabeads® is shown in Fig. 6.50, while Table 6.26 lists physical properties of Dynabeads®.

In addition to magnetic carriers in the micrometer size range, magnetic nanoparticles in the form of ferrofluids are often used for the labelling of target structures (usually cells) and as biodegradable or biocompatible carriers for biomedical applications. In contrast to micrometer-sized magnetic carriers, high-gradient magnetic separators are required to manipulate systems of magnetic nanoparticles.

In addition to its current progressive role in biosciences, novel developments in the biocompatible magnetic carrier technique are expected to have an enabling effect on capabilities in biological warfare defence, personnel monitoring and diagnostic and therapeutic treatments for military personnel. For instance, the Department of Defence (USA) believes that magnetic carriers are suited to provide mechanisms for monitoring and controlling biological activity [B45].

Magnetic separation, such as isolation and/or removal of specific cells, isolation of nucleic acids, proteins and heavy metals, is a common approach to exploit magnetic principles in biosciences. Biomagnetic techniques can be used in a variety of other applications. Magnetic support can be applied to immobilization of biologically active compounds and cells when magnetic carriers enable simple manipulation with immobilized structures.

Magnetic separation can also be used for modification of standard immunoassay analytical methods to determine concentrations of analytes in medical diagnostics. Ferrofluids have been studied in biomedical and clinical applications, often in connection with manipulation of drugs, viruses and tumours.

The separation of erythrocytes (red blood cells) from the remainder of the

Figure 6.51: A batch magnetic separator for biomedical applications (courtesy of Dynal Biotech ASA).

blood components using high-gradient magnetic separation has received considerable attention during the last two decades. The same technique has been successfully applied to the concentration of malarial parasites and bone marrow cells.

Magnetic separation has several advantages in comparison with standard separation procedures. Separation can often be performed directly in raw samples containing suspended solid material such as body fluids, food and clinical samples and waste water. As a result of their magnetic properties, target structures captured to specific surface-modified magnetic particles or magnetically modified affinity materials can be relatively easily and selectively removed from the sample in an external magnetic field.

A review of the current status of the application of magnetic methods in biosciences can be found, for instance, in an article by Šafaříková and Šafařík [S88].

6.5.2 Magnetic separators in biosciences

In recent years numerous magnetic separators have become available for applications in biosciences, ranging from small laboratory units based on permanent magnets, to automated cell sorters that are able to separate up to 10 million magnetically labelled cells per second. There are two basic types of laboratory magnetic separators for biomedical applications, namely batch and flow-through units [S88].

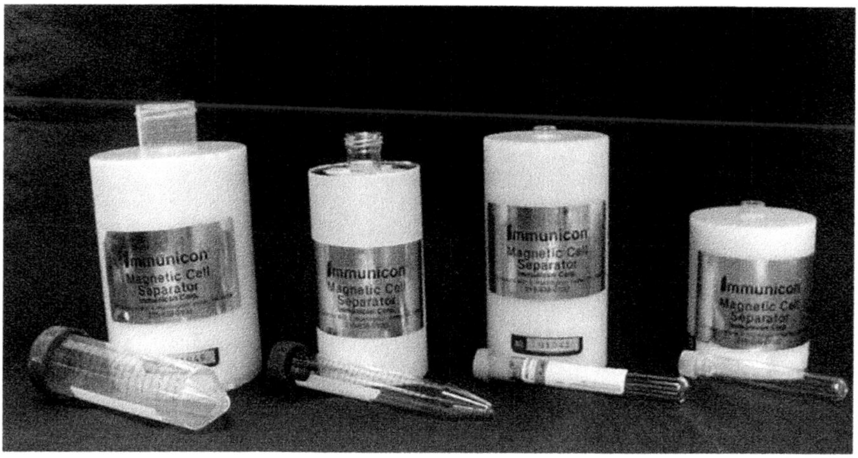

Figure 6.52: Quadrupole and hexapole batch magnetic separators for biomedical applications (courtesy of Immunicon, Inc.).

Batch magnetic separators

In most cases the isolation of magnetically labelled structures is performed in a batch mode. Test tube magnetic separators, an example of which is shown in Fig. 6.51, equipped with rare-earth permanent magnets, are used to separate magnetic microparticles from volumes ranging from $5\mu\ell$ to 50 $m\ell$. Magnetic nanoparticles require a higher magnetic force and can thus be separated in quadrupole or hexapole magnetic fields. Such QMGS and HMGS separators are shown in Fig. 6.52.

Flow-through magnetic separators

Flow-through separators are cyclic units consisting of a column packed with magnetizable stainless steel wool, placed between the poles of a permanent magnet or electromagnet. Magnetically labelled structures are passed through the column and are captured in the steel wool matrix. After the removal of the column from the magnetic field, the captured structures are removed from the matrix. High-gradient magnetic separators are used when working with sub-colloidal particles, particularly when separating magnetically labelled cells. Devices that can separate 10^{10} to 10^{11} cells in one cycle have become commercially available [S88].

Improved selectivity of separation of biological objects can be obtained when multipole d.c. magnetic fields or a.c. magnetic field are used. A flow-through quadrupole magnetic separator [Z2] generating a field gradient of 164 T/m and a hybrid a.c. magnetic device [G23] have been applied in various affinity-based separations.

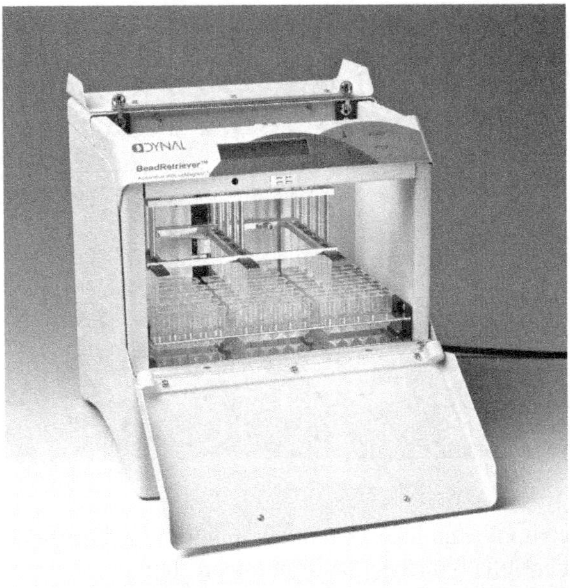

Figure 6.53: Bench-top BeadRetriever^{TM} instrument which performs automated immunomagnetic separation using Dynabeads® magnetic carriers (courtesy Dynal Biotech ASA).

6.5.3 The application of magnetic separation

Cell separation and isolation of nucleic acids

The isolation and removal of cells can be achieved using magnetic carriers as has been described above. The principle of separation is based on the fact that practically all body cells bear a specific cluster of differentiation (CD) receptors on their surface. These receptors determine specific functions of the cells in the body. By coating the magnetic particles with antibodies against respective CD receptors all types of cells can in principle be isolated from the corresponding body fluids or cell cultures. The magnetic carrier technology thus opens particularly interesting perspectives for the therapy and diagnosis of tumours. The quantitative isolation and determination of tumour cells from tissue samples can provide information on the progress of therapy. Magnetic carriers can also play a role in the separation of tumour cells during bone marrow transplants [M41].

The magnetic affinity separation is used particularly in molecular biology for isolation of nucleic acids (both RNA and DNA). The separated nucleic acids can be eluted from the beads and used for further applications [S88].

Immunomagnetic separation Immunomagnetic separation is the technique that uses antibodies immobilized on magnetic carriers, for capturing target cells. Immunomagnetic particles specific for a target organism are usually suspended in an aliquot of the mixed cell suspension. After an appropriate incubation period the immunomagnetic particles with bound target cells are separated from the suspension in a magnetic separator. The remaining suspension is removed and the magnetic particles are washed with a suitable solution. An important advantage of immunomagnetic separation is the ability to capture and detect dead or severely damaged microorganisms, which cannot be detected by standard cultivation processes [S89, Y10].

Immunomagnetic separation is often used in microbiology, cell biology and parasitology. Magnetic particles are used for the isolation of various cell subsets according to their CD receptors directly e.g. from tissue homogenates. Immunomagnetic removal of cancer cells from the blood and bone marrow of patients suffering from cancer is being practiced in numerous laboratories [S88]. A bench-top BeadRetriever^{TM} instrument that performs automated immunomagnetic separation is shown in Fig. 6.53.

Drug and radionuclide targeting

Although selective delivery of drugs to specific targets within an organism would be beneficial to many patients, conventional treatment is not able to achieve significant drug concentration in a target without distributing the drug throughout most other, usually healthy body parts and without developing drug resistance.

To circumvent these pitfalls, magnetic microspheres or nanospheres have been used for immobilization or encapsulation of drugs or radionuclides. These magnetic particles are then transported to a target position using an external magnetic field. Localized delivery of drugs at high concentration, for instance to tumours, a significant reduction of the toxic effects of most conventional anticancer drugs to healthy tissue has been achieved [P22, L26, J8, A43].

Transport of magnetic particles loaded with drugs can be achieved by magnetically guiding the particles using a high-gradient magnetic field. As a result of a short reach of such a magnetic field it is rather difficult to achieve sufficient focus and drug accumulation in a tissue that is located deep inside a body. A different principle of localizing a drug into the tumour area is illustrated in Fig. 6.54. A suspension of magnetic particles can be injected through a catheter into the artery feeding the tumour while a magnetic field is imposed onto the tumour area. The magnetic field can be conveniently generated by permanent magnets located outside the patient's body close to the tumour [K34].

Tumour regression through selective magnetic delivery of cytotoxic agents to the tumour site has been observed in numerous clinical cases, such as brain tumour and kidney and colon carcinoma. In order to increase the rate of drug delivery, the size of magnetic carriers should be as large as possible, compatible with the fact that the smallest blood vessels have a 5 μm inner diameter. Iron oxide particles are well tolerated and magnetite-based agents are approved for human use.

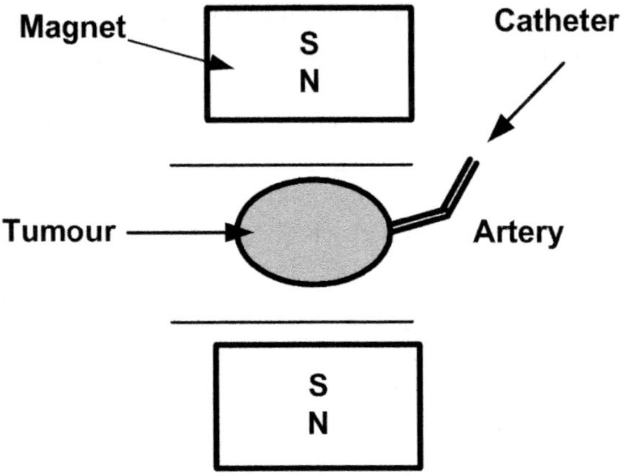

Figure 6.54: Schematic diagram of the introduction of drug-loaded magnetic carriers into a tumour (adapted from [K34]).

Another technique used to achieve tumour regression is based on the application of magnetorheological fluids that change their viscosity when exposed to an external magnetic field. Such a fluid can be injected into the blood vessels that supply the tumour. When a magnetic field is applied to the tumour site, the viscosity of the fluid increases and the fluid seals the blood vessels and inhibits blood supply and tumour necrosis develops [S94].

Hyperthermia

Local heating of certain organs or tissues to temperatures between 42 ^{0}C and 46 ^{0}C, particularly for cancer therapy, is called hyperthermia. The heating of the target tissue to this temperature range generally reduces the viability of cancer cells and increases their sensitivity to chemotherapy and radiation. The heating can be achieved by means of various external sources such as radio frequencies and microwaves. In order to optimize thermal homogeneity in the target volume, inductive heating of magnetic materials has been explored.

The use of ferrofluids, exposed to an external alternating electromagnetic field, for the localized non-invasive hyperthermia, has been thoroughly studied. Magnetic particles confined in the target volume dissipate the energy of the external field in the form of heat through several types of relaxation mechanisms. Multidomain particles exhibit hysteretic and Rayleigh losses while single-domain nanoparticles lose energy through Néel and Brownian relaxation [H34, R26]. Multidomain magnetic particles thus require large field amplitudes for extensive heating, while single-domain particles have large specific absorption rates at already lower field amplitudes. Under typical conditions of ferrofluid hyperthermia a magnetic induction of 0.04 T at 100 kHz is applied to a tumour for a period of 30 minutes.

Although hyperthermia with magnetic fluids has been studied for several decades, its performance has been demonstrated only on laboratory scale. Applications such as brain tumour, prostate carcinoma [J9], and AIDS therapy [M41] have been proposed and successfully investigated. Nevertheless, it appears that several problems must still be eliminated before this technique can be introduced into clinical practice [S88].

6.5.4 Separation of red blood cells by HGMS

The separation of erythrocytes (red blood cells) from whole blood using magnetic means has received considerable attention during the last twenty years. An innovative approach by a group of researchers from the Southampton University and other organizations [M42, M43,G24, G25, R27, R28] resulted in the development of a novel method of erythrocyte separation by high-gradient magnetic separation. This separation process is based on the fact that the paramagnetic properties of the reduced haemoglobin of erythrocytes allow capture of these cells under suitable conditions. By diluting fresh, anticoagulated blood in an isotonic reducing agent, the iron of haemoglobin is reduced to the paramagnetic Fe^{2+} form. The magnetic force exerted on these particles allows the red blood cells to be separated from the whole blood.

Red blood cells are flexible biconcave discoids with a diameter of 8.5 μm and a thickness of 2.3 μm. The volume magnetic susceptibility of red blood cells is estimated to be 3.9×10^{-6} (SI) when the haemoglobin is in the completely deoxygenated state. It can be shown [S90] that under conditions typical for HGMS of red blood cells, where the velocity of the red blood cells relative to the fluid of $v = 3\times10^{-4}$ m/s was usually employed [M42], a magnetic induction of 2 T and a field gradient greater than 10^4 T/m are required to render the magnetic force greater than the hydrodynamic drag. Such a magnetic field gradient can be obtained, for instance, using ferromagnetic stainless steel wire matrix of diameter smaller than 400 μm.

It was also demonstrated by Melville et al. [M43] that it was possible to collect selectively diamagnetic white blood cells and platelets onto the regions of diamagnetic capture of the matrix wire. By magnetically labelling various cell subpopulations, such as lymphocytes, in human blood and bone marrow it was possible to separate and concentrate magnetically these labelled cells at 99% purity and 68% recovery [R27].

Concentration of malarial parasites by HGMS

The malarial parasite, on entering the human body, spends an important part of its life cycle in the red blood cell. During this cycle the parasite progressively digests haemoglobin to leave oxidized haem product (pigment). The oxidized haem product is paramagnetic, in contrast to the diamagnetic oxygenated group found in oxyhaemoglobin.

This magnetic property can be exploited to separate malaria-parasitized red cells from a suspension of oxygenated normal cells at a short processing time (<

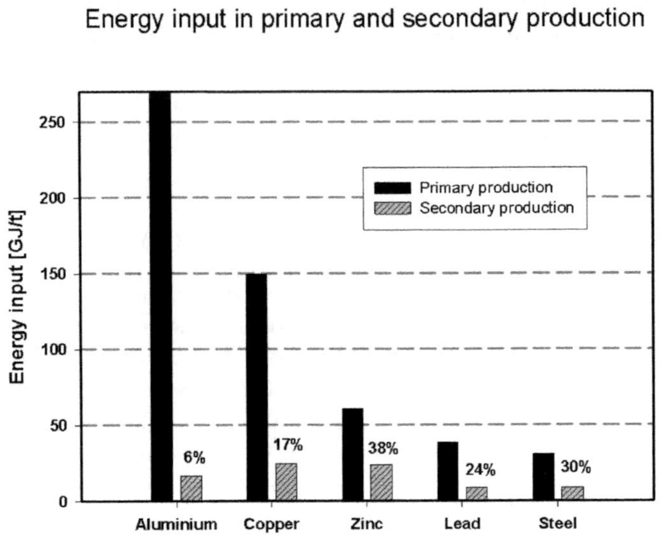

Figure 6.55: Comparson of the energy input in primary and secondary metal production (adapted from [N12]).

10 min) at a purity of the concentrate in excess of 90% [P23]. It was observed that in order to achieve percentage capture similar to the normal red cells case, the flow rate for parasitized cells had to be reduced by one order of magnitude.

6.6 Recovery of metals from wastes

Resource conservation by re-use, reclamation and recycling of materials has been practiced for millennia, and the recovery of metals and alloys is a prime example of such activities. The recovery of materials from waste and recycling of materials are of increasing importance due to a wide spectrum of factors [V11]. Increased scarcity of economically viable deposits of ores, energy saving, and increased emphasis on environmental control are some of the more important ones. Significant reduction of energy input in metal recycling compared to primary production is illustrated in Fig. 6.55, while Fig. 6.56 shows the relative costs of mining and extraction of various metallic products.

It is argued [V11] that it is the economic cost, rather than actual physical depletion, that will provide constraints on resource use in the future. It is likely that the only structural metals which will not be in short supply in the future are magnesium (from sea water), aluminium and iron [V11]. It is thus clear that recycling of materials is strongly associated with the economics of producing raw materials. This implies that for a financially self-sustaining system, the value of the recycled materials must exceed the cost of producing them [N13]. Recycling

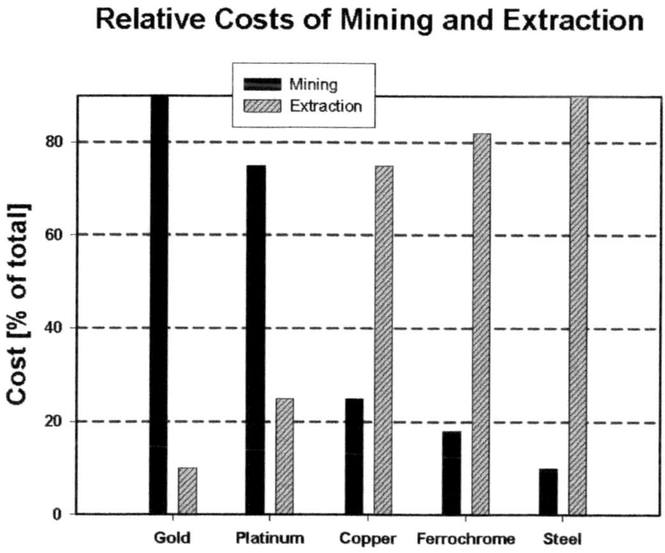

Figure 6.56: Comparison of relative costs of mining and extraction of various metallic products (adapted from [N12]).

will, therefore, take place only where the economic incentive is great enough.

6.6.1 The separation processes in metal recycling

In broad terms physical separation methods can be divided into dismantling, liberation, dry and/or wet separation and melting [N13, D21]. After liberation, a metal must be separated from a variety of materials. Materials have many unique physical and chemical properties, which may be applied in separation processes, ranging from colour and density to magnetic and electrical properties. Various dry and wet separation methods are used in practice, and their size ranges are listed in Tables 6.27 and 6.28.

Magnetic separation

Suspended magnets, magnetic pulleys and dry low-intensity drum magnetic separators are frequently used in metal recycling processes for concentration of ferromagnetic materials. These magnetic separators are usually situated directly after the primary shredder, or just after the air classifier, if used. Occasionally, secondary low-intensity magnetic separators are utilized to retrieve the 10% to 20% of ferrous metal missed by the primary separator. It is often necessary to re-shred and then air-classify a magnetic concentrate to free contaminants such as the contents of tin cans [V11]. A schematic diagram of major process steps

Table 6.27: Dry separation techniques for recycling (after Dalmijn et al. [D21]).

Dry separation	**Particle size range [mm]**
Gravity or density sorting	
Screening and sizing	1 - 100
Centrifuges and cyclones	0.05 - 20
Air classification	5 - 100
Friction separators, shaking tables	0.05 - 5
Magnetic and electrostatic separation	
Magnetic separation	0. 5 - 100
Eddy-current separation	1 - 100
Corona discharge separators	0.1 - 10
Triboelectrical separators	0.1 - 10
Particle-for-particle separation	
Sensors for material recognition	1 - 100
Image processing	1 - 100
Multi-sensor systems	1 - 100

Table 6.28: Wet separation techniques for recycling (after Dalmijn et al. [D21]).

Wet separation	**Particle size range [mm]**
Gravity or density separation	
Classification	1 - 100
Rising current separators	1 - 50
Centrifuges	1 - 50
Cyclones	0.05 - 0.5
Sink-float or DMS	1 - 100
Jigging	1 - 50
Flotation	0.05 - 0.5
Electromagnetic separation	
Magnetic separators	0.1 - 1
Eddy current or Magnus separators	0.05 - 5

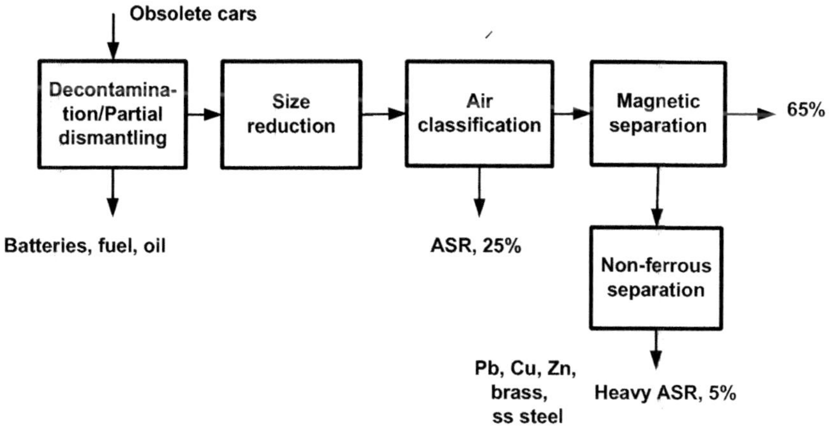

Figure 6.57: Major process operations in the mechanical processing of automobile scrap. ASR - autmotive shredder residue (adapted from Dalmijn et al. [D21]).

is shown in Fig. 6.57.

Dry drum magnetic separators equipped with rare-earth magnets are used for recovery of feebly magnetic stainless steel.

Wet low-intensity drum magnetic separators are used to recover magnetite and ferrosilicon from dense media plants prior to feeding scrap material to an eddy-current separation plant. High-intensity magnetic separators are also used to remove iron-aluminium, a complex material consisting of aluminium and magnetic parts mechanically connected, such as steel bolts [N13].

Eddy-current separation

Eddy-current separators are used in recycling in combination with a dense media separation plant to improve the grade and recovery of the metal products. Magnesium and aluminium are the first products from a dense media plant fed primarily with material from automobile shredders. This product still contains materials that responded to the same cut-point density at which DMS operated. These materials, such as stones, glass, copper wire and complex materials represent approximately 10% of the aluminium fraction with a size range of 10 to 65 mm [N13]. The aluminium concentrate can be obtained in a process shown in Fig. 6.55. The feed to an eddy-current separator is usually screened into two fractions, for instance - 32 + 8 mm and - 65 + 32 mm.

The heavy non-ferrous fraction obtained in DMS can also be processed by eddy-current separators. The current technology cannot, however, separate the heavy non-ferrous mixture into the final concentrates of individual components, such as zinc, lead and copper [N13]. However, eddy-current separators are useful to produce pre-concentrates for further sorting, for instance by image processing.

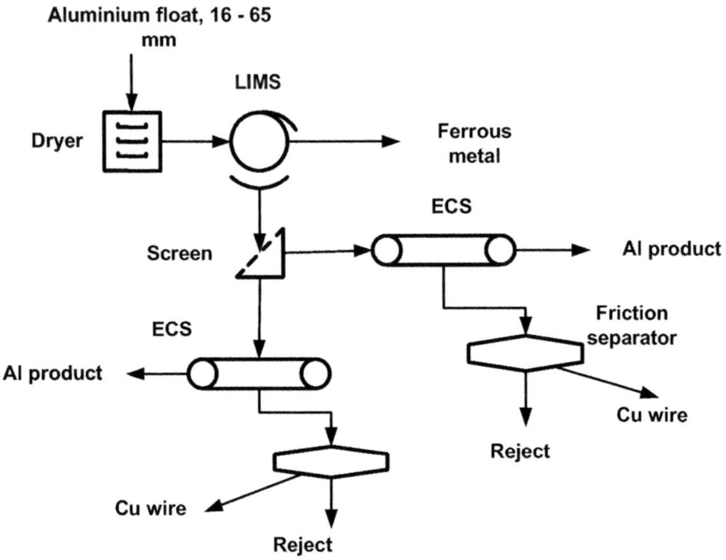

Figure 6.58: Schematic diagram of the processing of the aluminium float fraction with eddy-current (ECS) and friction separators (adapted from Nijkerk and Dalmijn [N13]).

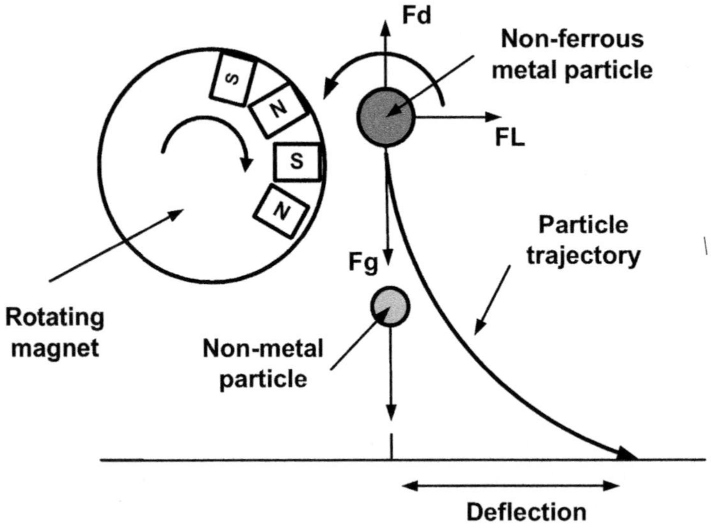

Figure 6.59: Principle of separation of non-ferrous metal particles from non-metal particles using the Magnus effect. FL = lift force.

Separation by Magnus-Robins Effect

It was demonstrated by Robins and later by Magnus that a particle, rotating either in water or air, does not fall strictly vertically, but, as a result of asymmetry of the flow and pressures around the rotating particle, a lift force is created that affects the particle trajectory [M45, B51, T13].

It has been suggested by Rem et al. [F36] that this effect can be applied to the separation of fine non-ferrous metal particles from a mixture. By exposing a mixture of vertically falling particles to a rotating magnetic field, the non-ferrous metal particles, in contrast to non-metallic particles, also start rotating. The Magnus lift force thus created deflects the metal particles from the vertical direction, as shown in Fig. 6.59.

A wet Magnus separator was applied, on pilot scale, to the separation of copper and aluminium particles from the bottom ash of a municipal waste incinerator [F36]. A production-size plant with a throughput of 50 t/h has been proposed [R31]. The wet Magnus separator can treat particles in the size range from 0.5 to 20 mm, while in a dry mode the size range is from 0.2 to 5 mm.

6.6.2 Sorting of heavy non-ferrous metals

Heavy non-ferrous metals include copper, lead, zinc, titanium, cobalt and precious metals such as silver, gold and platinum, and metal alloys such as brass (Cu+Zn) and bronze (Cu+Sn+Zn). Approximately 75% of the mixture of heavy non-ferrous metals comes from automobile scrap, the source of the rest being electronic and domestic waste. Brass and copper make up approximately 70% of the value of the mixture, as is illustrated in Fig. 6.60.

Hand sorting is generally the most cost-effective method for sorting of heavy metals. Labour rates for sorting in Asia are low and the costs of shipping scrap materials, from most parts of the world, to Asia are presently very attractive. Moreover, the Asian markets for non-ferrous metals are more attractive than, for instance, the European markets.

A rather recent entry into the sorting of heavy non-ferrous metals is colour sorting. High-speed, real-time processors interpret images from high-resolution cameras. The processors are pre-programmed to recognize certain metals by colour and geometry and compressed air is used to eject the recognized metal fragments. One metal type can be recovered per cycle per unit. The efficiency of this technique is limited by, for instance, insufficient liberation, contamination, dirt and corrosion. Recovery ranges from 80% for lead to 93% for zinc, while the purity of the concentrates is approximately 95% to 98%.

Sorting in a magnetic fluid

Separation in magnetic fluids, and particularly in ferrofluids (FHS) is a natural choice for sorting of non-ferrous metals into individual metal concentrates. Considerable effort was expended to develop the FHS for the treatment of automobile scrap [K16, K17, F21, S50, N4, R9, H35, A44], followed by the application

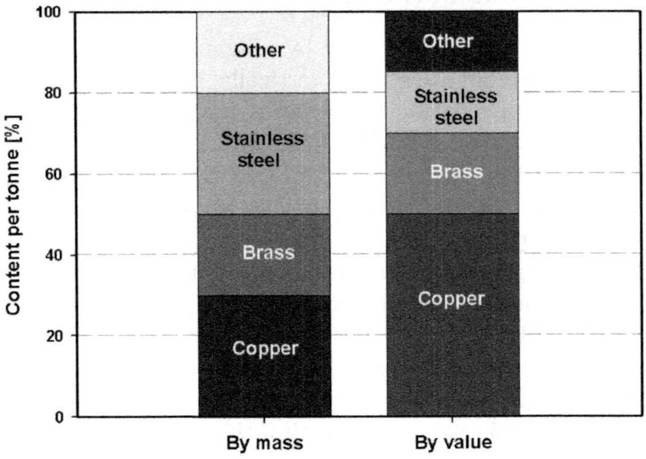

Figure 6.60: Distribution of heavy non-ferrous metals, per mass and per value, in a scrap metal mix (adapted from [S91]).

to electronic scrap treatment [B46, G26]. In view of large density differences between non-magnetic or weakly magnetic metals to be separated, only modest separation selectivity and a rather simple design of the FHS separator is required. It is fairly easy to design a ferrohydrostatic separator that can generate apparent densities exceeding 10 000 kg/m^3, required to float, for instance, lead, in a modestly magnetic (e.g. $J_s = 0.02$ T) ferrofluid.

A separator shown in Fig. 6.61 can treat up to 1 t/h of scrap material and it can be further scaled-up to double this throughput. Recovery of 95% of most heavy non-ferrous metals can achieved, with purity ranging from 98 to 100%. [M24]. A typical example of a copper concentrate obtained by FHS is shown in Fig. 6.62. Inclusion of ferrohydrostatic separation in the flowsheet for sorting of heavy non-ferrous metals is illustrated in Fig. 6.63.

As has been discussed in Section 5.6, there are two basic types of separators with magnetic fluids. A ferrohydrostatic separator with constant apparent density of ferrofluid along the vertical axis produces two products, namely float and sink. The selectivity and accuracy of separation in this type of separator can be very high. On the other hand, separators that use paramagnetic liquids or separators with ferrofluid with wedge-shaped pole-pieces generate a distribution of apparent density along the vertical. Although it is possible, at least in theory, to obtain several density products in one pass of a material through such a separator [e.g. A44], selectivity is inevitably limited. Interparticle interactions, distribution of density of the materials to be separated, and a continuum of the apparent density of the magnetic fluid along the vertical axis are some of the reasons that make this approach impractical.

While it is unlikely that the latter approach will find industrial applica-

Figure 6.61: A pilot-plant-scale ferrohydrostatic separator for sorting of heavy non-ferrous metals.

Figure 6.62: A photograph of the - 12 mm copper concentrate obtained by FHS as a sink product at the density of 8200 kg/m^3 and a float fraction at 10 300 kg/m^3[M24].

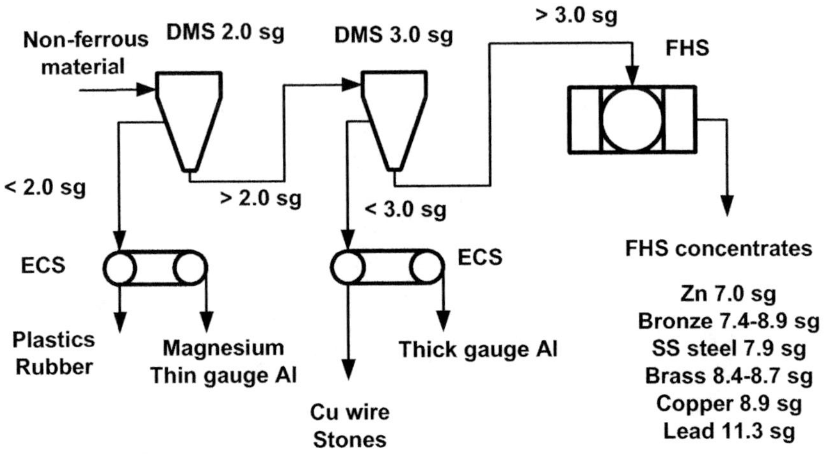

Figure 6.63: A flowsheet for sorting non-ferrous metals into individual metal concentrates, using FHS.

tion, sorting of non-ferrous metals by ferrohydrostatic separation holds a significant promise. A successful production-scale demonstration of this technology, availability of ferrohydrostatic separators with a throughput of 2 to 5 t/h, and availability of technical-grade ferrofluid at reasonable cost are prerequisites for introduction of this technology into the recycling process.

6.7 The applications of ferrohydrostatic separation

The range of potential applications of ferrohydrostatic separation is extensive and, in view of the unique capabilities of the FHS technology, this technique can change conventional practices of materials treatment. The main limitation is, however, the cost of ferrofluid, limited throughput of the existing separators and rather poor efficiency of separation of particles smaller than 500 μm. The viability of every application will be determined mainly by the losses of the fluid, its cost compared to the value of the products of separation, and by the efficiency of of the ferrofluid recycling, if it is used [S92].

The FHS technology is presently being used, on industrial scale, for the recovery of gold and other platinum-group metals and in the diamond industry. On laboratory and pilot-plant scale, the technology is being applied to the recovery of non-ferrous metals from scrap, as discussed in Section 6.6.2, and to the treatment of wastes and tailings, slags, coal analysis and recovery of base metals.

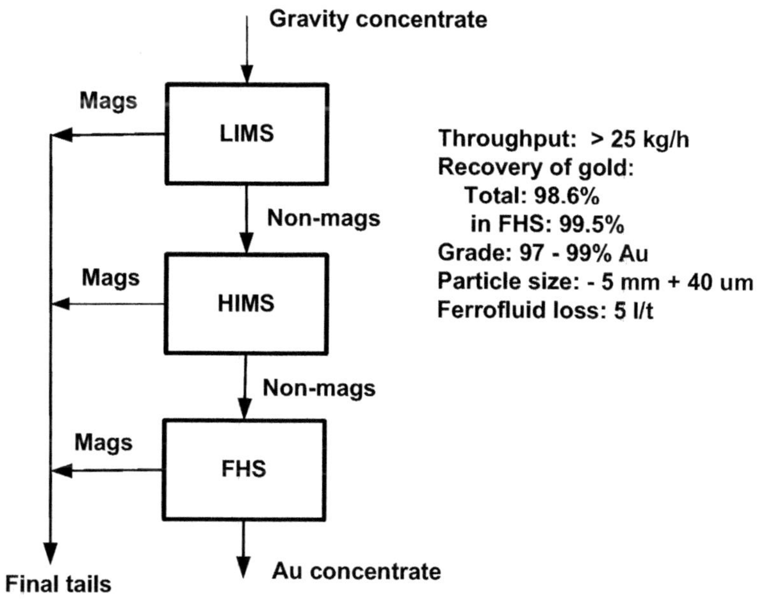

Figure 6.64: Recovery of gold by FHS at Aldanzoloto, Russia (adapted from [S55]).

Table 6.29: Beneficiation of gold-bearing alluvial sands by FHS [G27].

Product	Yield [%]	Grade [g/t Au]	Recovery [% Au]
Feed	100	473	100
Concentrate	0.1	452 000	95.5
Tailings	15.7	78	2.6
Magnetics	84.2	11	1.9

6.7.1 Recovery of gold and other platinum-group metals

At least fifty ferrohydrostatic separators, based either on electromagnets [S55], or permanent magnets [S54], are in operation in the territories of the former Soviet Union. The separators are used to recover gold from gravity concentrates and to treat previously accumulated tailings after the amalgamation of the gold concentrates. A typical flowsheet of the treatment of the gravity concentrate is shown in Fig. 6.64. Table 6.29 summarizes the results of the final concentration of black alluvial sand by an electromagnet-based GMUO FHS [G27].

Tables 6.30, 6.31 and 6.32 offer a detailed summary of the results of beneficiation of table gold concentrates in individual stages as outlined in Fig. 6.64.

Results of the concentration of gold from tailings in the Amur Region (Russia), using permanent-magnet based FHS, are shown in Table 6.33.

Table 6.30: Beneficiation of the gold gravity concentrate by a low-intensity magnetic separator (Severvostokzoloto Co.), B = 0.13 T.

Product	**Yield [%]**	**Grade [g/t Au]**	**Recovery [%Au]**
Feed	100	222146	100
Magnetics	31.81	1502	0.22
Non-magnetics	68.19	325074	99.78

Table 6.31: Beneficiation by a high-intensity magnetic separation of the non-magnetic fraction from LIMS (Severvostokzoloto Co.).

Product	**Yield [%]**		**Grade [g/t Au]**	**Recovery [%Au]**	
	per stage	overall		per stage	overall
Feed	100	68.19	325074	100	99.78
Magnetics	59.47	40.55	1864.6	0.34	0.34
Non-magnetics	40.53	27.64	799248	99.66	99.44

Table 6.32: Concentration of gold from the non-magnetic fraction of a high-intensity magnetic separator by FHS (Severvostokzoloto Co.).

Product	**Yield [%]**		**Grade [g/t Au]**	**Recovery [%Au]**	
	per stage	overall		per stage	overall
Feed	100	27.64	799248	100	99.44
Au concentrate	93.42	25.82		99.92	99.36
Tailings	6.58	1.82	9280.68	0.08	0.08

Table 6.33: Recovery of gold from tailings by permanent-magnet based FHS. Cut-point density: > 9000 kg/m3 [S54].

Size distribution	**Recovery [%Au]**
+ 0.315 mm, 8%	99.7
- 0.315 mm, 82%	71; 93.1 free gold
Full sample, 100%	89.1, 93.8 free gold

Table 6.34: Recovery of fine diamonds from the sorthouse feed at Yakutalmaz (Russia).

Parameter	Specification
Throughput	5 to 10 kg/h
Feed size	0.2 to 2 mm
Recovery of diamonds	99%
Input power	0.47 kW
Mass of the separator	450 kg

6.7.2 Applications in the diamond winning industry

Benefits of the unique capabilities of ferrohydrostatic separation for diamond winning were readily recognized and it was also realized that, in contrast to sorting of non-ferrous metals, separators with high selectivity operating at relatively low apparent densities (up to 4500 kg/m^3) were required. One of the first industrial applications of the technology was the installation of six GMUO FHS-40/3 units at Yakutalmaz Co. (Yakutia, Russia) in 1992. The separators were used at the sorthouse for the final recovery of fine diamonds. Typical performance data are summarized in Table. 6.34. Although the units operated satisfactorily, the FHS plant was later closed down, for reasons that were mainly of non-technical nature.

In order to meet the needs of the modern diamond-winning industry, De Beers Diamond Research Laboratory of De Beers Consolidated Mines (Pty.) Ltd. (South Africa) developed a range of sophisticated cost-effective FHS separators as has been discussed in Sections 2.8 and 5.6. In order to maintain accurate control of the apparent density, the separators employ a closed-loop control system. Such a control system automates the separation process and provides a user-friendly interface. An operator selects the desired cut-point density from a computer screen and the system automatically controls the plant at the set point.

In order to ensure high selectivity of separation, a proprietary detection system, which detects any changes in the density, is employed. The output of the detection system is used by the closed-loop controller to control the power supply output to maintain the ferrofluid at the constant apparent density.

Diamonds, by virtue of their high value, are highly sought after as an exchange medium in criminal activities. This results in significant pressure being applied to workers in the diamond mining industry to pilfer diamonds. It was therefore imperative to put into place mechanisms, which prevented people from accessing diamondiferous material during mining and recovery processes. Figures 2.108 and 6.65 show a two-stage automated and secure ferrohydrostatic separator built by De Beers for the recovery of diamonds [S23].

Secure and selective ferrohydrostatic separators can be used for the recovery of diamonds at various points of the diamond recovery process. Recovery of diamonds from DMS concentrates, X-ray tailings, X-ray concentrates and from

Figure 6.65: A two-stage ferrohydrostatic separator for the recovery of diamonds.

the sorthouse feed and tailings are some of the applications.

In addition to diamond recovery, ferrohydrostatic separators are used for the treatment of geological samples. Mass reduction of samples before hand-sorting, sorting of indicator minerals into individual narrow density fractions and replacement of toxic and expensive heavy liquids are the main benefits. The Rhomag® ferrohydrostatic separator, developed at De Beers and shown in Fig. 2.109, is used for treatment of exploration materials.

As a result of the ability of FHS to separate materials into narrow, well-defined density fractions over a wide range of densities, using a single fluid density, which can be changed by varying the magnetic field strength, the technique is particularly suitable for densimetric analysis. Densimetric analysis of the DMS feed and products by FHS allows to set the operating conditions of DMS and to assess its efficiency. Caution must, however, be exerted to take into account the magnetic properties of the material. As has been discussed in Section 3.8.4, particles with non-zero magnetic susceptibility exhibit an effective density different from their natural density. A failure to take into account this increment in density would result in distortion of the density distribution of the material.

6.7.3 The fractionation analysis of coal

Sink-and-float analysis of coal has been one of the areas of potential applications of FHS. Densimetric analysis of coal is usually carried out in bromoform

Table 6.35: Comparison of densimetric analysis of coal in heavy liquids and FHS [G17]. Concentration of impurities in FHS fractions is shown as a difference (in heavy liquids.

Separation density	**Size fraction [mm]**	**Concentration of impurities [%]**
2100 kg/m^3	- 100 + 50	0
	- 50 + 25	0
	- 25 + 13	0.7
	- 13 + 1	1.8
1800 kg/m^3	- 100 + 50	0
	- 50 + 25	0
	- 25 + 13	0.9
	- 13 + 1	1.4

and aqueous solutions of zinc chloride. Both heavy liquids are toxic and expensive to acquire and to dispose of. Such a hands-on operation, involving several liquids, to cover the entire density region from 1800 to 2400 kg/m^3 is tedious and environmentally unfriendly. The obvious advantage of using magnetic fluids to separate coal into narrow density fractions has been appreciated, over the last twenty years, by numerous researchers. Gubarevich [G17] employed a ferrohydrostatic separator with kerosene-based ferrofluid to investigate the fractionation of coal, ranging from 1 to 100 mm in size. The results, shown in Table 6.35, were compared with densimetric analysis obtained with heavy liquids.

It can be seen that while the best results were obtained with the coarse fractions, even in the fine fractions the concentration of impurities was low. The selectivity of separation was found to be high, with Ep ranging from 0.015 to 0.045, depending on the ferrofluid density. The high accuracy of separation is assisted by the fact that most coals are very feebly magnetic ($\chi_{coal} \cong 25\times10^{-9}$ m^3/kg). It was also observed that chemical interaction between the ferrofluid and organic and mineral matter in the coal was negligible.

Based on the above experiments, an industrial-scale continuous FHS separator (FGS-2) was developed by the GMUO Institute, Lugansk, the Ukraine. The separator, shown in Fig. 6.66, underwent several pilot-plant tests, with duration of each exceeding 600 hours. Typical results are summarized in Table 6.36. In spite of satisfactory performance, several attempts to commercialize this technology failed as a result of geopolitical changes in the former USSR at that time.

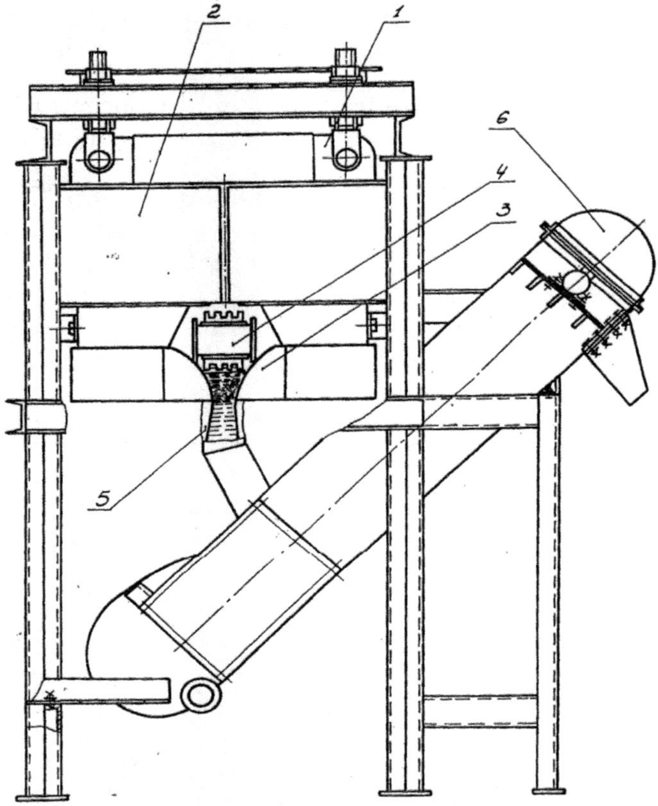

Figure 6.66: FGS-2 separator for fractionation of coal. 1-electromagnet, 2-coils, 3-pole-tips, 4-hopper, 5-bottomless separation chamber with hydraulic support, 6-lift.

Table 6.36: Typical results of pilot-plant-scale densimetric analysis of coal using FGS-2 [G17].

Density range	1300 to 2200 kg/m^3	
Cross-contamination [%] of FHS products	- 100 + 13 mm	0.5%
	- 13 + 1 mm	2%
Throughput	200 kg/h	
Carry-over of ferrofluid into products	1 to 4 kg/t	
Loss of ferrofluid after regeneration	0.4 kg/t	

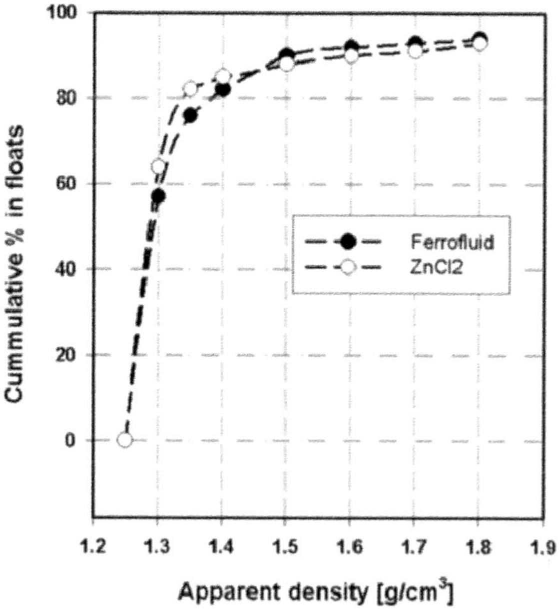

Figure 6.67: Comparison of densimetric analysis of coal using FHS and $ZnCl_2$. Particle size: - 3 + 0.3 mm (adapted from Fujita et al. [F34]).

Fujita et al. [F34] described the design of a batch ferrohydrostatic separator for densimetric analysis of coal. The results, an example of which is shown in Fig. 6.67 were encouraging. The separation curves for three size fractions, namely - 25 + 13 mm, - 13 + 3.3 mm and - 3.3 + 0.3 mm were similar for both separation methods, namely FHS and zinc chloride. The carry-over of the water-based ferrofluid into the products of separation was less than 5 kg per tonne of the feed material, in agreement with the value given by Gubarevich [G17]. Further regeneration of the ferrofluid reduced the loss to about 0.5 kg/t.

Ferrohydrostatic separators can thus be used as accurate tools for densimetric analysis of coal. FHS lends itself to automation, the rate of treatment is high and the detrimental effect of chemical and physical interaction of the heavy liquids with the coal matter is reduced. FHS thus possesses numerous advantages compared to conventional methods. As has been discussed by Svoboda [S93], further development of FHS technology that will result in more accurate and large-scale equipment for coal fractionation and in availability of cost-effective ferrofluids, combined with strong interest of the coal preparation fraternity, will inevitably result in the introduction of the FHS technique to the coal fractionation practice.

Chapter 7

Innovation and future trends in magnetic techniques of material treatment

7.1 Introduction

In the last two decades the mining and metallurgical industries have been addressing a variety of economic, environmental and social challenges through numerous technical and management programmes. These programmes include conservation of resources using more efficient mining and treatment processes, impounding and isolating wastes to minimize deterioration pending the day when economic use is found for these minerals, and facilitating the recycling of scrap and various types of wastes. These objectives of sustainable development can be met only if the mining and metallurgical operations employ the best available technology, improve energy efficiency and use renewable energy resources.

In forecasting future developments in material treatment, two approaches may be taken [A45]: the first one is a simple extrapolation from the present. The second approach is based on a long-term vision, on establishment of technology strategies and the timely identification of emerging science and technology.

If the first avenue is used, it appears that nothing really innovative can be expected during the next twenty years. The second, more risky avenue, is likely to produce some significant future developments as a result of the emergence of breakthrough technologies. Breakthrough, or 'disruptive' technologies based on intelligent materials, ceramics, nanostructured materials, high-temperature superconductors and others can allow mineral processing, and material treatment in general, to develop new processes and to explore future resources such

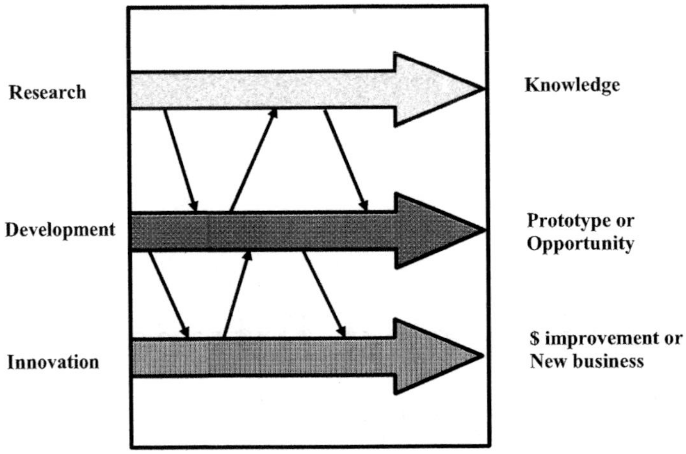

Figure 7.1: A possible connection between research, development and innovation (adapted from [B47]).

as deep-sea and extraterrestrial minerals. It is currently being speculated [W35] that it is possible to expect that significant developments will arise at all scientific and engineering levels in the next two decades.

The future of mining companies is thus based on sustainable competitive advantage in the changing and increasingly turbulent environment and continuous innovation is key to achieving this. The introduction of new technologies into the mining and mineral processing operations is, therefore, a strategic necessity. This necessity does not always appear to be part of strategic plans of major players in the minerals industry. It is often claimed that although even successful mineral companies are slow to adopt new research results, they are good at innovation [B47]. This view seems to contradict observations that many of the past innovations have actually come from the smaller, more entrepreneurial organizations. In addition, many of the large mining and processing companies are cutting back on their own research & development capabilities [N14, S96, B49].

7.2 Science and technological innovation

It is often argued, for example, by Batterham [B47], that research, development and innovation are separate processes. It is claimed that whilst the connections between these processes are frequent, as is illustrated in Fig. 7.1, they are on an 'as required' basis. It is even claimed that innovation can proceed without any research and development and it is innovation that generates the competitive advantage, not research or development *per se* [B47].

The view that research, development and innovation are separate often meets with strong disagreement in R&D circles [J10]. It also seems to contradict

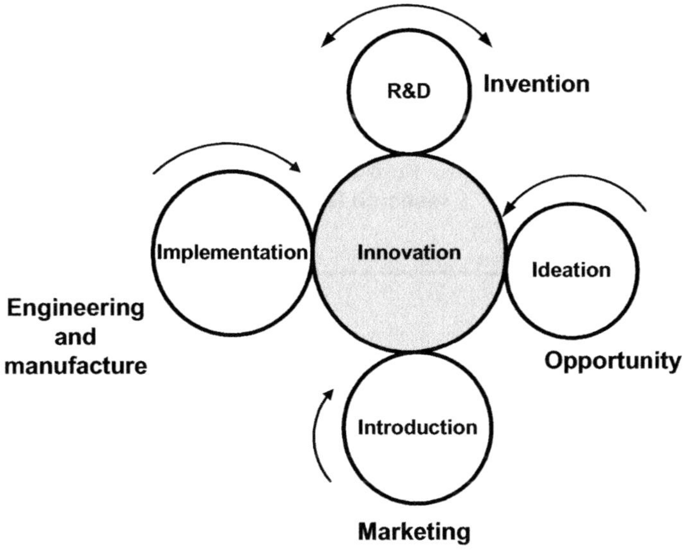

Figure 7.2: The elements of innovation (adapted from [R29]).

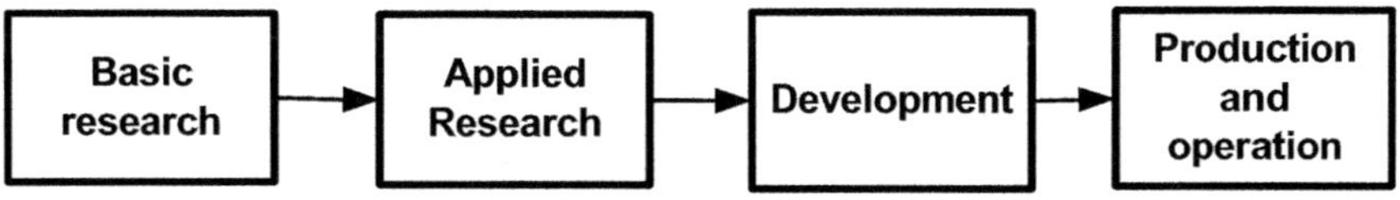

Figure 7.3: A linear model of scientific advances.

the actual history of science and technology. It can be postulated, and well supported by practical observations, that *innovation* is the process by which the *invention* is first brought into use [G28]. It involves the improvement or refinement of the invention, the initial design and production of prototypes, pilot plant testing and construction of production facilities. Innovation can be looked upon as the total process from the inception of an idea through to the manufacture of a product and finally to its ultimate application resulting in profit. It therefore includes invention and the many stages of implementation such as research, development, production and marketing [B48], as is illustrated in Fig. 7.2. In other words:

$$Innovation = Invention + Exploitation$$

In this equation invention is the generation of new knowledge, while innovation is the application of new knowledge for a purpose.

This dynamic non-linear model is in contradiction with the traditional linear model of research, development and technology transfer, a sequence of processes

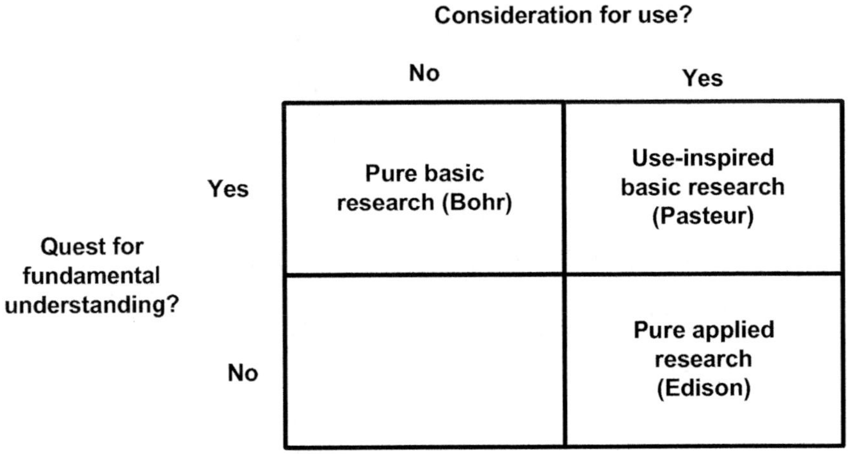

Figure 7.4: The quadrant model of scientific research (adapted from Stokes [S95]).

extending from basic research to new technology, as depicted in Fig. 7.3. This model is based on the belief that scientific advances are converted to practical use by a one-way flow from science to technology and that basic advances are the principal source of technological innovation. It thus transpires from this paradigm that each of the successive stages depends upon the preceding one.

This concept of a one-way linear flow is, however, in conflict with the history of science and technology and today has largely been abandoned [T13]. There are ample examples of the reverse flow from technology to science [S95] and it is no longer believed that pure, curiosity-driven basic science will itself guarantee the technology required to maintain competitive advantage.

It is also not possible to draw a sharp line between basic and applied research, as work directed toward applied goals can be highly fundamental in character. In order to take into account such a synthesis of the goals of understanding and of practical use, Stokes [S95] proposed a quadrant model of scientific research illustrated in Fig. 7.4.

The second quadrant (the upper left-hand cell) includes basic research that is guided solely by the quest for understanding without thought of practical use. Stokes called it Bohr's quadrant to illustrate how much Niels Bohr's quest of a model atomic structure was a pure voyage of discovery.

The fourth, Edison's, quadrant (the lower right-hand cell) includes research that is guided solely by applied goals without seeking a more general understanding of the phenomena of a scientific field.

The first quadrant in the upper right-hand cell includes basic research that seeks to extend frontiers of understanding but is also inspired by consideration for use. Or, conversely, it includes goal-orientated research which, due to its complexity, requires investigation of its fundamentals. Stokes called this section

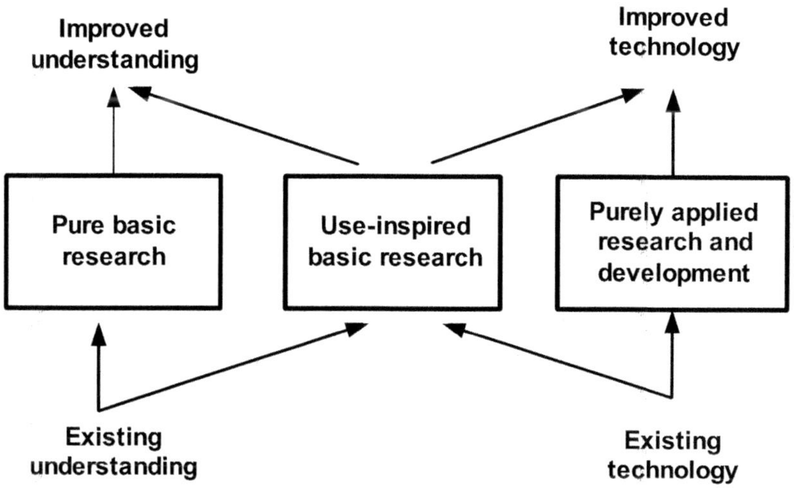

Figure 7.5: A dynamic model of science and innovation (adapted from Stokes [S95]).

Pasteur's quadrant to reflect Pasteur's drive towards understanding.

The history of science and technology is replete with examples of research, inventions and innovations that were directly influenced both by the quest for general understanding and by considerations of use. It follows from the analysis of these examples that the linear model that saw the advances of science as fully determining the development of technology must be replaced by a model that reflects the dual interactive character of basic and applied research and innovation. Such a model is shown schematically in Fig. 7.5.

Industrial magnetism is the scientific and engineering field that is particularly flush with such examples. Development of modern permanent magnetic materials, ferrofluids, magnetic carriers and of high-temperature superconductors are just a few examples of investigations undertaken either to gain new scientific and technical knowledge, directed primarily towards a specific practical aim or objective, or to furnish fundamental understanding of existing empirical knowledge.

7.3 Magnetic separation and innovation

Methods of treatment that use magnetic force offer a unique approach to material manipulation in a wide spectrum of industries. One of the main advantages of material treatment in a magnetic field is that this additional magnetic force can be applied in a controlled manner, in a wide range of values. Moreover, this force can be superimposed on other physical forces and several physical properties of materials can thus be exploited simultaneously.

In addition, magnetic separation in general can treat a wide range of types of materials, ranging from colloidal to large sizes and from "non-magnetic" to strongly magnetic. Magnetic separation is an environmentally friendly technique and it can operate in wet and dry modes, making it a technique of choice in arid and arctic regions.

7.3.1 Reliance on an empirical approach?

In spite of considerable progress, magnetic separation has not, so far, developed to its full potential. The fundamental problem appears to be a gap between scientific understanding and the engineering of this technology. Although numerous rigorous descriptions of assorted magnetic separation techniques have been introduced, their applicability to complicated engineering operations has been limited. Insufficient understanding of engineering practices, failure to appreciate the scale-up problems and incorrect assumptions often resulted in disappointments and sometimes in financial losses. Such a scenario has not helped to promote new technological advances in a conservative mining industry, which is slow in adopting new research results [A45, B47].

The production-scale innovation in magnetic methods of material manipulation has, therefore, been often left to plant engineers who had to rely on intuition and the empirical approach rather than on a working knowledge of the principles of magnetism. In 1975 Henry Kolm [K7] felt that physicists would have to learn more about mining because they were not likely to teach miners enough about magnetic technology.

It is doubtful whether this proposition has been implemented. It appears that only limited progress has been made in bridging the gap between "high technology" and "low technology" since the time when this challenge was identified [K7, S1]. This gap and lack of interaction among the many disciplines comprising magnetic separation has been delaying a progress in the application of sophisticated magnetic technology to the challenges facing magnetic separation. This is in spite of enormous progress in the development of new magnetic materials, modelling software and magnet design [S97].

7.3.2 A walk through innovation in magnetic separation

The innovation milestones

During the last fifty years of the 20th century, magnetic separation evolved from a simple technology to manipulate strongly magnetic coarse materials into a powerful technique for the treatment of weakly magnetic, finely dispersed particles. This is the result of development of high-intensity and high-gradient magnetic separators that used either resistive or superconducting electromagnets, or even permanent magnets.

As has been discussed in Section 1.1, the development of permanent magnetic materials and an improvement of their magnetic properties, particularly during the last thirty years, has been one of the main drivers of innovation in magnetic separation. The history of improvement in the energy product of permanent magnets is illustrated in Fig. 1.1.

Introduction of ferromagnetic bodies, or the matrix, into the magnetic field of a separator, in order to increase the field gradient and thus the magnitude of the magnetic force, was another significant driver that resulted in the development of sophisticated magnetic separators. This innovation extended the range of applicability of magnetic separation to many weakly paramagnetic or even diamagnetic materials of micrometer dimensions.

Introduction of superconductivity to the minerals industry is a significant milestone not only for magnetic separation but also for superconductivity itself. Although numerous superconducting magnetic separators are operating in mineral beneficiation plants, the potential of superconductivity still remains to be fully developed. The generation of a high magnetic field and high magnetic force, in large volumes, elimination of matrices, reduction in energy consumption, introduction of high-temperature superconductors, are some areas where superconductivity can become truly a breakthrough technology.

The role that these milestones played in an effort to improve the magnetic separation technique can be illustrated by numerous examples. The reasons for success or failure can also be identified.

Implementation of innovation: successes and failures

The reasons for failure to implement successfully a new technology often do not lie with the technology itself, but with the work system into which the technology is introduced [M44]. For instance, a new technology could have been technically too advanced for the level of skills available, or it may have required too radical a change in the way work was done. Possibly the need for a new technology was not strong enough. The cost might have been too high or it might have encountered resistance from the workforce. Premature application of an unproven technology, exaggerated expectations raised by an overzealous supplier and fear of job loss as a result of a new technology are additional factors that increase the risk of failure.

A review of the innovation landscape in magnetic separation shows that enormous effort has been expended over the years in order to convert a wealth of novel ideas into workable techniques and to introduce them into the material treatment operations. Successes and failures are almost equally represented in this process and they can serve as a lesson for future endeavours to introduce a new technology into the material processing industries [S96, S98].

Advent of rare-earth permanent magnets

One of the most successful innovations in magnetic separation resulted from the development of powerful rare-earth permanent magnets. An induced magnetic roll separator described in Section 2.3.2 and in Figs. 2.25, 2.26 and 7.6, has been traditionally used to treat dry weakly magnetic materials. There are several

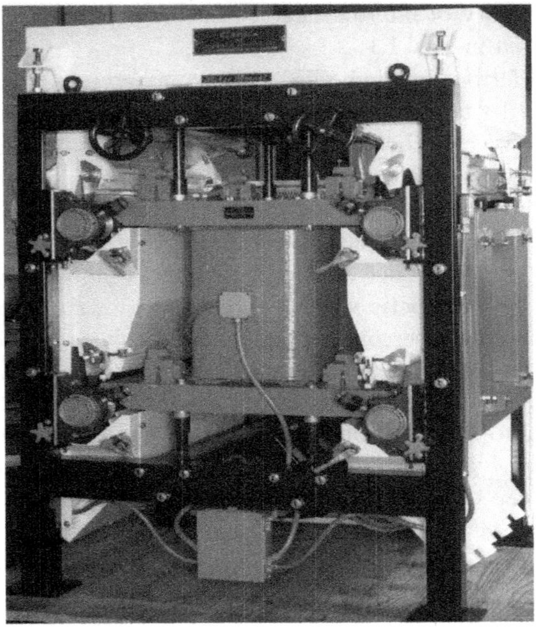

Figure 7.6: Carpco induced magnetic roll separator (courtesy of Carpco-Outokumpu Technology).

technical limitations to IMRs, namely a relatively low capacity per unit length and limited particle size range.

With the advent of rare-earth permanent magnetic materials it became possible to construct magnetic rolls that generate a magnetic field, gradient and, therefore, a magnetic force that exceeds that produced by an IMR separator. Although the magnetic field cannot be easily varied in such rolls, by a judicious selection of the permanent magnet material and of the thickness of the belt, and by optimizing the geometrical configuration of such a roll, it is possible to design rolls for treatment of materials of different size ranges and magnetic susceptibility distributions.

In addition to significantly lower energy consumption, there are numerous other advantages compared to an IMR. A permanent magnetic roll separator, shown in Figs. 2.30 and 7.7, has significantly lower mass and size. The absence of an air gap means that large particles, for instance as big as 25 mm or more, can be treated. A magnetic scalping stage is not needed since the belt detaches and easily discharges any strongly magnetic particles.

The considerable technical and commercial success of permanent magnet roll separators is a result of two factors. The first is the availability of a large variety of magnet grades and sizes, as well as decreasing cost and improved quality of the magnet material, particularly in terms of the energy product and thermal stability. The second aspect that played the decisive role is the

Figure 7.7: A permanent magnetic roll separator (courtesy of Roche Mining).

improved engineering reliability, such as better quality of belts, their tracking and an ease of replacement. At the present time, permanent magnetic rolls are manufactured by a variety of firms and are employed in a wide spectrum of mineral and chemical industries.

High-gradient magnetic separators

The introduction of a matrix into the circuits of magnetic separators resulted in a dramatic extension of the applicability of magnetic separation to materials that were previously considered too fine or too feebly magnetic. This significant development was achieved by Jones [J1] who combined Frantz's idea [F35] of a magnetized matrix with a high magnetic field of optimized orientation. The magnetic force was thus increased by several orders of magnitude. This idea initiated a long period of detailed academic research on the one hand and heuristic development on the other. This effort has been characterized by some remarkable achievements and, at the same time, disappointments.

High-gradient magnetic separators based on Jones's concept use electromagnets with an iron yoke to produce the magnetic field. For large-scale applications this approach suffers from two fundamental drawbacks. The maximum

Figure 7.8: Sala-Metso Minerals HGMS carousel high-gradient magnetic separator, model 480 (courtesy of Metso Minerals).

magnetic field is limited by the saturation magnetization of the yoke material and the scale-up is difficult. In order to treat a meaningful quantity of material, the working gap between the poles of an electromagnet must be large, which results in a significant drop in the magnetic field and the machine becomes massive. To generate a sufficiently large magnetic field at large volumes, an iron-clad solenoid must be used [M5], as has been discussed in Sections 2.4.3 and 4.4.

Sala HGMS While the design of a cyclic HGMS based on the iron-clad solenoid is a straightforward process, continuous solenoid separators require a more complex approach. Sala Magnetics, Inc. developed a carousel HGMS [O5] that employed a sophisticated saddle-shaped iron-clad solenoid surrounding the carousel ring. The separator shown in Figs. 2.64 to 2.67 and 7.8 can, in contrast to its predecessors, generate a background magnetic field as high as 2 T, allows a close control of flow velocity through the matrix and can be relatively easily scaled-up.

In spite of the obvious advantages of this second-generation HGMS, the Sala separator did not enjoy commercial success at that time. Although the separator was earmarked for several large-scale applications, the cost was apparently prohibitive and the market was reluctant to accept the product in spite of its progressive features.

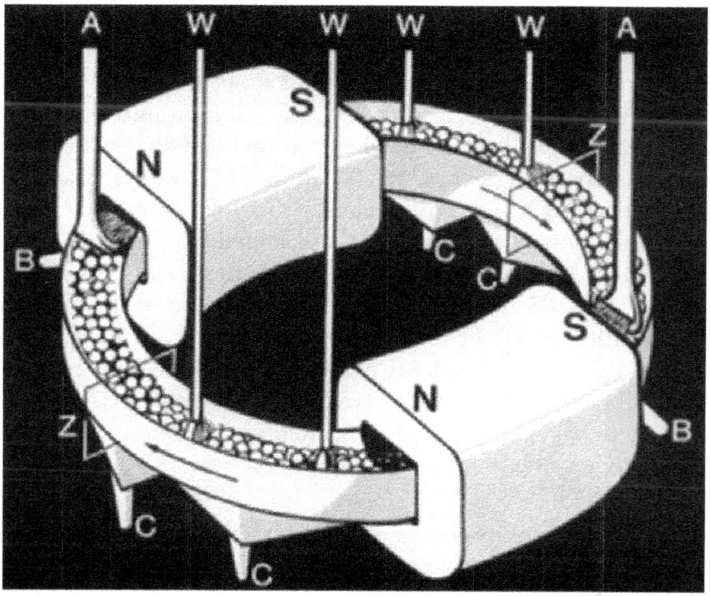

Figure 7.9: SOL high-gradient magnetic separator.

SOL HGMS In the SOL separator, designed and built by Krupp GmbH [S99, S1], the saddle-shaped solenoid was replaced by a simple axially orientated solenoid, as shown in Fig. 7.9. In order to provide a reasonably long retention time of the slurry in the magnetic field, the magnet had to be sufficiently long. The air gap of the magnet was, therefore, long and the magnetomotive force needed to generate the required magnetic field was very high. Moreover, because the coil was much longer that its diameter, the beneficial effect of the steel cladding was negligible.

As a result of this misconceived approach to the magnet design, the SOL separator could generate only a modest magnetic field at very high energy input per tonne of treated material. For these reasons the separator failed to gain acceptance by the industry.

VMS HGMS Probably the most onerous problem in the operation of HGMS separators is a permanent retention of particles in the matrix. Frequent shutdowns of the separators for cleaning or replacement of a blocked matrix severely impair the economic and metallurgical performance of the units. Numerous preventive measures that were investigated and applied in practice [S1] were found to be of interim efficacy only. The long-term solution was offered by machines that employed a back-flushing system.

The VMS magnetic separator developed by the Ore Research Institute, Prague, Czech Republic [C7], addressed the problem of matrix clogging by replacing the conventional horizontal rotor by a vertically rotating ring, as shown in Fig. 2.68. In addition to the reverse flush, the main advantages of the VMS separators, which employ an iron-clad solenoid, are excellent ability of scale-up and a high background magnetic field at modest input power.

The VMS separators were earmarked for incorporation into the Krivoy Rog and Kursk Magnetic Anomaly iron ore beneficiation plants in the former USSR. The proposed capacity of the plants was 30 Mt/a and it was envisaged that at least forty VMS separators, each with a capacity of 100 t/h, would be required.

Numerous difficulties were experienced, associated mainly with the disintegration of the Eastern Block in the late 1980s and early 1990s. In particular, the completion costs escalated as a result of rampant inflation in the former USSR during those transition years. In 1991 the Czechoslovak government decided to pull out from this joint venture and write off US$300 million already invested in the project. The beneficiation complex was never fully commissioned and political circumstances were, therefore, responsible for the failure of a promising innovative technology to be implemented on production scale.

SLON HGMS In spite of its commercial misfortune, the concept of the VMS separator proved beyond doubt that the use of an iron-clad solenoid and the application of the reverse flush of the matrix enhances the efficiency, scalability and availability of continuous high-gradient magnetic separators. The technical success of VMS separators prompted further innovative effort. The SLON HGMS developed at the Ganzhou Non-ferrous Metallurgy Research Institute, Ganzhou, China, and shown in Fig. 2.74, is based on the VMS concept [X1].

In order to improve the efficiency of separation even further, SLON employs an iron-clad solenoid located in the bottom section of the vertical ring carrying the matrix. This feature allows a pulsation of the slurry within the matrix to be introduced, which results in better selectivity of separation. The separator that is used, at the present time, to beneficiate weakly magnetic iron and ilmenite ores enjoys considerable commercial success in China and is making inroads into the international markets.

Magnetic separation and superconductivity

Superconductivity has been seen as a panacea in magnetic separation, particularly because of its ability to generate a magnetic field exceeding 2 T, the maximum field that can be generated by most resistive electromagnets. Additional advantages include low running costs, low mass and large volumes in which the high magnetic field can be generated.

Although numerous small-scale superconducting magnets were built in the late 1970s [S1], production-scale machines were slow to emerge. The first industrial superconducting separator was designed and built by the Institute for Refrigeration Engineering (VUPCHT), Prague, Czech Republic [F6]. This 5 T unit, shown in Fig. 2.83, was a reciprocating canister machine for beneficiation of kaolin.

The design of the magnetic system, the choice of the operating magnetic field strength and of the matrix were based on a theoretical description of particle

Figure 7.10: Installation of Cryofilter superconducting separators at Cadam, Brazil (courtesy of Outokumpu Technology, Inc.).

capture by a single magnetized wire. Such theoretical models, very popular in the late nineteen seventies and early eighties, failed to describe correctly the dependence of particle capture on key operational parameters. This approach sometimes resulted in costly and unnecessarily complicated design of magnetic separators and the VUPCHT separator was no exception.

Initially, the performance of the VUPCHT separator was encouraging. However, the subsequent deterioration in efficiency, and drop in throughput, as a result of matrix blockage, led to closure of the plant. The fate of the VUPCHT superconducting separator demonstrates how unproven academic reasoning and erroneous interpretation of laboratory data, with limited relevance to practical operation, can result in failure to implement a new technology.

Not surprisingly, the subsequent production-scale superconducting HGMS, developed by Eriez Magnetics, Inc. and shown in Fig. 2.80, was based on a conventional cyclic design [S16]. The magnet generated a 2 T magnetic field, sufficient to treat most minerals. The fast-ramp magnet allowed the maximum field to be reached in less than 60 s, which enabled a reasonable duty factor to be obtained without resorting to a mechanically more complicated reciprocating matrix system.

Further developments in superconducting technology and better understanding of mechanisms of HGMS led to a revival of the reciprocating matrix concept. After Carpco Inc. acquired, in 1991, the Magnetic Separation Group from Cryogenic Consultants Ltd., development of reciprocating canister superconducting magnets gained momentum. Several types of separators of different sizes were developed and installed at kaolin processing operations. The first large 5 T machine, with a diameter of the reciprocating tube of 500 mm and throughput of up to 50 t/h was delivered in 1993 to Caulim di Amazonia (Cadam), a Brazilian kaolin company. The machine requires 1000 litres of liquid helium per year, which is transferred annually. No liquid nitrogen is required and a small refrigerator, running at 10 kW, provides cooling at an intermediate temperature [W36]. The reduction in energy costs compared to conventional HGMS units amounts to about 95%. Further installations, including the largest separator of 1000 mm diameter, Cryofilter 5T/1000, under the Outokumpu Technology label after Outokumpu acquired Carpco Inc. in 1998, took place in several other kaolin operations. Figure 7.10 illustrates the two units delivered to Cadam.

Open-gradient magnetic separators

An unsuccessful attempt, potentially of breakthrough significance, to implement an innovative technique was the application of open-gradient magnetic separation to beneficiate pyroxenite ore. OGMS, instead of employing a matrix to generate a magnetic field gradient, uses the actual magnet windings to create the field gradient and magnetic force that deflect magnetizable particles. A schematic diagram of such a "falling curtain" superconducting OGMS is shown in Fig. 2.85.

As has been discussed in Section 2.5.3, Cryogenic Consultants Ltd. built a production-scale linear race-track magnet OGMS that was earmarked for the dry concentration of apatite from phlogopite [R19]. The magnet suffered from repeated quenches resulting from electromagnetic stresses [K31]. In spite of considerable expenses to resolve the problems and to salvage the project, the separator was never commissioned. This case illustrates the consequences of an attempt to introduce an immature technology into a production environment.

Separation in magnetic fluids

It has been discussed in Section 2.8.2 that AVCO Corp., NASA [M12], Hitachi [N4] and other organizations successfully demonstrated, in 1970s, that it was possible to achieve a very high selectivity of separation of non-ferrous metals from automobile scrap in ferrofluids at industrially meaningful separation rates. In spite of the positive outcomes of these campaigns, further production-scale implementation of the FHS technique did not happen. It can be speculated that the reason for non-acceptance of FHS by the market at that time was that the need for scrap recycling was not yet well developed. The perceived high cost of ferrofluid and the need to recycle it probably contributed to the fact that the introduction of FHS to the metal recycling industry was not successful.

At a somewhat later stage, a wide range of ferrohydrostatic separators was developed by the GMUO Institute in Lugansk, in the former USSR [G5]. The application of these units included the density separation of coal and non-ferrous metals and the recovery of diamonds, gold and other platinum-group metals.

Figure 7.11: The GMUO ferrohydrostatic separator FHS-40M.

Figure 7.11 shows an FHS-40M ferrohydrostatic separator designed for the recovery of fine diamonds from the final sorting concentrate.

In 1992 six of the FHS separators that had been supplied to Yakutalmaz in Yakutia, Russia, for the recovery of diamonds, underwent, successfully, detailed acceptance tests and it was decided to install them in production at Yakutalmaz to replace the toxic heavy liquid Clerici solution. It appears, however, that this new technology was treated with suspicion by the workforce. FHS was apparently seen as a threat to various benefits and perks. The management was forced to abandon this novel technique.

More fortunate was the GMUO FHS separator for the recovery of gold from placer gravity concentrates. It was reported that in 1993 this technology, which eliminates the need to use mercury amalgamation, was operating in nine Russian gold mining operations [S55]. Even more successful appears to be permanent magnet-based Geos ferrohydrostatic separators. It was reported that in 1995 these systems were in operation in 19 gold-processing plants in Russia [S54].

It can be seen that separation in ferrofluids, in spite of a few patches of misfortune, is on its way to successful implementation on production scale. The existing applications are in areas where a high density difference between the fractions to be separated allows a low level of sophistication of the equipment. Considerable progress has, however, been made in the development of selective FHS separators with automated density control with the main emphasis on application in the diamond-winning industry. Scale-up of the technology has also been addressed, technology for the recovery and recycling of ferrofluid has been developed, and low-cost ferrofluids are being produced on a production scale [S23]

These recent developments in FHS and new waste legislation in various First World countries open again the door for the application of FHS in the recycling of metals and metallic products and the recovery of non-ferrous metals [S92].

Concluding summary

Our walk through the innovation landscape in magnetic separation illustrates the enormous effort that was expended over the years in order to convert a wealth of novel ideas into workable techniques and to introduce them into material treatment operations. The cases discussed above illustrate that there have been numerous successful innovations in magnetic separation. Permanent magnetic roll separators, matrix separators as conceived by Jones, iron-clad solenoids as a spin-off of a search for the elusive magnetic monopole by the MIT scientists, the replacement of the Jones horizontal rotor in HGMS by a vertically rotating ring carrying the matrix and the introduction of eddy current techniques to the separation of non-ferrous metals are examples of the remarkable success of innovation and technology transfer.

On the other hand, there is an equal share of unmistakable failures. The reasons for these failures cover a wide spectrum of risks that accompany any effort to implement new technology. The derailers that were responsible for the failures in technology transfer can be identified from our analysis:

- The excessive cost of a product, in order to recover the research and development costs in the shortest possible time, not compatible with the market needs and with the availability of an alternative technique or product.
- Incorrect assessment of technical feasibility. A technique was not feasible or viable from first principles and in spite of this, it was still developed and offered to the industry.
- Political turmoil and changes on a regional level prevented implementation of a novel and sufficiently mature technology. In spite of financial losses and time delay, this setback became a basis for even further innovation.
- Premature application of unproven technology. Scale-up of the research results was attempted with limited consideration of the production-scale complexities.
- Poor engineering of an ambitious and complex technique.
- Insufficient market need in relation to the complexity and cost of the technique.

- Resistance to change from the workforce and fear of job and perk losses.

It is also interesting to note from the above analysis that the mining and material-handling companies that were trying to introduce innovation in their operations by implementing the magnetic separation technology usually did not carry out the research and development in-house, but rather preferred to procure proven, or seemingly proven, equipment or techniques. While this is a safe and cheaper way of acquiring a novel technology, it also carries some risks. The user becomes dependent on the supplier's expertise, credibility and integrity, and additional expenses might be incurred in order to align a novel technology with the work system into which it is being introduced.

7.4 The current status of magnetic separation technology

In order to review the current status of magnetic separation technology, it is instructive to apply Stokes's quadrant model of research, development and technology transfer, shown in Fig. 7.4. Without doubt, the ultimate objective of research and development of a novel technique should be to place it in the first, Pasteur's quadrant. A technique that finds itself in this quadrant is characterized by well developed theoretical foundations as well as by clearly outlined practical use or application.

The current status of magnetic separation technology, using Stokes's quadrant concept, is schematically depicted in Fig. 7.12. It can be seen that techniques such as low-intensity magnetic separation, permanent magnet-based high-intensity magnetic separation and large-particle eddy-current separation, while widely employed on production scale, are based on an empirical and heuristic approach rather than on good knowledge of the fundamentals [S97]. The magnetic aspects of the applications of magnetic techniques in biosciences are also of empirical nature rather than based on good understanding of the fundamentals of magnetism.

On the other hand, production-scale applications of open-gradient magnetic separation, dry HGMS, ferrohydrostatic separation and to some extent wet HGMS are rare, although the fundamental theoretical principles are reasonably well established. Flocculation of weakly magnetic materials and magnetic tagging, two techniques of potentially considerable industrial, environmental and social significance, find themselves only in the early stages of fundamental research.

7.5 What the future holds?

It transpires from our discussion that the future focus of magnetic techniques of material handling should be towards a good understanding of the physical principles, combined with clear goals for industrial applications.

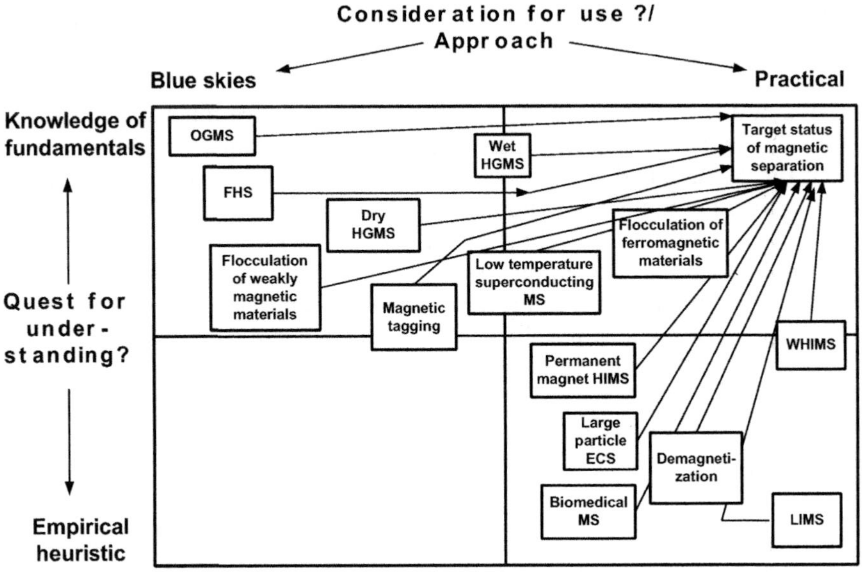

Figure 7.12: The current status of magnetic separation technology (adapted from Svoboda [S97]).

Presently, the magnetic techniques of manipulation of matter stand at a crossroads. The dominant drivers of the early 21st century differ markedly from those of the late 20th. The priorities of nuclear power, defence and beneficiation of mineral resources by brute force have been replaced by those associated with economic growth, discovery of new materials, environmental and knowledge management and health care. Processes and products from the last century cannot, therefore, meet the criteria of the 21st century and a new approach will have to be developed.

7.5.1 Mineral processing

The minerals industry faces problems of declining grades of ores, increasing proportions of deleterious impurities and a need to reduce the particle size in order to liberate the valuable components. New technologies capable of meeting these challenges have to be designed for environmental and sustainability reasons.

Magnetic separation, unlike many other technologies of mineral treatment, can operate in a dry mode. It can perform successfully in arid and arctic regions and even in extraterrestrial conditions. Problems of water consumption and effluent disposal can thus be largely eliminated. Using sophisticated magnet design or advanced permanent magnet materials, the energy consumption can be reduced significantly. In addition to lower operating costs, reduction of waste heat, CO_2 and other emissions will be, therefore, achieved.

In contrast to conventional flotation and flocculation, magnetic separation does not require chemicals that are responsible for steadily increasing operating costs and environmental concerns. Similarly, the use of heavy liquids, usually toxic and costly to acquire and to dispose of, in density separation and coal preparation, can be eliminated by using magnetism-based techniques.

7.5.2 Biomedical applications

There are three main areas in the biomedical field in which magnetic techniques are likely to play a role. The development of magnetic carriers and magnetic fluids as drug carriers and adsorbents is at the top of the list. Future applications also include the production of DNA sequences, development of magnetic affinity adsorbents for isolation and removal of selected biological compounds, the identification of unwanted toxic organisms in foodstuffs, and gene cloning. Drug delivery and the localization of magnetic fluids in target areas will benefit from new developments of powerful permanent magnet materials.

Although magnetic separation is the most common approach to exploit magnetic properties in biosciences, the design of existing commercially available separators is still based more on a trial-and-error approach than on specialized knowledge of the fundamentals of magnetism. Future magnetic separation systems will have to be able to discriminate between targets precisely, will have to be selective and will be required to leave targets functionally unharmed.

7.5.3 Recycling and waste processing

The recovery of materials from solid waste is technically, and in particular economically, complicated. Nevertheless, beneficial uses for stockpiled by-products are being found, and recycling is experiencing increasing emphasis. While the process design from the environmental point of view is based on short-term thinking, sustainability requires a longer view, perhaps 50 years into the future. It is a tough problem, in which recycling and waste treatment, in addition to renewability, dematerialization and product life extension, is an important part of the scenario.

The potential for improved success in the recovery of materials from solid waste increased considerably with the availability of eddy-current separators for small particles. The high intrinsic value of non-ferrous metals will inevitably lead to further improvements in the separation of fine and very fine particles of non-ferrous metals. This development, combined with availability of even more powerful permanent magnets that would allow the increase in the magnetic field strength in eddy-current separators, will extend the range of applicability of this technique and its viability [V11].

Magnetic carrier and tagging technology allow for rapid and selective separations of colloidal non-magnetic mineral particles and macromolecules using magnetic means. To date, this technology has been associated with small-scale high-value medical and analytical applications. However, with the development of new inexpensive magnetic support particles and genetic advances in mass production of biomaterials, such as antibodies, magnetic support technology has potential for large-scale and continuous separation in biological, mineral, effluent and soil cleaning processes.

The combination of magnetic support technology and processing by magnetic means may ultimately result in a technique which will not only compete with conventional separation techniques, but may also stimulate a radical change in overall plant design [M40].

Magnetic fluids

The use of magnetic fluids is attractive for separation of dense materials. Ferrofluids exposed to a non-homogeneous magnetic field acquire an apparent density exceeding densities obtained by conventional density-based methods. Ferrohydrostatic separation of copper, zinc, lead and precious metals is feasible on a production scale. The development of cost-effective, environmentally and user-friendly ferrofluids that can be easily recycled will enhance even further the significant potential of FHS.

A major contribution to the applicability of FHS on industrial scale would be the development of ferrofluids with high saturation polarization. Such fluids, with saturation polarization in the range from 0.1 T to 0.2 T have been prepared using particles of iron nitride or other ferromagnetic elements and alloys. However, the fluids lack stability in contact with the atmosphere. Efforts are under way to coat or otherwise protect the particle surfaces and enhance their usefulness [R30].

Present day magnetic fluids are black in colour. Clear or lightly coloured fluids would permit better visual control of the separation process. Recent advances in organic magnetic materials offer a hope in this line of investigation.

When magnets based on room-temperature superconductors become available, it will be possible to generate high magnetic fields at low cost and weakly magnetic fluids can replace the present-day ferrofluids. For instance, solutions of paramagnetic salts, because of their molecular make-up, are not appreciably subject to the inhomogeneities that can develop in sub-colloidal ferrofluids exposed to high gradient magnetic fields [R30]. Accuracy of density separation will thus be increased.

7.5.4 Prospects for superconductivity in magnetic separation

Superconducting magnetic separation has made a number of important technical advances in the last twenty years, The first large superconducting separator has been operating successfully in the USA since 1986 and a larger system, with twice the capacity, was installed in 1989. A novel design of a superconducting separator with a reciprocating canister system was installed and successfully operated for clay processing in 1989. Following this, a number of other cyclic and reciprocating separators have been installed for kaolin processing in various parts of the world. Furthermore, there appears to be an opportunity to retro-fit superconducting coils into conventional high-gradient magnetic separators [W36].

High-temperature superconductivity in magnetic separation

The discovery of high-temperature superconductors (HTS) in 1986 stimulated an enormous research effort into such materials. The discovery of the Y-Ba-Cu-O (YBCO) compounds with a superconducting transition temperature T_c as high as 92 K created a real possibility of practical devices operating in liquid nitrogen at 77 K. Such a development would imply that the capital and running costs, associated with the production and maintenance of the necessary conditions to operate these devices, would be considerably reduced, perhaps by as much as a factor of 10 [W37].

At present, and for the foreseeable future, high-T_c superconducting wires or tapes cannot compete economically and technically with low-T_c superconducting materials, as they are too expensive and the current density they can deliver is too small [W38, B50]. The limitation arises from the sensitivity of the high-T_c materials to their own magnetic field.

This situation will remain until there is a technical breakthrough in the processing and production of high-T_c materials in wire or tape form [W38]. Moreover, there are only limited economic advantages in using wire-wound high-T_c superconducting devices compared to low-temperature superconducting magnets. The wire-wound device will be of approximately the same cost as a low-T_c device. Since the cost of a large wire-wound superconducting system is usually dominated by the winding, which is 70% to 80% of the total cost, the saving will only be in the cryogenic areas [W37].

There is, however, a possibility of trapping the magnetic field in hollow tubes of a superconductor or in solid discs, without the necessity of winding a coil. Flux tubes are similar to solenoid permanent magnets and have a field trapped inside the tube. The field is produced by supercurrents flowing in the walls of the tube [W39, W40]. The system runs in the persistent mode. As the field in the flux tube decreases with time, due to flux creep, the field is topped up using a flux pump. Such a device is schematically shown in Fig. 7.13. The flux can be introduced into a pump through a flux gate 1, which is a region of a superconductor which can be switched thermally so that it can act in two modes. In these modes the flux gate can either prevent or allow the passage of magnetic flux. The flux introduced into the pump can be compressed with a mandrel and then it passes through flux gate 2 into the working region. This process can continue indefinitely.

This approach could provide the cheapest option for magnetic separation on a large scale [W41]. It has been estimated [W37] that the cost of manufacturing flux tubes, if it can be done by standard ceramics processing, could be reduced by a factor of 20 compared to an equivalent wire-wound solenoid. By including the cost reduction by the factor of 10 by operating the magnet at liquid nitrogen temperature, the economics of such devices would be improved significantly.

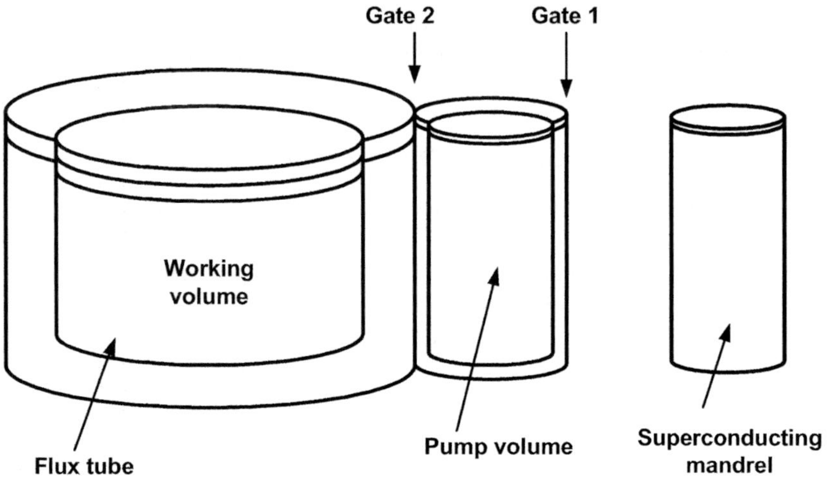

Figure 7.13: Flux tube with flux pump and superconducting mandrel (adapted from Watson [W38]).

And the likelihood of producing, in a short time, a high-T_c superconductor with high critical current J_c, of at least 5×10^2 A/mm^2 at 77 K, is much better than for a wire-wound solenoid [W38].

As yet, no flux tubes with flux pumps have been constructed from high-T_c material, but it appears to be potentially a rewarding direction of investigation that might lead to a breakthrough technology.

At the same time, in several recent projects, high-T_c superconducting coils have been used in small-scale magnetic separators. A collaboration between the Los Alamos National Laboratory and American Superconductor Corp. produced a small HTS magnetic separator with outer diameter of the magnet of 180 mm, inner diameter of 50 mm and height of 155 mm [D22]. The magnet, shown in Fig. 7.14, is made of 624 m of Ag/BSCCO superconducting wire and uses a pair of HTS current leads that are designed to operate with the warm ends at 75 K and the cold ends at 27 K. The system operates in vacuum and is cooled by a two-stage Gifford-McMahon cryocooler. The upper stage of the cryocooler cools the thermal shield and two heat pipe thermal intercepts attached to the current leads. The lower stage of the cryocooler cools the HTS magnet and the lower end of the HTS current leads. At 40 K the magnet can generate a magnetic induction of 2 T in 25 mm warm bore, at a current of 120 A .

A joint project between Aquafine Corp., Sumitomo Electric and Du Pont yielded an HTS magnet of dimensions similar to those of the LANL/ASC system [I7]. The magnet generated a maximum magnetic field of 4 T at 4.2 K, 3 T at 21 K and 0.65 T at 77 K. In the tests, the magnet was operated at 20 K, generating 2.5 T in 38 mm warm bore, and was cooled by a two-stage Gifford-McMahon cryocooler. The separator was applied to kaolin beneficiation and it

Figure 7.14: The LANL/American Superconductor Corp. high-temperature superconducting magnet for magnetic separation (courtesy of Los Alamos National Laboratory).

was observed that although an increase in the magnetic field from 2 T to 2.5 T did not produce any significant improvement in the quality of the concentrate, a higher throughput was achieved, without loss of quality.

An HTS magnetic separator built jointly by the University of Tsukuba, Iwate Industrial Promotion Centre and National Institute for Materials Science, Japan, has been applied to the removal of endocrine disrupting chemicals from waste water [M45]. The magnet generates a magnetic induction of 1.7 T in 200 mm inner bore. A photograph of the experimental set-up is shown in Fig. 7.15.

7.6 Research and development needs in magnetic separation

The current research and development initiatives and needs are shown in Fig. 7.16. Several important trends can be identified. As a result of these trends and drives, several magnetic techniques are likely to move, in the foreseeable future, into Pasteur's quadrant and become available for industrial applications.

Magnetic separation methods that have been, to a greater extent, conceived empirically and applied in practice, such as superconducting magnetic separation, HIMS, small-particle ECS and biomedical separation, are being studied from a more fundamental point of view and further progress can be expected in

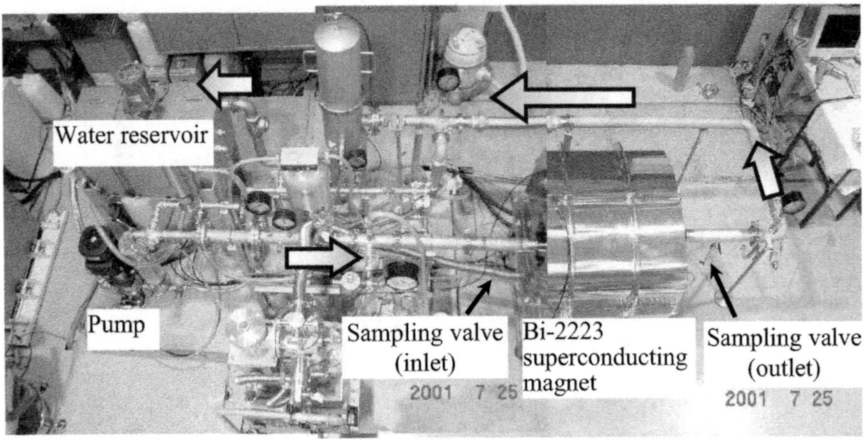

Figure 7.15: The Tsukuba/Iwate/NIMS (Japan) HTS magnetic separator for treatment of waste water.

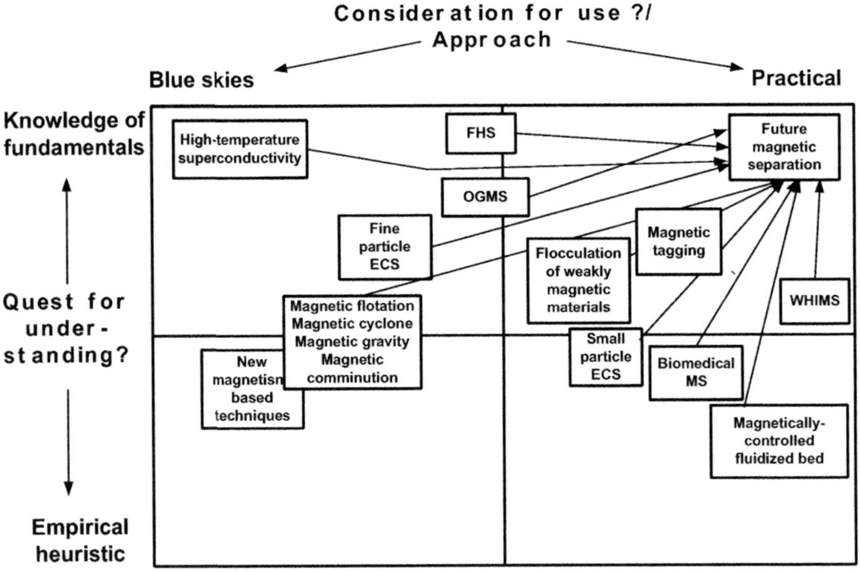

Figure 7.16: The current trends in the development of magnetic separation techniques (adapted from Svoboda [S97]).

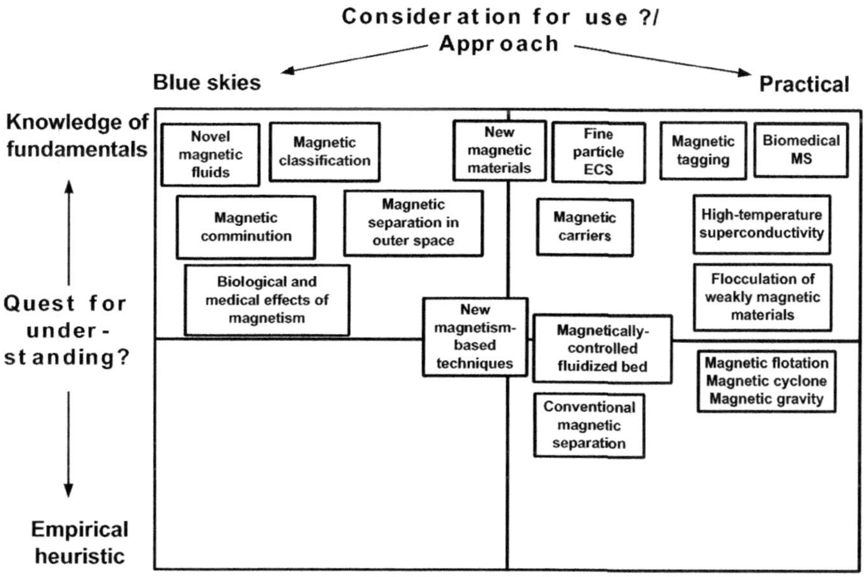

Figure 7.17: The future research and development focus in the magnetism-based methods of material treatment (adapted from Svoboda [S97]).

the near future.

On the other hand, methods such as fine-particle ECS, ferrohydrostatic separation, magnetic flocculation of weakly magnetic materials and magnetic tagging, that have received attention on an academic level, are likely to enter the development stage.

The application of high-temperature superconductivity to magnetic separation and novel magnetism-based techniques are also being explored, either theoretically, or empirically (magnetic flotation or gravity separation).

Of particular promise for rapid incorporation into wide-scale industrial use are the following areas that are currently enjoying strong focus:

- An improved understanding of the theoretical and operational principles of HGMS/WHIMS and possible resurrection of open-gradient magnetic separation, based on correct assessment of the needs of the industry.

- The judicious and justified incorporation of low-temperature superconductivity into magnetism-based methods of materials handling.

- Further development of advanced permanent magnetic materials and their incorporation into sophisticated magnet design.

To predict the future focus of research and development is a risky business; nevertheless some well-defined trends can be identified. These trends allow the author to venture a possible scheme of future developments. These conjectures are illustrated in Fig. 7.17. The application of high-temperature superconductivity, magnetic flocculation of weakly magnetic materials, magnetic tagging and carriers are likely to dominate the field of technological needs. It can also be speculated that new magnetism-assisted techniques, such as magnetic gravity separation, magnetic flotation, magnetic comminution and classification will take advantage of having a much wider control over these processes as a result of the presence of this additional external force.

However, for successful development and implementation of such novel methods of materials treatment, a research and development process that combines vision, deep knowledge of scientific fundamentals, incentives and a clearly defined action plan for a production environment, are essential.

List of Symbols

a	radius of matrix element [m]
a_p	acceleration of particle [m/s^2]
A	Hamaker constant [J]
A	cross-sectional area [m^2]
A_c	cross-sectional area of cladding [m^2]
A_m	cross-sectional area of magnet [m^2]
A_p	cross-sectional area of particle [m^2]
b	radius of particle [m]
B	magnetic flux density (magnetic induction) [T]
B_c	critical magnetic induction [T]
B_g	magnetic induction in air gap [T]
B_m	magnetic induction in magnet [T]
B_0	magnetic induction on surface of magnet [T]
B_p	magnetic induction for flocculation into primary minimum [T]
B_r	remanent magnetic induction [T]
B_s	magnetic induction for flocculation into secondary minimum [T]
BH	energy product [J/m^3]
$(BH)^i$	intrinsic energy product [J/m^3]
$(BH)_{\max}$	maximum energy product [J/m^3]
c	concentration of particles [m^{-3}]
c_{in}	influent concentration [m^{-3}]
c_{out}	effluent concentration [m^{-3}]
c_{col}	concentration of colliding particles [m^{-3}]
C	throughput [t/h]
d_p	diameter of particle [m]
D	diameter [m]
e	electronic charge [1.6022×10^{-19} C]
Ep	mean probable error
f	frequency [s^{-1}]
f	filling factor
F	force [N]

$F(\alpha,\beta)$	field factor
F_c	centrifugal force [N]
F_d	hydrodynamic drag [N]
F_e	electric force [N]
F_f	force of friction [N]
F_g	force of gravity [N]
F_h	force of capillary attraction [N]
F_i	inertial force [N]
F_L	Lorentz force [N]
F_{Lp}	Lorentz force on particle [N]
F_p	force on particle [N]
F_{pg}	force of gravity on particle [N]
F_{pm}	magnetic force on particle [N]
F_{pgb}	buoyancy force on particle [N]
F_{pmb}	magnetically induced buoyancy force [N]
F_r	radial component of force [N]
F_θ	azimuthal component of force [N]
g	acceleration of gravity (9.807 m/s^2)
G	weight [N]
G_b	weight of burden
h	distance [m]
H	magnetic field strength (intensity) [A/m]
H_0	external magnetic field strength [A/m]
H_c	coercive force [A/m]
H_{ci}	intrinsic coercive force [A/m]
H_F	magnetic field strength of flocculation [A/m]
H_s	saturation magnetic field strength [A/m]
I	electric current [A]
I_m	ionic strength of medium [N/C^2]
j	current density [A/m^2]
j_s	surface current density [A/m]
J	magnetic polarization [T]
J_f	magnetic polarization of fluid [T]
J_r	remanent polarization [T]
J_s	saturation polarization [T]
k	Boltzmann constant (1.3806×10^{-23} J/K)
K_1	leakage factor
K_2	loss factor
l	length [m]
L	filter length [m]
L	matrix loading
L_e	inductance [H]
L_m	length of magnet [m]
L_g	length of air gap [m]
m	mass [kg]
m_p	mass of particle [kg]

M	magnetization [A/m]
M_f	magnetization of fluid [A/m]
M_p	magnetization of particle [A/m]
M_s	saturation magnetization [A/m]
N	number of revolutions [s^{-1}]
N	number of turns in a coil
N	demagnetization factor
N_0	initial concentration of particles [m^{-3}]
N_s	saturation concentration of particles [m^{-3}]
$N(x,\tau)$	number of particles captured at position x and time τ
P	permeance [Tm^2/A]
P_g	permeance of air gap [Tm^2/A]
P_m	permeance of magnet [Tm^2/A]
ΔP	pressure drop [Pa]
Q	flowrate [m^3/s]
r	radius vector [m]
R	radius [m]
R	resistance [Ω]
R	recovery
R_c	capture radius [m]
R_{ca}	normalized capture radius
s	shape factor
S	cross-sectional area [m^2]
S_v	coefficient of magnetic viscosity
t	time [s]
$t_{1/2}$	half-time of flocculation [s]
T	temperature [K, ^0C]
T	torque [Nm]
T_c	critical temperature [K]
U	magnetic potential [A]
U_a	perimeter of part a [m]
v	velocity [m/s]
v_0	superficial velocity [m/s]
v_r	radial component of velocity [m/s]
v_θ	azimuthal component of velocity [m/s]
V	voltage [V]
V_A	London-van der Waals interaction energy [J]
V_b	volume of burden [m^3]
V_g	volume of air gap [m^3]
V_m	volume of magnet [m^3]
V_M	magnetic interaction energy [J]
V_R	interaction energy of electric double layer [J]
V_T	total interaction energy [J]
w	width [m]
W	input power [W]
W	stability ratio

X	Debye-Hückel factor [m^{-1}]
α	temperature coefficient [$^0C^{-1}$]
γ	surface tension [N/m]
γ_c	correction factor
δ	skin depth [m]
ϵ	porosity
ϵ_0	permittivity of vacuum (8.854×10^{-12} F/m)
ϵ_d	porosity of deposit
ϵ_i	porosity of clean matrix
ϵ_r	dielectric constant (relative permittivity)
η	dynamic viscosity [Pa×s]
θ_c	critical angle
κ	volume magnetic susceptibility
κ_f	volume magnetic susceptibility of fluid
κ_p	volume magnetic susceptibility of particle
λ	filling factor
λ	filter coefficient [m^{-1}]
λ	periodicity of magnetic system
Λ	reluctance [H^{-1}]
Λ_g	reluctance of air gap [H^{-1}]
Λ_m	reluctance of magnet [H^{-1}]
μ_0	magnetic permeability of vacuum ($4\pi\times10^{-7}$ H/m)
μ_d	coefficient of dynamic friction
μ_m	relative magnetic permeabilty of medium
μ_M	magnetic moment [Am^2]
μ_p	relative magnetic permeability of particle
μ_s	coefficient of static friction
ρ	density [kg/m^3]
ρ_a	apparent density [kg/m^3]
ρ_b	bulk density [kg/m^3]
ρ_{cp}	cut-point density [kg/m^3]
ρ_{eff}	effective density [kg/m^3]
ρ_m	density of medium [kg/m^3]
ρ_p	density of particle [kg/m^3]
ϱ	resistivity [Ωm]
ϱ_p	resistivity of particle [Ωm]
σ_0	initial surface charge density [C/m^2]
σ_p	specific electric conductivity of particle [S/m]
σ_s	surface charge density [C/m^2]
τ	characteristic time [s]
τ_0	shear stress [$kg/m/s^2$]
χ	mass (specific) magnetic susceptibility [m^3/kg]
χ_f	mass magnetic susceptibility of fluid [m^3/kg]
χ_p	mass magnetic susceptibility of particle [m^3kg]
ψ	surface potential [V]
ω	angular velocity [s^{-1}]

ω_c critical angular velocity [s^{-1}]
ω_p angular velocity of particle [s^{-1}]
ω_r angular velocity of roll [s^{-1}]
Ω angular velocity of particle rotation [s^{-1}]

Bibliography

[A1] U. Andres: *Magnetohydrodynamic and Magnetohydrostatic Methods of Mineral Separation.* John Wiley and Sons, New York and Israel University Press, Jerusalem 1976

[A2] N. Amin, S. Arajs and E. Matijevic: Magnetic properties of submicronic $\alpha-Fe_2O_3$ particles of uniform size distribution at 300 K. *Phys. Stat. Sol.* (a) **104** (1987), K65

[A3] M. Aubrey and R.M. Funk: The quadrupole magnetic separator - a new concept in magnetic separation equipment. In: *Proc. 32nd Ann. Min. Proc. Symp.,* Duluth, Minn., USA (1971), 55

[A4] J.R. Andreachi and O.E. Palasvirta: Magnetic separator. *Canadian Patent no. 711.991* (1965)

[A5] B.R. Arvidson, J.A. Oberteuffer and I. Wechsler: Continuous HGMS pilot plant: machine description and mineral processing results. *AIME Reprint No. 76B-341* (1976)

[A6] Anon: Novel high-gradient magnetic separator solves the problem. *SA Mining, Coal, Gold & Base Minerals,* July 1991, p. 53

[A7] Anon: Superconducting magnetic separator abandoned. *Magn Electr. Sep.* **3** (1991), 68

[A8] Anon: Superconducting magnets: systems for research and industry. *Oxford Instruments Brochure 1998*

[A9] N.R. Allen: Low intensity rotating magnetic field separation. In: *Proc. Int. Con. Min. Proc. Extr. Metal. MINPREX 2000*, Melbourne, Australia 2000, p.303

[A10] N.R. Allen: The concept of magnetic mineral separation by particle rotation. *Magn. Electr. Sep.* **11** (2002), 33

[A11] N.R. Allen: Mineral particle rotation measurements for magnetic rotation separation. *Magn. Electr. Sep.* **11** (2002), 155

[A12] U. Andres: Magnetohydrostatic separation. *Zh. Prikl. Met. Tekh. Fiz.*, No. 3 (1966)

[A13] B.R. Arvidson: Recent developments of rare-earth magnetic roll and drum separators. In: *Proc. SME Ann. Meeting*, Salt Lake City, UT, USA (2000), Preprint 00-140

[A14] U. Andres, A.L. Devernoe and M.S. Walker: Apparatus and method employing magnetic fluids for separating particles. *US Patent 4,594,149* (1986)

[A15] Anon: A manual published by Magnetic Equipment Standards Council, USA

[A16] S. Arajs, C.A. Moyer, R. Aidun and E. Matijevic: Magnetic filtration of submicroscopic particles through a packed bed of spheres. *J. Appl. Phys.* **57** (1985), 4286

[A17] N.R. Allen: unpublished results (2003)

[A18] U. Andres and W. O'Reilly: Selectivity in the magnetic separation of minerals. *Powder Technol.* **79** (1994), 147

[A19] U. Andres: Magnetohydrodynamic and magnetohydrostatic separator: A new prospect for mineral separation in the magnetic field. *Min. Sci. Eng.* **7** (2) (1975), 99

[A20] U. Andres and W. O'Reilly: Separation of minerals by selective magnetic fluidization. *Powder Technol.* **69** (1992), 279

[A21] N.R. Allen: The rotating magnetic field separation of minerals. *Ph.D. Thesis, University of Tasmania, Hobart* (1999)

[A22] N.R. Allen: Small particle eddy current separation. In: *Proc. Int. Congress Min. Process. Extr. Metall. MINPREX 2000,* Melbourne, Australia, p.293

[A23] B.R. Arvidson and B. Dillé: Improved high-intensity magnetic separation for industrial minerals. *Ind. Minerals Supplement,* September 1996, p. 29

[A24] B.R. Arvidson: The many uses of rare-earth magnetic separators for heavy minerals sands processing. In: *Proc. Int. Heavy Minerals Conf.,* Freemantle, Australia (2001), 1

[A25] B.R. Arvidson: Advances in rare-earth magnetic drum separators for heavy mineral sands processing. In: *Int. Heavy Minerals Conf.*, Durban, South Africa (1999)

[A26] B.R. Arvidson and L. Rademeyer: Rare-earth magnetic separators for mineral sands applications. In: *Int. Heavy Minerals Conf.*, Johannesburg, South Africa (1997), 129

[A27] M.A. Aramante, A. Botelho de Sousa and M. Machado Leite: Beneficiation of a feldspar ore for application in the ceramic industry. In: *Proc. Int. Conf. "Minerals & Materials '96*, Somerset West, South Africa (1996), p.168

[A28] D.M. Allen et al.: Treatment of combined sewer overflows by HGMS. *EPA Report EPA-600/2/77/015* (1977)

[A29] A. Aharoni: Traction force on paramagnetic particles in magnetic separators. *IEEE Trans. Magn.* **12** (1976), 234

[A30] B. R. Arvidson: Metallurgical and economical advantages of HGMS. In: *Proc. 12th Int. Min. Proc. Congress,* Sao Paulo, Brazil (1977), 1

[A31] N.R. Allen: Effects of roasting temperature on the magnetism of ilmenite. *Phys. Sep. Sci. Eng.* **12** (2003),

[A32] N.R. Allen: The effect of roasting temperature on ilmenite particle rotation characteristics in a rotating magnetic field. *Phys., Sep. Sci. Eng.* **12** (2003),

[A33] U.T. Andres: Potentialities of magnetohydrostatic separation using solution of salts of rare earth elements. *J. S.A. Inst. Min. Metal.*, October 1975, p. 113

[A34] Anon.: Magnetic hydrocyclone thickener - a new mineral processing tool. *Min. J.*, January 7 (1983), 6

[A35] J.W. Anderson, T.J. Veasey and I.S. Wells: The magnetic hydrocyclone. In: *Solid-Liquid Separation Practice III* **113**, Chapter 22 (1989), 17

[A36] S.C. Amarante, A.C. Araujo, R. Meillus and R.A. Merwin: Ferrous Wheel magnetic concentration of iron ore fines: project implementation in the Brazilian iron ore industry. Presented at *MEGS '01 Conference,* Falmouth, UK (2001)

[A37] Ü. Atalay, G. Özbayoğlu and Z. Doğan: Concentration of chromite gravity tailings by wet high intensity magnetic separation. In: *Proc. XVI Int. Min. Proc. Congress,* Stockholm, Sweden (1988), 1305

[A38] L.R. Avens et al.: Magnetic separation for soil decontamination. In: Proc. Waste *Management Symposium "Working Towards a Cleaner Environment"*, Tucson, Arizona, USA (1993), 787

[A39] K.R. Alikhanov et al.: HGMF model tests for heat carrier cleaning on Leningrad nuclear power plant. *IEEE Trans. Magn.* **28** (1992), 668

[A40] D.M. Allen, R.L. Sargent and J.A. Obertueffer: Treatment of combined sewer overflows by HGMS. *EPA Report* EPA-600/2-77-015 (1977)

[A41] D.M. Allen: Treatment of combined sewer overflows by HGMS. On-site testing with mobile pilot plant trailer. *EPA Report* 600/2-78-209 (1978)

[A42] N.J. Anderson and A.J. Priestley: Colour and turbidity removal with reusable magnetite particles - V. Process development. *Water Res.* **17** (1983), 1227

[A43] Ch. Alexiou et al. :Magnetic drug targeting: biodistribution and dependency on magnetic field strength. *J. Magn. Magn. Mater.* **252** (2002), 363

[A44] U. Andres: Separation of non-ferrous metals. *Aufbereitungstechnik* **35** (1994), 71

[A45] N. Arbiter: Mineral processing - past, present and future. In: *Proc. Symp. "Advances in Mineral Processing: A Half-Century of Progress in Application of Theory to Practice",* New Orleans, La., USA (1986), 3

[B1] J. Billan: Materials. In: *Proceedings of CERN Accelerator School: Magnetic Measurements and Alignment.* Geneva 1992, p.17

[B2] R.M. Bozorth: *Ferromagnetism.* IEEE Press, New York 1993

[B3] W.F. Brown, Jr.: *Magnetostatic Principles in Ferromagnetism.* North-Holland, Amsterdam 1962

[B4] V.I. Bagin et al.: Study of magnetic properties and Mössbauer effect at the transformation temperature of siderite. *Fiz. Zemli* No. 11 (1971), 71

[B5] L.F. Bates: *Modern Magnetism.* Cambridge University Press, London 1963

[B6] A.E. Berkowitz and E. Kneller: *Magnetism and Metallurgy.* Vol. I, Academic Press, New York 1969

[B7] D. Brown et al.: Developments in the processing and properties of NdFeB-type permanent magnets. *J. Magn. Magn. Mater.* **248** (2002), 432

[B8] J.A. Bartnik, H.D. Wasmuth and W.H. Zabel: Production of coarse grained iron ore sinter feed using Jones WHIMS. *Aufbereitungstechnik,* September 1982, p. 490

[B9] R.R. Birss, M.R. Parker and M.K. Wong: Modeling of fields in magnetic drum separators. *IEEE Trans. Magn.* **15** (1979), 1305

[B10] R.R. Birss and M.R. Parker: High-intensity magnetic separation. In: *Progress in Filtration and Separation* (R.J. Wakeman, Editor), Vol. 2, Elsevier Science Publishers, Amsterdam (1981), 171

[B11] G.K. Batchelor: *An Introduction in the Fluid Dynamics.* Cambridge University Press, Cambridge 1967

[B12] R.R. Birss, R. Gerber and M.R. Parker: Theory and design of axially ordered filters for high intensity magnetic separation. *IEEE Trans. Magn.* **12** (1976), 892

[B13] V. Badescu, V. Murariu, O. Rotariu and N. Rezlescu: Optimization of the recovery efficiency in an axial HGMF cell with bounded flow field. *J. Phys. D: Appl. Phys.* **29** (1996), 2515

[B14] R.R. Birss, M.R. Parker and T.J. Sheerer: Statistics of particle capture in HGMS. *IEEE Trans. Magn.* **16** (1980), 830

[B15] R.R. Birss and M.R. Parker: Magnetic separation of diamagnetic particles. *IEEE Trans. Magn.* **15** (1979), 1523

[B16] J. Boehm, R. Gerber, D. Fletcher and M.R. Parker: Deflection of weakly magnetic materials by superconducting OGMS. *IEEE Trans. Magn.* **24** (1988), 1674

[B17] J. Boehm: Particle transport in OGMS. *IEEE Trans. Magn.* **27** (1991), 5381

[B18] R.I. Berdichevskiy and V.V. Karmazin: Flocculation of magnetite concentrates during flotation in magnetic field. Obog. Rud **73**/1 (1968), 41

[B19] J.A. Bartnik and R.W. Schively: separation of solids and fluids by magnetic flocculation. In: *Proc. 1971 AIME Cent. Ann. Meeting,* New York (1971), Preprint No. 71-B-44

[B20] W.H. Benson, J.A. Bartnik and G.D. Rose: Demagnetizing coils and magnetic flocculators used in the magnetite beneficiation industry. In: *Proc. 29th Ann. Min. Symp.,* Duluth, Minn., USA (1968), 139

[B21] J.A. Bartnik and C.F. Giermak: Chemical-magnetic flocculation process. *CIM Bull.,* March 1969, p. 263

[B22] R.C. Bunge and D.W. Fuerstenau: Separation characteristics of a magnetogravimetric separator. In: *Proc. 17th Int. Min. Process. Congress,* Dresden, Germany (1991), 31

[B23] R.C. Bunge and D.W. Fuerstenau: Magnetogravimetric separation in a rotational device. *Magn. Electr. Sep.* **7** (1996), 163

[B24] F. Bitter: The design of powerful electromagnets. Part I: The use of iron. *Rev. Sci. Instr.* **7** (1936), 479

[B25] K.J. Binns and P.J. Lawrenson: *Analysis and Computation of Electric and Magnetic Field Problems.* Pergamon Press, Oxford

[B26] A.A. Berezin: Electrification of Solid Materials. In: *Handbook of Electrostatic Processes* (Jen-Shih Chang, A.J. Kelly and J.M. Crowley, Editors), Marcel Dekker Inc., New York (1995), 25

[B27] A.G. Bailey: Electrostatic phenomena during powder handling. *Powder Tech.* **37** (1984), 71

[B28] W.J. Bronkala: How to select wet magnetic separators for heavy media plants. *Eng. Min. Int.* **164** (1963), 98

[B29] A.S. Baran and J. Svoboda: unpublished results (1999)

[B30] J.A. Bartnik and R.W. Schively: Separation of solids and fluids by magnetic flocculation. In: *Proc. 1971 AIME Cent. Ann. Meeting*, New York (1971), Preprint No. 71-B-44

[B31] W.H. Benson, J.A. Bartnik and G.D. Rose: Demagnetizing coils and magnetic flocculators used in the magnetite beneficiation industry. In: *Proc. 29th Ann. Min. Symp.*, Duluth, Minn. (USA), (1968), 139

[B32] B.M. Berkovsky, V.F. Medvedev and M.S. Krakov: *Magnetic Fluids: Engineering Applications.* Oxford University Press, Oxford 1993

[B33] P.R. Bissel et al.: Long term stability measurement on magnetic fluids. *IEEE Trans. Magn.* **20** (1984), 1738

[B34] G. Blendulf, N. Dawson and D. James: Utilisation of Readings wet high intensity magnetic separators at Richards Bay Minerals. In: *Australia: A World Source of Ilmenite, Rutile, Monazite and Zircon.* Australasian Inst. Min. Metal., Parkville, Victoria (1986), 81

[B35] J. Babůrek: The preparation of kaolin by means of NIMS. *Interceram* **21** (1972), 54

[B36] W.P. Barnett and Th. J. Feeley, III: Status of advanced coal cleaning as a compliance technology. *Mining. Eng.*, October 1992, p. 1225

[B37] D.A. Butcher and N.A. Rowson: Coal cleaning in high intensity magnetic fields. *Aufbereitungstechnik* **36** (1995), 218

[B38] D.D. Bluhm, G.E. Fanslow and S. O. Nelson: Enhanced magnetic separation of pyrite from coal after microwave heating. *IEEE Trans. Magn.* **22** (1986), 1887

[B39] D.A. Butcher and N.A. Rowson: Microwave pretreatment of coal prior to magnetic separation. *Magn. Electr. Sep.* **6** (1995), 87

[B40] J. Broomberg, S. Gélinas, J.A. Finch and Z. Xu: Review of magnetic carrier technologies for metals ion removal. *Magn. Electr. Sep.* **9** (1999), 169

[B41] G. Bitton and R. Mitchell:The removal of Escherichia Coli-bacteriophage T_7 by magnetic filtration. *Water Res.* **8** (1974), 549

[B42] G. Bitton, J.L. Fox and H.G. Strickland: Removal of algae from Florida lakes by magnetic filtration. *Appl. Microbiol.* **30** (1975), 905

[B43] G. Bitton, R. Mitchell, C. de Latour and E. Maxwell: Phosphate removal by magnetic filtration. *Water Res.* **8** (1974), 107

[B44] A.S. Bahaj, D.C. Ellwood and J.H.P. Watson: Extraction of heavy metals using microorganisms and high gradient magnetic separation. *IEEE Trans. Magn.* **27** (1991), 5371

[B45] V.M. Browning: Potential Department of Defence (DoD) applications of magnetic carriers. In: *Proc. Int. Conf. "Scientific and Clinical Applications of Magnetic Carriers"*, Tallahassee, Florida, USA (2002), 13

[B46] M. Brozek, K. Nowakowski, W. Pilch and A. Siwiec: The possibility of applying the magnetic separator with ferrofluid for separation of nonferrous scrap. In: *Proc. XVII Int. Min. Proc. Congress,* Dresden, Germany (1991), **VII**, 81

[B47] R.J. Batterham and S.H. Algie: The role of technology in the minerals industry. In: *Proc. XIX Int. Min. Proc. Congress,* San Francisco, CA, USA (1995), **3,** 19

[B48] M.M.J. Berry and J.H. Taggart: Managing technology and innovation: A review. *R&D Manag.* **24** (1994), 341

[B49] R.B. Batterham: The mine of the future: will it be visible? In: *Proc. XXII Int. Min. Proc. Congress,* Cape Town, South Africa (2003), 42

[B50] L. Bolt and J.H.P. Watson: Magnetic separation using high-T_c superconductors. *Supercond. Sci. Technol.* **11** (1998), 154

[C1] J. Crangle and M. Gibbs: Units and unity in magnetism: a call for consistency. *Physics World,* November 1994, p. 31

[C2] *Proceedings of CAS - CERN Accelerator School: Magnetic Measurements and Alignment.* Geneva 1992

[C3] J.J. Croat, J.F. Herbst, R.W. Lee and F.E. Pinkerton: Pr-Fe and Nd-Fe based materials: a new class of high-performance permanent magnets. *J. Appl. Phys.* **55** (1984), 2078

[C4] D.F. Cygan and M.J. Mc Nallan: Corrosion of NdFeB permanent magnets in humid environments at temperatures up to 150^0C. *J. Magn. Magn. Mater.* **139** (1995), 131

[C5] P.W. Chu: High-temperature superconductors. *Sci. Amer.*, September 1995, p. 162

[C6] H.E. Cohen and J.A. Good: The principles and operation of a very high intensity magnetic mineral separator. *IEEE Trans. Magn.***12** (1976), 552

[C7] J. Cibulka et al.: A new concept of high-gradient magnetic separators. In: *Proc. XV Int. Min. Proc. Congress,* Cannes (France), 1985, p. 363

[C8] I.J. Corrans and J. Levin: Wet high-intensity magnetic separation for the concentration of Witwatersrand gold-uranium ores and residues. *J.S. Afr. Inst. Min. Metall.* **79** (1979), 210

[C9] I. Cakir, A.J.W. Rozelaar and I.S. Wells: Preliminary analysis of the mechanical forces acting on a particle on the rotor of an electrical separator. In: *Proc. XII Int. Min. Proc. Congress,* Sao Paulo, Brazil (1977), Vol. 1, p. 47

[C10] C. Cowen, F.J. Friedlaender and R. Jaluria: Single wire model of high gradient magnetic separation process-I. *IEEE Trans. Magn.* **12** (1976), 466

[C11] D.L. Cummings, D.C. Prieve and G.J. Powers: The motion of small paramagnetic particles in HGMS. *IEEE Trans. Magn.* **12** (1976), 471

[C12] J.T. Cookson: Removal of submicron particles in packed beds. *Environ. Sci. Technol.* **4** (1970), 128

[C13] H.K. Collan et al.: Superconducting open-gradient magnetic separator. *Trans. Int. Min. Metall. (Sect. C: Min. Process. Extr. Metall.)* **91** (1982), C5

[C14] D.Y. Chan and D. Henderson: The orientation-averaged interaction of a pair of magnetic dipolar spheres in an external magnetic field. *J. Coll. Interface Sci.* **101** (1984), 419

[C15] D.Y. Chan, D. Henderson, J. Barojas and A.M. Homola: The stability of a colloidal suspension of coated magnetic particles in an aqueous solution. *IBM J. Res. Develop.* **29** (1985), 11

[C16] P. Campbell: *Permanent Magnet Materials and their Applications.* Cambridge University Press, Cambridge 1994

[C17] B. Collins, T.J. Napier-Munn and M. Sciarone: The production, properties and selection of ferrosilicon powders for heavy-medium separation.J.S. Afr.Inst. *Min. Metall.* **75** (1974), 103

[C18] I.J. Corrans and J. Svoboda: Magnetic separator. *South African Patent* 83/3323 (1983)

[C19] I.J. Corrans: Magnetic separators. *US Patent* 4,260,477 (1981)

[C20] G. Cavanough: Measurement of magnetic susceptibility in titanium minerals processing. *Ph.D. Thesis,* University of Queensland, Australia (2003)

[C21] Z. Cui, Q. Liu and T.H. Etsell: Magnetic properties of ilmenite, hematite and oil sand minerals after roasting. *Min. Eng.* **15** (2002), 1121

[C22] S.W. Charles and J. Popplewell: Ferromagnetic liquids. In: *Ferromagnetic Materials,* Vol. 2 (E.P. Wohlfarth), North-Holland Pucl. Co., Amsterdam (1980), 509

[C23] S. Chikazumi, S. Taketomi, M. Ukita, M. Mizukami, H. Miyajima, M. Setogawa and Y. Kurihara: Physics of magnetic fluids. *J. Magn. Magn. Mater.* **65** (1987), 245

[C24] R.W. Chantrell, J. Sidhu, P.R. Bissell and P.A. Bates: Dilution induced instability in ferrofluids. *J. Appl. Phys.* **53** (1982), 8341

[C25] F. Chovanec, I. Hlásnik, J. Pitel and V. Hencl: The removal of sulphur and ash contamination from lignite coal by superconducting OGMS. In: *Proc. 10th ICEC Conference,* Otaniemi, Finland (1984)

[D1] R.S. Dean and C.W. Davis: *Magnetic Separation of Ores.* US Bureau of Mines Bulletin 425, Washington D.C.,1941, 417 pp

[D2] D.C. Dahlin and A.R. Rule: Magnetic susceptibility of minerals in high magnetic fields. *Report RI 9449,* US Bureau of Mines (1993)

[D3] Du-Xing Chen et al.: Demagnetising factors for cylinders. *IEEE Trans. Magn.* **27** (1991), 3601

[D4] V.G. Derkach: *Special Methods of Beneficiation of Minerals.* Nedra, Moscow 1966 (in Russian)

[D5] D.J. Dunlop: Magnetic properties of fine-particle hematite. *Ann.Geophys.* **27** (1971), 269

[D6] K. Dwight: Experimental techniques with general applicability for the study of magnetic phenomena.*J. Appl. Phys.* **38** (1967), 1505

[D7] H. Desportes: Three decades of superconducting magnet development. *Cryogenics* **34**, ICEC Suppl. (1994), 47

[D8] S.J. Dobney: Magnetic separation: New developments. *Ind. Miner.,* July 1998, p. 50

[D9] P.G. de Gennes and P.A. Pincus: Pair correlation in a ferromagnetic colloid. *Phys. Kondens. Mater.* **11** (1970), 189

[D10] B.V. Derjaguin and L.D. Landau: Theory of molecular attraction. *Acta Physikokhim (USSR)* **14** (1941), 633

[D11] D. De Klerk: *The Construction of High-Field Electromagnets.* Newport Instruments, Newport Pagnell 1960

[D12] J. Danoczi: Magnetic tracers. *Min. Eng.* **14** (2001), 797

[D13] K.A. Dardis: The design and operation of heavy medium recovery circuits for improved medium recovery. In: *Proc. Dense Medium Operators Conf.,* Brisbane, Australia (1987), 157

[D14] M. Dunglison, T.J. Napier-Munn and F.N. Shi: The rheology of ferrosilicon dense medium suspensions. *Min. Proc. Ext. Met. Rev.* **20** (1999), 183

[D15] D.J. Dunlop: Theory of magnetic viscosity of lunar and terrestrial rocks. *Rev. Geophys. Space Phys.* **11** (1973), 855

[D16] S. Dumbu: Some aspects of ferrohydrostatic separation of minerals and the recycling of ferrofluid. *M.Sc. Dissertation,* University of Stellenbosch (South Africa), 2001

[D17] J. Domenico: How to teach an old dog new tricks: Increasing the profitability of industrial mineral operations using complementary processing techniques. In: *Proc. IMIL Proc. Profit Conf.*, Atlanta, Georgia, USA (2001)

[D18] H. Dinçer et al.: Removal of sulphur and mineral matter from coal slurry by a pilot scale wet magnetic separator. In: *Proc. XXI Int. Min. Proc. Congress,* Rome, Italy (2000), **C7**-15

[D19] C. de Latour: Magnetic separation in water pollution control. *IEEE Trans. Magn.* **9** (1973),314

[D20] C. de Latour and H. Kolm: Magnetic separation in water pollution control II. *IEEE Trans. Magn.* **11** (1975), 1570

[D21] W.L. Dalmijn, M.A. Reuter, T.P.R. de Jong and U.M.J. Boin: The optimization of resource cycle impact of the combination of technology, legislation and economy. In: *Proc. XXII Int. Min. Proc. Congress,* Cape Town, South Africa (2003), 80

[D22] M.A. Dougherty et al.: HTS high gradient magnetic separation system. *IEEE Trans. Appl. Supercond.* **7** (1997), 650

[E1] I. Eisenstein: Magnetic traction force in an HGMS with an ordered array of wires. *IEEE Trans. Magn.* **14** (1978), 1148

[E2] J.A. Ewing: *Magnetic Induction in Iron and Other Metals.* The Electrician Printing and Publishing Co., Ltd., London 1900, 3rd edition

[E3] T. England, P.E. Hand, D.C. Michael, L.M. Falcon and A.D. Yell (Editors): *Coal Preparation in South Africa.* The South African Coal Processing Society, Johannesburg 2002

[E4] S. Endoh and K. Yamaguchi: Characteristics of pressure drop across spongy nickel filter in magnetic separation process. *Powder Tech.* **45** (1985), 35

[E5] A. Einstein: *Ann. Phys.* **19** (1906), 289

[E6] B. Eisenbraum: Magnetic iron recovery improvements at EVTAC Mining ulilizing the Metso mineral separator technology. In: *Proc. Ann. SME Convention,* Duluth, Minn., USA (2002)

[E7] D.E. Erasmus: Dry magnetic and electrostatic beneficiation of a Gravelotte heavy mineral spiral concentrate. In: *Proc. Heavy Minerals Conference*, Johannesburg, South Africa (1997)

[E8] S.C. Everett, I.S. Schache, G.F. Balderson and P.A. Bannister: Pooncarie project: The technical challenges. In: *Proc. "Heavy Minerals 2003" Conference*, Cape Town, South Africa (2003)

[E9] J. Elder: The role of physical processing enhancing the quality of industrial minerals. In: *Proc. 14th Indust. Miner. Int. Congress*, Denver, Colorado, USA (2000)

[E10] B.B. Emory: Nuclear power reactor application of HGMS. *IEEE Trans. Magn.* **18** (1982), 1686

[F1] F. Fraas: Alternating current matrix-type magnetic separator. *US Bureau of Mines Report of Investigations 7746 (1973)*

[F2] S. Foner: Review of magnetometry. *IEEE Trans. Magn.* **7** (1981), 3358

[F3] S. Foner: Vibrating sample magnetometer. *Rev. Sci. Instr.* **30** (1959), 549

[F4] R.H.J. Fastenau and E.J. van Loenen: Applications of rare earth permanent magnets. *J. Magn. Magn. Mater.* **157** (1996), 1

[F5] P.J. Ford and G.A. Saunders: High-temperature superconductivity - ten years on. *Contemp. Phys.* **38** (1997), 63

[F6] J. Fojtek and Z. Kaiser: Superconducting magnetic separator for kaolin purification. In: *Proc. Int. Conf. "Kryogenika"*, Prague, Czechoslovakia (1984), 27

[F7] S.G. Frantz: Magnetic separation method. *US Patent 2,056,426* (1936)

[F8] T. Fujita: Separation of nonmagnetic particles with magnetic fluid. In: *Magnetic Fluids and Applications Handbook* (B. Berkovski and V. Bashtovoy, Editors), Begell House, Inc., New York (1996)

[F9] F.J. Friedlaender, M. Takayasu, T. Nakano and W.H. Neese: Diamagnetic capture in single-wire HGMS. *IEEE Trans. Magn.* **15** (1979), 1526

[F10] F.J. Friedlaender, M. Takayasu, J.B. Rettig and C.P. Kentzer: Particle flow and collective process in single wire HGMS studies. *IEEE Trans. Magn.* **14** (1978), 1158

[F11] F.J. Friedlaender, R. Gerber, W. Kurz and R.R. Birss: Particle motion near and capture on single spheres in HGMS. *IEEE Trans. Magn.* **17** (1981), 2801

[F12] F.J. Friedlaender, R. Gerber, H.P. Henkel and R.R. Birss: Particle build-up on single spheres in HGMS. *IEEE Trans. Magn.* **17** (1981), 2804

[F13] N. Fuchs: On the stability and charging of aerosol. *Z. Physik* **89** (1934), 736

[F14] T. Fujita and M. Mamiya: Interaction forces between nonmagnetic particles in the magnetized magnetic fluid. *J. Magn. Magn. Mater.* **65** (1987), 207

[F15] D. Fletcher, R. Gerber, P. Lawson and J. Boehm: Eddy-current separation of non-ferrous conductors and non-conductors: theory and initial experiments. *IEEE Trans. Magn.* **27** (1991), 5375

[F16] D. Fletcher and R. Gerber: Electromagnetic separation: the prediction and measurement of conductor separability. *IEEE Trans. Magn.* **29** (1993), 3255

[F17] D. Fletcher, R. Gerber and T. Moore: The electromagnetic separation of non-ferrous metals from each other. *IEEE Trans. Magn.* **31** (1995), 4184

[F18] M. Fofana: unpublished results

[F19] M.J. Flanagan: Simplified dilute medium circuit and improved medium recovery with Permax separator. In: *Proc. 7th Samancor Symp. Dense Media Separation,* Zimbali Lodge, South Africa 2000, p. 87

[F20] M. Fofana and J. Svoboda: unpublished results

[F21] T. Fujita, K. Nakatsuka and J. Shimoiizaka: Sink and float separation of heavy nonmagnetic metals with magnetic fluid. *Metal. Rev. MMIJ* **1** (1984), 38

[F22] K.S.E. Forssberg and N.R. Kostkevicius: Comparative pilot scale tests with WHIMS separators. *Erzmetall* **35** (1982), 285

[F23] M. Franzreb, U. Habich and G. Resch: Development of a new permanent magnet-based separator for recycling of ferrous microparticles from aqueous effluents. In: *TMS Fall 2002 Extraction & Processing Division Meeting on Recycling & Waste Treatment in Mineral and Metal Processing. Technical & Economical Aspects.* Luleå, Sweden (2002)

[F24] F. Fraas: Magnetization delay in the separation of minerals. *US Bureau of Mines Report* **RI 6411** (1964)

[F25] J. Farkas: A solids concentration pilot-plant process using ferromagnetic fluid as the variable density medium. *Sep. Sci. Technol.* **18** (1983), 701

[F26] J. Farkas: A pilot-plant process for manufacturing kerosene-base ferromagnetic fluid. *Sep. Sci. Technol.* **18** (1983), 787

[F27] J. Farkas and B. Hargitay: Recovery and reconstitution of ferromagnetic fluids. *Sep. Sci. Technol.* **18** (1983), 917

[F28] A.G. Fricker: Magnetic hydrocyclone separator. *Trans. Inst. Min. Metall., Sect. C: Min. Process. Extr. Metall.* **94** (1985), C158

[F29] R.J. Freeman, N.A. Rowson, T.J. Veasey and I.R. Harris: The progress of the magnetic hydrocyclone. *Magn. Electr. Sep.* **4** (1993), 139

[F30] R.J. Freeman, N.A. Rowson, T.J. Veasey and I.R. Harris: The development of a magnetic hydrocyclone for processing finely-ground magnetite. *IEEE Trans. Magn.* **30** (1994), 4665

[F31] G. Ferrara and G.D. Schena: Design criteria and control strategies for dynamic dense media separation processes treating fine ores. In: *Proc. XVI Int. Min. Proc. Congress*, Stockholm, Sweden (1988), 885

[F32] D.E. Freeman, H. Aral and L.K. Smith: The potential for the recovery of chromite sands from the Murray Basin, Australia. In: *Proc. "Heavy Minerals 2003" Conference,* Cape Town, South Africa (2003)

[F33] S. Fawell: Magnetic separation in sand processing. *Glass,* December 1997, p. 467

[F34] T. Fujita, S. Mori, M. Mamiya and J. Shimoiizaka: An improved sink-float testing apparatus for coal preparation using a water-based magnetic fluid. In: *Proc. 11th Int. Coal Prep. Congress,* Tokyo, Japan (1990), 109

[F35] S.G. Frantz: Method of magnetic separation. *US Patent 2,074,085* (1937)

[G1] R. Gerber and R.R. Birss: *High Gradient Magnetic Separation.* Research Studies Press, John Wiley & Sons, Chichester 1983

[G2] R.B. Goldfarb: Panel discussion on units in magnetism. *Magn. Electr. Sep.* **6** (1995), 105

[G3] J.A. Good and K. White: Superconducting magnet for 60 tonne/hour mineral separator with closed cycle 4 K refrigeration. *J. Physique, Coll. C1, Suppl. 1,* **45** (1984), C1

[G4] V.N. Gubarevich and S.V. Vidsota: Theoretical principles, present status and prospects for development of material separation in magnetic fluids. *Magn. Electr. Sep.* **5** (1994), 169

[G5] R. Gerber: Magnetic separation. In: *Applied Magnetism (*R. Gerber, C.D. Wright and G. Asti, Editors), Kluwer Academic Publishers, NATO ASI Series, Series E: Applied Sciences, Vol. 253, Dordrecht, The Netherlands 1994

[G6] R. Gerber: Theory of particle capture in axial filters for high gradient magnetic separation. *J. Phys. D: Appl. Phys.* **11** (1978), 2119

[G7] H. Greiner and H. Hoffmann: The magnetic field and force in a perpendicularly magnetized matrix consisting of parallel ordered ferromagnetic wires. *J. Magn. Magn. Mater.* **38** (1983), 187

[G8] R. Gerber and M.H. Watmough: Superconducting and conventional OGMS: Comparative study. *J. Magn. Magn. Mater.* **83** (1990), 490

[G9] R. Gerber and M.H. Watmough: Monte Carlo simulation of collision limited OGMS process. *IEEE Trans. Magn.* **24** (1988), 1677

[G10] R. Gerber and M.H. Watmough: A design of a superconducting split-coil open-gradient magnetic separator. *IEEE Trans. Magn.* **21** (1985), 2053

[G11] J. Gerhold: Superconducting magnet design for open gradient magnetic separation. *Cryogenics* **26** (1986), 523

[G12] A. Gupta, K.K. Mukhopadhaya and A.H. Cory: A new method of magnetic separation of ferro- and para-magnetic minerals from feebly magnetic and non-magnetic minerals. *Proc. Australas. Inst. Min. Metall.* **274** (1980), 23

[G13] O. Goodluck, J. Finch and M. Leroux: Magnetic separation in alternating fields. *IEEE Trans. Magn.* **23** (1987), 1909

[G14] V.V. Gogosov, R.D. Smolkin and O.P. Saiko: Computation and development of a separator with magnetic fluids. *Magn. Hydrodynam. (Riga)* No. 2 (1988), 99 (in Russian)

[G15] V.V. Gogosov, R.D. Smolkin, Yu.M. Garin, O.P. Saiko and V.S. Krokhmal: Design characteristics of separators with magnetic fluids using magnetic systems without a plate. In: *Proc. XI Riga Conf. on Magnetohydrodynamics,* Salaspils, Vol. 3 (1984), 191 (Inst. of Physics, Academy of Sciences of the Latvian SSR)

[G16] V.V. Gogosov, R.D. Smolkin, Yu. M. Garin, V.S. Krokhmal and G.A. Shaposhnikova: Some theoretical and practical problems of separation in magnetic fluids. *J. Magn. Magn. Mater.* **39** (1983), 169

[G17] V.N. Gubarevich: *Separation of Materials in Magnetic Fluids.* Nedra, Moscow 1987 (in Russian)

[G18] G.W. Govier, C.A. Shook and E.O. Lilge: The rheological properties of water suspensions of finely subdivided magnetite, galena and ferrosilicon. *CIM Trans.* **LX** (1957), 147

[G19] V.N. Gubarevich: A method of regeneration of magnetic fluid. *Author's Certificate (USSR)* SU 1708421 (1987)

[G20] V.N. Gubarevich, N.D. Kravchenko, A.V. Dubinin and A.I. Alipov: M : Magnetic separator. *Authors' Certificate (USSR)* SU 1710133 (1990)

[G21] V.N. Gubarevich, M.P. Nerush and A.I. Alipov: Magnetic Separator. *Authors' Certificate (USSR)* SU 1715429 (1990)

[G22] Gang Shen and J.A. Finch: Theoretical analysis of multipole magnetic hydrocyclones. Can. Metall. *Quart.* **29** (1990), 171

[G23] A.N. Ghebremeskel and A. Bose: A continuous hybrid field-gradient device for magnetic colloid-based separations. *J. Magn. Magn. Mater.* **261** (2003), 66

[G24] M.D. Graham: Efficiency comparison of two preparative mechanisms for magnetic separation of erythrocytes from whole blood. *J. Appl. Phys.* **52** (1981), 2578

[G25] M.D. Graham: Comparison of volume and surface mechanisms for magnetic filtration of blood cells. *J. Physique, Coll. C!, Suppl. No. 1* **45** (1984), C1-779

[G26] V.N. Gubarevich: Magnetic separators for complex composite scrap processing of ferrous and non-ferrous metals and other mixtures. In: *Int. Min. Proc. Congress,* Dresden, Germany (1991), **VII**, 121

[G27] V.N. Gubarevich: Method and equipment for final concentration of the gold-bearing alluvial sands. *Magn. Electr. Sep.* **6** (1995), 229

[G28] L.A. Girifalco: *Dynamics of Technological Change.* Van Nostrand Reinhold (1991)

[H1] K. Honda: The thermomagnetic properties of elements. *Ann. Phys. (Lpz.)* **32** (1910), 1048

[H2] H.E. Hawkes: *US Geol. Survey Bull.* **A973** (1951)

[H3] K.N. Henrichsen: Classification of magnetic measurement methods. In: *Proc. CERN Accelerator School: Magnetic Measurements and Alignment.* Geneva 1992, p. 70

[H4] J.F. Herbst and J.J. Croat: Neodymium-iron-boron permanent magnets. *J. Magn. Magn. Mater.* **100** (1991), 57

[H5] H. Harada and M. Tokunaga: Samarium-cobalt-based rare-earth magnets. In: *Magnetic Materials in Japan.* Elsevier, Oxford 1991, p. 276

[H6] S. Hirosawa: Nd-Fe-B permanent magnets. In: *Magnetic Materials in Japan.* Elsevier, Oxford 1991

[H7] D.A. Huse et al.: Are superconductors really superconducting? *Nature* **358** (1992), 553

[H8] G.P. Hatch and R.E. Stelter: Magnetic design considerations for devices and particles used for biological high-gradient magnetic separation systems. *J. Magn. Magn. Mat.* **225** (2001), 262

[H9] V. Hencl et al.: High-gradient drum magnetic separator for wet beneficiation of weakly magnetic minerals. *Czechoslovak Patent 194464* (1982)

[H10] V. Hencl: Current developments in the application of high-gradient magnetic separation. *Rudy* **30** (1982), 41

[H11] U. Habich: Non-ferrous metals separator. *Sales Representative Convention (Steinert Elektromagnetbau GmbH)*, 2000

[H12] D.M. Hopstock: Fundamental aspects of design and performance of low-intensity dry magnetic separators. *Trans. AIME/SME* **258** (1975), 222

[H13] A.W. Hamilton and E.S. Twichell: *EMJ,* September 1971, p.142

[H14] K. Hayashi and S. Uchiyama: On particle trajectory and capture efficiency around many wires. *IEEE Trans. Magn.* **16**(1980), 827

[H15] J.P. Herzig, D.M. Leclerc and P. De Goff: Flow of suspensions through porous media - application to deep bed filtration. *Ind. Eng. Chem.* **62** (1970), 8

[H16] V. Hencl and J. Svoboda: The possibility of magnetic flocculation of weakly magnetic minerals. In: *Proc. XIII Int. Min. Proc. Congress,* Warsaw, Poland (1979), 209

[H17] R. Hogg, T.W. Healy and D.W. Fuerstenau: Mutual coagulation of colloidal dispersions. *Trans. Farad. Soc.* **62** (1966), 1638

[H18] D. Hadfield (Editor): *Permanent Magnets and Magnetism.* Ilife Books Ltd., London 1962

[H19] C.D. Hendricks: Charging macroscopic particles. In: *Electrostatics and its Applications* (A.D. Moore, Editor), John Wiley & Sons, New York (1973), p. 57

[H20] M.S. Hunt, J.O. Hansen and A.T. Davy: The influence of ferrosilicon properties on dense medium separation plant consumption. *Bull. Proc. Australas. Inst. Min. Metall.* **291** (1986), 73

[H21] W. Howell and P. Mellin: Magnetite recovery in fine-coal heavy-media cyclone circuits. *Eriez Magnetics unpublished report* (1981)

[H22] V. Hencl and G. Šebor: Optimization of the parameters of HGMS of complex siderite ore. In: *Proc. 15th Int. Min. Proc. Congress,* Cannes (1985), p.354

[H23] D. M. Hopstock: A reexamination of the performance of demagnetizing coils on finely ground natural magnetite. *Int. J. Min. Proc.* **59** (2000), 45

[H24] H. Heep, T. Heyer and H.-D. Wasmuth: The commercial application of the new two-stage Jones WHIMS process for martitic iron ores at the Fábrica Mine of Ferteco Mineração SA, Brazil. In: *Proc. XVIII Int. Min. Proc. Congress,* Sydney, Australia (1993), 411

[H25] C. Hahn, H. Cordes and H.-D. Waldhecker: Separation by wet high intensity magnetic separation WHIMS JONES and new pneumatic flotation system PNEUFLOT. In: *Innov. Min. Coal Process.*, Balkema, Rotterdam (1998), 695

[H26] S. Holtzhausen and P.C. Engelbrecht: Magnetic separation as a means of increasing the treatment rate of a sulphide flotation plant. In: *African Mining '95 Conference,* Windhoek, Namibia (1995), 251

[H27] E.C. Hise, I. Wechsler and J.M. Doulin: Continuous separation of dry crushed coal at one ton per hour by HGMS. *Oak Ridge National Laboratory Report ORNL-5763* (1981)

[H28] A.S. Holman, E.C. Hise and J.E. Jones: Initial investigation of open-gradient magnetic separation. *Oak Ridge National Laboratory Report ORNL-5764* (1982)

[H29] K. Harding and W. Baxter: Application of HGMS to ferric hydroxide filtration. *IEEE Trans. Magn.* **17** (1981), 2795

[H30] J.R. Harland, L. Nilsson and M. Wallin: Pilot-scale high-gradient magnetic filtration of steel mill waste water. *IEEE Trans. Magn.* **12** (1976), 904

[H31] H.G. Heitmann: Studies of the application of electromagnetic filters in power plants. In: *Industrial Applications of Magnetic Separation* (Ed. Y.A. Liu), IEEE Publ. 78 CH 1447-2MAG (1979), 115

[H32] H.G. Heitmann, V. Schneider, K. Fröhlich and G. Rupp: Investigations of feed water purification in nuclear power plants by electromagnetic filtration. In: *Proc. "Electrical and Magnetic Separation and Filtration Technology" Conference,* Antwerp (1984), 73

[H33] V. Hencl and P. Mucha: The application of high-gradient magnetic separation to water treatment by means of chemically precipitated magnetite. *Magn. Electr. Sep.* **5** (1994), 155

[H34] R. Hergt et al.: Physical limits of hyperthermia using magnetite fine particles. *IEEE Trans. Magn.* **34** (1998), 3745

[H35] G. Hartfeld: Magnetohydrostatic separators state of development. *Aufbereitungstechnik* **4** (1985), 224

[I1] Y. Ishikawa and S. Chikazumi: Design of high power electromagnets. *Jap. J. Appl. Phys.* **1** (1962), 155

[I2] J. Iannicelli: New developments in magnetic separation. *IEEE Trans. Magn.* **12** (1976), 436

[I3] J. Iannicelli, N. Millman and W.J.D. Stone: Process for improving the brightness of clay. *US Patent 3,471,011* (1969)

[I4] K.J. Ives: Capture mechanisms in filtration. In: *The Scientific Basis of Filtration (*K.J. Ives, Editor), Noordhoff, Leiden, 1975, p. 183

[I5] H. Isogami, N. Saho, M. Morita and T. Takagi: Sewage treatment performance of a continuous superconducting magnetic separator. *JSME Int. J., Series B,* **44** (2001), 675

[I6] J. Iannicelli: High extraction magnetic filtration of kaolin clay. *Clays Clay Min.* **24** (1976), 64

[I7] J. Iannicelli et al.: Magnetic separation of kaolin clay using a high temperature superconducting magnet system. *IEEE Trans. Appl. Supercond.* **7** (1997), 1061

[J1] G.H. Jones: Wet magnetic separator for feebly magnetic minerals. I. Description and theory. In: *Proc. 5th Int. Miner. Proc. Congress,* London 1960, p. 717

[J2] Jack Ji-Nong Sun: Magnetic barrier-a new concept in magnetic separation. *IEEE Trans. Magn.* **12** (1976), 483

[J3] T. Janowski and S. Kozak: Efficiency of a superconducting OGMS separator. *IEEE Trans. Magn.* **26** (1990), 1864

[J4] T. Janowski and S. Kozak: The superconducting OGMS separator optimization. *IEEE Trans. Magn.* **29** (1993), 3261

[J5] P.G. Jordan: Association phenomena in a ferromagnetic colloid. *Mol. Phys.* **25** (1973), 961

[J6] P.J. Jonker: The operational efficiency of an Eriez WHIMS at Buffalo Fluorspar Mine. Presented at: *Colloq. Mag. Sep.,* Council for Mineral Technology, Randburg, South Africa (1986)

[J7] J.A. Jirestig and E. Forssberg: HGMS - a new dry high gradient magnetic separator. *Aufbereitungstechnik* **36** (1995), 257

[J8] J. Johnson, T. Kent, J. Koda, C. Peterson, S. Rudge and G. Tapolsky: The MTC technology: A platform technology for the site-specific delivery of pharmaceutical agents. In: *Proc. 4th Int. Conf. "Scientific and Clinical Applications of Magnetic Carriers",* Tallahassee, Florida, USA (2002), 22

[J9] A. Jordan: Presentation of a new magnetic field therapy system for the treatment of human solid tumours with magnetic field hyperthermia. *J. Magn. Magn. Mater.* **225** (2001), 118

[J10] M. Jacob: Training benefits from basic research. *CERN Courier* **48**, October 2003, p. 54

[K1] H.H. Kolm: The future of superconducting technology. *Cryogenics* **15** (1975), 63

[K2] V.I. Karmazin and V.V. Karmazin: *Magnetic Methods of Beneficiation.* Nedra, Moscow 1984 (in Russian)

[K3] V.V. Karmazin and V.I. Karmazin: *Magnetic and Electrical Methods of Beneficiation.* Nedra, Moscow 1988 (in Russian)

[K4] Y. Kaneko: Highest performance of Nd-Fe-B magnet over 55 MGOe. *IEEE Trans. Magn.* **36** (2000), 3275

[K5] H.H. Kolm: Process for magnetic separation. *US Patent 3,676,337* (1972)

[K6] O. Kolář, M. Horáček and J. Cibulka: Magnetic separators with permanent magnets. In: *Proc. 12th Natl. Miner. Proc. Conf.,* Prague 1977, p. 49 (in Czech)

[K7] H.H. Kolm: The large-scale manipulation of small particles. *IEEE Trans. Magn.* **11** (1975), 1567

[K8] Z. Kaiser et al.: Magnetic separator with a superconducting magnet and reciprocating matrix. *Sci. Tech. Froid.* No. 2 (1986), 45

[K9] F.S. Knoll, R. Allen and D. Richards: Superconducting magnetic separators in mineral processing. *Preprint 97-150, SME* (1997)

[K10] J. Kopp and J.A. Good: The physics of high-intensity dry magnetic separation. *IEEE Trans. Magn.* **18** (1982), 833

[K11] J. Kopp: The physics of "falling curtain" dry magnetic separation. *Int. J. Min. Proc.* **10** (1983), 297

[K12] P. Krist, R. Gerber, D. Fletcher and J. Boehm: The diamagnetic mode of OGMS. *IEEE Trans. Magn.* **33** (1997), 4248

[K13] J. Kopp: Performance limits of open-gradient superconducting magnets. *IEEE Trans. Magn.* **23** (1987), 2761

[K14] V.V. Karmazin, V.I. Karmazin and V.A. Binkevich: *Magnetic Regeneration and Separation in the Beneficiation of Ores and Coal.* Nedra, Moscow 1968 (in Russian)

[K15] M. Koizumi: Improvements in electromagnetic separators for the separation of minerals. *US Patent 254 030* (1925)

[K16] S.E. Khalafalla: Beneficiation with magnetic fluids - magnetic separation of the second kind. *Min. Proc. Extr. Metall. Rev.* **2** (1985), 21

[K17] S.E. Khalafalla and G.W. Reimers: Beneficiation with magnetic fluids. *Bureau of Mines Report of Investigations RI 8532,* US Dept. of Interior 1981

[K18] R. Kaiser and L. Mir: Hyperbolic magnet poles for sink-float separators. *US Patent 3,898,156* (1975)

[K19] L. Krüger, J. Svoboda, B. Dale and V. Murariu: unpublished results (2002)

[K20] J.D. Krige: Heavy medium separation at Iscor Sishen iron-ore mine. In: *Proc. Dense Medium Operators Conf.,* Brisbane, Australia (1987), p. 65

[K21] V.I. Karmazin et al.: Finishing tungsten-tin slimes on the 2VK-3 high-intensity magnetic separator. *Tsvetn. Metal.* **10** (1969), 35

[K22] L. Kuzev and S. Stoev: Vibromagnetic separation of iron ores. *World Min. Equip.,* December 1986, p. 84

[K23] L. Krüger: unpublished results

[K24] R. Kaiser and G. Miskolczy: Magnetic properties of stable dispersions of subdomain magnetic particles. *J. Appl. Phys.* **41** (1970), 1064

[K25] N.D. Kravchenko, A.V. Dubinin and G.A. Denisov: Investigation of the temperature regime of removal and regeneration of the ferromagnetic fluid from the products of FHS. *Obogasch. Rud (Leningrad),* n. 4 (1988), 25

[K26] T. Kojovic: Application of *Magstream* in mineral sands separation. *Magn. Electr. Sep.* **5** (1994), 231

[K27] U.A. Khvatov et al.: The utilization of the tailings from wet magnetic separation at the Krivoy Rog GOK. *Magn. Electr. Sep.* **6** (1995), 179

[K28] Y.S. Kim, T. Fujita, S. Hashimoto and J. Shimoiizaka: The removal of Cu sulphide minerals from Pb flotation concentrate of black ore by high gradient magnetic separation. In: *Proc. XV Int. Min. Proc. Congress,* Cannes, France (1985),381

[K29] V.I. Karmazin, Yu. S. Mostika and E.N. Savluk: Investigation of the removal of ash, sulphur and other contaminants from coal by wet magnetic separation. *Magn. Electr. Sep.* **6** (1995), 1

[K30] J.K. Kindig: The Magnex process: Review and current status. In: *Industrial Applications of Magnetic Separation.* IEEE Publ. 78CH1447-2MAG (1979), 99

[K31] H. Koca, S. Koca and O.M. Kockar: Upgrading of Kutahya region lignites by mild pyrolysis and high intensity magnetic separation. *Min. Eng.* **13** (2000), 657

[K32] M.A. Kokkala: Water treatment by means of magnetic separation. *Ph.D. Thesis,* Helsinki University of Technology (1982)

[K33] L.O. Kolarik: Colour and turbidity removal with reusable magnetite particles -IV. Alkali activated magnetite- a new solid, reusable coagulant-adsorbent. *Water Res.* **17** (1983), 141

[K34] A.A. Kuznetsov et al.: Ferro-carbon particles: Preparation and clinical applications. In: *Scientific and Clinical Applications of Magnetic Carriers (*U. Häfeli, W. Schütt, J. Teller and M. Zborowski, Editors), Plenum Press, New York (1997), 379

[K35] J. Kopp: Superconducting magnetic separators. *Magn. Electr. Sep.* **3** (1991), 17

[K36] U. Kohaupt: Selection of magnetic separators for removing tramp iron. *Aufbereitungstechnik* **44** (2003), 32

[L1] E. Laurila: On the magnetic permeability of mixtures containing ferromagnetic particles and porous ferromagnetic materials. *Ann. Acad. Sci. Fenn., Series A, VI. Physica no. 7,* Helsinki 1961

[L2] E. Laurila: On the fields and permanent magnets in magnetic pulleys and separators. *Acta Polytech. Scand.* AP 312 (1962)

[L3] F.J. Lowes: *Proc. Roy. Soc. London* **A337** (1974), 555

[L4] K. Lileg and B. Schnizer: Influence of particle shape on forces in magnetic separators. *IEEE Trans. Magn.* **25** (1989), 4292

[L5] L.D. Landau and E.M. Lifshitz: *Electrodynamics of Continuous Media.* Pergamon Press, Oxford 1984, 2nd Edition, p. 42

[L6] F.E. Luborsky and B.J. Drummond: High gradient magnetic separation: theory versus experiment. *IEEE Trans. Magn.* **11** (1975), 1696

[L7] W.F. Lawson, W.H. Simons and R.P. Treat: The dynamics of a particle attracted by a magnetized wire. *J. Appl. Phys.* **48** (1977), 3213

[L8] Zhengnan Li and J.H.P. Watson: Vortex magnetic separation (VMS). *IEEE Trans. Magn.* **30** (1994), 4662

[L9] H. Lantto: The effect of magnetic flocculation in the beneficiation of magnetic materials. *Acta Polytech. Scand.* **Ch133** (1977)

[L10] E. Laurila: On the behaviour of magnetic powder in a rotating field. *Ann. Acad. Sci. Fenn.* **AVI34** (1959)

[L11] S. Lu, S. Song and Z. Dai: The hydrophobic and magnetic combined aggregation of paramagnetic minerals. In: *Proc. XVI Int. Min. Proc. Congress,* Stockholm, Sweden (1988), p. 999

[L12] J.E. Lawver and W.P. Dyrenforth: Electrostatic separation. In: *Electrostatics and its Applications* (A.D. Moore, Editor), John Wiley & Sons, New York (1973), p.221

[L13] T. Lindberg: Experience with the Jones separator at Buchanan. *Min. Mag.*, Sept. 1979, p. 199

[L14] F.E. Luborsky: HGMS for removal of sulphur from coal. *Report EPA-600/7-78-208* (1978)

[L15] Li Zhengnan and T.J. Veasey: An improved process for china clay beneficiation using high-gradient magnetic separation. *Min. Eng.* **1** (1988), 311

[L16] E.A. Laurila: Magnetic flocculation and demagnetization. In: *SME Mineral Processing Handbook* (N.L. Weiss, Editor), SME, New York (USA), 1985, p. 6-43

[L17] I.J. Lin and N.P. Finkelstein: The application of magnetohydrostatic separation to metallurgical problems. *J. S.A. Inst. Min. Metal.*, October 1975), 111

[L18] I.J. Lin and T. Fujita: Classification in a hydrocyclone using magnetic fluids. *Magn. Gidrodyn.* no. 1, January-March 1991, p. 44

[L19] I.J. Lin, T. Fujita and M. Mamiya: Separation of particles in hydrocyclone using magnetic fluids. In: *Proc. XVI Int. Min. Proc. Congress*, Stockholm, Sweden, (1988), 915

[L20] I.J. Lin, M. Krush-Bram and G. Rosenhouse: The beneficiation of minerals by magnetic jigging, Part 1. Theoretical aspects. *Int. J. Min. Proc.* **50** (1997), 143

[L21] I.J. Lin, M. Krush-Bram and G. Rosenhouse: The beneficiation of minerals by magnetic jigging, Part 2. Identification of the parameters and verification of the mathematical model for the theoretical analysis of the mineral particles motion in the magnetic jig. *Int. J. Min. Proc.* **54** (1998), 29

[L22] I.J. Lin, M. Krush-Bram and G. Rosenhouse: The beneficiation of minerals by magnetic jigging, Part 3. The bed effects and the multifrequency magnetic jig. *Int. J. Min. Proc.* **55** (1998), 61

[L23] L.A. Lomovtsev, K.V. Nikolaenko, Yu. A. Davydov, A.P. Kazantsev and A.L. Tarasenko: Recovery of manganese from slimes by high-intensity magnetic separation. *Magn. Electr. Sep.* **6** (1995), 99

[L24] S. V. Lesnikov and A.P. Kazantsev: The application of progressive methods of beneficiation at the Ordzhonikidze GOK. *Metall. Gordnorud. Prom.* **4** (1993), 47

[L25] C.H. Lofthouse and C.W. Scobie: Beneficiation of kaolin using commercial high-intensity magnetic separator. In: *Proc. XIII Int. Min. Proc. Congress*, Round Table Seminar "Beneficiation of Clay Raw Materials", Warsaw, Poland (1979), 163

[L26] A.S. Lübbe, C. Bergemann, J. Brock and D.G. McClure: Physiological aspects in magnetic drug-targeting. *J. Magn. Magn. Mater.* **194** (1999), 149

[M1] H. Mauritsch et al.: Comparison of the hysteresis characteristics of synthetic samples with different magnetite and haematite contents. *Phys. Earth Planet. Inter.* ***46*** (1987), 93

[M2] A. Mookherji and S.C. Mathur: Magnetic studies of siderite. *J. Phys. Soc. Japan* ***20*** (1965), 1336

[M3] P. Mitchell: Corrosion protection of NdFeB magnets. *IEEE Trans. Magn.* **26** (1990), 1933

[M4] D.B. Montgomery: The generation of high magnetic fields. *Rep. Progr. Phys.* **26** (1963), 69

[M5] P.G. Marston: Magnetic design and its effects on the economics of HGMS process. In: *Proc. HGMS Symposium,* MIT 1973, p.25

[M6] E.G. Maximov: High-temperature superconductivity: the current state. *Physics - Uspekhi* **43** (2000), 965

[M7] D.G. Morgan and W.J. Bronkala: The selection and application of magnetic separation equipment. Part I. *Magn. Electr. Sep.* **3** (1991), 5

[M8] D.G. Morgan: Magnetic separators for protection of process equipment. *Powder Bulk Eng.* **2,** No. 7, (1988)

[M9] P.G. Marston and J.J. Nolan: Moving matrix magnetic separator. *US Patent 3,920,543* (1975)

[M10] V. Murariu and J. Svoboda: The applicability of the Davis tube tests to ore separation by drum magnetic separators. *Phys. Sep. Sci. Eng.* **12** (2003),

[M11] J. McAndrew: Calibration of a Frantz isodynamic separator and its application to mineral beneficiation. *Proc. Aust. IMM* **181** (1957), 59

[M12] L. Mir, C. Simard and S.D. Grana: In: *Proc. 3rd Urban Technol. Conf. Tech. Display, Boston (USA),* AIAA Paper no. 73-959 (1973)

[M13] M. Marinescu et al.: New permanent magnetic separator with NdFeB meets theoretical predictions. *IEEE Trans. Magn.* **25** (1989), 2732

[M14] V. Murariu: unpublished results

[M15] C. Moyer, M. Natenapit and S. Arajs: Particle capture by an assemblage of spheres in high-gradient magnetic separation. *J. Appl. Phys.* **55** (1984), 2589

[M16] C. Moyer, M. Natenapit and S. Arajs: Magnetic filtration of particles in laminar flow through a bed of spheres. *J. Magn. Magn. Mater.* **44** (1984), 99

[M17] R.I. Mackie, R.M.W. Horner and R.J. Jarvis: Dynamic modelling of deep-bed filtration. *AIChE J.* **33** (1987), 1761

[M18] V. Murariu and J. Svoboda: The effect of magnetic susceptibility on the motion of a particle in a ferrohydrostatic separator. *Magn. Electr. Sep.* **11** (2002), 51

[M19] M. McCaig: *Permanent Magnets in Theory and Practice.* Pentech Press, London 1977

[M20] L.R. Moskowitz: *Permanent Magnet Design and Application Handbook.* R.E. Krieger Publ. Co., Malabar, Florida 1986

[M21] D.B. Montgomery: Iron magnet design. In: *High Magnetic Fields (*H. Kolm et al., Editors), M.I.T., Cambridge, Mass., USA (1962), p. 180

[M22] D.B. Montgomery: *Solenoid Magnet Design.* Wiley-Interscience, New York 1969

[M23] P.G. Marston: Magnetic design and its effects on the economics of HGMS process. In: *Proc. HGMS Symp.*, MIT, Mass., USA (1973), p. 25

[M24] V. Murariu and J. Svoboda: unpublished results

[M25] V. Murariu, M. Fofana and J. Svoboda: unpublished results

[M26] D.G. Morgan and W.J. Bronkala: The selection and application of magnetic separation equipment. Part II. *Electr. Magn. Sep.* **4** (1993), 151

[M27] E. Madai: On the problem of matrix adaptation in HGMS operation. Aufbereitungstechnik **24** (1983), 99

[M28] V.M. Malyi: A study of matrices for high-intensity magnetic separators. *Magn. Electr. Sep.* **6** (1995), 31

[M29] D.J. Miller, D.G. James and J.H. Turner: Recovery of - 100 μm hematite by wet high intensity magnetic separation. In: *Proc. 18th Int. Min. Proc. Congress,* Sydney, Australia (1993), 397

[M30] I. Makarow, Th. Agrafiotis and J. A. Finch: Magnetizing roast for chalcopyrite: effect of particle size and dilution. *Mag. Sep. News* **1** (1984), 115

[M31] I.K. Myburgh: Experimental investigation into the application of a magnetic dense medium cyclone in a production environment. *M. Eng. Dissertation,* University of Potchefstroom, South Africa (2001)

[M32] E. Martinez: The gravity-magnetic separator: A new development for recovery of weakly magnetic minerals. In: *Advances in Mineral Processing: A half-century of progress in application of theory to practice.* March 1986, New Orleans, La., USA, Chapter 32, p. 545

[M33] V.M. Malyi, T.B. Gansenko, G.I. Notovich and I.M. Savchenko: Development of high-intensity magnetic separation for weakly magnetic ores. In: *Proc. XX Int. Min. Proc. Congress,* Aachen, Germany (1997), 611

[M34] V. Malyi, L. Lomovtsev, A. Kostin, O. Dimchenko and V. Myasoyedov: Full-scale plant for beneficiation of fine-disseminated oxidized iron ores. In: *Proc. XVII Int. Min. Proc. Congress*, Dresden, Germany (1991), Vol. III, p. 161

[M35] S.E. Male: Magnetic susceptibility and separation of inorganic material from UK coals. *J. Phys. D: Appl. Phys.* **17** (1984), 155

[M36] S.E. Male: The magnetic beneficiation of coal. *Magn. Sep. News* **2** (1985), 1

[M37] M.J. Mortimer: Magnetic susceptibility measurements on fission product insolubles. *IEEE Trans. Magn.* **17** (1981), 2819

[M38] M.J. McNallan, K.C. Russel and J.A. Oberteuffer: High-gradient magnetic filtration of steel plant waste water. *Iron & Steel Eng.* **53** (1976), 40

[M39] O.I. Martynova and A.S. Kopylov: The use of electromagnetic filters for removing ferromagnetic impurities from water. *Teploenergetika* **19** (1972), 67

[M40] G. Moffat, R.A. Williams, C. Webb and R. Stirling: Selective separations in environmental and industrial processes using magnetic carrier technology. *Min. Eng.* **7** (1994), 1039

[M41] D. Müller-Schulte: Novel magnetic micro- and nanoparticles for biomedical separation and as means for a new approach to AIDS therapy. *Magn. Electr. Sep.* **10** (2000), 141

[M42] D. Melville, F. Paul and S. Roath: High gradient magnetic separation of red cells from whole blood. *IEEE Trans. Magn.* **11** (1975), 1701

[M43] D. Melville, F. Paul and S. Roath: Fractionation of blood components using high gradient magnetic separation. *IEEE Trans. Magn.* **18** (1982), 1680

[M44] A.S. Macfarlane: The implementation of new technology in southern African mines: Pain or panacea. *J. SAIMM* **101** (2001), 115

[M45] K. Mitsuhashi et al.: Purification of endocrine disrupter-polluted water using HTS magnetic separation system. *Phys. Sep. Sci. Eng.***12** (2004),

[N1] T. Nagata: *Rock Magnetism.* Maruzen Company Ltd., Tokyo 1961

[N2] J.E. Nesset and J.A. Finch: A static model of HGMS based on force within the fluid boundary layer. In: *Proc. Int. Conf. "Industrial Applications of Magnetic Separation"* (Y.A. Liu, Ed.), IEEE Publication No. 78 CH 1447-2 MAG, New York 1979

[N3] A.A. Nijkerk and W.L. Dalmijn: *Handbook of Recycling Techniques.* Nijkerk Consultancy, The Hague, The Netherlands 1995

[N4] S. Nogita et al.: Recovery of non-ferrous metals using magnetic fluid techniques. *Hitachi Rev.* **26** (1977), 139

[N5] J.E. Nesset and J.A. Finch: A loading equation for HGMS and application in identifying the fine size limit of recovery. In: *Fine Particles Processing* (P. Somasundaran, Ed.) Vol. 2, AIME, New York 1980, p. 1217

[N6] B.W. Ninham: On progress in forces since the DLVO theory. *Adv. Coll. Interface Sci.* **83** (1999), 1

[N7] T.J. Napier-Munn, T. Kojovic, I.A. Scott, F. Shi, J.H. Masinja and P.J. Baguley: Some causes of medium loss in dense medium plants. *Min. Eng.* **8** (1995), 659

[N8] T.J. Napier-Munn and I.A. Scott: The effect of demagnetisation and ore contamination on the viscosity of the medium in a dense medium cyclone plant. *Min. Eng.* **3** (1990), 607

[N9] A. Newns and R.D. Pascoe: Influence of path length nand slurry velocity on the removal of iron from kaolin using a high gradient magnetic separator. *Min. Eng.* **15** (2002), 465

[N10] M.I. Nasr and M.A. Youssef: Optimization of magnetizing reduction and magnetic separation of iron ores by experimental design. *ISIJ Int.* **36** (1996), 631

[N11] J. Nell and P. den Hoed: Separation of chromium oxides from ilmenite by roasting and increasing the magnetic susceptibility of Fe_2O_3-$FeTiO_3$ (ilmenite) solid solutions. In:*Proc. "Heavy Minerals 1997" Conf.*, Johannesburg, South Africa (1997), 75

[N12] M.J. Nicol: New advances in metallurgical processes: Techniques for enhancing cost-effectiveness through improved efficiency in the processing of minerals. In: *Proc. Conf. "Developing Advanced Techniques to Address the New role of the Mine Manager"*, Midrand (South Africa), 1993

[N13] A. A. Nijkerk and W. L. Dalmijn: *Handbook of Recycling Techniques.* Nijkerk Consultancy, The Hague (2001)

[N14] M.J. Nicol: Hydrometallurgy into the next millennium. In: *Proc. MINPREX 2000 Conf.*, Melbourne, Australia (2000), 11

[O1] M. Owen: Magnetochemical investigations. The thermomagnetic properties of elements II. *Ann. Phys. (Lpz.)* **37** (1912), 657

[O2] W. O'Reilly: *Rock and Mineral Magnetism.* Blackie & Son, Ltd., Glasgow and London 1984

[O3] J.A. Oberteuffer: HGMS: basic principles, devices and applications. In: *Proc. Conf. "Industrial Applications of Magnetic Separation"* (Y.A. Liu, Editor), IEEE Publ. No. 78 CH1447-2 MAG,, New York 1979, p. 3

[O4] J.A. Oberteuffer: Engineering development of HGMS. *IEEE Trans. Magn.* **12** (1976), 444

[O5] J.A. Oberteuffer: Magnetic separation: a review of principles, devices and applications. *IEEE Trans. Magn.* **10** (1974), 223

[O6] M. Ozaki, H. Suzuki, K. Takahashi and E. Matijevic: Reversible ordered agglomeration of hematite particles due to weak magnetic interactions. *J. Coll. Interface Sci.* **113** (1986), 76

[O7] M. Ozaki, T. Egami, N. Sugiyama and E. Matijevic: Agglomeration in colloidal hematite dispersions due to weak magnetic interactions. II. The effects of particle size and shape. *J. Coll. Interface Sci.* **126** (1988), 212

[O8] J.A. Oberteuffer and I. Wechsler: Recent advances in HGMS. In: *Int. Symp. Fine Particle Process.* (ed. P. Somasundaran), Las Vegas, Nevada, USA (1980), 1178

[O9] T. Okuda, I. Sugano and T. Tsuji: Removal of heavy metals from waste water by ferrite co-precipitation. *Filtr. Sep.* **12** (1975), 472

[P1] N.G. Petrova: Investigation of the pulverised magnetite. *Izv. AN SSSR, Ser. Geogr. Geophys.* **12** (1948)

[P2] J.M. Pastrana and D.M. Hopstock: Magnetic properties of natural hematite and goethite. *Trans. AIME* **262** (1977), 1

[P3] H.E. Powell and C.K. Miller: Magnetic susceptibility of siderite. *US Bureau of Mines RI 6224* (1963)

[P4] R.J. Parker: *Advances in Permanent Magnetism.* John Wiley & Sons, New York 1990

[P5] S.S. Papell: Low viscosity magnetic fluid obtained by colloidal suspension of magnetic particles. *US Patent 3,215,572* (1965)

[P6] J. Pitel: Analytical approach to the Newton's motion equation solution of a single particle in superconducting OGMS. In: *Proc. 1st Italo-Czechoslovak Symp. Supercond.,* CIL Lecco, Italy (1989)

[P7] J. Pitel: Methodics of the design of the superconducting magnet system for a linear deflecting separator to process dry grain mixtures. In: *Proc. 1st Japanese-Czechoslovak Seminar on Applied Electromagnetic in Materials,* Piestany, Czechoslovakia (1992), B5-3

[P8] M.R. Parker, R.P.A.R. van Kleef, H.W. Myron and P. Wyder: Particle aggregation in colloids in high magnetic fields. *J. Magn. Magn. Mater.* **27** (1982), 250

[P9] M.R. Parker et al.: Kinetics of magnetic flocculation in colloidal dispersions. *J. Coll. Interface Sci.* **101** (1984), 314

[P10] R.J. Pugh, Y. Wang and E. Forssberg: The influence of magnetic and surface forces on the coagulation of hematite and chromite. *Min. Metall. Proc.,* September 1994, p. 133

[P11] R.J. Pugh: Selective coagulation of colloid mineral particles. In: *Colloid Chemistry in Mineral Processing* (J.S. Laskowski and J. Ralston, Editors), Elsevier, Amsterdam 1992, p.243

[P12] R.J. Parker and R.J. Studders: *Permanent Magnets and Their Applications.* John Wiley and Sons, New York 1962

[P13] J. Parsonage: Selective magnetic coating for mineral separation. *Trans. Inst. Min. Metall., Sec. C: Min. Process. Extr. Metall.* **93** (1984), C37

[P14] P. Parsonage: Small-scale separation of minerals by use of paramagnetic liquids. *Trans. Instn. Min. Metall., Sect. B* **86** (1977), B43

[P15] B. Portillo, A. Perez, D. Norrgran and J. Wernham: Iron ore recovery from tailings ponds using high gradient permanent magnetic separators. In: *Proc. XIX Int. Min. Proc. Congress,* SME (1995), Chapter 34, p. 179

[P16] B.K. Parekh, A.E. Bland, D.A. Norrgran and R.A. Merwin: Dry coal cleaning using a rare earth magnetic separator. In: *Proc. 4th Pittsburgh Coal Conference,* Pittsburgh, PA, USA (1987)

[P17] J. Pitel and F. Chovanec: The application of superconducting magnet systems to dry magnetic separation of coal. *Magn. Electr. Sep.* **4** (1992), 19

[P18] H.E. Plumblee et al.: Development of an antinide magnetic separation system for enhancement of the TRUclean system. In: *Proc. WM'93 Symposium on Waste Management "Working Towards a Cleaner Environment" Conference,* Tucson, Arizona, USA (1993)

[P19] V.V. Pereverzev et al.: Magnetic extraction of radioactive contamination from Chernobyl NPP regions. *IEEE Trans. Magn.* **28** (1992), 678

[P20] P. Parsonage: Principles of mineral separation by selective magnetic coating. *Int. J. Min. Proc.* **24** (1988), 269

[P21] P. Parsonage: Coating and carrier methods for enhancing magnetic and flotation separations. In: *Colloid Chemistry in Mineral Processing* (Ed. J.S. Laskowski and J. Ralston), Elsevier, Amsterdam (1992), 331

[P22] S.K. Pulfer and J.M. Gallo: Targeting magnetic microspheres to brain tumors. In: *Scientific and Clinical Applications of Magnetic Carriers* (U. Häfeli, W. Schutt, J. Teller and M. Zborowski, Editors), Plenum Press, New York (1997), 445

[P23] F. Paul, D. Melville, S. Roath and D.C. Warhurst: A bench top magnetic separator for malarial parasite concentration. *IEEE Trans. Magn.* **17** (1981), 2822

[P24] W.A.P.J. Premaratne and N.A. Rowson: Development of a magnetic hydrocyclone separation for the recovery of titanium from beach sands. *Phys. Sep. Sci. Eng.* **12**/4 (2004)

[P25] C. Prenger: Personal communication (2003)

[R1] R.E. Rosensweig: Material separation using ferromagnetic liquid techniques. *US Patent 3,483,969* (1969)

[R2] M.G. Rylatt and G.M. Popplewell: Diamond processing at Ekati in Canada. *Min. Eng.*, February 1999, p. 19

[R3] P.W. Riley and D. Hocking: A reciprocating canister superconducting magnetic separator. *IEEE Trans. Magn.* **17** (1981), 3299

[R4] D. Richards and R. Allen: Superconducting magnets: the "users" choice. In: *Proc.. Third Processing for Profit Conf.,* Antwerp, Belgium 1999

[R5] R.A. Rikers, H.J. Glass and W.L. Dalmijn: The Frantz barrier separator: merits and characterization of polluted soils. In: *Proc. XX IMPC,* Aachen Germany (1997), 649

[R6] P.C. Rem, S. Zhang, E. Forssberg and T. P.R. de Jong: The investigation of separability of particles smaller than 5 mm by eddy-current technology. Part II: Novel design concepts. *Magn. Electr. Sep.* **10** (2000), 85

[R7] R.E. Rosensweig: Buoyancy and stable levitation of a magnetic body immersed in a magnetizable fluid. *Nature* **120** (1966), 6

[R8] R.E. Rosensweig: Fluidmagnetic buoyancy. *AIAA J.* **4** (1966), 1751

[R9] G.W. Reimers, S.A. Rholl and H.B. Dahlby: Density separations of nonferrous scrap metals with magnetic fluids. In: *Proc. 5th Miner. Waste Util. Symp.*, Chicago (USA), 1976

[R10] G.W. Reimers and S.E. Khalafalla: Production of magnetic fluids by peptization techniques. *US Patent 3,843,540* (1972)

[R11] V.E. Revnetsev: *Soviet Inform. Bull.* **P4** (1982)

[R12] G. Reger, R. Gerber, F.J. Friedlaender and H. Hoffmann: The efficiency of particle capture by an infinite array of ferromagnetic wires at low Reynolds numbers. *J. Magn. Magn. Mater.* **49** (1985) 291

[R13] G. Reger, H. Hoffmann and F.J. Friedlaender: Ordered HGMS filters in the transverse configuration. *Appl. Phys.* **A44** (1987), 161

[R14] R. Rajagopalan and Chi Tien: The theory of deep bed filtration. In: *Progress in Filtration and Separation* (R.J. Wakeman, Editor), **Vol. 1,** Elsevier Sci. Publ. Co., Amsterdam !979, p. 179

[R15] Ya. Rabinovich and N.V. Tchukaryev: Effect of electromagnetic retardation on the force of molecular attraction. *Koll. Zh.* **41** (1979), 468

[R16] R.E. Rosensweig: *Ferrohydrodynamics.* Cambridge University Press, Cambridge 1985

[R17] P.C. Rem, P.A. Leest and A.J. van den Akker: A model for eddy current separation. *Int. J. Min. Proc.* **49** (1997), 193

[R18] P.C. Rem, E.M. Beunder and A.J. van der Akker: Simulation of eddy-current separators. *IEEE Trans. Magn.* **34** (1998), 2280

[R19] E.H. Roux, J.G. Goodey, E.F. Wepener and K.R. Hodierne: Industrial-scale dry beneficiation of phosphate-bearing pyroxenite ore. In: *Proc. Int. Conf. MINTEK 50,* Johannesburg 1984, Paper A6/2

[R20] J.G. Rayner and T.J. Napier-Munn: A mathematical model of recovery of dense medium magnetics in the wet drum magnetic separator. *Int. J. Min. Proc.* **69** (2003), 157

[R21] P.W. Riley and J.H.P. Watson: The use of paramagnetic matrices for magnetic separations. *Filtr. Separ.* July/August 1977, p. 1

[R22] R.J. Rots: *Chern. Metallurg.* No. 4/936 (1983), 29

[R23] J.G. Rayner and T.J. Napier-Munn: The mechanism of magnetic capture in the wet drum magnetic separator. *Min. Eng.* **13** (2000), 277

[R24] R.E. Rosensweig: Magnetic fluids. *Scientific American* **247**/4 (1982), 124

[R25] D. Richards, J. Boehm and R. Allen: Commercial acceptance of superconducting magnetic separation. In: *Proc. XX Int. Min. Proc. Congress,* Aachen, Germany (1997), 639

[R26] T. Rheinländer, R. Kötitz, W. Weitschies and W. Semmler: Magnetic fluids: Biomedical applications and magnetic fractionation. *Magn. Electr. Sep.* **10** (2000), 179

[R27] A.J. Richards et al.: Improved high gradient magnetic separation for positive selection of human blood mononuclear cells using ordered wire filters. *J. Magn. Magn. Mater.* **122** (1993), 364

[R28] A.J. Richards et al.: The mechanisms of high gradient magnetic separation of human blood and bone marrow. *IEEE Trans. Magn.* **32** (1996), 459

[R29] V.E. Ross: A formula for successful innovation. In: *Proc. SAIChE 2000 Conf. "Innovation for Growth"* Secunda, South Africa (2000), 128

[R30] R.E. Rosensweig: Past, present and future of magnetic fluids. In: *Proc.9th Int. Conf. Magn. Fluids ICMF9*, Bremen, Germany (2001), 1

[R31] B. Radebe and J. Svoboda : unpublished results

[S1] J. Svoboda: *Magnetic Methods for the Treatment of Minerals.* Elsevier, Amsterdam 1987

[S2] J. Svoboda: The effect of the magnetic field strength on the efficiency of magnetic separation. *Min. Eng.* **7** (1994), 747

[S3] L.J. Swartzendruber: Properties, units and constants in magnetism. *J. Magn. Magn. Mater.* **100** (1991), 573

[S4] P.C. Scholten: Which SI? *J. Magn. Magn. Mater.* **149** (1995), 57

[S5] F.D. Stacey and S.K. Banerjee: *The Physical Principles of Rock Magnetism.* Elsevier, Amsterdam 1974

[S6] F.C. Schwerer and W. Gundaker: Magnetic properties of natural chromites: mechanical and thermal effects. *Trans. AIME* **258** (1975), 88

[S7] J. Sammer et al.: Automatic measurement of nonuniform magnetic fields with industrial Hall probes. *IEEE Trans. Magn.* **26** (1990), 2064

[S8] J. Svoboda: unpublished results

[S9] K.J. Strnat and R.M.W. Strnat: Rare-earth-cobalt permanent magnets. *J. Magn. Magn. Mater.* **100** (1991), 38

[S10] M. Sagawa, S. Fujimura, N. Togawa, H. Yamamoto and Y. Matsuura: New material for permanent magnets based on Nd and Fe. *J. Appl. Phys.* **55** (1984), 2083

[S11] M. Sagawa, S. Fujimura, H. Yamamoto, Y. Matsuura and S. Hirosawa: Magnetic properties of rare-earth-iron-boron permanent magnetic materials. *J. Appl. Phys.* **57** (1985), 4094

[S12] J. Svoboda: Optimization of the design of a magnetic roll. *MTC Report S102* (1990)

[S13] R.T. Sundberg: Wet low intensity magnetic separation of iron ore. In: *Innov. Miner. Coal Process.*, Balkema, Amsterdam 1998, p. 711

[S14] J. Svoboda: Development of a linear high-gradient magnetic separator. *IEEE Trans. Magn.* **24** (1988), 749

[S15] Z.J.J. Stekly: A superconducting high-intensity magnetic separator. *IEEE Trans. Magn.* **11** (1975), 1594

[S16] J.A. Selvaggi, P. Vander Arend and J. Colwell: An industrial superconducting high gradient magnetic separator. *Adv. Cryog. Eng.* **33** (1988), 53

[S17] A.A. Stadtmuller, J.A. Good and N.J. Riches: Developments in superconducting magnetic separation for commercial application. In: *8th Ind. Miner. Int. Congress*, Boston, USA, 1988, p. 163

[S18] A. Stadtmuller et al.: Superconducting magnetic separation. *Ind. Miner.*, October 1997, p.80

[S19] K. Schönert, A. Supp and H. Dörr: Solenoid-pile separator, a new high intensity magnetic separator with super conductive coils. In: *Proc. XII Int. Min. Proc. Congress,* Sao Paulo, Brazil 1977, paper 4/1

[S20] N.F. Schulz: Determination of the magnetic separation characteristics with the Davis magnetic tube. *Trans. SME-AIME* **229** (1964), 211

[S21] H.J. Steinert and A. Boehm: Prediction of the performance of low-intensity wet-magnetic separators in the processing of partly altered magnetite ores. In: *Proc. XXI Int. Min. Proc. Congress,* Vol. A, A7-35-41, Rome, Italy (2000)

[S22] H. Schubert: Eddy-current separation: fundamentals, separators, applications. *Aufbereitungstechnik* **35** (1994), 553

[S23] J. Svoboda: Separation in magnetic fluids: time to meet the technological needs. In: *Proc. Int. Congr. Miner. Proc. Extr. Metall. MINPREX 2000,* Melbourne, Australia (2000), 297

[S24] A. W. Stradling: The physics of open-gradient dry magnetic separation. *Int. J. Min. Proc.* **39** (1993), 1

[S25] J. Svoboda: Extraction of ferromagnetic particles by a suspended magnet. *Magn. Electr. Sep.* **4** (1993), 223

[S26] T.J. Sheerer, M.R. Parker, F.J. Friedlaender and R. R. Birss: Theory of capture of weakly magnetic particles in random matrices in the longitudinal configuration in HGMS. *IEEE Trans. Magn.* **17** (1981), 2807

[S27] J. Svoboda, F.J. Friedlaender, Hua Fu and S.W. Luan: Single wire particle collection with magnetic field components along the wire. *IEEE Trans. Magn.* **24** (1988), 2419

[28] J. Svoboda: Vortex magnetic separation: An illusion or reality? *Min. Eng.* **8** (1995), 571

[S29] A. Stadtmuller and S. Fawell: HI filters to superconducting HGMS: The new generation. In: *Proc. XXI Int. Min. Proc. Congress,* Rome, Italy (2000), A7-78

[S30] W.H. Simons and R.P. Treat: Particle trajectories in a lattice of parallel magnetized fibres. *J. Appl. Phys.* **51** (1980), 578

[S31] J. Svoboda and V.E. Ross: Particle capture in the matrix of a magnetic separator. *Int. J. Min. Proc.* **27** (1989), 75

[S32] J. Svoboda: The effect of the magnetic field strength on the efficiency of magnetic separation. *Min. Eng.* **7** (1994), 747

[S33] J. Svoboda: A realistic description of the process of high-gradient magnetic separation. *Min. Eng.* **14** (2001), 1493

[S34] J. Svoboda: A theoretical approach to the magnetic flocculation of weakly magnetic minerals. *Int. J. Min. Proc.* **8** (1981), 377

[S35] J. Svoboda: Magnetic flocculation and treatment of fine weakly magnetic minerals. *IEEE Trans. Magn.* **18** (1982), 796

[S36] J. Svoboda and J. Žofka: Magnetic flocculation in secondary minimum. *J. Coll. Interface Sci.* **94** (1983), 37

[S37] J. Svoboda: The selective magnetic flocculation of fine weakly magnetic minerals. In: *Proc. Int. Conf. MINTEK 50,* Sandton, South Africa (1984), 329

[S38] J. Svoboda: The influence of surface forces on magnetic separation. *IEEE Trans. Magn.* **18** (1982), 862

[S39] J. Škvarla: Influence of polar surface forces on magnetic coagulation of fine paramagnetic minerals: I. Theory. *Magn. Electr. Sep.* **3** (1992), 171

[S40] J. Škvarla: Influence of polar surface forces on magnetic coagulation of fine paramagnetic minerals. II. Experiment. *Magn. Electr. Sep.* **18** (1992), 185

[S41] J. Škvarla and F. Zeleňák: Magneto-hydrophobic coagulation of paramagnetic minerals: a correlation of theory with experiment. *Int. J. Min. Proc.* **68** (2003), 17

[S42] P.C. Scholten and D.L.A. Tjaden: Mutual attraction of superparamagnetic particles. *J. Coll. Interface Sci.* **73** (1980), 254

[S43] J.H. Schenkel and J.A. Kitchener: A test of the DLVO theory with colloidal suspensions. *Trans. Farad. Soc.* **56** (1960), 161

[S44] J. Svoboda: A contribution to the theory of separation in a rotating ferrofluid. *Min. Eng.* **9** (1996), 743

[S45] E. Schlömann: Separation of nonmagnetic metals from solid waste by permanent magnets. I. Theory. *J. Appl. Phys.* **46** (1975), 5012

[S46] E. Schloemann: Eddy-current separation methods. In: *Progress in Filtration and Separation (*R.J. Wakeman, Editor), Elsevier, Amsterdam 1979, p. 29

[S47] J.A. Stratton: *Electromagnetic Theory.* McGraw-Hill Book Co., New York 1941, p.118

[S48] J.A. Selvaggi and J.P. Selvaggi: External and internal magnetic field equations for domain oriented permanent magnets. In: *Proc. Magnet Technology Conference MT-10,* Boston, Mass., USA (1987)

[S49] J. Svoboda and V. Murariu: unpublished results

[S50] J. Shimoiizaka, K. Nakatsuka, T. Fujita and A. Kounosu: Sink-float separators using permanent magnets and water based magnetic fluid. *IEEE Trans. Magn.* **16** (1980), 368

[S51] T. Shima: Separation of non-magnetic metals using a magnetic fluid. *J. Soc. Powder Technol. Japan* **20** (1983), 32

[S52] A.B. Solodenko: Analytical method of calculation of magnetic field parameters in a working zone of MG-separators with permanent magnets. *Izv. Vysch. Ucheb. Zaved. Tsvet. Metall.,* No. 2 (1991), 5 (in Russian)

[S53] A.B. Solodenko: Scientific basis of development of technology for beneficiation of minerals in ferromagnetic colloids. *D.Sc. Dissertation,* Moscow Institute of Steel and Alloys (1992)

[S54] A.B. Solodenko: Personal communication (1999)

[S55] R.D. Smolkin, V.S. Krokhmal and O.P. Saiko: Commercial equipment designed to recover gold from gravitational concentrates by means of magnetic separation and the separation in magnetic fluids. In: *Proc. XVIII Int. Min. Proc. Congress,* Sydney, Australia (1993), 425

[S56] J. Svoboda: Ferrohydrostatic separation method and apparatus. *US Patent 6,026,966 (2000), South African Patent 96/9288, Australian Patent 44342/97*

[S57] J. Svoboda: Ferrohydrostatic separation method and apparatus. *South African Patent 99/1255, Namibian Patent 2001/0049,* other countries patent pending

[S58] J. Svoboda: The recovery of ferrochrome metallics from slag. *MTC Report S111 (1992)*

[S59] G.D. Sealy and W.F. Howell: Magnetic recovery of medium from heavy media circuits. *World Coal,* June 1977, p. 35

[S60] J.A. Selvaggi and D. Cottrell: Operational characteristics of superconducting magnetic separators. *Filtr. Separ.,* March 1995, p.219

[S61] J. Svoboda: The selection of a matrix for the recovery of uranium by WHIMS. *Report M204,* Council for Mineral Technology (South Africa), 1985

[S62] E.A. Sultanovich and V.I. Karmazin: Kaolin beneficiation in a high-gradient magnetic separator with a ball matrix. *Magn. Electr. Sep.* **5** (1993), 1

[S63] J. Svoboda: HGMS: A search for a matrix material. *Int. J. Min. Proc.* **8** (1981), 165

[S64] J. Svoboda and I.J. Corrans: The removal of particles from the matrix of a high-gradient magnetic separator. *IEEE Trans. Mag.* **21** (1985), 53

[S65] L.A. Spielman: Flow through porous media and fluid-particle hydrodynamics. In: *The Scientific Basis of Filtration.* (Ed. K.J. Ives), Noordhoff, Leyden (1975), 129

[S66] I. Spevakova, L. Ruditser, V. Eroshenko and G. Nikolskiy: New magnetic filter for chemical industry. *Magn. Electr. Sep.* **5** (1993), 17.

[S67] J. Svoboda, R.N. Guest and W.J.C. Venter: Recovery of copper and lead minerals from Tsumeb flotation tailings. *J. South Afr. Min. Metall.* **88** (1988), 9

[S68] Y. Shao, T.J. Veasey and N.A. Rowson: Wet high intensity magnetic separation of iron minerals. *Mag. Electr. Sep.* **8** (1996), 41

[S69] J. Svoboda: The effect of particle size and colloid stability on the WHIMS of uranium from cyanidation residues. *Report M236,* Council for Mineral Technology, Randburg, South Africa (1986)

[S70] P. Somasundaran: Principles of flocculation, dispersion and selective flocculation. In: *Fine Particles Processing* (Ed. P. Somasundaran) Vol. **2** , AIME, New York (1980), 947

[S71] A.B. Solodenko and V.V. Karmazin: Magnetogravimetric separation of ores of non-ferrous metals. In: *Proc. Conf. "Theoretical and Practical Problems of Magnetic Separation of Raw Materials",* Košice, Czechoslovakia (1989), 118

[S72] J. Svoboda: The investigation of the effect of colloid stability in magnetic separation. *Aufbereitungstechnik* **24** (1983), 520

[S73] J. Svoboda, I.J. Corrans and M.H.E. Spitze: The effect of pH on the recovery of uranium and gold by HGMS. *Int. J. Min. Proc.* **17** (1986), 83

[S74] F. Špaček and M. Dočkal: The application of pulse magnetic field in iron ore treatment. *Rudy (Prague)* **34** (1986), 45

[S75] J. Svoboda, P. Mangaya and S. Tsilevhulevhu: The effect of magnetic viscosity on magnetic separation. *Powder Technol.* **57** (1989), 241

[S76] J. Svoboda, M. Lazer and W.A.M. te Riele: Selective magnetic separation of uranium and gold. *IEEE Trans. Mag.* **23** (1987), 283

[S77] V.D. Samygin, L.A. Barsky and S.M. Angelova: Mechanism of mutual flocculation of particles differing in size. *Kol. Zh. (USSR)* **30** (1968), 581

[S78] Z. Sun and Z. Li: A study of vibration of HGMS. In: *Proc. 15th Int. Min. Proc. Congress,* Cannes (1985), 410

[S79] J. Svoboda: Magnetic flocculation and treatment of fine feebly magnetic minerals. *Rudy* **29** (1981), 321

[S80] J. Svoboda and L. Krüger: unpublished results

[S81] J. Svoboda and S. Dumbu: unpublished results

[S82] J. Svoboda, S. Dumbu, L. Krüger, M. Fofana and N. Monteiro: Recent developments in ferrohydrostatic separation as applied to material treatment at De Beers. In: *Proc. De Beers METCON,* Johannesburg (South Africa), 2000

[S83] E.M. Savitzky et al.; Magnetic hydrocyclone. *Authors' Certificate (USSR)* 5259961/22-03 (1982)

[S84] J. Svoboda and Q.P. Campbell: Magnetic cyclone and method of operating it. *South African Patent* 97/3440 (1996), *Australian Patent* 731513 (2001), *Canadian Patent* CA 2205339 (2001)

[S85] J. Svoboda, C. Coetzee and Q.P. Campbell: Experimental investigation into the application of a magnetic cyclone for dense medium separation. *Min. Eng.* **11** (1998), 501

[S86] T. Shuttleworth, C. S. Bon and D. Milton: Magnetic separation techniques for SAG grinding circuits. In: *SAG 2001*, Vancouver. B.C. Canada (2001), III-393

[S87] A.M. Saeid, D.A. Butcher and D.A. Rowson: Coal desulphurisation and ash removal in intensified magnetic fields. *Magn. Electr. Sep.* **4** (1993), 107

[S88] M. Šafaříková and I. Šafařík: The application of magnetic techniques in biosciences. *Magn. Electr. Sep.* **10** (2001), 223

[S89] I. Šafaříková, I. Šafařík and S.J. Forsythe: The application of magnetic separations in applied microbiology. *J. Appl. Bacter.* **78** (1995), 575

[S90] J. Svoboda: Separation of red blood cells by magnetic means. *J. Magn. Magn. Mater.* **220** (2000), L103

[S91] J. Staudinger and G. Keoleian: Management of end-of-life vehicles (ELVs) in the US. *Bureau of Int. Recycling* (2001)

[S92] J. Svoboda: Separation of materials in magnetic fluids: Principles, applications and the future developments. In: *Proc. 3rd Int. Symposium "Waste Processing and Recycling in Mineral and Metallurgical Industries III"* Calgary , Canada (1998), 31

[S93] J. Svoboda: The fractionation analysis of coal using ferrofluids. In: *XIV Int. Coal Prep. Congress,*, Johannesburg, South Africa (2002), 47

[S94] R. Sheng, G.A. Flores and J. Liu: In vitro investigation of a novel cancer therapeutic method using embolizing properties of magnetorheological fluids. *J. Magn. Magn. Mater.* **194** (1999), 167

[S95] D.E. Stokes: *Pasteur Quadrant: Basic Science and Technological Innovation.* Brookings Institution Press, Washington D.C. (1997)

[S96] J. Svoboda: Innovation in electromagnetic techniques of material treatment. *Resources Process. (Japan)* **48** (2001), 1

[S97] J. Svoboda: Magnetic methods of material treatment: technology at crossroads. In: *Proc. 4th Int. Conf. Mater. Resources*, Akita, Japan (2001), 131

[S98] J. Svoboda and T. Fujita: Innovation in magnetic techniques of material treatment: Technology at crossroads. In: *Proc. XXII Int. Min. Proc. Congress,* Cape Town, South Africa (2003), 261

[S99] W. Süsse: SOL separator: A new development in the field of high-intensity wet magnetic separation. *Erzmetall* **32** (1979), 321

[S100] J. Svoboda and T. Fujita: Recent developments in magnetic methods of material separation. *Min. Eng.* **16** (2003), 785

[T1] D.H. Tarling and F. Hrouda: *The Magnetic Anisotropy of Rocks.* Chapman & Hall, London 1993, 217 pp.

[T2] A.M. Turkenich: Theory and performance of a slurry flow in form of a film in the 6-ERM-35/135 separator with a high-gradient magnetic field. *Magn. Electr. Sep.* **8** (1996), 53

[T3] S.J. Truscott: *A Text-Book of Ore Dressing*, MacMillan and Co., Ltd., London 1923, p. 534

[T4] M. Takayasu, F.J. Friedlaender, Y. Hiresaki and D.R. Kelland: Matrices for selective diamagnetic HGMS. *IEEE Trans. Magn.* **17** (1981), 2810

[T5] C. Tsouris and T.C. Scott: Coagulation of paramagnetic particles in a magnetic field. *J. Coll. Interface Sci.* **171** (1995), 319

[T6] R.K. Tenzer: Estimating leakage factors for permanent magnets from geometry of magnetic circuit. *Electr. Manuf.*, February 1957, p. 94

[T7] M. Troy: Study of magnetic filtration applications to the primary and secondary systems of PWR plants. *EPRI NP-514 Report TPS-76-665 (1978)*

[T8] S.C. Trindade, M. Saddy and J.L.F. Monteiro: Magnetic recovery of sulphur from coal in Brazil. *IEEE Trans. Magn.* **12** (1976), 355

[T9] K. Takino, T. Tankaka and T. Schichiri: Operation of a full-scale device of high-gradient magnetic filter for treatment of gas scrubber water from vacuum degassing process in a steel mill. In: *Industrial Applications of Magnetic Separation* (Ed. Y.A. Liu), IEEE Publ. 78 1447-2MAG (1979), 137

[T10] M. Tsuge, J. Yano, E. Schichi, K. Kawashima and S. Matsumoto: Treatment of cold rolling coolant for steel by HGMS. *IEEE Trans. Magn.* **23** (1987), 2764

[T11] M. Troy: Study of magnetic filtration applications to the primary and secondary systems of PWR plants. *EPRI Report* **NP-514** (1978)

[T12] Y. Terashima, H. Ozaki and M. Sekine: Removal of dissolved heavy metals by chemical coagulation, magnetic seeding and high gradient magnetic filtration. *Water. Res.* **20** (1986), 537

[T13] J.C. Tsang et al.: Science, technology and economic growth. In: *American Physical Society: Panel on Public Affairs* (2003), www.aps.org/public_affairs/popa/reports/steg

[U1] K.H. Unkelbach: *Magnetic separators: Mode of operation and applicability for the separation of materials.* KHD Humboldt Wedag AG, Cologne 1990

[U2] R.S. Ulubabov and V.I. Karmazin: Development and performance of the 6-ERM-35/135 separator with a high-intensity magnetic field. *Magn. Electr. Sep.* **3** (1992), 77

[U3] R.S. Ulubabov et al.: Simultaneous increase in the grade and the recovery of iron in HGMS by controlling magnetic susceptibility of the separation medium. *Magn. Electr. Sep.* **5** (1994), 223

[U4] S. Uchiyama, S. Kondo and M. Takayasu: Performance of parallel stream type magnetic filter for HGMS. *IEEE Trans. Magn.* **12** (1976), 895

[U5] G.G. O.O. Uwadiale: Magnetizing roasting of iron ores. *Min. Proc. Extract. Metall. Rev.* **11** (1992), 1

[V1] E.J.W. Verwey and J.T.G. Overbeek: *Theory of the Stability of Lyophobic Colloids.* Elsevier, Amsterdam 1948

[V2] J. Visser: The concept of negative Hamaker constant.*Adv. Coll. Interface Sci.* **15** (1981), 157

[V3] H.J.L. van der Valk, W.L. Dalmijn and W.P.C. Duyvesteyn: Eddy-current separation methods with permanent magnets for the recovery of non-ferrous metals and alloys.*Erzmetall* **41** (1988), 266

[V4] H.J.L. van der Valk, B.C. Braam and W.L. Dalmijn: Eddy-current separation by permanent magnets. Part I: Theory. *Resources & Conservation* **12** (1986), 233

[V5] R.P.A.R. van Kleef, H.W. Myron, P. Wyder and M.R. Parker: Continuous flow separation: an application of selective magnetosedimentation. *J. Physique* **45**, C1, Suppl. no. 1 (1984), C1-763

[V6] V. Vand: Viscosity of solutions and suspensions. *J. Phys. Coll. Chem.* **52** (1948), 277

[V7] L. Vatta: unpublished results

[V8] L.L. Vatta, R. Kekana, B. Radebe, I. Myburgh and J. Svoboda: The effect of magnetic field on the performance of a dense medium separator. *Phys. Sep. Sci. Eng.* **12** (2003),

[V9] N. Vorobyev, L. Lomovtsev, Yu. Khvatov, A. Dobrynin and F. Žůrek: Application of a technology for a combined beneficiation of low-grade manganese ore slimes. In: *Proc. XVII Int. Min. Proc. Congress,* Dresden Germany (1991), Vol. V, p. 195

[V10] M.V. Verkhoturov, Yu. M. Emel'yashin and V.P. Tarasenko: Moisture of air and efficiency of dry magnetic separation. *Sov. Min. Sci.* **22** (1987), 508

[V11] T.J. Veasey, R. J. Wilson and D. M. Squires: *The Physical Separation and Recovery of Metals from Wastes.* Gordon and Breach Publishers (1993)

[W1] M.S. Walmer et al.: A new class of Sm-TM magnets for operating temperatures up to 550 ^{0}C. *IEEE Trans. Magn.* **36** (2000), 3376

[W2] M.N. Wilson: *Superconducting Magnets.* Clarendon Press, Oxford 1983

[W3] M.S. Walker, A.L. Devernoe and W.S. Urbanski: Separation of non-magnetic minerals using magnetic fluids in a flow-through MHS rotor. *Miner. Metall. Process.* **7** (1990), 209

[W4] W. Windle: A method of, and apparatus for, separating magnetisable particles from a fluid in which they are suspended. *British Patent 1,469,765* (1977)

[W5] A.J. Winters and J.A. Selvaggi: Large-scale superconducting separator for kaolin processing. *Chem. Eng. Progress,* January 1990, p. 36

[W6] H.D. Wasmuth et al.: New medium intensity drum-type permanent magnetic separator PERMOS and its practical applications for the processing of industrial minerals and martitic iron ores. *Magn. Electr. Sep.* **6** (1995), 201

[W7] M.S. Walker, A.L. Devernoe, R.W. Stuart and W.S. Urbanski: A new method for the commercial separation of particles of differing densities using magnetic fluids. In: *Proc. XV Int. Min. Proc. Congress,* Cannes, France (1985), 307

[W8] J.H.P. Watson: Magnetic filtration. *J. Appl. Phys.* **44** (1973), 4209

[W9] J.H.P. Watson: Theory of capture of particles in magnetic high-intensity filters. *IEEE Trans. Magn.* **11** (1975), 1597

[W10] J.H.P. Watson: HGMS at moderate Reynolds numbers. *IEEE Trans. Magn.* **15** (1979), 1538

[W11] J.H.P. Watson: Theoretical and single-wire studies of vortex magnetic separation. *Min. Eng.* **5** (1992), 1147

[W12] J.H.P. Watson and Z. Li:Vortex magnetic separation: a highly selective process with high recovery and a high processing rate. *Min. Eng.* **8** (1995), 574

[W13] J.H.P. Watson and S.J.P. Watson: The ball matrix magnetic separator. *IEEE Trans. Magn.* **19** (1983), 2698

[W14] J.H.P. Watson: Approximate solutions of the magnetic separator equations. *IEEE Trans. Magn.* **14** (1978), 240

[W15] M.E. Weber and D. Paddock: Interceptional and gravitational collision efficiencies of single collectors at intermediate Reynolds numbers. *J. Coll. Interface Sci.* **94** (1983), 328

[W16] J.H.P. Watson: Magnetic separation at high magnetic fields. In: *Proc. 6th Int. Conf. Cryog. Eng.,* Grenoble, France (1976), 223

[W17] Y. Wang, R.J. Pugh, E. Forssberg: The influence of interparticle surface forces on the coagulation of weakly magnetic mineral ultrafines in a magnetic field. *Col. Surf. A: Physiochem. Eng. Asp.* **90** (1994), 117

[W18] Y. Wang and E. Forssberg: Aggregation between magnetite and hematite ultrafines utilizing remanent magnetization. *Min. Eng.* **5** (1992), 895

[W19] Y. Wang, E. Forssberg, R.J. Pugh and S.A. Elming: Magnetic aggregation in dispersions of mineral ultrafines. *J. Disp. Sci. Tech.* **16** (1995), 137

[W20] M.S. Walker, A.L. Devernoe and W.S. Urbanski: Separation of non-magnetic minerals using magnetic fluids in a flow-through MHS rotor. *Min. Metal. Process.* **7** (1990), 209

[W21] M.S. Walker, A.L. Devernoe, W.S. Urbanski and G.R. Morrow: Mainly "gravity" separations using magnetic fluids under rotation. In: *Proc. 17th Int. Min. Proc. Congress,* Dresden, Germany (1991), 105

[W22] M.S. Walker and A.L. Devernoe: Mineral separations using rotating magnetic fluids. *Int. J. Min. Proc.* **31** (1991), 195

[W23] J.H.P. Watson, D. Rassi and A.S. Bahaj: The recovery of gold and uranium from gold leached residues by HGMS. *IEEE Trans. Mag.* **19** (1983), 2136

[W24] I. Wechsler, J. Doulin and R. Eddy: Coal preparation using magnetic separation. Vol. 3, *EPRI Report CS-1517* (1980)

[W25] Y. Wang, R.J. Pugh and E. Forssberg: Carrier coagulation of chromite fines in wet magnetic separation. *Mag. Electr. Sep.* **5** (1993), 33

[W26] J.H.P. Watson and K.B.M. Yusof: Magnetic flocculation of paramagnetic particles in high magnetic field. In: *Digest Intermag Conf.*, Grenoble (France), 1981, p.14-3

[W27] J.L. Watson and K. Amoako-Gyampah: Cycloning in magnetic fields. *SME Preprint* 83-335 (1983)

[W28] J.L. Watson and P. Gardner: Multiforce dewatering for magnetic waste materials. *Min. Eng.* **8** (1995), 191

[W29] J.L. Watson and Zhicheng Li: Application of magnetic forces to disk vacuum filtration in the laboratory and plant. *Min. Eng.* **12** (1999), 1253

[W30] Y. Wang and E. Forssberg: Development and application of magnetic separation techniques in processing Swedish minerals. In: *Proc. Int. Conf. "Minerals & Materials '96",* Sommerset West, South Africa (1996), Vol. I: Minerals.

[W31] W.L. Wilkinson: Filtration techniques in the nuclear fuel industry. *Filtr. Sep.* January/February 1982, p. 17

[W32] J.A. Williams, and C.M. Leslie: HGMS in the nuclear fuel cycle. *IEEE Trans. Magn.* **17** (1981), 2790

[W33] J.A. Williams et al.: The application of HGMS in nuclear fuel reprocessing. *Nucl. Technol.* **52** (1981), 284

[W34] J.H.P. Watson and D.C. Ellwood: Biomagnetic separation and extraction process. *IEEE Trans. Magn.* **23** (1987), 3751

[W35] R. A. Williams: The impact of fundamental research on minerals processing operations of the future. In: *Proc. XXII Int. Min. Proc. Congress,* Cape Town, South Africa (2003), 2

[W36] J.H.P. Watson: Status of superconducting magnetic separation in the minerals industry. *Min. Eng.* **7** (1994), 737

[W37] J.H.P. Watson and L. Bolt: High temperature superconductors for magnetic separation. In: *Proc. XXI Int. Min. Proc. Congress,* Rome, Italy (2000), **A7**, 105

[W38] J.H.P. Watson: High temperature superconducting permanently magnetised discs and rings: Prospects for use in magnetic separation. *Min. Eng.* **12** (1999), 281

[W39] J.H.P. Watson: High magnetic field production with ceramic superconductors. *Physica* C **153-155** (1988), 1401

[W40] S.F. Wipf and H.L. Laquer: Superconducting permanent magnets. *IEEE Trans. Magn.* **25** (1989), 1877

[W41] J.H.P. Watson: The design for a high-T_c superconducting magnetic separator. *Supercond. Sci. Technol.* **5** (1992), 694

[X1] Xiong Da-he: New development of the SLON vertical ring and pulsation HGMS separator. *Magn. Electr. Sep.* **5** (1994), 211

[X2] Xiong Da-he: Development and commercial test of SLON-2000 vertical ring and pulsating high-gradient magnetic separator. *Magn. Electr. Sep.* **8** (1997), 89

[X3] Xiong Da-he: Research and commercialization of treatment of fine ilmenite with SLON magnetic separators. *Magn. Electr. Sep.* **10** (2000), 121

[X4] Xiong Da-he: A large scale application of SLON magnetic separator in Meishan iron ore mine. *Magn. Electr. Sep.* **11** (2002), 1

[X5] Xiong Da-he: SLON magnetic separator applied to upgrading the iron concentrate. *Phys. Sep. Sci. Eng.* **12** (2003), 63

[Y1] I. Yaniv: Magnetic separators. *South African Patent Application* 814271 (1981)

[Y2] Yang Peng, Liu Shuyi and Chen Jin: The separation performance of the pulsating high-gradient magnetic separator. *Magn. Electr. Sep.* **4** (1993), 211

[Y3] Yuan Xuemin, He Pingbo and Sun Zhongyuan: Studies on vibrating high gradient magnetic separation of mixed sulphide flotation concentrates. In: *Proc. 16th Int. Min. Proc. Congress,* Stockholm (1988), 1065

[Y4] Yu Kang Chun: The removal of iron from hard pulverized kaolin by dry high-gradient magnetic separation. *Mag. Electr. Sep.* **6** (1995), 171

[Y5] P.P. Yurov et al.: Beneficiation of roasted ores in a reconstructed section of the CGOK beneficiation plant. *Chernaya Metallurg.* No. 9 (1970), 3

[Y6] P.P. Yurov et al.: The application of magnetic hydrocyclones in the circuits for beneficiation of iron ores. *Gorniy Zh.* No. 4 (1986), 34

[Y7] T. Yalcin: Magnetic-column flotation. In: *Innovation in Mineral and Coal Processing,* Balkema Publ., Rotterdam (1998), 213

[Y8] I. Yildirim, Güven Önal and M. S. Celik: Desulphurization of low-rank lignitic coals using semicoking and dry magnetic techniques. In: *Proc. XX Int. Min. Proc. Congress,* Aachen, Germany (1997), 631

[Y9] Jun-ichi Yano and I. Eguchi: Application of HGMS for water treatment in steel industry. In: *Industrial Applications of Magnetic Separation* (Ed. Y.A. Liu), IEEE Publ. 78 CH 1447-2MAG (1979), 134

[Y10] Hao Yu: Use of immunomagnetic carriers for the rapid detection of virulent bacteria from biological samples. In: *Scientific and Clinical Applications of Magnetic Carriers* (U. Häfeli, W. Schütt, J. Teller and M. Zborowski, Editors), Plenum Press, New York (1997), 341

[Z1] H. Zijlstra: *Experimental Methods of Magnetism.* North-Holland, Amsterdam 1967

[Z2] M. Zborowski et al.: Continuous cell separation using novel magnetic quadrupole flow sorter. *J. Magn. Magn. Mater.* **194** (1999), 224

[Z3] V. Žežulka and F. Žůrek: Continuous high-gradient magnetic separator for kaolin treatment. In: *XVI Proc. Int. Min. Proc. Congress* (E. Forssberg, Editor), Elsevier, Amsterdam 1988, p. 1075

[Z4] S. Zhang, P.C. Rem and E. Forssberg: Particle trajectory simulation of two-drum eddy current separators. *Res. Conserv. Recycl.* **26** (1999), 71

[Z5] S. Zhang, P. C. Rem and E. Forssberg: The investigation of separability of particles smaller than 5 mm by eddy-current separation technology. Part I: Rotating type eddy-current separators. *Magn. Electr. Sep.* **9** (1999), 233

[Z6] G. Zabel: Deposition of aerosol flowing past a cylindrical fiber in a uniform electric field. *J. Colloid Sci.* **20** (1965), 522

[Z7] Y. Zimmels, I. Yaniv, I.J. Lin and U. Levite: On the design and application of magnetohydrostatic laboratory systems. *IEEE Trans. Magn.* **14** (1978), 269

[Z8] Y. Zhang, S. Liu and W. Fu: Study of separation of refractory sulphide minerals by HGMS. In: *Proc. XXI Int. Min. Proc. Congress,* Rome, Italy (2000), **C7** - 22

Index

GPSR Compliance

The European Union's (EU) General Product Safety Regulation (GPSR) is a set of rules that requires consumer products to be safe and our obligations to ensure this.

If you have any concerns about our products, you can contact us on ProductSafety@springernature.com

In case Publisher is established outside the EU, the EU authorized representative is:

Springer Nature Customer Service Center GmbH
Europaplatz 3
69115 Heidelberg, Germany

Batch number: 10755248

Printed by Printforce, the Netherlands